SEEDS: BIOLOGY, DEVELOPMENT AND ECOLOGY

SEEDS: BIOLOGY, DEVELOPMENT AND ECOLOGY

Edited by

Steve W. Adkins

School of Land and Food Sciences, The University of Queensland, Australia

Sarah E. Ashmore

Centre for Forestry and Horticultural Research, Griffith University, Australia

Sheldon C. Navie

School of Land and Food Sciences, The University of Queensland, Australia

www.cabi.org

CABI is a trading name of CAB International

CABI Head Office
Nosworthy Way
Wallingford
Oxfordshire OX10 8DE
UK
Tel: +44 (0)1491 832111
Fax: +44 (0)1491 833508
E-mail: cabi@cabi.org
Website: www.cabi.org

CABI North American Office
875 Massachusetts Avenue
7th Floor
Cambridge, MA 02139
USA
Tel: +1 617 395 4056
Fax: +1 617 354 6875
E-mail: cabi-nao@cabi.org

A catalogue record for this book is available from the British Library, London, UK.

A catalogue record for this book is available from the Library of Congress, Washington, DC.

ISBN-13: 978 1 84593 197 1

Typeset by SPi, Pondicherry, India.
Printed and bound by CPI Group (UK) Ltd, Croydon, CR0 4YY

Contents

Contributors

S.W. Adkins, *Integrated Seed Research Unit, School of Land and Food Sciences, The University of Queensland, St Lucia, Queensland 4072, Australia*

L.J. Armstrong, *School of Computer and Information Science, Edith Cowan University, Mt Lawley, Western Australia 6050, Australia*

N. Ashwath, *School of Biological and Environmental Sciences, Central Queensland University, Rockhampton, Queensland 4700, Australia*

H. Bar, *Department of Weed Research, Agricultural Organization, Newe-Ya'ar Research Center, Ramat Yishai 30095, Israel*

C. Bechetoille, *Centre National de la Recherche Scientifique (CNRS)–Bayer CropScience Joint Laboratory, Bayer CropScience, 14–20 rue Pierre Baizet, 69263 Lyon, Cedex 9, France*

M. Belghazi, *Institut National de la Recherche Agronomique, Unité Mixte de Recherche N° 6175, Service de Spectrométrie de Masse Pour la Protéomique, Nouzilly, France*

S.M. Bellairs, *School of Science and Primary Industries, Charles Darwin University, Darwin, Northern Territory 0909, Australia*

P. Berjak, *School of Biological and Conservation Sciences, University of KwaZulu-Natal, Durban 4041, South Africa*

K.J. Bradford, *Department of Plant Sciences, Seed Biotechnology Center, One Shields Avenue, University of California, Davis, CA 95616-8780, USA*

W. Brandt, *Department of Molecular and Cell Biology, University of Cape Town, Rondebosch 7700, South Africa*

C.M. Bray, *Faculty of Life Sciences, University of Manchester, Oxford Road, Manchester, UK*

P.H. Brown, *Tasmanian Institute of Agricultural Research, School of Agricultural Science, University of Tasmania, Private Bag 98 Hobart, Tasmania 7001, Australia*

J. Buitink, *UMR 1191 Physiologie Moléculaire des Semences (Université d'Angers/INH/INRA), 16 Bd Lavoisier, 49045 Angers, France*

C. Calistru, *School of Biological and Conservation Sciences, University of KwaZulu-Natal, Durban 4041, South Africa*

J. Catusse, *Centre National de la Recherche Scientifique (CNRS)–Bayer CropScience Joint Laboratory, Bayer CropScience, 14–20 rue Pierre Baizet, 69263 Lyon, Cedex 9, France*

D. Côme, *Physiologie Végétale Appliquée, Université Pierre et Marie Curie-Paris 6, 75252 Paris, Cedex 05, France*

F. Corbineau, *Physiologie Végétale Appliquée, Université Pierre et Marie Curie-Paris 6, 75252 Paris, Cedex 05, France*

F. Coste, *Laboratoire d'Ecophysiologie Végétale et Agronomie, Ecole Supérieure d'Agriculture, 55 Rue Rabelais, BP 30748, 49007 Angers, Cedex 01, France*

Y. Crozat, *Laboratoire d'Ecophysiologie Végétale et Agronomie, Ecole Supérieure d'Agriculture, 55 Rue Rabelais, BP 30748, 49007 Angers, Cedex 01, France*

H. Dassen, *Laboratory of Plant Physiology, Wageningen University and Research Centre, Arboretumlaan 4, 6703 BD Wageningen, The Netherlands*

R.A. de Maagd, *Plant Research International, Bornsesteeg 65, 6708 PD Wageningen, The Netherlands*

C.J. Delpratt, *Department of Resource Management, The University of Melbourne, Burnley Campus, Richmond, Victoria 3121, Australia*

A.B. Downie, *Department of Horticulture, University of Kentucky, Lexington, KY 40546, USA*

J.M. Dunwell, *School of Plant Sciences, The University of Reading, Whiteknights, Reading RG6 6AS, UK*

C. Dürr, *INRA Agronomie LRM, Rue Fernand Christ, 02007 Laon Cedex, France*

M. Dutt, *Department of Horticulture, University of Kentucky, Lexington, KY 40546, USA*

S. Eggers, *School of Biological and Conservation Sciences, University of KwaZulu-Natal, Durban 4041, South Africa*

D. Erdey, *School of Biological and Conservation Sciences, University of KwaZulu-Natal, Durban 4041, South Africa*

J.M.R. Faria, *Laboratory of Plant Physiology, Wageningen University, Arboretumlaan 4, 6703 BD Wageningen, The Netherlands; Departamento de Ciências Florestais, Universidade Federal de Lavras, CP 3037, CEP 37200-000, Lavras, Minas Gerais, Brazil*

J.M. Farrant, *Department of Molecular and Cell Biology, University of Cape Town, Rondebosch 7700, South Africa*

U. Fischer, *KWS SAAT AG, Grimsehlstr. 31, 37555 Einbeck, Germany*

E.A. Fitch, *Department of Biology, Middle Tennessee State University, Murfreesboro, TN 37132, USA*

M.E. Foley, *US Department of Agriculture–Agricultural Research Service (USDA–ARS), Biosciences Research Laboratory, Fargo, ND 58105-5674, USA*

J.R. Fu, *School of Life Sciences, Sun Yat-Sen University, Guangzhou 510275, China*

A. Geard, *Tasmanian Institute of Agricultural Research, School of Agricultural Science, University of Tasmania, Private Bag 98 Hobart, Tasmania 7001, Australia*

T. Gendler, *Ben-Gurion University of the Negev, Desert Research Institutes, Sede Boker Campus, 84990 Israel*

R.L. Geneve, *Department of Horticulture, University of Kentucky, Lexington, KY 40546, USA*

M.P. González-García, *Departamento de Fisiología Vegetal, Centro Hispano-Luso de Investigaciones Agrarias, Facultad de Biología, Universidad de Salamanca, Plaza de los Doctores de la Reina s/n, 37007 Salamanca, Spain*

A.J. Gracie, *School of Agricultural Science, University of Tasmania, Hobart, Tasmania, Australia*

T. Groom, *Botanical Resources Australia Pty. Ltd., Industrial Drive, Ulverstone, Tasmania, Australia*

S.P.C. Groot, *Plant Research International, Wageningen University and Research Centre, PO Box 16, 6700 AA Wageningen, The Netherlands*

X.-Y. Gu, *Department of Plant Sciences, North Dakota State University, Fargo, ND 58105, USA*

Y. Gutterman, *Ben-Gurion University of the Negev, Desert Research Institutes, Sede Boker Campus, 84990 Israel*

B. Hess, *Institute for Biology II, Botany/Plant Physiology, Albert-Ludwigs-University Freiburg, Schänzlestr. 1, D-79104 Freiburg i. Br., Germany*

J.R. Hewitt, *Department of Horticulture, Oregon State University, Corvallis, OR 97331, USA*

S.N. Hidayati, *Department of Biology, Middle Tennessee State University, Murfreesboro, TN 37132, USA*

H.W.M. Hilhorst, *Laboratory of Plant Physiology, Wageningen University, Arboretumlaan 4, 6703 BD Wageningen, The Netherlands*

C. Hlongwane, *CSIR Bio/Chemtek, Pretoria 0001, South Africa*

T.M. Homrichhausen, *Department of Chemical Engineering, Oregon State University, Corvallis, OR 97331, USA*

Z.-Y. Huang, *Ben-Gurion University of the Negev, Desert Research Institutes, Sede Boker Campus, 84990 Israel; Institute of Botany, The Chinese Academy of Sciences, Beijing 100093, Peoples Republic of China*

R. Huguet, *Centre National de la Recherche Scientifique (CNRS)–Bayer CropScience Joint Laboratory (UMR CNRS 2847), Bayer CropScience, 14/20 rue Pierre Baizet, BP 9163, F-69263 Lyon Cedex 09, France*

A. Jagonak, *KWS SAAT AG, Grimsehlstr. 31, 37555 Einbeck, Germany*

L. Jiang, *Department of Biology, The Chinese University of Hong Kong, Shatin, New Territories, Hong Kong, China*

C. Job, *Centre National de la Recherche Scientifique (CNRS)–Bayer CropScience Joint Laboratory (UMR CNRS 2847), Bayer CropScience, 14/20 rue Pierre Baizet, BP 9163, F-69263 Lyon, Cedex 09, France*

D. Job, *Centre National de la Recherche Scientifique (CNRS)–Bayer CropScience Joint Laboratory (UMR CNRS 2847), Bayer CropScience, 14/20 rue Pierre Baizet, BP 9163, F-69263 Lyon, Cedex 09, France*

D.M. Joel, *Department of Weed Research, Agricultural Organization, Newe-Ya'ar Research Center, Ramat Yishai 30095, Israel*

S.F. Kianian, *Department of Plant Sciences, North Dakota State University, Fargo, ND 58105, USA*

J. Kodde, *Laboratory of Plant Physiology, Wageningen University, Arboretumlaan 4, 6703 BD Wageningen, The Netherlands*

N. Koizuka, *Department of Horticulture, Oregon State University, Corvallis, OR 97331, USA*

B. Kucera, *Institute for Biology II, Botany/Plant Physiology, Albert-Ludwigs-University Freiburg, Schänzlestr. 1, D-79104 Freiburg i. Br., Germany*

M. Lammers, *Laboratory of Plant Physiology, Wageningen University and Research Center, Arboretumlaan 4, 6703 BD Wageningen, The Netherlands*

P.A. Lane, *School of Agricultural Science, University of Tasmania, Hobart, Tasmania, Australia*

G.G. Leubner-Metzger, *Institute for Biology II, Botany/Plant Physiology, Albert-Ludwigs-University Freiburg, Schänzlestr. 1, D-79104 Freiburg i. Br., Germany*

J. Leymarie, *Physiologie Végétale Appliquée, Université Pierre et Marie Curie-Paris 6, 75252 Paris, Cedex 05, France*

W. Ligterink, *Laboratory of Plant Physiology, Wageningen University and Research Centre, Arboretumlaan 4, 6703 BD Wageningen, The Netherlands*

G.G. Lindsey, *Department of Molecular and Cell Biology, University of Cape Town, Rondebosch 7700, South Africa*

X.Q. Li, *The Vegetable Research Centre, Academy of Agricultural Sciences of Beijing, China*

P.-P. Liu, *Department of Horticulture, Oregon State University, Corvallis, OR 97331, USA*

O. Lorenzo, *Departamento de Fisiología Vegetal, Centro Hispano-Luso de Investigaciones Agrarias, Facultad de Biología, Universidad de Salamanca, Plaza de los Doctores de la Reina s/n, 37007 Salamanca, Spain*

Y. Lovigny, *Centre National de la Recherche Scientifique (CNRS)-Bayer CropScience Joint Laboratory (UMR CNRS 2847), Bayer CropScience, 14/20 rue Pierre Baizet, BP 9163, F-69263 Lyon, Cedex 09, France*

P.J. Lutman, *Rothamsted Research, Harpenden, Hertfordshire, AL5 2JQ, UK*

D.D. Mackenzie, *Department of Renewable Resources, University of Alberta, Edmonton, Alberta T6G 2H1, Canada*

R.C. Martin, *US Department of Agriculture–Agricultural Research Service (USDA-ARS), National Forage Seed Production Research Center (NFSPRC), 3450 SW Campus Way, Corvallis, OR 97331, USA*

A.M. Mayer, *Department of Botany, The Hebrew University of Jerusalem, Jerusalem 91904, Israel*

M.B. McDonald, *Department of Horticulture and Crop Science, The Ohio State University, Columbus, OH 43210, USA*

J. Meinhard, *KWS SAAT AG, Grimsehlstr. 31, 37555 Einbeck, Germany*

V. Merhar, *Electron Microscope Unit, University of KwaZulu-Natal, Durban 4041, South Africa*

Y. Miao, *Department of Biology, The Chinese University of Hong Kong, Shatin, New Territories, Hong Kong, China*

L. Miché, *Centre National de la Recherche Scientifique (CNRS)–Bayer CropScience Joint Laboratory (UMR CNRS 2847), Bayer CropScience, 14/20 rue Pierre Baizet, BP 9163, F-69263 Lyon, Cedex 09, France*

A.H. Millar, *School of Biomedical, Biomolecular and Chemical Sciences, Faculty of Life and Physical Sciences, University of Western Australia, Western Australia 6009, Australia*

P. Moreau-Valancogne, *Laboratoire d'Ecophysiologie Végétale et Agronomie, Ecole Supérieure d'Agriculture, 55 Rue Rabelais, BP 30748, 49007 Angers, Cedex 01, France*

L. Mtwisha, *Department of Molecular and Cell Biology, University of Cape Town, Rondebosch 7700, South Africa*

K. Müller, *Institute for Biology II, Botany/Plant Physiology, Albert-Ludwigs-University Freiburg, Schänzlestr. 1, D-79104 Freiburg i. Br., Germany*

A.J. Murdoch, *Seed Science Laboratory, Department of Agriculture, The University of Reading, Earley Gate, Reading RG6 6AR, UK*

M.A. Naeth, *Department of Renewable Resources, University of Alberta, Edmonton, Alberta, T6G 2H1 Canada*

T.S. Nair, *Tasmanian Institute of Agricultural Research, School of Agricultural Science, University of Tasmania, Hobart, Tasmania 7001, Australia*

C. Nicolás, *Departamento de Fisiología Vegetal, Centro Hispano-Luso de Investigaciones Agrarias, Facultad de Biología, Universidad de Salamanca, Plaza de los Doctores de la Reina s/n, 37007 Salamanca, Spain*

G. Nicolás, *Departamento de Fisiología Vegetal, Centro Hispano-Luso de Investigaciones Agrarias, Facultad de Biología, Universidad de Salamanca, Plaza de los Doctores de la Reina s/n, 37007 Salamanca, Spain*

H. Nonogaki, *Department of Horticulture, Oregon State University, Corvallis, OR 97331, USA*

M.K.J. Ooi, *Institute for Conservation Biology, University of Wollongong, Wollongong, NSW 2522, Australia; Biodiversity Conservation Science Section, Department of Environment and Conservation (NSW), Hurstville, NSW 2220, Australia*

N.W. Pammenter, *School of Biological and Conservation Sciences, University of KwaZulu-Natal, Durban 4041, South Africa*

E. Pestsova, *Institut für Entwicklungs und Molekularbiologie der Pflanzen, Heinrich-Heine-Universität, Universitätsstr, 1, 40225 Düsseldorf, Germany*

L. Rajjou, *Centre National de la Recherche Scientifique (CNRS)–Bayer CropScience Joint Laboratory (UMR CNRS 2847), Bayer CropScience, 14/20 rue Pierre Baizet, BP 9163, F-69263 Lyon, Cedex 09, France*

R.P. Rawnsley, *Tasmanian Institute of Agricultural Research, Burnie, Tasmania, Australia*

D. Reyes, *Departamento de Fisiología Vegetal, Facultad de Biología, Universidad de Salamanca-Centro Hispano Luso de Investigaciones Agrarias, Plaza de los Doctores de la Reina s/n, 37007 Salamanca, Spain*

D. Rodríguez, *Departamento de Fisiología Vegetal, Centro Hispano-Luso de Investigaciones Agrarias, Facultad de Biología, Universidad de Salamanca, Plaza de los Doctores de la Reina s/n, 37007 Salamanca, Spain*

J. Sadowski, *Department of Genetics and Plant Breeding, University of Technology and Agriculture, Al. Kaliskiego 7, 85-789 Bydgoszcz, Poland*

S.K. Shepherd, *Seed Science Laboratory, Department of Agriculture, The University of Reading, Earley Gate, Reading RG6 6AR, UK*

E. Sliwinska, *Department of Genetics and Plant Breeding, University of Technology and Agriculture, Al. Kaliskiego 7, 85-789 Bydgoszcz, Poland*

Y. Soeda, *Technology Development Centre, Sumika Agrotech Company, Ltd, 5-1 Sobiraki-cho, Niihama 792-0001, Japan*

S.Q. Song, *Xishuangbanna Tropical Botanical Garden, The Chinese Academy of Sciences, Menglun, Mengla 666303, Yunnan, China; School of Life Sciences, Sun Yat-Sen University, Guangzhou 510275, China*

C.J. Spurr, *Tasmanian Institute of Agricultural Research, School of Agricultural Science, University of Tasmania, Private Bag 98 Hobart, Tasmania 7001, Australia*

K.J. Steadman, *School of Plant Biology, Faculty of Natural and Agricultural Sciences, University of Western Australia, Western Australia 6009, Australia; School of Pharmacy, The University of Queensland, Queensland 4072, Australia*

J.-M. Strub, *Laboratoire de Spectrométrie de Masse Bio-Organique, ECPM, 25 rue Becquerel, 67087 Strasbourg, France*

X. Tian, *School of Life Sciences, Jishou University, Jishou 416000, China; School of Life Sciences, Sun Yat-Sen University, Guangzhou 510275, China*

Y.C. Tse, *Department of Biology, The Chinese University of Hong Kong, Shatin, New Territories, Hong Kong, China*

A.H.M. van der Geest, *Plant Research International, Wageningen University and Research Centre, PO Box 16, 6700AA Wageningen, The Netherlands*

A. van Dorsselaer, *Laboratoire de Spectrométrie de Masse Bio-Organique, ECPM, 25 rue Becquerel, 67087 Strasbourg, France*

V. Verdoucq, *Department of Horticulture, Virginia Tech, Blacksburg, VA 24061, USA*

D. Vinel, *Physiologie Végétale Appliquée, Université Pierre et Marie Curie-Paris 6, 75252 Paris, Cedex 05, France*

J.L. Walck, *Department of Biology, Middle Tennessee State University, Murfreesboro, TN 37132, USA*

J. Wang, *Department of Biology, The Chinese University of Hong Kong, Shatin, New Territories, Hong Kong, China*

W.M. Waterworth, *Faculty of Life Sciences, University of Manchester, Oxford Road, Manchester, UK*

G. Welbaum, *Department of Horticulture, Virginia Tech, Blacksburg, VA 24061, USA*

P. Westhoff, *Institut für Entwicklungs und Molekularbiologie der Pflanzen, Heinrich-Heine-Universität, Universitätsstr, 1, 40225 Düsseldorf, Germany*

J. Westwood, *Department of Plant Pathology, Physiology and Weed Science, Virginia Tech, Blacksburg, VA 24061, USA*

S. Wilson, *Tasmanian Institute of Agricultural Research, School of Agricultural Science, University of Tasmania, Hobart, Tasmania 7001, Australia*

Y. Xu, *The Vegetable Research Centre, Academy of Agricultural Sciences of Beijing, China*

X.Y. Zheng, *The Vegetable Research Centre, Academy of Agricultural Sciences of Beijing, China*

Preface

The Eighth International Workshop on Seeds was held in Brisbane, Australia, in May 2005, following the earlier workshops in this series in Jerusalem (1980), Wageningen (1985), Williamsburg (1989), Angers (1992), Reading (1995), Merida (1999) and Salamanca (2002). In the tradition of each of these workshops, the proceedings contain papers presented at the Brisbane Workshop. It was not possible to include all of the 221 papers and posters presented, but the 43 papers included in this volume are representative of the exciting advances being made in all areas of seed biology, ecology and technology. Particularly promising are the contributions of younger scientists using new technologies to investigate seed biology, opening the way for continuing advancement of the field.

The Workshop was attended by more than 250 participants, including 50 students from 35 countries around the world. In addition to the scientific sessions, the participants enjoyed a day experiencing some of Queensland's historic and cultural delights. The success of the Workshop can be attributed to the hard work and dedication of the Local Organizing Committee: Steve Adkins (Chairperson), Sarah Ashmore, Sheldon Navie, Sally Allen and Ron McLean. They were assisted in the scientific planning of the Workshop by the Executive Committee of the International Society for Seed Science (ISSS).

The ISSS was established following the Sixth Workshop in Merida in 1999, and now has many members worldwide. The Eighth Workshop in Brisbane was the second to be held under the auspices of the ISSS when Professor Ralph Obendorf was its second president. The current president, J. Derek Bewley, and President-elect Patricia Berjak lead this growing professional organization for seed science (more information available at: www.SeedSciSoc.org/). The Workshop organizers thank the ISSS and the following sponsors for their assistance and support:

Australian Society of Plant Scientists
Bayer CropScience
Centre for Forestry and Horticultural Research, Griffith University

Millennium Seed Bank Project, Kew
The University of Queensland

The ISSS will hold the Ninth International Conference on Seed Biology in Olsztyn, Poland, in 2008. Professor Ryszard Gorecki will chair the organizing committee and information will be posted on the ISSS website as the arrangements are made. We invite all those interested in seed biology to become members of the ISSS and join us in Olsztyn in 2008.

Steve Adkins
Sarah Ashmore
Sheldon Navie

1 Desiccation Tolerance in the -omics Era: New Tools for an Old Enigma?*

J. Buitink

UMR 1191 Physiologie Moléculaire des Semences (Université d'Angers/ INH/INRA), 16 Bd Lavoisier, 49045 Angers, France

Abstract

Desiccation tolerance (DT), or the capacity to survive nearly complete drying, is considered to be a multifactorial trait, involving the protection of macromolecules against water removal and regulation of the metabolism to prevent oxidative stress and the production of by-products to toxic levels. To obtain a general overview of the processes, molecules and regulating pathways involved in the acquisition of DT, transcriptome profiling was carried out on desiccation-tolerant and desiccation-sensitive stages of *Medicago truncatula* Gaertn. seeds using 16K microarrays. In order to discriminate DT from other developmental pathways, differences in gene expression were determined for two different stages: (i) between desiccation-sensitive embryos at 14 days after pollination (DAP) and those that had acquired their DT at 20 DAP; and (ii) between desiccation-sensitive radicles of germinated seeds and radicles of those germinated seeds that had received an osmotic treatment, resulting in the re-establishment of DT. Functional analysis of genes that were at least twofold upregulated or downregulated in both desiccation-tolerant stages revealed that the main groups containing upregulated genes belonged to defence and seed storage reserves. Groups containing typically downregulated genes were those involved in cell cycle and DNA processing as well as primary and energy metabolism. The main findings are discussed and compared with other desiccation-tolerant organisms.

Introduction

In the later stages of maturation, seeds of many species acquire the capacity to withstand removal of the majority of their water (i.e. they become desiccation-tolerant). This characteristic allows for prolonged survival in the dry state. During seed imbibition, water is taken up and metabolic processes resume, leading to the emergence of the radicle through the seed coat, concomitant with the loss of

*This chapter has been chosen to be the first in this book as it was presented at the workshop as the inaugural 'Michael Black Founders Lecture'.

desiccation tolerance (DT). In these radicles, DT can be re-established by treating the germinated seeds in an osmoticum at −1.5 to −1.7 MPa (Bruggink and van der Toorn, 1995; Buitink *et al.*, 2003). This re-establishment is possible only during a small developmental window after germination, and when radicles grow above a certain size full DT can no longer be reinduced (Buitink *et al.*, 2003).

The mechanisms of DT are thought to involve the synthesis of protective molecules, such as non-reducing sugars, late embryogenesis abundant (LEA) proteins, heat shock proteins (HSP) and various stress proteins (see Bartels and Salamini, 2001; Hoekstra *et al.*, 2001 for reviews). In addition, it has been hypothesized that a controlled downregulation of metabolism might be essential to prevent the production of detrimental reactive oxygen species (Leprince *et al.*, 2000; Hoekstra *et al.*, 2001). Considering the multifactorial trait of DT, it is possible that several regulatory pathways act in parallel and interact with one another to induce DT. Abscisic acid (ABA) and ABA-related signalling pathways play an important role in the acquisition of DT (Bartels and Salamini, 2001). Much of the research that has focused on elucidating signalling in DT has been performed on non-seed organisms, such as the so-called resurrection plants (Neale *et al.*, 2000; Bartels and Salamini, 2001; Ramanjulu and Bartels, 2002; Collett *et al.*, 2004) or on the rehydration of desiccation-tolerant bryophytes such as *Tortula ruralis* (Hedw.) Gaertn. (Oliver *et al.*, 2004). In seeds, much of the progress on the elucidation of signalling pathways and developmental programmes controlling DT has come from the isolation and analysis of viviparous mutants and mutants impaired in embryogenesis and maturation, such as the *fus3*, *lec1* and *abi3* mutants of *Arabidopsis* (*Arabidopsis thaliana* (L.) Heynh.), three master seed development regulators (Raz *et al.*, 2001). In *Arabidopsis* and maize (*Zea mays* L.), some of the target genes of these activators are genes proposed to have a role in DT, e.g. small HSPs (Wehmeyer and Vierling, 2000) and LEA proteins (Delseny *et al.*, 2001). However, progress in unravelling the signal transduction pathways for acquisition of DT has been impeded by the difficulty in discriminating them from those involved in other overlapping developmental programmes, such as reserve accumulation and acquisition of germinability. For instance, the desiccation-sensitive *abi3/fus3* mutant of *Arabidopsis* exhibits an array of pleiotropic defects ranging from decreased accumulation of storage proteins to reduced dormancy and vivipary (Delseny *et al.*, 2001; Raz *et al.*, 2001).

Transcriptome analysis is a useful tool for investigating the regulatory and protective mechanisms involved in the acquisition of DT and may shed light on shifts in the metabolic and cellular processes necessary during preparation for the dry state. Comparison of the genes induced or repressed upon the re-establishment of DT in germinated radicles, with those coinciding with the acquisition of DT during maturation, might help to uncouple the different developmental and stress pathways and put emphasis on those genes that are related to DT. The legume *M. truncatula* Gaertn., which is closely related to economically relevant species such as lucerne (*M. sativa* L.), soybean (*Glycine max* (L.) Merr.) and pea (*Pisum sativum* L.), is a major target species for the understanding of seed biology and storage reserve accumulation in proteinaceous seeds (Bell *et al.*, 2001). Seeds of *M. truncatula* are relatively large and can be easily selected after germination according to their radicle length to obtain a synchronized and reproducible curve for the re-establishment of DT (Buitink *et al.*, 2003).

With the expression data obtained from 16,086 probe oligo microarrays of *M. truncatula* (Buitink *et al.*, 2006), this study focuses on highly expressed genes in relation to DT. It also examines two stages of seed development – before and after the acquisition of DT – as well as upon the re-establishment of DT in germinated radicles after an osmotic treatment. Further characterization of these candidate genes is expected to contribute to the elucidation of mechanisms involved in seed survival in the dry state.

Materials and Methods

Biological material

M. truncatula 'Paraggio' (Seedco, Hilton, Australia) seeds were imbibed in distilled water at 20°C in the dark. In order to reinduce DT in germinated, desiccation-sensitive radicles of seeds of *M. truncatula*, 20 h imbibed germinated seeds with a protruded radicle length of 2.7–2.9 mm (referred to as 3 mm) were selected and submitted to an osmotic treatment by incubation in a polyethylene glycol (PEG) 8000 solution (−1.7 MPa) at 10°C in the dark (Buitink *et al.*, 2003). In order to test DT, germinated seeds with 3 mm long protruded radicles without PEG treatment, or after 72 h of PEG incubation, were dried for 2–3 days at 20°C under an air flow at 42% relative humidity (RH) (Buitink *et al.*, 2003). Seeds were considered desiccation-tolerant when the radicle resumed growth upon reimbibition. Each sample contained 50 seeds and the treatment was replicated at least three times. Water contents were determined gravimetrically on three replicates of 25 radicles. To determine the acquisition of DT during maturation, *M. trunculata* plants were grown under a 16 h light regime (250 μmol/m^2/s photosynthetic photon flux density) at 23/21°C and flowers were labelled when pollination occurred. At different time points, 100 seeds were removed from the pods and dried for 2–3 days at 20°C under an air flow at 42% RH to determine the percentage of DT. Water contents were determined gravimetrically on three replicates of five seeds.

RNA extraction, label preparation and hybridization of microarrays

Total RNA was isolated according to Verwoerd *et al.* (1989) from 50 embryos at two maturation stages (i.e. 14 and 20 DAP) and from 300 radicles (3 mm long emerged) before and after 72 h of PEG incubation. After excision, the material was directly frozen in liquid nitrogen to prevent a wounding response. Total RNA aliquots (400–800 μg) were used to purify mRNA using the PolyATract mRNA isolation system III kit according to the manufacturer's protocol (Promega, Madison, Wisconsin, USA). The quality of mRNA was assessed using an Agilent 2100 bioanalyser (Agilent, Palo Alto, California, USA). A biological replicate consisted of two separate mRNA extractions. Cy3- and Cy5-labelled cDNAs were prepared using the CyScribe cDNA post-labelling kit (Amersham Pharmacia Biotech). The two replicate mRNA extractions from the PEG-treated radicles and embryos aged for 14 DAP were each labelled individually with Cy3. The reference mRNA (i.e. untreated radicles or 20 DAP) was labelled with Cy5 (see Buitink *et al.*, 2006).

Mt16kOLI1 microarrays contain 16,086 70-mer oligonucleotide probes (Qiagen) representing all tentative consensus sequences (TCs) of The Institute for Genomic Research (TIGR) *M. truncatula* Gene Index 5 (http://www.tigr.org/tdb/mtgi) (Hohnjec *et al.*, 2005). Duplicate spots are present in the same grids throughout Mt16kOLI1 microarrays. Each Cy3-labelled sample was mixed with an equal amount of Cy5-labelled sample, preincubated with yeast tRNA and poly A^+RNA, denatured (i.e. 2 min at 100°C, 30 min at 37°C) and hybridized to the microarrays. Two independent hybridizations were performed for each RNA sample, leading to four hybridizations per sample. After 16 h of hybridization at 42°C, microarrays were washed, dried and treated with Dyesaver (Genisphere, Hatfield, Pennsylvania, USA).

Image acquisition and data analysis

Hybridized arrays were scanned by fluorescence confocal microscopy on a ScanArray Express Scanner (PerkinElmer, Boston, Massachusetts, USA) at laser power ranging from 75% to 100% and photo-multiplier tube gain settings ranging from 65% to 100%. Measurements were obtained separately for each fluorochrome at 10 μm/pixel resolution. Signal intensities were extracted with the GenePix Pro 5.0 image analysis software (Axon Instruments, USA). Image processing was performed using MADSCAN software (Le Meur *et al.*, 2004). This dynamic procedure validates the quality of the raw data points and of the microarray, corrects systematic and random biases by normalizing the filtered data, and detects outliers and validates statistically the expression level of each probe. For the comparisons between embryos of 20 vs 14 DAP, and radicles of 3 mm treated for 72 h with PEG vs the untreated radicles, genes were retained when at least four out of the eight replicates were valid and with a false discovery rate of 0.008% using a statistical analysis of microarray (SAM) one class response *t*-test. Genes were considered differently expressed when the genes were at least twofold differentially expressed.

Results

Differentially expressed genes between desiccation-sensitive and desiccation-tolerant tissues were identified using the microarrays available for *M. truncatula*, containing 16,086 genes that represent all the contigs present in the MtGI5 database (www.tigr.org) (Buitink *et al.*, 2006). Differential gene expression was related to DT using two comparisons: one between desiccation-sensitive embryos of 14 DAP and 20-day-old embryos that had acquired their DT, evident from the high percentage of survival after rapid drying (Fig. 1.1a and c); and a second one between desiccation-sensitive 3 mm radicles from germinated seeds and those from germinated seeds that had received a PEG treatment, which led to the re-establishment of DT (Fig. 1.1b and d) (see Buitink *et al.*, 2003 for details). Each slide was hybridized with a mixture of Cy3-labelled cDNA from tolerant tissues and Cy5-labelled cDNA from sensitive tissues.

In total, over 10,000 of the 16,086 genes were expressed in the tissues at some stage (Buitink *et al.*, 2006). Here, differentially expressed genes were selected when

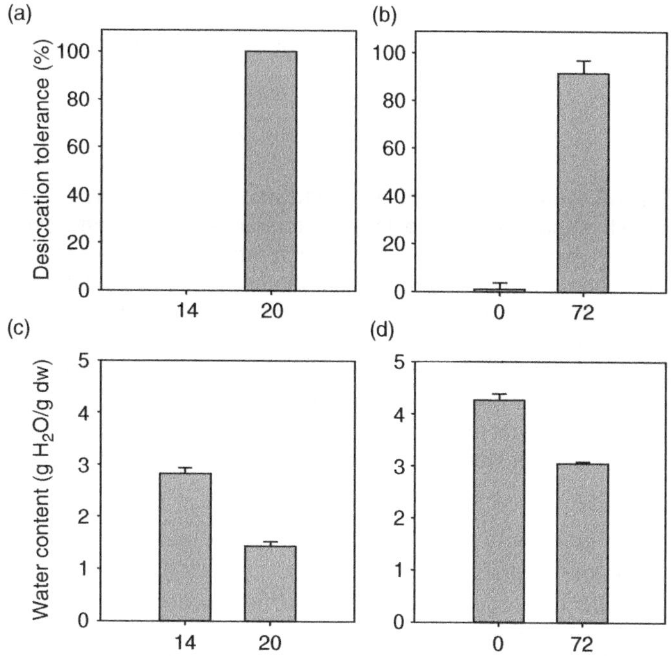

Fig. 1.1. (a and c) Desiccation tolerance (DT) and water content during maturation and (b and d) the re-establishment of DT in *Medicago truncatula* seeds. (a) DT was determined at 14 and 20 days after pollination (DAP) by counting the percentage of seeds that germinated after drying and subsequent rehydration. Seeds were dried for 3 days at 20°C in an air flow of 42% relative humidity (RH) and were considered to have germinated when the radicle protruded through the seed coat. DT was determined on a population of 100 seeds. (c and d) Water content (g H_2O/g dw) was determined gravimetrically on triplicate of five seeds (maturation in terms of DAP) or 25 radicles (with or without polyethylene glycol (PEG) treatment). Mean ± standard error (SE). (b) Mean DT was determined on germinated seeds with a radicle length of 3 mm that were either dried directly (i.e. 0 h) or incubated for 72 h at 10°C in a PEG 8000 solution (–1.7 MPa) in the dark prior to drying (for 3 days at 20°C in an air flow of 42% RH). DT was described as above on triplicates of populations of 50 seeds (mean ± SE).

the expression between the two stages was at least twofold upregulated or downregulated. For the comparison between 14- and 20-day-old embryos, 677 genes were upregulated and 411 downregulated upon the acquisition of DT (DEV in Fig. 1.2). Likewise, 1123 and 1264 genes were upregulated or downregulated, respectively, upon the re-establishment of DT in the radicles (PEG in Fig. 1.2). A Venn diagram was constructed between both desiccation-tolerant data-sets (DEV and PEG, acquisition and re-establishment of DT during maturation and PEG treatment, respectively). In all, 227 upregulated genes and 107 downregulated genes were found to be common between the two data-sets (Fig. 1.2).

Subsequently, the differentially regulated genes that were common were classified into functional groups. The relative representation of the genes in the different

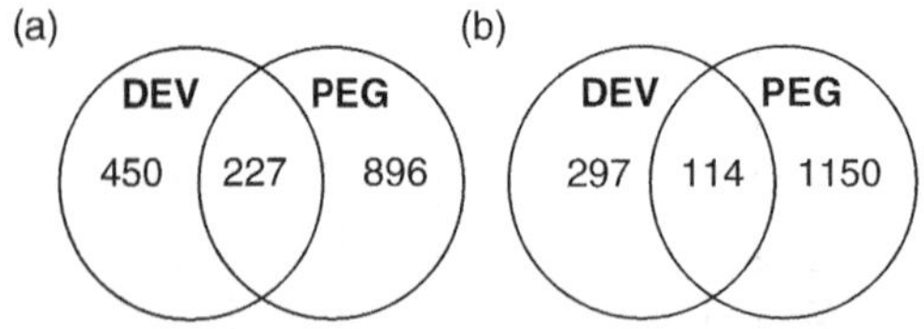

Fig. 1.2. Venn diagrams of (a) upregulated and (b) downregulated genes expressed upon the acquisition of desiccation tolerance (DT) and after the re-establishment of DT in germinated seeds. Expression ratios were obtained from comparison of desiccation-tolerant, 20-day-old embryos with desiccation-sensitive 14-day-old embryos harvested during seed development (DEV) and from the comparison of desiccation-tolerant, 3 mm long protruded radicles from germinated seeds in which DT was re-established after 72 h of PEG treatment (−1.7 MPa, 10°C) with similar radicles prior to polyethylene glycol (PEG) treatment. Genes were selected when they were at least twofold differentially expressed with a false discovery rate of 0.008% using a statistical analysis of microarrays (SAM) one class response *t*-test.

groups is shown in pie charts (Fig. 1.3). The groups representing the largest number of known genes were those related to cellular defence (21%) and metabolism (12%) for the upregulated genes. For the downregulated genes, the largest number of genes was related to metabolism (25%) and cell cycle (12%). Certain groups were represented by typically upregulated or downregulated genes. For instance, groups that contained mostly downregulated genes were those related to nucleotide and lipid metabolism, cell cycle, and DNA processing and transport (Fig. 1.3). Alternatively, groups that contained mainly upregulated genes in relation to DT were those related to cellular defence, transcription and cellular communication, and storage reserves, such as lipids and proteins.

In order to obtain an overview of the genes that are strongly induced or repressed in both tolerant and sensitive comparisons, genes were selected that were at least threefold upregulated or downregulated, and are listed in Table 1.1. The indicated names are based on the closest homology of the basic local alignment search tool (BLAST) search of the contigs.

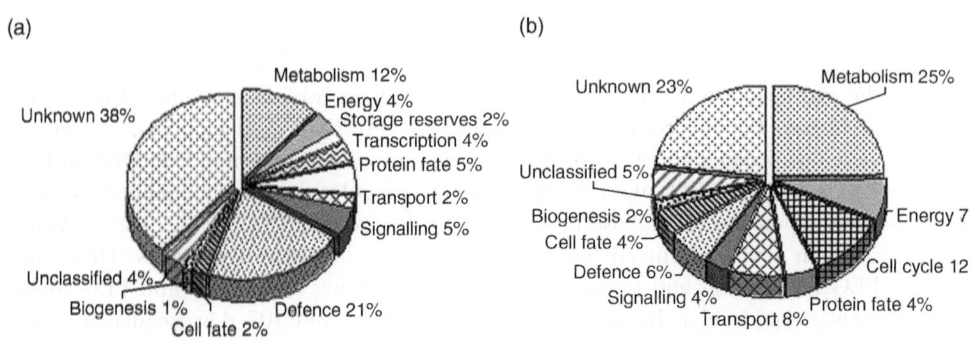

Fig. 1.3. Distribution of genes associated with desiccation tolerance (DT) in functional groups. Selected genes are over twofold (a) upregulated or (b) downregulated both during the acquisition and re-establishment of DT.

Table 1.1. Classification of genes that are more than threefold upregulated or downregulated upon acquisition and re-establishment of desiccation tolerance (DT). Indicated names are based on the closest homology of the basic local alignment search tool (BLAST) search of the contigs.

Classification	Identification
Upregulated	
Metabolism	Homocysteine *S*-methyltransferase, aldose reductase, allantoinase ADP-glucose pyrophosphorylase, cinnamoyl-CoA reductase, UDP-glycose
Energy	Pyrophosphate-fructose-6-phosphate-1-phospho-transferase, seed maturation protein PM34, aconitate hydratase, succinic semialdehyde dehydrogenase, aldehyde dehydrogenase
Storage reserves	Caleosin (2), steroleosin
Transcription	bHLH transcription factor
Protein fate	Cysteine proteinase precursor, trypsin inhibitor precursor
Transport	Pore protein homologue, outer membrane lipoprotein, ferritin precursor
Signalling	RING finger protein, SnRK1 subunit SNF4b, protein phosphatase type 2C, fiber annexin, calcineurin B-like protein 4
Defence/protection	BURP domain-containing protein, ethylene-responsive protein, metallothionein-like protein, 20.5 kDa heat shock protein, dormancy-associated protein (3), metallothionein II, 1-cys peroxiredoxin, embryonic abundant protein gD-34, LEA 4, seed maturation protein PM27, seed biotin-containing protein SPB65, maturation protein PM3, maturation protein PM32, CapLea1, PM10, cold acclimation responsive protein BudCAR5, Em protein, embryonic abundant protein 59K, dehydrin-like protein, dehydrin, 35 kDa seed maturation protein, ripening related protein, cationic peroxidase, phospholipid hydroperoxide glutathione peroxidase, lipoxygenase, isoflavone 3′-hydroxylase
Unclassified/unknown	22
Downregulated	
Metabolism	Argininosuccinate synthase, aspartate transaminase, beta-1,3-glucanase, sucrose synthase, enoyl-ACP reductase precursor, acetyl-CoA carboxylase, hydroxymethylglutaryl-CoA reductase
Energy	Fumarate hydratase, malate dehydrogenase, ferredoxin [2Fe-2S] fd1 precursor
Cell cycle	Beta tubulin, tubulin beta-8 chain, cdc2-like protein kinase, tubulin alpha-5 chain
Protein synthesis	Ribosomal protein S7
Protein fate	Serine proteinase
Transport	Aquaporin TIP-type, aquaporin PIP1-3, phosphoenolpyruvate/phosphate translocator, H+ATPase, dicarboxylate/tricarboxylate carrier, integral membrane protein
Signalling	Protein kinase, receptor protein kinase PERK1
Defence	L-ascorbate oxidase precursor
Biogenesis	Extensin-like protein, arabinogalactan protein
Unclassified/unknown	10

Discussion

The comparison of gene expression in seeds during maturation and the re-establishment of DT allows for the identification of genes that are potentially involved in DT. However, gene expression does not necessarily lead to the translation of the gene product or enzyme activity, mRNA stability could have been altered and mRNA can be stored in mRNPs to function in the rehydration phase (Wood and Oliver, 1999); thus caution is required in the interpretation of the data. It is unlikely that all the regulatory mechanisms leading to the re-establishment of DT in germinated radicles are similar to those involved in the acquisition of DT during maturation. For instance, downregulated genes could be related to the arrest of radicle growth or germination processes that have not yet been initiated in developing seeds. For example, no specific transcription factor (TF) was found that was downregulated both during acquisition and re-establishment of DT (Fig. 1.3), illustrating the difference between the developmental stage of the two tissues prior to the induction of DT. A closer look at the genes that are only upregulated during the PEG treatment shows that some of them do correspond to genes activated during seed development. For instance, genes encoding seed storage proteins (e.g. legumin, albumin, globulin, vicilin and seed storage protein SSP2) or oleosins and caleosins as well as genes expressed at the later stages of maturation (e.g. stachyose synthase, raffinose synthase and oleosins) (Buitink *et al.*, 2006). These genes were not all picked up as being differentially expressed during maturation. Most likely, they are expressed before or after the time interval of 14–20 DAP. The multitude of genes reactivated in the PEG-treated radicles which are originally expressed during seed development is consistent with the idea of a partial return to the development of the germinated radicles, as was previously observed for drying of germinating pea axes (Lalonde and Bewley, 1986). Interestingly, for both maturation and germination, the induction of DT was accompanied by a decrease in water content (wc). The decrease was 1.4 H_2O g/g dw for the embryos after the acquisition of DT during maturation (Fig. 1.1c) and 1.3 g/g for the radicles that were submitted to the PEG treatment (Fig. 1.1d). Previously, Black *et al.* (1999) showed that a reduction in wc comparable with that measured in this study promotes the acquisition of DT in developing embryos of wheat (*Triticum aestivum* L.). It is possible that the change in wc is responsible for the activation of comparable signalling pathways both during acquisition and re-establishment of DT.

A number of genes activated in the early stages of dehydration in resurrection plants are similar to those expressed in the desiccating seed of most plant species (Neale *et al.*, 2000; Ramanjulu and Bartels, 2002; Collett *et al.*, 2004) or even upon rehydration of dried desiccation-tolerant bryophytes (Oliver *et al.*, 2004) (Table 1.1). Indeed, most of the screens for genes involved in DT were carried out on resurrection plants, and a comparison with our data allows for the detection of possible common pathways or protective mechanisms involved in DT. It would be particularly interesting to compare genes of unknown function that are upregulated in this study (85 genes) to those expressed in resurrection plants during drying or in bryophytes upon rehydration. Some genes were identified that share homology with known genes in other DT screens. For instance, a gene that is upregulated during the acquisition and re-establishment of DT, albeit at a lower level during maturation than PEG, is encoding a suppressor of initiation (SUI) protein, a translation initiation factor (Buitink *et al.*, 2006). A homologue of this gene, *eIF1*, was detected previously in *Sporobolus stapfianus* Gand. (Neale *et al.*, 2000). In these plants, high levels of *eIF1* transcripts were detected during the mid-stage of dehydration and the transcript levels were highest in completely dehydrated plants, suggesting a role in the reinitiation of metabolic activity upon rehydration (Neale *et al.*, 2000).

A functional group that contains many upregulated as well as downregulated genes is that of transport, including water transport (Fig. 1.3). Aquaporins are a complex family of channel proteins that facilitate the transport of water along the transmembrane water potential gradients (Maurel and Chrispeels, 2001), and their regulation of expression is mediated by dehydration or water stress. Using double *Arabidopsis* antisense lines for a *PIP1* and *PIP2* isoform, Martre *et al.* (2002) found that these plasma membrane intrinsic protein (PIP) aquaporins played a significant role during recovery from water deficit. In our study, we found a gene encoding a putative aquaporin, a homologue to PIP1–3 (Table 1.1; Buitink *et al.*, 2006). A PIP aquaporin in *Xerophyta humilis* (Bak.) Dur. and Schinz was also found to be downregulated upon dehydration (Collett *et al.*, 2004). It is not known whether these homologues have a comparable function to those studied in *Arabidopsis* or why they need to be downregulated. Another gene encoding an aquaporin, a tonoplast intrinsic protein (TIP), was also strongly downregulated (Table 1.1). This TIP gene, *MtAQP1*, was induced in symbiosis when compared to non-mycorrhizal control roots, and injection of *in vitro* transcribed RNA into *Xenopus* oocytes revealed that the encoded MtAQP1 protein specifically facilitates water transport (Krajinski *et al.*, 2000). The authors suggested that MtAQP1 might play a role in buffering osmotic fluctuations in the highly compartmental vacuole of arbuscule cells. Only one gene encoding a TIP alpha is two- to threefold upregulated during the induction of DT (Buitink *et al.*, 2006). The expressed sequence tag (EST) libraries in which this gene is found are all seed-specific. This protein shows homology to a TIP of *Phaseolus vulgaris* L., a seed-specific and highly conserved protein (Johnson *et al.*, 1990). This TIP is located in protein storage vacuoles in seeds and transports small metabolites between the storage vacuoles and cytoplasm of seed storage tissues.

Protection

The stress and defence class was the one containing the most upregulated genes (Fig. 1.3). Indeed, many of the highly expressed genes encode LEA proteins (14) and HSPs, as well as genes involved in detoxification (Table 1.1). In *S. stapfianus*, functional classification of 55 upregulated cDNAs showed that a large number of cDNAs had a protective function (Collett *et al.*, 2004). In their study, LEA proteins accounted for 31% of the identified dehydration upregulated genes. In our study, 14 out of a total of the 19 known LEA proteins present in the EST database were strongly induced, and one gene sharing homology to a 35 kDa maturation protein showed a >200-fold increase (Buitink *et al.*, 2006). LEA proteins are low-complexity, highly hydrophilic and mostly unordered proteins in the hydrated state and heat-stable after boiling (Cuming, 1999; Wise, 2003). They are classified into at least five groups by virtue of similarity in their amino acid sequences (Cuming,

1999; Wise, 2003), and those detected in this DT screen included members of all five groups. Generally, the presence of LEA proteins correlates well with DT and they are thought to play a protective role during dehydration. *In vitro*, several LEA proteins were found to protect enzymes against nearly complete loss of water brought about by rapid evaporation or vacuum drying (Goyal *et al.*, 2003; Grelet *et al.*, 2005). However, their role in DT *in vivo* has not yet been clearly identified. All desiccation-tolerant organisms appear to accumulate LEA proteins, ranging from bacteria and nematodes to bryophytes, pollen, seeds and resurrection plants (Blackman *et al.*, 1991; Battista *et al.*, 2001; Wolkers *et al.*, 2001; Ramanjulu and Bartels, 2002; Goyal *et al.*, 2003; Collett *et al.*, 2004). Other genes encoding proteins with a protective role that were identified in our study are small heat shock proteins (sHSPs) (Table 1.1). Quantitation of sHSPs in desiccation-intolerant seeds of *abi3–6*, *fus3–3* and *lec1–2 Arabidopsis* mutants showed that all had <2% of wild-type HSP17.4 levels, correlating to a reduction in sHSPs with desiccation sensitivity (Wehmeyer and Vierling, 2000).

Genes that were classified in detoxification encode enzymes that are mostly involved in detoxification of reactive oxygen species, such as 1-Cys peroxiredoxin (Prx) and peroxidases (Table 1.1), confirming the importance of free radical processing systems in DT. 1-Cys Prx is an antioxidant that is also present in *X. viscosa* Baker (Mowla *et al.*, 2002). Recently, it was suggested that 1-Cys Prx is employed to sense and/or react to seed environmental conditions, thus preventing germination under unfavourable conditions (Haslekås *et al.*, 2003). In this group we also found metallothionein genes that are known to be upregulated upon drought (Bray, 2002) and in the desiccation-tolerant resurrection plant *X. humilis* (Collett *et al.*, 2004). These proteins act by sequestering intracellular iron, which could be involved in the generation of reactive hydroxyl radicals through a Fenton reaction (Bajaj *et al.*, 1999). Genes encoding these proteins were also found in the rehydration transcriptome of the desiccation-tolerant bryophyte *T. ruralis* (Oliver *et al.*, 2004). It is possible that their mRNA abundance during rehydration in *T. ruralis* might indicate that they are accumulating in the seed tissues prior to drying but play a role in the early rehydration phase in seed tissues.

One gene that was upregulated in relation to DT encodes aldose reductase (Table 1.1). This gene is closely linked to a barley (*Hordeum vulgare* L.) gene that was shown to be associated with the acquisition of DT in barley embryos (Bartels *et al.*, 1991) and an aldose reductase gene from *X. viscosa* (Mundree *et al.*, 2000). Ectopic expression of a homologue of this gene in tobacco (*Nicotiana tabacum* L.) resulted in improved tolerance to oxidative stress and dehydration, and physiological analysis suggests that it probably functions by reducing the level of aldehydes (Oberschall *et al.*, 2000). This confirms that mechanisms exist to avoid the build-up of metabolites to toxic levels during drying in desiccation-tolerant tissues.

Cell cycle and metabolism

One of the functional groups with genes mainly downregulated during the acquisition and re-establishment of DT is related to cell cycle and DNA processing (Fig. 1.3 and Table 1.1). Growth of the radicles was completely arrested during the

PEG incubation (Buitink *et al.*, 2003), explaining the downregulation of a cdc2-like protein kinase (Magyar *et al.*, 1997) (Table 1.1). The mRNA of the homologue in *M. sativa, cdc2MsF,* was found to peak during G2–M phases, in parallel with their protein abundance, and Mtcdc2MsF interacted with a G2- to M-phase-related kinase complex (Magyar *et al.*, 1997). Indeed, Faria *et al.* (2005) found that imbibed *M. truncatula* radicles contained relatively high 4C DNA contents, suggesting that during germination the cell cycle may be at the G2/M boundary. In addition, several α and β tubulin genes were strongly downregulated (Table 1.1). The incubation of germinated *M. truncatula* seeds in a PEG solution for 3 days caused no changes in the microtubular cytoskeleton in the radicles (Faria *et al.*, 2005), but when they were dehydrated the microtubular cytoskeleton was partially dismantled. It is possible that the strong decrease in α and β tubulin transcripts (Table 1.1) is an indication that the radicles are preparing themselves for the dry state.

Overall, the induction of DT goes together with a massive repression of genes involved in primary and energy metabolism (e.g. those involved in amino acid metabolism, carbon metabolism, lipid metabolism, secondary metabolism, tricarboxylic acid (TCA) cycle and electron transport) (Buitink *et al.*, 2006). Depression of metabolism is a key survival strategy for organisms that undergo long periods of anoxia or freezing (MacDonald and Storey, 1999a) and is hypothesized to play a role in the DT of seeds as well (Leprince *et al.*, 2000). Indirect evidence for this comes from the reduced rate of respiration after the re-establishment of DT in radicles of germinated cucumber (*Cucumis sativus* L.) seeds (Leprince *et al.*, 2000). Also, in *X. humilis*, a large number of the downregulated genes were involved in metabolism (25%) as well as photosynthesis (25%) (Collett *et al.*, 2004), and it was suggested that this might be related to the controlled shutdown of metabolism and photosynthesis in preparation for the desiccated state. Interestingly, genes encoding H+ATPases were downregulated at the end of the re-establishment of DT (Table 1.1), as was found for drought stress (Seki *et al.*, 2002). Several studies have shown that regulated suppression of activities of membrane ion channels and ion pumps is a key factor in achieving the energy conservation that results in metabolic rate depression (Hochachka *et al.*, 1997; MacDonald and Storey, 1999b).

Regulation

A relatively large number of TFs were expressed upon the induction of DT (Buitink *et al.*, 2006). It would be tempting to speculate that analysis of the so far uncharacterized TFs might provide further clues about the regulatory mechanisms that are involved in DT. On the other hand, several upregulated TFs and signalling molecules were identified that share homology to known genes previously studied in other species, thereby giving insights into putative regulatory and signalling pathways that might be involved in DT. A review of the functions of these genes, based on their homology to known genes, reveals that they are either related to developmental and maturation processes occurring in seeds or involved in regulating responses to abiotic stresses. For instance, a basic leucine zipper (bZIP) DNA-binding protein shares homology with *ROM2* (regulator of MAT2), a repressor of maturation (MAT)-specific and LEA-specific gene expression in dicots, acting during seed desiccation (Chern *et al.*, 1996). Another Myb TF encoded a tuber-specific and sucrose-responsive element-binding factor. The expression of a homologue of this TF from *Arabidopsis* (i.e. MYBR1) is affected by the *fus3*, *lec1* and *abi3* mutations (Kirik *et al.*, 1998). In our study, *abi3* was present but not differentially expressed, and the homologue of *fus3* could not be identified. Furthermore, one gene was expressed in relation to DT that shows homology to drought-responsive element (DRE)-binding proteins, involved in the ABA-independent pathway of drought tolerance (Liu *et al.*, 1998) (data not shown). This gene is known to be induced in drought-stressed tissues. The expression of this gene during the acquisition of DT indicates that partial overlap might exist between the regulatory pathways involved in drought and DT. This is also suggested by the upregulation in our study of a homologue of a Myb-related gene, *cpm10*, from the desiccation-tolerant resurrection plant *Craterostigma plantagineum* Hochst. (Iturriga *et al.*, 1996) (data not shown). Overexpression of this gene in *Arabidopsis* led to drought and salt tolerance of transgenic lines (Villalobos *et al.*, 2004).

Only a small number of signalling genes with high transcript levels were found in common between both data-sets (Table 1.1). One example is *SFN4b*, a putative activating subunit of the sucrose non-fermenting-related kinase (SnRK1) complex, whose expression was already found to be correlated to DT (Buitink *et al.*, 2004). The SnRK1 complex in seeds seems to play an important role in regulating carbon and energy metabolism (Thelander *et al.*, 2004). Whether SNF4b is indeed involved in activating one of the SnRK1 complexes in seeds is currently under investigation in our laboratory. Furthermore, several Ca^{2+}-related signalling molecules were discovered, such as a calcineurin B-like (CBL) protein (Table 1.1). These proteins are part of a family of Ca^{2+} sensors that are involved in plant signal transduction processes in response to stress conditions, and function by interacting with other signalling proteins, such as the CBL-interacting protein kinases (CIPKs), also known as the SnRK3 family (Kim *et al.*, 2000). Another signalling gene found to be strongly upregulated was phosphatase 2C (PP2C) (Table 1.1). Plant PP2Cs are represented by a large family, and the different phosphatases appear to have specific functions as regulators of signal transduction pathways and are also involved in development (Schweighofer *et al.*, 2004). The PP2C identified in this study does not share homology with the group A or B PP2Cs, involved in the negative regulation of ABA signalling or regulation of mitogen-activated protein kinase (MAPK) signalling, respectively. It shares homology with the *Arabidopsis* PP2C At3g15260 that belongs to group F, which has not been characterized so far (Schweighofer *et al.*, 2004).

Conclusions

A transcriptomic analysis on seeds of *M. truncatula* using 16K probe oligo microarrays identified genes that are expressed both upon the acquisition of DT during maturation and after the re-establishment of DT by PEG incubation of germinated seeds. Many genes previously identified in relation to protection were upregulated, whereas genes involved in cell cycle and growth were downregulated. Furthermore, a large number of genes involved in primary and secondary metabolism, as well as transport, were also differentially expressed. The large shift of the transcriptome accompanying the acquisition of DT suggests that tissues enter a more quiescent state in a regulated manner that allows them to survive the dry state. Further characterization of the candidate genes involved in signalling and regulatory molecules in relation to DT will bring us a step closer to the understanding of DT mechanisms.

Acknowledgements

We thank B. Jettner (Seedco Australia Co-operative Ltd, Hilton, Australia) for the generous gift of *M. truncatula* 'Paraggio' seeds. Dr J. Leger, I. Guisle and Dr N. Le Meur are gratefully acknowledged for help with the microarray experiments. This work was supported in part by grants from Contrat Etat-Région 2000–2006 and the French Institute for Agronomy Research (INRA).

References

Bajaj, S., Targolli, J., Liu, L.F., Ho, T.H.D. and Wu, R. (1999) Transgenic approaches to increase dehydration-stress tolerance in plants. *Molecular Breeding* 5, 493–503.

Bartels, D. and Salamini, F. (2001) Desiccation tolerance in the resurrection plant *Craterostigma plantagineum*: a contribution to the study of drought tolerance at the molecular level. *Plant Physiology* 127, 1346–1353.

Bartels, D., Engelhardt, K., Roncarati, R., Schneider, K., Rotter, M. and Salamini, F. (1991) An ABA and GA modulated gene encodes an aldose reductase-related protein. *The EMBO Journal* 10, 1037–1043.

Battista, J.R., Park, M.-J. and McLemore, A.E. (2001) Inactivation of two homologues of proteins presumed to be involved in the desiccation tolerance of plants sensitizes *Deinococcus radiodurans* R1 to desiccation. *Cryobiology* 43, 133–139.

Bell, C.J., Dixon, R.A., Farmer, A.D., Flores, R., Inman, J., Gonzales, R.A., Harrison, M.J., Paiva, N.L., Scott, A.D., Weller, J.W. and May, G.D. (2001) The *Medicago* genome initiative: a model legume database. *Nucleic Acids Research* 29, 114–117.

Black, M., Corbineau, F., Gee, H. and Côme, D. (1999) Water content, raffinose, and dehydrins in the induction of desiccation tolerance in immature wheat embryos. *Plant Physiology* 120, 463–472.

Blackman, S.A., Wettlaufer, S.H., Obendorf, R.L. and Leopold, A.C. (1991) Maturation proteins associated with desiccation tolerance in soybean. *Plant Physiology* 96, 868–874.

Bray, E.A. (2002) Classification of genes differentially expressed during water-deficit stress in *Arabidopsis thaliana*: an analysis using microarray and differential expression data. *Annals of Botany* 89, 803–811.

Bruggink, T. and van der Toorn, P. (1995) Induction of desiccation tolerance in germinated seeds. *Seed Science Research* 5, 1–4.

Buitink, J., Ly Vu, B., Satour, P. and Leprince, O. (2003) A physiological model to study the re-establishment of desiccation tolerance in germinated radicles of *Medicago truncatula* Gaertn. seeds. *Seed Science Research* 13, 273–286.

Buitink, J., Thomas, M., Gissot, L. and Leprince, O. (2004) Starvation, osmotic stress and desiccation tolerance lead to expression of different genes of the regulatory β and γ subunits of the SnRK1 complex in germinating seeds of *Medicago truncatula*. *Plant, Cell and Environment* 27, 55–67.

Buitink, J., Leger, J.J., Guisle, I., Vu, B.L., Wuillème, S., Le Bars, A., Le Meur, N., Becker, A., Küster, H. and Leprince, O. (2006) Transcriptome profiling uncovers metabolic and regulatory processes occurring during the transition from desiccation sensitive to tolerant stages in *Medicago truncatula* seeds. *The Plant Journal* 47, 735–750.

Chern, M.S., Bobb, A.J. and Bustos, M.M. (1996) The regulator of MAT2 (ROM2) protein binds to early maturation promoters and represses PvALF-activated transcription. *The Plant Cell* 8, 305–321.

Collett, H., Shen, A., Gardner, M., Farrant, J.M., Denby, K.J. and Illing, N. (2004) Towards transcript profiling of desiccation tolerance in *Xerophyta humilis*: construction of a normalized 11k *X. humilis* cDNA set and microarray expression analysis of 424 cDNAs in response to dehydration. *Physiologia Plantarum* 122, 39–53.

Cuming, A.C. (1999) LEA proteins. In: Shewry, P.R. and Casey, R. (eds) *Seed Proteins*. Kluwer Academic Publishers, Dordrecht, The Netherlands, pp. 753–780.

Delseny, M., Bies-Etheve, N., Carles, C., Hull, G., Vicient, C., Raynal, M., Grellet, F. and Aspart, L. (2001) Late embryogenesis abundant (LEA) protein gene regulation during *Arabidopsis* seed maturation. *Journal of Plant Physiology* 158, 419–427.

Faria, J.M.R., Buitink, J., van Lammeren, A.A.M. and Hilhorst, H.W.M. (2005) Changes in DNA and microtubules during loss and re-establishment of desiccation tolerance in germinating *Medicago truncatula* seeds. *Journal of Experimental Botany* 56, 2119–2130.

Goyal, K., Tisi, L., Basran, A., Browne, J., Burnell, A., Zurdo, J. and Tunnacliffe, A. (2003) Transition from natively unfolded to folded state induced by desiccation in an anhydrobiotic nematode protein. *Journal of Biological Chemistry* 278, 12977–12984.

Grelet, J., Benamar, A., Teyssier, E., Avelange-Macherel, M.-H., Grunwald, D. and Macherel, D. (2005) Identification in pea seed mitochondria of a later embryogenesis abundant protein able to protect enzymes from drying. *Plant Physiology* 137, 157–167.

Haslekås, C., Viken, M.K., Grini, P.E., Nygaard, V., Nordgard, S.H., Meza, T.J. and Aalen, R.B. (2003) Seed 1-cysteine peroxiredoxin antioxidants are not involved in dormancy, but contribute to inhibition of germination during stress. *Plant Physiology* 133, 1148–1157.

Hochachka, P.W., Land, S.C. and Buck, L.T. (1997) Oxygen sensing and signal transduction in metabolic defense against hypoxia: lessons from vertebrate facultative anaerobes. *Comparative Biochemistry and Physiology* 118A, 23–29.

Hoekstra, F.A., Golovina, E.A. and Buitink, J. (2001) Mechanisms of plant desiccation tolerance. *Trends in Plant Science* 5, 431–438.

Hohnjec, N., Vieweg, M.F., Pühler, A., Becker, A. and Küster, H. (2005) Overlaps in the transcriptional profiles of *Medicago truncatula* roots inoculated with two different *Glomus* fungi provide insights into the genetic program activated during arbuscular mycorrhiza. *Plant Physiology* 137, 1283–1301.

Iturriga, G., Leyns, L., Villegas, A., Gharabeih, R., Salamini, F. and Bartels, D. (1996) A family of novel *myb*-related genes from the resurrection plant *Craterostigma plantagineum* are specifically expressed in callus and roots in response to ABA or desiccation. *Plant Molecular Biology* 32, 707–716.

Johnson, K.D., Hofte, H. and Chrispeels, M.J. (1990) An intrinsic tonoplast protein of protein storage vacuoles in seeds is structurally related to a bacterial solute transporter (GIpF). *The Plant Cell* 2, 525–532.

Kim, K.-N., Cheong, Y.H., Gupta, R. and Luan, S. (2000) Interaction specificity of *Arabidopsis* calcineurin B-like calcium sensors and their target kinases. *Plant Physiology* 124, 1844–1853.

Kirik, V., Kölle, K., Miséra, S. and Bäumlein, H. (1998) Two novel MYB homologues with changed expression in late embryogenesis-defective *Arabidopsis* mutants. *Plant Molecular Biology* 37, 819–827.

Krajinski, F., Biela, A., Schubert, D., Gianinazzi-Pearson, V., Kaldenhoff, R. and Franken, P. (2000) Arbuscular mycorrhiza development regulates the mRNA abundance of *Mtaqp1* encoding a mercury-insensitive aquaporin of *Medicago truncatula*. *Planta* 211, 85–90.

Lalonde, L. and Bewley, J.D. (1986) Patterns of protein synthesis during the germination of pea axes, and the effects of an interrupting desiccation period. *Planta* 167, 504–510.

Le Meur, N., Lamirault, G., Bihouée, A., Steenman, M., Bédrine-Ferran, H., Teusan, R., Ramstein, G. and Léger, J.J. (2004) A dynamic, web-accessible resource to process raw microarray scan data into consolidated gene expression values: importance of replication. *Nucleic Acids Research* 32, 5349–5358.

Leprince, O., Harren, F.J.M., Buitink., J., Alberda, M. and Hoekstra, F.A. (2000) Metabolic dysfunction and unabated respiration precede the loss of membrane integrity during dehydration of germinating radicles. *Plant Physiology* 122, 597–608.

Liu, Q., Kasuga, M., Sakuma, Y., Abe, H., Miura, S., Yamaguchi-Shinozaki, K. and Shinozaki, K. (1998) Two transcription factors, DREB1 and DREB2, with an EREBP/AP2 DNA binding domain separate two cellular signal transduction pathways in drought- and low-temperature-responsive gene expression, respectively, in *Arabidopsis*. *The Plant Cell* 10, 1391–1406.

MacDonald, J.A. and Storey, K.B. (1999a) Cyclic AMP-dependent protein kinase: role in anoxia and freezing tolerance of the marine periwinkle *Littorina littorea*. *Marine Biology* 133, 193–203.

MacDonald, J.A. and Storey, K.B. (1999b) Regulation of ground squirrel Na+K+-ATPase activity by reversible phosphorylation during hibernation. *Biochemical and Biophysical Research Communications* 254, 424–429.

Magyar, Z., Meszaros, T., Miskolczi, P., Deak, M., Feher, A., Brown, S., Kondorosi, E., Athanasiadis, A., Pongor, S., Bilgin, M., Bako, L., Koncz, C. and Dudits, D. (1997) Cell cycle phase specificity of putative cyclin-dependent kinase variants in synchronized alfalfa cells. *The Plant Cell* 9, 223–235.

Martre, P., Morillon, R., Barrieu, F., North, G.B., Nobel, P.S. and Chrispeels, M.J. (2002) Plasma membrane aquaporins play a significant role during recovery from water deficit. *Plant Physiology* 130, 2101–2110.

Maurel, C. and Chrispeels, M.J. (2001) Aquaporins: a molecular entry into plant water relations. *Plant Physiology* 125, 135–138.

Mowla, S.B., Thomson, J.A., Farrant, J.M. and Mundree, S.G. (2002) A novel stress-inducible antioxidant enzyme identified from the resurrection plant *Xerophyta viscosa* Baker. *Planta* 215, 716–726.

Mundree, S.G., Whittaker, A., Thomson, J.A. and Farrant, J.M. (2000) An aldose reductase homolog from the resurrection plant *Xerophyta viscosa* Baker. *Planta* 211, 693–700.

Neale, A.D., Blomstedt, C.K., Bronson, P., Guthridge, K., Evans, J., Gaff, D.J. and Hamill, J.D. (2000) The isolation of genes from the resurrection grass *Sporobolus stapfianus* which are induced during severe drought stress. *Plant, Cell and Environment* 23, 265–277.

Oberschall, A., Deak, M., Torok, K., Sass, L., Vass, I., Kovacs, I., Feher, A., Dudits, D. and Horvath, G.V. (2000) A novel aldose/aldehyde reductase protects transgenic plants against lipid peroxidation under chemical and drought stresses. *The Plant Journal* 24, 437–446.

Oliver, M.J., Dowd, S.E., Zaragoza, J., Mauget, S.A. and Payton, P.R. (2004) The rehydration transcriptome of the desiccation-tolerant bryophyte *Tortula torturalis*: transcript classification and analysis. *BMC Genomics* 5, 89.

Ramanjulu, S. and Bartels, D. (2002) Drought- and desiccation-induced modulation of gene expression in plants. *Plant, Cell and Environment* 25, 141–151.

Raz, V., Bergervoet, J.H. and Koornneef, M. (2001) Sequential steps for developmental arrest in *Arabidopsis* seeds. *Development* 128, 243–252.

Schweighofer, A., Hirt, H. and Meskiene, I. (2004) Plant PP2C phosphatases: emerging functions in stress signalling. *Trends in Plant Science* 9, 236–243.

Seki, M., Narusaka, M., Ishida, J., Nanjo, T., Fujita, M., Oono, Y., Kamiya, A., Nakajima, M., Enju, A., Sakurai, T., Satou, M., Akiyama, K., Taji, T., Yamaguchi-Shinozaki, K., Carninci, P., Kawai, J., Hayashizaki, Y. and Shinozaki, K. (2002) Monitoring the expression profiles of 7000 *Arabidopsis* genes under drought, cold and high-salinity stresses using a full-length cDNA microarray. *The Plant Journal* 31, 279–292.

Thelander, M., Olsson, T. and Ronne, H. (2004) Snf1-related protein kinase 1 is needed for growth in a normal day-light cycle. *The EMBO Journal* 23, 1900–1910.

Verwoerd, T.C., Dekker, B.M. and Hoekema, A. (1989) A small-scale procedure for the rapid isolation of plant RNAs. *Nucleic Acids Research* 17, 2362.

Villalobos, M.A., Bartels, D. and Iturruaga, G. (2004) Stress tolerance and glucose insensitive phenotypes in *Arabidopsis* overexpressing the *CpMYB10* transcription factor gene. *Plant Physiology* 135, 309–324.

Wehmeyer, N. and Vierling, E. (2000) The expression of small heat shock proteins in seeds responds to discrete developmental signals and suggests a general protective role in desiccation tolerance. *Plant Physiology* 122, 1099–1108.

Wise, J. (2003) LEAping to conclusions: a computational reanalysis of the late embryogenesis abundant proteins and their possible roles. *BMC Bioinformatics* 4, 52–71.

Wolkers, W.F., McCready, S., Brandt, W.F., Lindsey, G.G. and Hoekstra, F.A. (2001) Isolation and characterization of a D-7 LEA protein from pollen that stabilizes glasses *in vitro*. *Biochimica et Biophysica Acta* 1544, 196–206.

Wood, A.J. and Oliver, M.J. (1999) Translational control in plant stress: the formation of messenger ribonucleoprotein particles (mRNPs) in response to desiccation of *Tortula ruralis* gametophytes. *The Plant Journal* 18, 359–370.

2 Recent Progress Towards the Understanding of Desiccation Tolerance

P. Berjak and N.W. Pammenter

School of Biological and Conservation Sciences, University of KwaZulu-Natal, Durban 4041, South Africa

Abstract

Desiccation tolerance (DT), which is a remarkable survival mechanism, has been identified in one or other life cycle stage in all the major plant taxa, and is a property found across a variety of other life forms. The present contribution considers the commonality of some of the characteristics of DT across the spectrum of organisms, focusing particularly on the involvement of proteins – generally late embryogenesis abundant (LEA) proteins – with sugars in the intracellular glassy state; free radicals, active oxygen species and antioxidants; and recent evidence that, even under ideal conditions, stored seeds have extended, but not infinite, lifespans.

Introduction

Desiccation tolerance (DT), which is just one mechanism of drought tolerance (Alpert, 2005), pertains to organisms that will survive dehydration to an overall water concentration ≤0.1 g/g dw, or water potentials (Ψ) ≤ −100 MPa (Vertucci and Farrant, 1995; Walters *et al.*, 2005a). At intracellular water concentrations ≤0.1 g/g, macromolecules are no longer surrounded by a water monolayer; thus enzymatic activity and metabolism are precluded (Billi and Potts, 2002). However, unique water properties pertain at such low intracellular concentrations, so this is not the same thing as complete dryness. Hence the terms anhydrobiotic and anhydrobiosis are avoided by the present authors. In macromolecules, water may be considered a structural component, the complete removal of which would probably result in conformational changes (Billi and Potts, 2002).

DT in seeds is the outcome of the interaction of a suite of genetically controlled traits (Pammenter and Berjak, 1999), considered to be intrinsic cell characteristics (Walters *et al.*, 2005a). However, the absence or presence, and degree of expression and interaction, of those traits present, must account for differences in response to dehydration among highly desiccation-tolerant, less tolerant and

 Seeds: Biology, Development and Ecology (eds S. Adkins, S. Ashmore and S.C. Navie)

variously desiccation-sensitive organisms, or life cycle stages of species (Pammenter and Berjak, 1999). It must be stressed that DT necessitates not only the means to withstand extreme water loss, but also the ability to survive for prolonged periods in the dehydrated state.

Is There Commonality Facilitating Desiccation Tolerance Among Organisms?

Besides being a trait of orthodox seeds, DT occurs in bacteria, terrestrial microalgae, lichens, bryophytes, resurrection angiosperms and five animal phyla (Alpert, 2005).

Resurrection plants

Desiccation-tolerant resurrection plants are able to remain viable despite considerable dehydration, resuming metabolic activity when water once again becomes available. As is the case for orthodox seeds, DT of the vegetative tissues is based on a spectrum of relatively complex protection mechanisms that accompany dehydration (Illing *et al.*, 2005).

Late embryogenesis abundant (LEA) proteins, sucrose and certain oligosaccharides accumulate coincidently with the acquisition of DT during orthodox seed development (Buitink *et al.*, 2002), and particular antioxidant enzymes become prominent (Bailly, 2004). The expression of at least 16 different LEA genes (identified from a survey of only 425 cDNAs) (J.M. Farrant, 2005, University of Cape Town, South Africa, personal communication) has been found to occur in the leaves of the xerotolerant resurrection plant, *Xerophyta humilis* (Baker) Dur. and Schinz during dehydration (Collett *et al.*, 2004). The antioxidant 1-cys-peroxiredoxin, which had previously been considered to be seed-exclusive, was found to be abundantly expressed in tissues of the resurrection plants *X. humilis* and *X. viscosa* (Baker). Illing *et al.* (2005) also reported that sucrose accumulates only in the tissues of the desiccation-tolerant *Eragrostis nindensis* (Ficalho & Hiern), and not in related sensitive *Eragrostis* species indicating a further commonality between desiccation-tolerant seeds in general and vegetative tissues of resurrection plants.

Lichens

The result of the lichen-forming symbiosis between a fungus (i.e. the mycobiont) and a green alga or cyanobacterium (i.e. the photobiont) is that neither of the partners remains constrained to the cryptic habitats that would be obligatory for either one alone (Kranner *et al.*, 2005). In the lichen *Cladonia vulcani* (Savicz), those same authors have shown this to be the outcome of significant upregulation of both photo-protective and antioxidant mechanisms, resulting in the lichen thallus being desiccation-tolerant (which is a prerequisite for its occurring above ground).

Desiccation-tolerant animals

DT has been documented for nematodes (Browne *et al.*, 2002) and bdelloid rotifers (Lapinski and Tunnacliffe, 2003) and, classically, for the encysted embryos of the brine shrimp, *Artemia* species (see Clegg, 2005). In nematodes, accumulation of a 'novel heat-stable protein' was reported as a response to desiccation (Solomon *et al.*, 2000), as was the upregulation of a gene coding for a LEA protein in response to desiccation stress in the desiccation-tolerant nematode, *Aphelenchus avenae* (Bastian) (Browne *et al.*, 2002). In parallel, the latter authors found that *A. avenae* accumulates trehalose in response to dehydration, leading to the suggestion that the LEA–trehalose combination might act synergistically in the formation of a glass, as is currently favoured for LEA–sucrose-based glasses in plant cells (see later). A protein–trehalose synergism has also been demonstrated *in vitro* in the highly desiccation-tolerant encysted embryos of *Artemia* (Viner and Clegg, 2001; Clegg, 2005). However, neither of two species of bdelloid rotifers expressing a LEA-type protein were found to accumulate trehalose, or other simple sugars other than glucose (Lapinski and Tunnacliffe, 2003), the implications of which will be considered later.

Microorganisms

Prokaryotes in soil, which might periodically become very dry, need to be able to protect against the consequences of dehydration, and survival of bacteria in the dry state has important implications for health issues (Billi and Potts, 2002). In the stationary phase under salt-stress conditions, *Bacillus subtilis* is characterized by considerable accumulation of a protein, GsiB, which bears a marked similarity to LEA Group 1 proteins (Stacy and Aalen, 1998).

The Intracellular Glassy State

During the developmental drying phase (i.e. maturation drying) of desiccation-tolerant seeds a marked increase in cytoplasmic viscosity occurs as tissue water concentrations decline below 0.3 g/g (Buitink and Leprince, 2004), which is indicative of the formation of the intracellular glassy state. Although vitrification below the glass transition temperature (T_g) will occur in any supercooled liquid that remains uncrystallized (Burke, 1986), dehydration is the principal factor resulting in glass formation during maturation drying of orthodox seeds. Hence the glassy state of the cytomatrix is essentially a super-saturated solution of extremely high viscosity. Glasses are described as amorphous solids lacking regular structure, but in which the molecules are interconnected, so that mobility is restricted within the matrix (Walters, 1998). However, like liquids, glasses lack specific molecular orientation, resulting in free volume (i.e. 'holes') within the matrix, which allows the diffusion of small molecules (Walters, 1998) or other chemical species (e.g. free radicals) (Hendry, 1993; Benson and Bremner, 2004) through the matrix.

The original realization that the intracellular solution within orthodox seeds at low water concentrations assumes the glassy state (Koster and Leopold, 1988;

Williams and Leopold, 1989; Koster, 1991) provided an explanation of the basis of seed survival in the desiccated state. The metastability of the glassy state could also account for the fact that seed storage lifespan is limited. However, the original conceptualization was that glass formation in seeds occurred as a consequence of their sucrose and oligosaccharide content, and other possible contributors to the intracellular glasses were largely overlooked.

LEAs with sugars as the basis of intracellular glasses

Walters *et al.* (1997) first showed that for wheat (*Triticum aestivum* L.) ~50% of the sugars from the LEA protein–sugar fraction appeared to be tightly protein-associated. Later, Buitink *et al.* (2000) showed that the intrinsic stability of the intracellular glassy state did not change after osmo-priming, which was accompanied by decreased oligosaccharide and increased sucrose contents, compared with the original (i.e. non-primed) condition. However, the longevity of the primed seeds was considerably reduced compared with non-primed seeds stored under the same conditions, despite the fact that water concentration was similar in both (Buitink *et al.*, 2000). These observations indicate that factors other than sugar composition and relative concentrations, as well as the physical properties of the intracellular glasses per se, affect the seed storage lifespan. However, in view of the abundance of non-reducing sugars in seeds, it is inevitable that they must participate in the glassy state, perhaps by filling free volume between large molecules, thereby increasing the density of the glass (Buitink and Leprince, 2004).

Wolkers *et al.* (2001) isolated a D-7 LEA from desiccation-tolerant pollen of *Typha latifolia* L., and showed it to have an unordered, random coil conformation in solution, as has been demonstrated for other LEAs (Goyal *et al.*, 2003; Wise and Tunnacliffe, 2004). However, the conformation of the protein changed considerably upon drying: when the LEA protein was dehydrated (either slowly or rapidly) in the presence of sucrose, it assumed an α-helical conformation (Wolkers *et al.*, 2001). The most interesting result emanated from a comparison of pure sucrose glasses with glasses containing both sucrose and the LEA protein in different ratios. The presence of the protein was found to increase both the T_g of the glasses, as well as the average strength of hydrogen bonding (Wolkers *et al.*, 2001), leading to the suggestion that the protein acted synergistically with the sucrose in the formation of the glassy matrix. This attributes functions to both the LEA protein and the disaccharide in conferring long-term stability on intracellular glasses in desiccation-tolerant organisms in the dry state. There is, however, additional significance of the accumulation of sucrose at the expense of monosaccharides that accompanies maturation drying in orthodox seeds, in that respiratory substrates become increasingly unavailable. This is not a new idea, as Rogerson and Matthews (1977) correlated the fall in respiratory rate with a decline in respiratory substrate in developing pea (*Pisum sativum* L.) seeds.

Goyal *et al.* (2003) have characterized a LEA-like protein (AavLEA1) from the desiccation-tolerant nematode *A. avenae*, which is natively unfolded, but has the propensity to develop a significant α-helical component upon dehydration. These authors speculated that, upon dehydration, AavLEA1 could form coiled coils similar in morphology to cytoskeletal intermediate filaments (IFs), a concept further explored by Wise and Tunnacliffe (2004). Significantly, IF morphology confers intracellular flexibility, strength and resilience. Goyal *et al.* (2003) speculated that LEA protein filaments could act synergistically with sugar glasses, thus increasing tensile strength. This is an attractive idea, providing a functional role for both LEAs and sucrose in seeds (or trehalose in desiccation-tolerant animal cells) in providing a relatively stable dehydrated intracellular matrix.

Wise and Tunnacliffe (2004) suggest that the function of LEA proteins has been obscure mainly because of their natively unfolded structure in the hydrated state. They also maintain that the apparently low relatedness between LEAs and other polypeptides is probably a consequence of LEAs containing regions of low sequence complexity that are routinely masked out during conventional searches for sequence similarity. Therefore, Wise and Tunnacliffe (2004) have developed a novel computational approach, the protein or oligonucleotide probability profile (POPP), which facilitates analyses based on similarities in peptide composition of proteins, rather than on sequence similarities. As a consequence, they have proposed a revised, and essentially less complex, system for LEA classification. The findings about the similarities among IFs and the LEAs from *T. latifolia*, and the nematode *A. avenae*, are described as fitting in well with predictions made by POPP analysis (Wise and Tunnacliffe, 2004).

It is apparent that views about the intracellular amorphous state in desiccation-tolerant organisms have moved from the concept of sugar glasses to that of glasses in which proteins, especially LEAs, play a dominant role. However, there is a multiplicity of other cytomatrical components that will be caught up in, and may contribute to, the intracellular glass. Intraorganellar solutions will also vitrify at sufficiently low water concentrations, and these are likely to differ compositionally from the cytoplasmic solution (Buitink and Leprince, 2004). Furthermore, the interior of organelles such as mitochondria and plastids, as well as the milieu of the chromatin might represent localized pools of higher water concentration (Berjak *et al.*, 1986; Rinne *et al.*, 1999), making a variable local density of 'intracellular glass' possible. Further, it might be conjectured that there are intracellular 'pockets' of sufficient residual water to preclude localized glass formation, accounting for the progressive action of endo- and exo-nucleases in desiccated seeds (Elder *et al.*, 1987) and enzyme activity, which has been suggested to be maintained in dehydrin-associated local pools of water, in the otherwise dehydrated cells of birch (*Betula pubescens* Ehrh.) buds (Rinne *et al.*, 1999). The evidence and conjecture therefore support the concept of non-homogeneous intracellular glass of variable density, probably with pockets that are sufficiently fluid to permit some chemical reactions in dehydrated, desiccation-tolerant tissues. This implies that dehydrated, desiccation-tolerant tissues are not at thermodynamic equilibrium. This suggested state of disequilibrium is maintained not by the expenditure of energy but simply by the rate at which processes leading to equilibrium occur. Perhaps the achievement of a state of equilibrium coincides with death.

Is the accumulation of simple sugars imperative for desiccation tolerance?

Lapinski and Tunnacliffe (2003) have demonstrated that bdelloid rotifers withstand desiccation although they do not contain trehalose or any equivalent disaccharide.

The rotifers do, however, accumulate an LEA-related protein upon dehydration to the water concentration 0.06–0.10 g/g, at which glasses must have formed.

Sugars, however, appear to be ubiquitous in desiccated seeds and, apparently, resurrection plants (Illing *et al.*, 2005). Bryant *et al.* (2001) have suggested that large solutes (perhaps the LEAs) may well be excluded from narrow intermembrane spaces, where glasses incorporating only small solutes (e.g. sugars) might provide a buffer reducing physical stresses and the associated increase in fluid-to-gel transition temperature (T_m) of the lipid that would prevail were sugars absent. This proposition does not detract from the suggested LEA–sugar synergism that would presumably prevail in cytomatrical locations outside of narrow intermembrane spaces.

Free Radicals, Active Oxygen Species and Antioxidants

The highly reactive free radicals, along with hydrogen peroxide (H_2O_2), are collectively termed active/reactive oxygen species (AOS/ROS). While the intracellular glassy state must curtail molecular and AOS mobility, and thus interaction, during the process of desiccation, intracellular structures are highly vulnerable as conditions for radical generation are enhanced (Vertucci and Farrant, 1995; Pammenter and Berjak, 1999; Walters *et al.*, 2005a).

The most frequently discussed AOS in the context of intracellular phenomena are the superoxide radical ($O_2^{-}\cdot$), H_2O_2 and the hydroxyl radical ($\cdot OH$). While $O_2^{-}\cdot$ itself is relatively short-lived, it may be dismutated to form H_2O_2, which, if not reduced to water, can be involved in a major route of $\cdot OH$ formation by reaction with Fe^{2+} (Hendry, 1993; Benson and Bremner, 2004), the major source of which is thought to be the reduction of Fe^{3+} by $O_2^{-}\cdot$ (Hendry, 1993). The $\cdot OH$ is considered to be the most damaging AOS in biological tissues, reacting with nucleic acids, lipids and, indirectly, with proteins (Benson and Bremner, 2004).

Possession and effective operation of a suite of antioxidants is of prime importance during dehydration of orthodox seeds, and again as soon as water uptake commences by the desiccated cells (Pammenter and Berjak, 1999). Additionally, non-enzymic antioxidants are likely to confer protection during the desiccated state in seeds (Bailly, 2004) and in the cells of any organism or structure capable of surviving extreme desiccation. Most studies on AOS and antioxidants have been focused on the final stages of seed desiccation, but in his review Bailly (2004) draws together the limited information available for various stages of seed development (and during germination). The work by Aalen (1999) and Stacy *et al.* (1999) makes a good start to this extension.

Although the family of peroxiredoxins (which reduce H_2O_2 and $\cdot OH$) is wide-ranging across organisms generally (Aalen, 1999), prior to the studies of Illing *et al.* (2005) the expression of 1-cys-peroxiredoxin genes was thought to be unique to seeds. It takes place in barley (*Hordeum vulgare* L.) only in those tissues that survive desiccation (Stacy *et al.*, 1999). Most interestingly, 1-cys-peroxiredoxin has been localized to the nuclei of imbibed, dormant barley embryo and aleurone cells, where it is held to afford antioxidant protection to the DNA, perhaps by preventing free radical attack during desiccation (Stacy *et al.*, 1999). If localized water pools do occur *in absentia* in the milieu of the chromatin (see earlier), it is highly possible that 1-cys-peroxiredoxin can and does function to protect the genome against AOS in the desiccated seed.

What is the situation during seed development? From work on developing seeds, Bailly (2004) noted that the activity of the major antioxidant enzymes changes with development. High activity of superoxide dismutase (SOD; $O_2^{-}\cdot \rightarrow H_2O_2$) and ascorbic peroxidase (APX; $H_2O_2 \rightarrow H_2O$) characterizes the desiccation-sensitive stage in bean (*Phaseolus vulgaris* L.) seeds, when the activity of catalase (CAT; $H_2O_2 \rightarrow H_2O$) and glutathione reductase (GR) is low. In contrast, the desiccation-tolerant condition is characterized by high CAT and GR activity, and low SOD and APX activity. According to de Tullio and Arrigoni (2003) neither ascorbate nor APX are essential for the desiccation phase, particularly as the presence of ascorbate in association with transition metals at that stage could lead to ·OH formation. As discussed by Bailly (2004), a trend of increasing CAT activity during seed desiccation in various species seems to be emerging, leading to the conjecture that transcription of the catalase gene might be finely regulated by water loss.

During orthodox seed dehydration, from 0.45 to <0.25 g/g (i.e. Ψ −3 to <−11 MPa), unregulated metabolism is held to occur accompanied by the potential for free radical generation (Vertucci and Farrant, 1995; Walters *et al.*, 2005a). It is thus particularly necessary that the components of the antioxidant system appropriate to operation in this water concentration range are fully expressed and active.

Stored seeds at water concentrations <0.1 g/g remain vulnerable to free radical production by autoxidation (Walters *et al.*, 2005a). Considering the prevalence of the glassy state and low Ψ (<−100 MPa), conditions are not considered to be conducive to enzymic antioxidant operation, and non-enzymic mechanisms are suggested to prevail (Bailly, 2004). However, as the same arguments about restricted molecular mobility must apply to both sorts of antioxidants, the appropriate scavengers must be located where they are required, and localized conditions must permit their reactivity. This may well be the case (as for 1-cys-peroxiredoxin in nuclei) for a spectrum of non-enzymic antioxidants, especially if localized water pools do exist within organelles (Berjak *et al.*, 1986; Rinne *et al.*, 1999). Nevertheless, this is a situation that should not preclude localized antioxidant enzyme activity, given that the appropriate enzymes are present. However, to establish the situation, microscopical immunocytochemical (ICC) methodology should be widely used to locate specific molecules in freeze-substituted specimens, thus avoiding spurious results that may be generated by aqueous-based analytical approaches. As an indication of the benefits, Stacy *et al.* (1999) localized 1-cys-peroxiredoxin to nuclei by means of ICC and confocal fluorescence microscopy.

Despite putative control mechanisms, if free radicals are generated while seeds are in the desiccated condition they may accumulate, especially during prolonged storage under poor conditions (Hendry, 1993), making essential the mobilization of appropriate antioxidants at the start of imbibition. It appears that H_2O_2 is produced early during imbibition in a variety of seeds, particularly when non-enzymic antioxidants come into play (Bailly, 2004 and references therein) and increased activity of antioxidant enzymes, particularly GR and CAT, occurs prior to radicle protrusion in sunflower (*Helianthus annuus* L.) seeds (Bailly *et al.*, 2002).

Although much progress has been made in the area of AOS and antioxidants in seeds, and recognizing the positive aspects of AOS (Bailly, 2004), there are still gaps that are required to be filled before an overall appreciation of the fine details can be obtained.

The brief overview of AOS and antioxidants in relation to DT would not be complete without a consideration of the intriguing findings on the lichen *C. vulcani* by Kranner *et al.* (2005). These authors showed that in the lichenized condition, both the photo-protective and antioxidant systems of the alga were upregulated relative to photo-assimilation – properties that were lost to the alga in the separated state. The mycobiont apparently benefits from the association in terms of antioxidant defence, showing only an essentially ineffective glutathione-based system if separated. Kranner *et al.* (2005) hypothesized that the resilience of the lichen to desiccation is the consequence of mutual stimulation by both partners of the photo-protective and antioxidant capacities, which facilitates not only enhanced DT but also rapid and complete recovery upon rehydration.

The Mortality of Stored Seeds

There are a few biological specimens that tolerate overall water concentrations below 0.01 g/g without incurring any obvious damage. Encysted embryos of *Artemia* constitute one example (Clegg, 1986), while spores of *Aspergillus niger* showed no reduction in viability when stored at 1% RH (Walters *et al.*, 2005a). Seeds of *Welwitschia mirabilis* Hook. f. (which is endemic to the Namib Desert) survived after drying to 0.007 g/g (Whitaker *et al.*, 2004) and showed no reduction in viability when stored at this low water concentration in sealed foil containers for 2 years at 5.5°C or −20°C. However, in all these cases the structures are heavily encapsulated, and it is feasible that the living tissues were actually at water concentrations somewhat higher than the surrounding non-living structures. Nevertheless, from the results of ultra-dry seed storage experiments, it seems unlikely that viability could be retained in the complete absence of intracellular water (Walters and Engels, 1998).

From a wide-ranging study across species, Walters *et al.* (2005b) found that even when stored under near-ideal gene-banking conditions, the viability of desiccated seeds declines. Characteristic P_{50} values were not correlated with content of sucrose or oligosaccharides (Walters *et al.*, 2005b), supporting the view that the stability of the intracellular glassy state is not primarily dependent on these compounds (Buitink and Leprince, 2004).

It is, however, apparent that the properties of intracellular glasses do change at very low water concentrations (Buitink and Leprince, 2004), supporting Walters' (1998) identification of temperature-dependent critical water contents, below which the deterioration rate of dry seeds increases. As relaxation (i.e. movement) within the glassy matrix may occur with time, Walters (1998) also concluded that the nature and kinetics of chemical deteriorative processes might be expected to change, which could underlie the kinetics of seed ageing. Walters *et al.* (2004) concluded that, even below −130°C, ageing reactions are enabled because molecules remain sufficiently mobile, which has serious implications for cryostorage of desiccated material. It is pertinent that free radicals could persist within the glassy matrix, despite the significant extension of lifespan afforded by cryostorage (Benson and Bremner, 2004).

The nature of deteriorative reactions in seeds at water concentrations or Ψ below the critical level does change (Vertucci and Farrant, 1995; Walters *et al.*, 2005a), which, at water concentrations ≤0.08 g/g (i.e. Ψ −150 to −1000 MPa), include free radical production via autoxidation, evolution of carbonyls and destabilization of protein and membrane structure. Concomitantly, increased seed ageing rates that typify ultra-dry seed storage occur (Walters *et al.*, 2005a).

What might occur at water concentrations ≤0.08 g/g, to account for increased molecular mobility, change in the nature of deleterious events that occur and increased rate of deterioration that characterizes ultra-dry seeds? Water is suggested to play a critical role in the integrity of the glassy matrix by hydrogen bonding between components: on withdrawal of water below the critical level, the structural continuity of the glass is suggested to be perturbed, and, as suggested by Walters (1998), structural integrity of macromolecules may be compromised.

Loss of glass integrity could result in increased porosity, with the potential for greater molecular or free radical mobility. Hence, the rate of deleterious reactions might well increase while the desiccated state prevails, the consequences of which would be cumulative with time. Additionally, if free radicals do persist in the glassy state (Benson and Bremner, 2004), the increase in water in the seed during post-storage imbibition is also potentially hazardous.

Recent progress has furthered the understanding of DT, the commonality of certain traits across a range of organisms, and the statics and dynamics of the desiccated state. In the context of orthodox seeds, the major challenge is translating this progress into best practices for plant germplasm conservation.

Acknowledgements

We are pleased to acknowledge helpful discussions with Tony Ford (University of KwaZulu-Natal, Durban, South Africa), Jill M. Farrant (University of Cape Town, South Africa) and Chris Walters (National Center for Genetic Resources Preservation, Fort Collins, Colorado, USA).

References

Aalen, R.B. (1999) Peroxiredoxin antioxidants in seed physiology. *Seed Science Research* 9, 285–295.

Alpert, P. (2005) The limits and frontiers of desiccation-tolerant life. *Integrative and Comparative Biology* 45, 685–695.

Bailly, C. (2004) Active oxygen species and antioxidants in seed biology. *Seed Science Research* 14, 93–107.

Bailly, C., Bogatek-Leszczynska, R., Côme, D. and Corbineau, F. (2002) Changes in activities of antioxidant enzymes and lipoxygenase during growth of sunflower seedlings and seeds of different vigour. *Seed Science Research* 12, 47–55.

Benson, E.E. and Bremner, D. (2004) Oxidative stress in the frozen plant: a free radical point of view. In: Fuller, B.J., Lane, N. and Benson, E.E. (eds) *Life in the Frozen State.* CRC Press, Boca Raton, Florida, pp. 205–241.

Berjak, P., Dini, M. and Gevers, H.O. (1986) Deteriorative changes in embryos of long-stored, uninfected maize caryopses. *South African Journal of Botany* 52, 109–116.

Billi, D. and Potts, M. (2002) Life and death of dried prokaryotes. *Research in Microbiology* 153, 7–12.

Browne, J., Tunnacliffe, A. and Burnell, A. (2002) Anhydrobiosis: plant desiccation gene found in a nematode. *Nature* 416, 38.

Bryant, G., Koster, K.L. and Wolfe, J. (2001) Membrane behaviour in seeds and other systems at low water content: the various effects of solutes. *Seed Science Research* 11, 17–25.

Buitink, J. and Leprince, O. (2004) Glass formation in plant anyhdrobiotes: survival in the dry state. *Cryobiology* 48, 215–228.

Buitink, J., Hemminga, M.A. and Hoekstra, F.A. (2000) Is there a role for oligosaccharides in seed longevity? An assessment of intracellular glass stability. *Plant Physiology* 122, 1217–1224.

Buitink, J., Hoekstra, F.A. and Leprince, O. (2002) Biochemistry and biophysics of tolerance systems. In: Black, M. and Pritchard, H.W. (eds) *Desiccation and Survival in Plants: Drying without Dying*. CAB International, Wallingford, UK, pp. 293–318.

Burke, M.J. (1986) The glassy state and survival of anhydrous biological systems. In: Leopold, A.C. (ed.) *Membranes, Metabolism and Dry Organisms*. Comstock Publishing Associates, Ithaca, New York, pp. 358–363.

Clegg, J.S. (1986) The physical properties and metabolic status of *Artemia* cysts at low water contents: the 'water replacement hypothesis'. In: Leopold, A.C. (ed.) *Membranes, Metabolism and Dry Organisms*. Comstock Publishing Associates, Ithaca, New York, pp. 169–187.

Clegg, J.S. (2005) Desiccation tolerance in encysted embryos of the animal extremophile, *Artemia*. *Integrative and Comparative Biology* 45, 715–724.

Collett, H., Shen, A., Gardner, M., Farrant, J.M., Denby, K.J. and Illing, N. (2004) Towards transcript profiling of desiccation tolerance in *Xerophyta humilis*: construction of a normalized 11k *X. humilis* cDNA set and microarray expression analysis of 242 cDNAs in response to dehydration. *Physiologia Plantarum* 122, 39–53.

de Tullio, M.C. and Arrigoni, O. (2003) The ascorbic acid system in seeds: to protect and to serve. *Seed Science Research* 13, 249–260.

Elder, R.H., Dell'Aquila, A., Mezzina, M., Sarasin, A. and Osborne, D.J. (1987) DNA ligase in repair and replication in the embryos of rye. *Mutation Research* 181, 61–71.

Goyal, K., Tisi, L., Basran, A., Browne, J., Burnell, A., Zurdo, J. and Tunnacliffe, A. (2003) Transition from natively unfolded to folded state induced by desiccation in an anydrobiotic nematode protein. *Journal of Biological Chemistry* 278, 12977–12984.

Hendry, G.A.F. (1993) Oxygen and free radical processes in seed longevity. *Seed Science Research* 3, 141–153.

Illing, N., Denby, K., Collett, H., Shen, A. and Farrant, J.M. (2005) The signature of seeds in resurrection plants: a molecular and physiological comparison of desiccation tolerance in seeds and vegetative tissues. *Integrative and Comparative Biology* 45, 771–787.

Koster, K.L. (1991) Glass formation and desiccation tolerance in seeds. *Plant Physiology* 96, 302–304.

Koster, K.L. and Leopold, A.C. (1988) Sugars and desiccation tolerance in seeds. *Plant Physiology* 88, 829–832.

Kranner, I., Cram, W.J., Zorn, M., Wornik, S., Yoshimura, I., Stabenheiner, E. and Pfeifhofer, H.W. (2005) Antioxidants and photoprotection in a lichen as compared with its isolated symbiotic partners. *Proceedings of the National Academy of Sciences of the United States of America* 102, 3141–3146.

Lapinski, J. and Tunnacliffe, A. (2003) Anhydrobiosis without trehalose in bdelloid rotifers. *FEBS Letters* 553, 387–390.

Pammenter, N.W. and Berjak, P. (1999) A review of recalcitrant seed physiology in relation to desiccation-tolerance mechanisms. *Seed Science Research* 9, 13–37.

Rinne, P.L.H., Kaikuranta, P.L.M., van der Plas, L.H.W. and van der Schoot, C. (1999) Dehydrins in cold-acclimated apices of birch (*Betula pubescens* Ehrh.): production, localization and potential role in rescuing enzyme function during dehydration. *Planta* 209, 377–388.

Rogerson, N.E. and Matthews, S. (1977) Respiratory and carbohydrate changes is developing pea (*Pisum sativum*) seeds in relation to their ability to withstand desiccation. *Journal of Experimental Botany* 28, 304–313.

Solomon, A., Salomon, R., Paperna, I. and Glazer, I. (2000) Desiccation stress of entomopathogenic nematodes induces the accumulation of a novel heat-stable protein. *Parasitology* 121, 409–416.

Stacy, R.A.P. and Aalen, R.B. (1998) Identification of a sequence homology between the internal hydrophilic repeated motifs of Group 1 late-embryogenesis-abundant proteins in plants and hydrophilic repeats of the general stress protein GsiB of *Bacillus subtilis*. *Planta* 206, 476–478.

Stacy, R.A.P., Nordeng, T.W., Culiáñez-Maciá, F.A. and Aalen, R.B. (1999) The dormancy-related peroxiredoxin antioxidant PER1 is located to the nucleus in barley embryo and aleurone. *The Plant Journal* 19, 1–8.

Vertucci, C.W. and Farrant, J.M. (1995) Acquisition and loss of desiccation tolerance. In: Kigel, J. and Galili, G. (eds) *Seed Development and Germination*. Marcel Dekker, New York, pp. 237–271.

Viner, R.I. and Clegg, J.S. (2001) Influence of trehalose on the molecular chaperone activity of p26, a small heat shock protein. *Cell Stress Chaperones* 6, 126–135.

Walters, C. (1998) Understanding the mechanisms and kinetics of seed aging. *Seed Science Research* 7, 223–244.

Walters, C. and Engels, J. (1998) The effects of storing seeds under extremely dry conditions. *Seed Science Research* 8, 3–8.

Walters, C., Ried, J.L. and Walker-Simmons, M.K. (1997) Heat-soluble proteins extracted from wheat embryos have tightly bound sugars and unusual hydration properties. *Seed Science Research* 7, 125–134.

Walters, C., Wheeler, L. and Stanwood, P.C. (2004) Longevity of cryogenically stored seeds. *Cryobiology* 48, 229–244.

Walters, C., Hill, L.M. and Wheeler, L.M. (2005a) Dying while dry: kinetics and mechanisms of deterioration in desiccated organisms. *Integrative and Comparative Biology* 45, 751–758.

Walters, C., Wheeler, L.M. and Grotenhuis, J. (2005b) Longevity of seeds stored in a genebank: species characteristics. *Seed Science Research* 15, 1–20.

Whitaker, C., Berjak, P., Kolberg, H. and Pammenter, N.W. (2004) Responses to various manipulations, and storage potential, of seeds of the unique desert gymnosperm, *Welwitschia mirabilis* Hook. fil. *South African Journal of Botany* 70, 621–629.

Williams, R.J. and Leopold, A.C. (1989) The glassy state in corn embryos. *Plant Physiology* 89, 977–981.

Wise, M.J. and Tunnacliffe, A. (2004) POPP the question: what do LEA proteins do? *Trends in Plant Science*, 9, 13–17.

Wolkers, W.F., McCready, S., Brandt, W.F., Lindsey, G.G. and Hoekstra, F.A. (2001) Isolation and characterization of a D-7 LEA protein from pollen that stabilizes glasses *in vitro*. *Biochimica et Biophysica Acta* 1544, 196–206.

3 Gene Expression in Relation to Seed Development and Longevity

S.P.C. Groot,[1] Y. Soeda[2] and A.H.M. van der Geest[1]

[1]*Plant Research International, Wageningen University and Research Centre, PO Box 16, 6700 AA Wageningen, The Netherlands;* [2]*Technology Development Centre, Sumika Agrotech Company, Ltd, 5-1 Sobiraki-cho, Niihama 792-0001, Japan*

Abstract

During maturation seeds acquire stress tolerance and increase in longevity, which are lost completely or partially during germination and priming. Slow drying can partly restore the longevity of primed seeds. Cabbage (*Brassica oleracea* L.) seeds were used as a model to analyse expression of genes during these processes. Six genes for which the expression profiles correlated with the longevity level were identified. Two of these genes are discussed in more detail, *EM6* and *RAB18*, next to the kinetics of two cell cycle genes, histone *H1-1* and histone *H3*. These genes, and the proteins they encode, might serve as markers for optimization of seed processing protocols to produce seeds with a high level of longevity.

Introduction

Most seeds have the remarkable ability to survive in dry conditions. However, there is large variation in longevity among desiccation-tolerant seeds, and even seeds from the same species can differ in longevity, depending on their genotype and physiological conditions. When seeds are harvested prematurely, they may not even survive subsequent drying, but once the developing seeds have acquired desiccation tolerance, they continue to increase in longevity, and seeds most often reach maximum vigour at the moment of natural dispersal. When seeds start their germination process they lose their storability (Hong and Ellis, 1992). Since seed priming includes the initiation of germination, this contributes to the reduced storability of primed seeds in general. Upon radicle protrusion seeds also lose their desiccation tolerance.

In commercial seed production, seeds need to be harvested before shedding to prevent yield losses and this produces seeds that are not fully matured. Additionally,

 Seeds: Biology, Development and Ecology (eds S. Adkins, S. Ashmore and S.C. Navie)

many seed crops have an extended period of flowering, resulting in seed lots containing seeds of different maturities. Seed cleaning techniques, such as sieving and density sorting, are required to get rid of most of the immature and less mature seeds, but they are not always efficient.

Jalink *et al.* (1998) developed a system that measures the level of chlorophyll fluorescence (CF) of individual seeds and subsequently sorts them based on their chlorophyll level. Many seeds are green during development but the chlorophyll is degraded during the seed maturation phase. Measurement of CF of individual seeds has been shown to be a very sensitive method for analysing the relative maturity of different seed types. When coupled to a sorting device, it enables sorting a single seed lot into fractions of different maturities. For research purposes this has the advantage that subsamples of different maturity can be compared which have the same history of seed production, processing and storage. Cabbage (*Brassica oleracea* L.) seeds with a relatively low level of CF were shown to be more tolerant in a controlled deterioration (CD) test compared with seeds with a relatively high level of CF (Jalink *et al.*, 1998). CF levels are inversely correlated with seed maturity, which confirms that mature seeds have better storability.

To increase the rate and uniformity of seedling emergence and to enable seed germination under a broader range of environmental conditions, seed companies may apply priming treatments to seeds. Different methods of seed priming have been developed recently, but the main principle is that seeds can take up enough water to start part or all of their germination processes under favourable conditions. To avoid loss of desiccation tolerance, radicle protrusion is prevented during the priming process either through restriction of the maximum water uptake or through limitations in the period of moist incubation. The major disadvantage of priming is a reduction in the storability of the seeds, which can be severe, depending on the species and the priming protocol used. Bruggink *et al.* (1999) developed a 'shelf-life treatment' to increase the storability of primed seeds and showed that the drying method used after priming has a large effect on the subsequent storability. The principle of the treatment is slow drying of the primed seeds for a few days, which can be combined with moderately elevated temperatures, followed by fast drying.

The general picture that emerges from these observations is that a metabolically active maturing seed prepares itself for survival in a dry state, through induction of protective mechanisms. Adequate protection is important as repair of damage requires enzyme activity, which in itself requires a minimum amount of water. Since repair cannot be performed under dry conditions, it is critical to prevent damage as much as possible. It is the other way around during germination. When the germination process is initiated, the seeds become metabolically and transcriptionally active, and proteins are synthesized from stored and newly produced mRNAs. The protective mechanisms, such as tight packaging of the DNA that preserved the cells during dry storage, will need to be removed in order to activate these processes, making the seed vulnerable to sudden desiccation and storage, as with primed seeds. Slow drying may reimpose part of these protective mechanisms.

We are interested in the genes and proteins that contribute to the acquisition of seed longevity and protect the seed during storage. Cabbage seeds are used as a model system. The close taxonomic relationship between cabbage and *Arabidopsis* (*Arabidopsis thaliana* (L.) Heynh.) enables the use of the *Arabidopsis* gene sequence database. Previously, we have reported on an mRNA microarray analysis of about 850 different genes, using seeds of diverse longevity and seeds during germination and priming (Soeda *et al.*, 2005). Here, we concentrate on some genes for which the expression correlates with seed longevity. Their mRNA levels increase during seed maturation, decrease during seed germination or priming and increase again during a shelf-life treatment. Details of the samples and analyses are published elsewhere (Soeda *et al.*, 2005).

Materials and Methods

Production of seed samples at various physiological stages

To obtain seeds of different degrees of maturation, a non-commercial cabbage seed lot (*B. oleracea* 'Bartolo') was sorted into a fully mature and a less mature fraction using a SeedScan CF sorter (Plant Research International B.V., Wageningen, The Netherlands). An independent, high-quality seed lot (*B. oleracea* 'Bartolo'), containing mostly mature seeds, was used for priming and germination stage samples. Germination was performed at 20°C and 8 h of light with four replicates of 50 seeds. Seeds were considered germinated when radicles emerged. Evaluation of seedling quality was performed according to the rules of the International Seed Testing Association (ISTA, 2003). Seed germination parameters as the days required to obtain 50% of total germination (T_{50}) were calculated with the software package Seed Calculator 2.1 (Plant Research International B.V., Wageningen, The Netherlands). Seed priming was done in −1.0 MPa polyethylene glycol 6000 (PEG) for 7 days. Primed seeds were briefly washed, blotted dry and dried either quickly (fast drying) under circulating air above a saturated salt solution (32% RH) or slowly (slow drying) using a 'shelf-life treatment' (3 days in standing air at 30°C and 75% RH, followed by fast drying). Seed longevity was analysed either by CD tests (85% RH and 40°C) or by storage at more natural deteriorating conditions (66% RH and 30°C) (Soeda *et al.*, 2005). Flow cytometric analysis of cell cycle activation was performed by analysing the relative frequency of nuclei with 2C or 4C DNA levels (Jing *et al.*, 1999).

Gene-expression was analysed using a dedicated microarray with cDNAs obtained from developing and germinating rape (*B. napus* L.) seeds, representing about 850 individual genes. RNA was isolated from the primed and dried seeds, from germinating seeds (15, 30 and 45 h of imbibition) and from 7-day primed seeds (−1.0 MPa PEG) without drying. The latter served as a reference in the microarray analysis.

Results and Discussion

Characterization of the seed samples

The seed samples used in the analysis differed in vigour and longevity. The least mature seeds from the non-commercial seed lot had reduced vigour compared with the most mature seeds, expressed by slower germination (T_{50} 2.59 ± 0.11 days vs

2.18 ± 0.05 days), a reduced frequency of germinating seeds (73.0% ± 2.4% vs 96.0% ± 0.8%) and fewer normal seedlings (60.0% ± 2.2% vs 95.5% ± 1.0%). A CD treatment of the less mature seeds strongly reduced the frequency of normal seedlings (35.0% ± 2.7% vs 87.5% ± 2.1%), illustrating their vulnerability to stress as compared with the mature seeds.

Priming accelerated the germination of the cabbage seeds at temperatures ranging from 12°C to 30°C. With the exception of 12°C, slow-dried seeds germinated in about half a day faster than fast-dried seeds at these temperatures. This difference in T_{50} between slow- and fast-dried primed seeds is striking, at 24°C the T_{50} was 0.20 ± 0.09 days and 0.68 ± 0.13 days, respectively. It is not likely that the relatively faster germination of the slow-dried seeds is due to a further progression of germination related processes during drying, since the seeds had already been primed for 7 days. More likely is a relative delay in germination of the fast-dried primed seeds due to a need for repair. In seeds with DNA damage, initiation of DNA replication is delayed while deteriorated seeds also display a delay of protein synthesis (see Delpratt, Chapter 34, this volume). The primed and dried seeds were only briefly stored prior to the germination experiments; so if the delay is related to repair of damage, this damage has most probably occurred either during drying or during reimbibition of the seeds. There was no significant difference between slow- and fast-dried primed seeds in an electro-conductivity assay (data not shown), which indicates that relative retardation of germination speed from fast-dried vs slow-dried primed seeds is not related to cell membrane damage.

Additionally, fast-dried primed seeds were more sensitive to ultra drying (using silica gel) followed by cold imbibition (10°C) than slow-dried primed seeds (Groot *et al.*, 2003). After that treatment the fast-dried seeds produced about 74% normal seedlings compared with 91% normal seedlings for the slow-dried seeds. This is in accordance with previous observations by Hong and Ellis (1992), who reported an increase in sensitivity to extreme drying during germination of barley (*Hordeum vulgare* L.) and mung bean (*Vigna radiata* L.) seeds. Reduced longevity of primed seeds and a difference in this respect between fast- and slow-dried primed seeds were also apparent when the seeds were stored under more natural conditions (i.e. 66% RH and 30°C, resembling non-protected storage conditions frequently seen in the tropics). There was a good correlation between the data obtained after CD storage and those obtained after this more natural storage (r = 0.93 and 0.95 for total germination and frequency of normal seedlings, respectively).

Gene expression correlated with longevity

From these seed physiological analyses, it is clear that the fast- and slow-dried primed seed samples differed in stress tolerance and seed longevity. In addition to these primed and dried seeds, samples from non-treated seeds, non-dried primed seeds and water-imbibed seeds were used. From all these samples RNA was extracted and relative RNA levels were determined for approximately 850 genes on the cDNA microarray.

The general results of the gene-expression analyses have been published (Soeda *et al.*, 2005). The majority of the genes were upregulated during osmopriming and germination in water, and this was related to the resumption of metabolic activity. A group of nine genes was identified as being upregulated during seed maturation, parallel to the increase in longevity, and downregulated during seed germination or priming (Fig. 3.1). Six of these genes also had higher mRNA levels in slow-dried primed seeds compared with fast-dried primed seeds, thus showing a correlation with seed longevity.

Not surprisingly, a number of these genes show homology to genes previously identified as coding for late embryogenesis abundant (LEA) proteins, which are by definition accumulated during seed maturation. LEA proteins form a varied group of proteins. Although several authors have made classifications based on biochemical properties or sequence similarities, the function of most LEA proteins remains unknown. Wise (2003) classified the LEA proteins according to their protein or oligonucleotide probability profile (POPP) and their similarities with proteins of known function.

One of the genes that is highly expressed during late seed maturation is the *B. oleracea* homologue of *ATEM6* gene (At2G40170). Hardly, any transcript is present in imma-

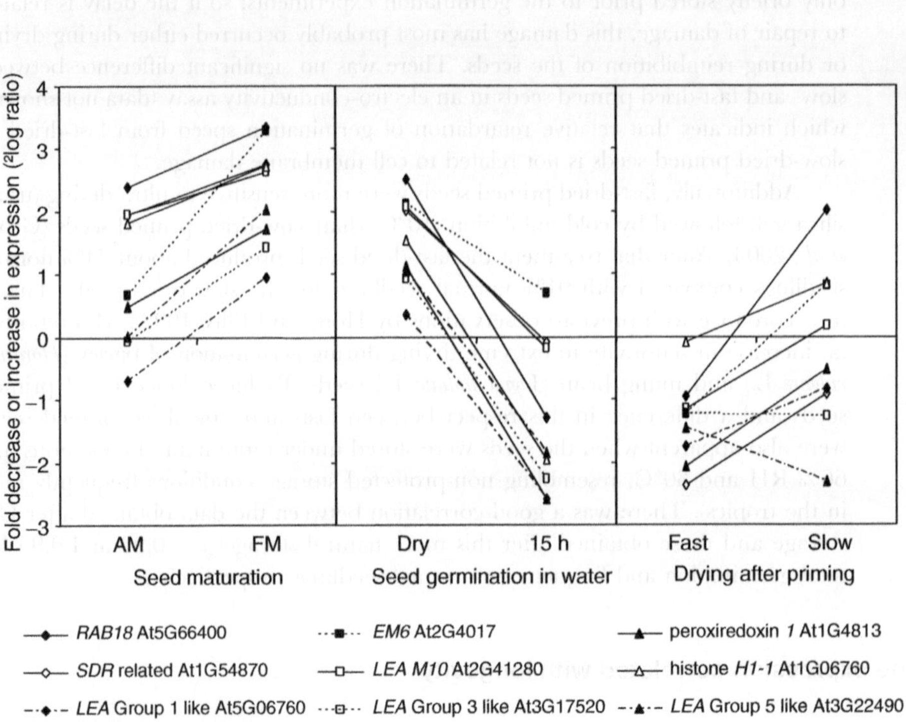

Fig. 3.1. Relative expression of nine genes during seed maturation, germination and drying of primed seeds. The sample from primed non-dried seeds was used as a reference (equal expression to the reference would result in a ^{2}log value of zero). The genes are named after their closest homologue from *Arabidopsis*. Physiological stages analysed: AM = almost mature seeds (high chlorophyll fluorescence signal); FM = fully mature seeds (low chlorophyll fluorescence signal); dry = dry mature seeds; 15 h = mature seeds imbibed for 15 h in water; fast = mature seeds primed for 7 days in −1 MPa PEG with a subsequent fast-drying treatment; slow = mature seeds primed for 7 days in −1 MPa PEG with a subsequent slow-drying treatment.

ture cabbage seeds and some is present in the less mature seed fraction (Fig. 3.2). In the most mature seed fraction, the *EM6* transcript level was approximately six times higher compared with that in the less mature seeds. There was hardly any effect of storage on the relative level of *EM6* transcript, but the level of this transcript strongly decreased during imbibition in water or during osmopriming, indicating active degradation or turnover. The *EM6* gene is a group 1 *LEA* gene (Dure *et al.*, 1989), whose expression is highly responsive to, and dependent on, ABA biosynthesis (Gaubier *et al.*, 1993). In *Arabidopsis*, *ATEM6* transcript accumulates later during seed development and is rapidly degraded during seed germination (Gaubier *et al.*, 1993; Bies *et al.*, 1998). *BNEM6* from *B. napus* follows the same expression pattern (Vicient *et al.*, 1998). We showed that in cabbage seeds, *EM6* gene-expression is reinduced during the slow-drying process (Soeda *et al.*, 2005). Interestingly, such a reinduction of *EM6* gene-expression was also found by Buitink and co-workers in experiments where they reinduced desiccation tolerance with germinated *Medicago truncatula* L. seeds (see Buitink, Chapter 1, this volume).

The novel LEA protein classification based on their POPP puts the class 1B LEA proteins (to which EM6 belongs) into the LEA superfamily 6 (Wise, 2003). The proteins in this superfamily share characteristics with proteins associated with dsRNA binding, DNA gyrases, DNA breakage, calmodulin-like protein (CLP) and ATP binding as well as with potential functions in nucleic acid unwinding, nucleic acid repair or molecular chaperone activity. This potential molecular chaperone activity has been tested in wheat (*Triticum aestivum* L.) using citrate synthase as a

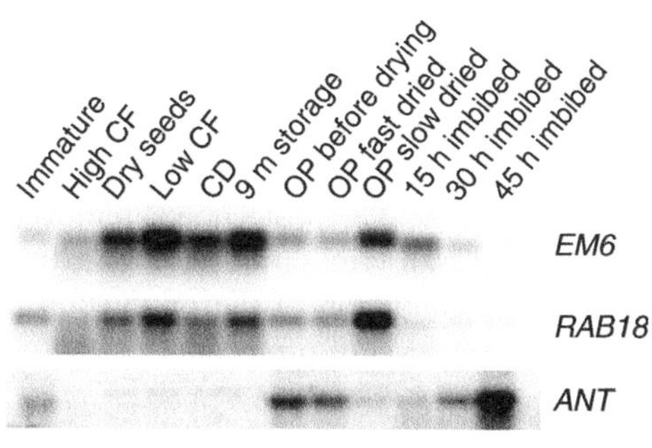

Fig. 3.2. Gene-expression during seed development, storage, osmopriming, drying and imbibition. RNA blot of three genes: *EM6* (homologous to At2G40170), *RAB18* (homologues to At5G66400) and an *Adenine nucleotide translocator* (*ANT*) gene (homologous to At5G13490). The *ANT* gene is shown as example of genes upregulated during germination. Physiological stages analysed: immature = bent cotyledon stage developing seeds; high CF = almost mature seed with high levels of chlorophyll fluorescence; dry seeds = non-sorted mature seeds; low CF = full mature seed with low levels of chlorophyll fluorescence; CD = seeds after a 5-day controlled deterioration treatment; 9 m storage = seeds stored for 9 months at 66% RH and 30°C; OP before drying = osmoprimed seeds in −1 MPa PEG before drying; OP fast-dried = osmoprimed seeds after fast drying; OP slow-dried = osmoprimed seeds after slow drying; 15 h imbibed = germinating seeds 15 h after start of imbibition on water; 30 h imbibed = germinating seeds 30 h after start of imbibition on water; 45 h imbibed = germinating seeds 45 h after start of imbibition on water. The data have previously been published in Soeda *et al.* (2005) and are copyright by ASPB.

model (Goyal *et al.*, 2005). The wheat EM LEA protein, which has a high protein sequence similarity to ATEM6, was unable to independently prevent heat-induced aggregation of citrate synthase, but exhibited a protective effect in the presence of trehalose, a sugar known to accumulate during seed maturation (Goyal *et al.*, 2005). Additionally, wheat EM was able to protect citrate synthase from aggregation due to desiccation and freezing. It is plausible that in cabbage seeds EM6 protein performs a similar protective function upon drying of the seeds.

A second gene, with expression correlated to the level of cabbage seed stress tolerance, was homologous to the *RAB18* gene from *Arabidopsis* (At5G66400). The RAB18 protein belongs to the group 2 LEA proteins and is an ABA inducible dehydrin expressed in developing seeds (Lang *et al.*, 1994). Based on the consensus POPP classification of LEA proteins (Wise, 2003), the RAB18 protein sequence has characteristics related to ATP binding, DNA topoisomerase activity, protein biosynthesis, repair, DNA binding, transcriptional regulation, intermediate filament, keratin, chaperone, homeobox, coiled coil, high mobility group (HMG) box domain, cytoskeletal, etc. Many of these keywords refer to the interaction with DNA and suggest a molecular chaperoning role in the nucleus.

The *ATRAB18* mRNA accumulates in *Arabidopsis* plants that are exposed to low temperature, water stress or exogenous ABA (Lang and Palva, 1992; Nylander *et al.*, 2001). Overexpression of the *RAB18* or *COR47* (a second dehydrin gene) alone had no effect on stress tolerance of transgenic *Arabidopsis* plants, but overexpression of both the genes together strongly increased the freezing tolerance in *Arabidopsis* plants (Puhakainen *et al.*, 2004). Immunohistochemical localization studies demonstrated the presence of the RAB18 protein in the nuclei of stomatal guard cells, consistent with the nuclear localization predicted from the POPP classification, whereas ABA treatment (inducing *RAB18* expression) resulted in staining of the whole guard cells (Nylander *et al.*, 2001). Therefore, RAB18 may play a role in protecting DNA during stress. The protection is removed during germination or priming to allow repair and transcription of the DNA. In contrast to slow drying, fast drying may not allow reinitiation of the production of the protein and thus confer the DNA with a lower level of protection.

Cell cycle activity

Flow cytometric analysis of radicle tip nuclei showed that cell cycle activity was initiated in the seeds during germination and priming treatment. The frequency of 4C and 8C nuclei has strongly increased prior to radicle protrusion, with no difference between fast- and slow-dried primed seeds (Table 3.1). Histones play an important role in DNA compaction. The DNA is wrapped around the nucleosomes and histone octamer, with two copies each of the histones H2A, H2B, H3 and H4. Histone H1 is a linker histone. On our microarray, we could follow the activity of the histones *H1* and *H3* transcription. Histone H1 proteins are linker histones and in most eukaryotic cells several distinct histone H1 variants occur, of which the expression is developmentally regulated (Jerzmanowski *et al.*, 2000). Histone H1 proteins can modulate the DNA structure by bending, and the presence of histone H1 protein in the nucleosome provides protection against nuclease activity in *in vitro*

Table 3.1. Influence of priming and drying conditions on nuclear DNA levels in radicle tip cells from cabbage seeds. Average frequency of nuclei with 2C, 4C or 8C DNA levels are presented (± standard errors) from three replicates of five root tips each.

Treatment	2C (%)	4C (%)	8C (%)
Control (non-primed)	86.9 ± 1.5	12.3 ± 1.3	0.8 ± 0.3
Primed and fast-dried	16.6 ± 1.4	64.4 ± 0.6	19.0 ± 2.0
Primed and slow-dried	13.9 ± 1.1	63.4 ± 1.9	22.7 ± 1.4

assays (Jerzmanowski *et al.*, 2000). *Arabidopsis* has three histone *H1* genes: *H1-1* and *H1-2*, which are expressed in a variety of tissues and which make up the majority of linker histones; and *H1-3*, a drought-inducible linker histone that was shown to bind the *RAB18* gene DNA during drought stress (Ascenzi and Gantt, 1999). Transgenic *Arabidopsis* plants with more than 90% reduction in histone *H1* expression – as a result of silencing of the three genes simultaneously – exhibit a spectrum of aberrant developmental phenotypes, correlating with specific methylation patterns of the DNA (Wierzbicki and Jerzmanowski, 2005). A histone *H1-1* clone from *Arabidopsis* was spotted on our microarray. We obtained a clear hybridization signal in dry cabbage seeds, with no significant difference between less mature and mature seeds (Fig. 3.3). During germination in water the histone *H1-1* signal initially strongly decreased, but some recovery of the signal was observed after radicle protrusion and after the onset of DNA replication. In 7-day primed seeds the histone *H1* signal was higher than at 15 h with water-imbibed seeds. It is not clear if this was due to a smaller decrease, or due to a reinitiation of transcription in relation to cell cycle activation during osmopriming.

Histone *H3* gene expression was low in dry seeds and increased strongly during seed germination in water, along with the start of DNA replication. In primed seeds the histone *H3* signal was similar to that observed in dry seeds; apparently the osmotic inhibition also blocked expression of this histone gene. No mitosis occurred during osmopriming in contrast to the situation with water-imbibed seeds, where mitosis occurred after radicle protrusion.

Nuclei in dry maize (*Zea mays* L.) embryos contain relatively large amounts of histone H1 proteins, about two molecules of histone H1 per nucleosome, which drop rapidly after imbibition to one molecule per nucleosome, the general ratio for plant and animal tissues (Ivanov and Zlatanova, 1989). Our histone *H1-1* expression data are in accordance with these observations. It might be that the relatively large amount of histone H1 proteins in dry seeds plays a role in extra DNA compaction, which might aid in the protection of the DNA protein complex during dry storage.

Conclusions

Our study confirms the importance of controlling the drying process of seeds to obtain maximum seed quality during seed production and after seed priming. Seeds that are harvested prematurely and quickly dried cannot finish their

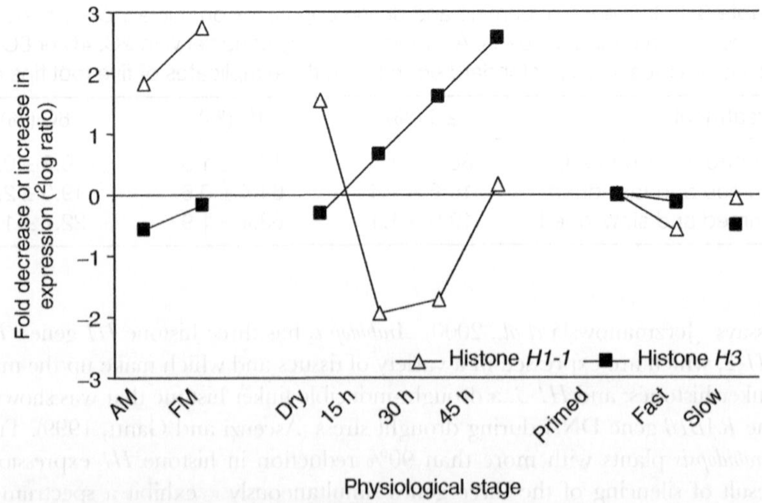

Fig. 3.3. Relative expression of two histone genes during seed maturation, germination and drying of primed seeds. The sample from primed non-dried seeds was used as a reference (equal expression to the reference would result in a ²log value of zero). The genes are named after their closest homologue from *Arabidopsis*. Physiological stages analysed: AM = almost mature seeds (high chlorophyll fluorescence signal); FM = full mature seeds (low chlorophyll fluorescence signal); dry = dry mature seeds; 15h = mature seeds imbibed for 15h in water; 30h = mature seeds imbibed for 30h in water; 45h = mature seeds imbibed for 45h in water; primed = mature seeds primed for 7 days in −1 MPa PEG; fast = primed seeds with a subsequent fast-drying treatment; slow = primed seeds with a subsequent slow-drying treatment.

developmental programme required for implementation of protective mechanisms; consequently, such seeds have a lower viability compared with seeds that are harvested fully mature after slow, natural drying on the mother plant. Similarly, when a primed or germinated seed is quickly dried it might lack some of its protective mechanisms. This might play a role in the decreased longevity of primed seeds and lack of desiccation tolerance after radicle protrusion. The observation that slow drying of primed seeds, or temporary incubation at an elevated temperature, can reinduce at least part of the storability of the seeds (Bruggink *et al.*, 1999) indicates that these treatments might reimpose protective mechanisms.

Maintenance of DNA integrity is extremely important and it can be expected that one or more protection mechanisms will be operational in the dry seed. When the germination process is initiated and metabolic activity resumes, DNA transcription starts and proteins are synthesized from stored and newly produced mRNA. Cell cycle activity is also initiated, including DNA replication. DNA repair, transcription and replication require access to the DNA and, most likely, removal of the protection mechanisms.

We believe that the protective mechanisms employed during seed maturation share common elements with those of slow-dried primed seeds, and involve expression of the three genes, *EM6*, *RAB18* and histone *H1-1*, discussed in this chapter. The genes identified in our studies, or the proteins they are coding for, might serve as markers for optimization of seed treatments that are aimed at producing primed seeds with optimal germination characteristics and shelf life.

References

Ascenzi, R. and Gantt, J.S. (1999) Subnuclear distribution of the entire complement of linker histone variants in *Arabidopsis thaliana*. *Chromosoma* 108, 345–355.

Bies, N., Aspart, L., Carles, C., Gallois, P. and Delseny, M. (1998) Accumulation and degradation of Em proteins in *Arabidopsis thaliana*: evidence for post-transcriptional controls. *Journal of Experimental Botany* 49, 1925–1933.

Bruggink, G.T., Ooms, J.J.J. and van der Toorn, P. (1999) Induction of longevity in primed seeds. *Seed Science Research* 9, 49–53.

Dure, L. III, Crouch, M., Harada, J., Ho, T.-H.D., Mundy, J., Quatrano, R., Thomas, T. and Sung, Z.R. (1989) Common amino acid sequence domains among the LEA proteins of higher plants. *Plant Molecular Biology* 12, 475–486.

Gaubier, P., Raynal, M., Hull, G., Huestis, G.M., Grellet, F., Arenas, C., Pages, M. and Delseny, M. (1993) Two different Em-like genes are expressed in *Arabidopsis thaliana* seeds during maturation. *Molecular General Genetics* 238, 409–418.

Goyal, K., Walton, L.J. and Tunnacliffe, A. (2005) LEA proteins prevent protein aggregation due to water stress. *Biochemical Journal* 388, 151–157.

Groot, S.P.C., Soeda, Y., Stoopen, G., Konings, M.C.J.M. and van der Geest, A.H.M. (2003) Gene-expression during loss and regaining of stress tolerance at seed priming and drying. In: Nicolas, G., Bradford, K.J., Côme, D. and Pritchard H.W. (eds) *The Biology of Seeds: Recent Research Advances*. CAB International, Wallingford, UK, pp. 279–287.

Hong, T.D. and Ellis, R.H. (1992) The survival of germinating orthodox seeds after desiccation and hermetic storage. *Journal of Experimental Botany* 43, 239–247.

ISTA (2003) International rules for seed testing. The International Seed Testing Association, Bassersdorf, Switzerland.

Ivanov, P.V. and Zlatanova, J.S. (1989) Quantitative changes in the histone content of the cytoplasm and the nucleus of germinating maize embryo cells. *Plant Physiology and Biochemistry* 27, 925–930.

Jalink, H., van der Schoor, R., Frandas, A., van Pijlen, J.G. and Bino, R. (1998) Chlorophyll fluorescence of *Brassica oleracea* seeds as a non-destructive marker for seed maturity and seed performance. *Seed Science Research* 8, 437–443.

Jerzmanowski, A., Przewloka, M. and Grasser, K.D. (2000) Linker histones and HMG1 proteins of higher plants. *Plant Biology* 2, 586–597.

Jing, H.C., van Lammeren, A.A.M., de Castro, R.D., Bino, R.J., Hilhorst, H.W.M. and Groot, S.P.C. (1999) β-tubulin accumulation and DNA synthesis are sequentially resumed in embryo organs of cucumber (*Cucumis sativus* L.) seeds during germination. *Protoplasma* 208, 230–239.

Lang, V. and Palva, E.T. (1992) The expression of a rab-related gene, *rab18*, is induced by abscisic acid during the cold acclimation process of *Arabidopsis thaliana* (L.) Heynh. *Plant Molecular Biology* 20, 951–962.

Lang, V., Mantyla, E., Welin, B., Sundberg, B. and Palva, E.T. (1994) Alterations in water status, endogenous abscisic acid content, and expression of rab18 gene during the development of freezing tolerance in *Arabidopsis thaliana*. *Plant Physiology* 104, 1341–1349.

Nylander, M., Svensson, J., Palva, E.T. and Welin, B.V. (2001) Stress-induced accumulation and tissue-specific localization of dehydrins in *Arabidopsis thaliana*. *Plant Molecular Biology* 45, 263–279.

Puhakainen, T., Hess, M.W., Mäkelä, P., Svensson, J., Heino, P. and Palva, E.T. (2004) Overexpression of multiple dehydrin genes enhances tolerance to freezing stress in *Arabidopsis*. *Plant Molecular Biology* 54, 743–753.

Soeda, Y., Konings, M.C.J.M., Vorst, O., van Houwelingen, A.M.M.L., Stoopen, G.M., Maliepaard, C.A., Kodde, J., Bino, R.J., Groot, S.P.C. and van der Geest, A.H.M. (2005) Gene expression programs during *Brassica oleracea* seed maturation, osmopriming and germination are indicators of progression of the germination process and the stress tolerance level. *Plant Physiology* 137, 354–368.

Vicient, C.M., Roscoe, T.J. and Delseny, M. (1998) Characterization of an Em-like gene from *Brassica napus*. *Journal of Experimental Botany* 49, 1061–1062.

Wierzbicki, A.T. and Jerzmanowski, A. (2005) Suppression of histone *H1* genes in *Arabidopsis* results in heritable developmental defects and stochastic changes in DNA methylation. *Genetics* 169, 997–1008.

Wise, M.J. (2003) LEAping to conclusions: a computational reanalysis of late embryogenesis abundant proteins and their possible roles. *BMC Bioinformatics* 4, 52–71.

4 The Influence of Harvest Method on Seed Yield, Seed Size and Germination Capacity of *Bulbine bulbosa* (R. Br.) Haw. (*Liliaceae*)

C.J. Delpratt

Department of Resource Management, The University of Melbourne, Burnley Campus, Richmond, Victoria 3121, Australia

Abstract

The yield, size and germination capacity of seeds harvested from intact inflorescences of the Australian perennial forb *Bulbine bulbosa* (R. Br.) Haw. (*Liliaceae*) were compared with those harvested from inflorescences that were detached from the parent plant and allowed to dry at 20°C. Inflorescences were detached when the first, third and sixth capsules, respectively, reached harvest maturity, and seeds were harvested as capsules opened during storage. Capsules on intact inflorescences were harvested from the plant as they opened. The influence of capsule position within an inflorescence was assessed by separately analysing seeds harvested from the proximal, central and distal third of all inflorescences. Seed yield from intact inflorescences was highest, but daily harvests were required and spanned an average of 33 days per inflorescence, approximately twice as long as detached inflorescences. There was no significant difference in the number of seeds harvested per capsule, but there was a higher proportion of small seeds harvested from detached inflorescences than from intact inflorescences. Following 8 months of dry storage, germination, 21 days after imbibition, was >70% for all harvest treatments. Removing and storing inflorescences, when 1–3 capsules reach harvest maturity, improve harvest efficiency of *B. bulbosa* seeds.

Introduction

The grasslands and grassy understoreys are among the most threatened plant communities of eastern and southern Australia. The restoration of these communities depends on readily available supplies of high-quality seed from a diverse range of locally indigenous grasses and forbs. Rather than relying solely on seed collected from remnant vegetation, an increasing number of species (e.g. *Bulbine bulbosa* (R. Br.) Haw.) are being cultivated, often for the first time, in seed production areas (Delpratt, 1997).

 Seeds: Biology, Development and Ecology (eds S. Adkins, S. Ashmore and S.C. Navie)

The objective of the cultivation of plants for seed production has been defined as 'the maximization of the production of seed with suitable germination capacity' (Bowring *et al.*, 1980). However, the structure of a species' inflorescence, the number and uniformity of inflorescences on an individual plant and the uniformity of flowering between plants can strongly influence the timing and uniformity of seed maturation within a seed crop and reduce the amount of seed that can be harvested economically. For instance, to minimize loss of seed from opening capsules, Desai *et al.* (1997) recommended that onion (*Allium cepa* L.) seed crops be harvested when 1–3% of the umbels in the field have mature seeds (i.e. when capsules are open showing black seeds). Seed heads are harvested intact, with a short (unspecified) length of scape attached. The intact umbels are then gradually dried for a period of days, after which the seeds are extracted by threshing.

For seed production, Copeland and McDonald (1995) described three general stages of seed development. Approximately 80% of seed growth occurs during the first stage, which begins with fertilization of the ovule, and is characterized by numerous cell divisions and differentiation of organs, acquisition of assimilates from the parent plant and a substantial increase in seed mass. The second stage involves the beginning of desiccation, the degeneration of the funicle and the consequent separation of the seed from the parent plant. At this stage, termed 'physiological maturity', the seed attains its maximum dry weight. The third stage involves further desiccation until the seed reaches a moisture content of usually 15–20%. If high-quality seed is to be harvested directly from the plant, the harvest process must be delayed sufficiently to allow the individual seed to reach harvest maturity, but not to the point where the mature seed is shed from the parent plant.

One consequence of differences in seed maturity at harvest is a lack of uniformity in seed size. All seed lots contain a range of seed sizes, and uniformity in seed size within a seed lot is considered to be one measure of seed quality (Thomson, 1979). Non-uniform seed lots are difficult to sow mechanically and small seed size may indicate a small embryo or reduced endosperm reserves. However, the causes of heterogeneity in seed size within a seed lot include differences in moisture content, genotypic differences between plants for seed size, seed position within a multiseeded fruit or complex inflorescence and seed maturity. The impact of small (relative to its cohorts) seed size on subsequent germination, seedling establishment and plant growth is complex and will be influenced by the factors that contributed to the size differences. However, it is a common practice to screen seeds during cleaning to improve the uniformity in seed size of the processed seed lot.

When harvesting seeds of a heterogeneous, undomesticated species, considerable variability can be expected in plant development and morphological factors that affect seed size. Many wild species have adaptations that spread the production of their seeds over time. Since only a small proportion of their potential seed production is mature at one time, they are difficult to harvest efficiently.

B. bulbosa is a perennial lily native to many grasslands and understorey communities in eastern and southern Australia. It has a basal tuft of succulent, linear leaves of 30 cm, growing from a compressed stem or bulb. In natural communities it produces leaves in autumn, followed in spring and early summer by one to numerous unbranched inflorescences. Although usually dormant from mid-summer to mid-autumn, in cultivation it will retain its leaves and extend flower and seed production if it has adequate water (Delpratt, 1997). The fruit is a globose to obovoid capsule (Conran and Walsh, 1994). Capsules generally contain 18 ovules and, in cultivation, mature capsules produce 1–13 black, angular seeds that are 2–3 mm long (author's observations). Capsules mature sequentially from the base of the inflorescence upwards. It is common to have dehiscing capsules, immature capsules, two or three open flowers and unopened buds in one inflorescence at the same time. In cultivation, single plants commonly produce ten or more inflorescences during spring, summer and autumn (author's observations).

Hand-harvesting of *B. bulbosa* every 1–3 days, and picking capsules just prior to opening, achieves a high recovery of mature seeds. However, this process is tedious and time-consuming. An alternative may be to harvest intact inflorescences, allow them to dry and then recover the seed. With the range of reproductive stages on a given inflorescence, it is not obvious when to harvest an inflorescence, or whether seeds will continue to develop and mature once an inflorescence is detached from the plant.

This experiment investigated the influence of harvest method of cultivated plants of *B. bulbosa* on total seed yield, seed size and germination capacity. The conventional hand-harvesting method was compared to storing of harvested inflorescences of various ages in a controlled environment.

Materials and Methods

Harvest treatments

In early summer, 40 container-grown *B. bulbosa* plants of similar age, each with an inflorescence at a specific and similar stage of maturity, were selected for the experiment. The criteria for inflorescence selection were: (i) the inflorescence had already set at least 20 capsules; and (ii) the oldest capsule was approaching harvest maturity as defined by the lightening in colour of the capsule sutures, just prior to splitting.

Plants were randomly allocated to one of four harvest treatments, with ten plants in each harvest. The four harvest treatments were: (i) Harvest 1 – the inflorescences remained on the plant and capsules were harvested individually when each reached harvest maturity; (ii) Harvest 2 – the inflorescences were harvested completely when the first (proximal) capsule reached harvest maturity; (iii) Harvest 3 – the inflorescences were harvested completely when the third capsule reached harvest maturity; and (iv) Harvest 4 – the inflorescences were harvested completely when the sixth capsule reached harvest maturity.

During the experiment, the plants were grown on an outdoor nursery bench, in a randomized complete block design. They were hand-watered once a day to avoid wetting of the inflorescences with overhead irrigation.

For Harvest 1, each capsule that reached harvest maturity was counted as part of the harvest. This treatment represented hand-harvesting, a thorough but time-consuming method. For inflorescences in Harvests 2, 3 and 4, the first, the first three and the first six capsules, respectively, were not counted as part of the harvest. However, the seeds from these capsules were collected and counted as part of total seed numbers produced per inflorescence because they represented mature seeds potentially lost from the harvest. These latter harvests investigated whether there was an optimum harvest time for the destructive harvesting of inflorescences, an approach to harvesting that had the potential to improve harvest efficiency. Each harvested inflorescence was dried slowly under controlled conditions (see method below) and seeds were harvested as capsules reached harvest maturity.

At harvest, the information recorded for each inflorescence was the number of unopened buds, the number of open flowers, the number of capsules and the number of pedicels of aborted buds, flowers or capsules.

At the same time, each inflorescence was divided with a twist tie into three segments: proximal, central and distal. Each segment contained one-third of the capsules that had set at that time. When the number of capsules that had set did not divide into three equal portions, the additional capsules were included in the proximal and central segments, in that order.

For Harvest 1, all open flowers and buds (above the distal segment) were retained on the inflorescence and were allowed to develop normally. Any capsule that set seed above the distal segment was counted separately but included in the harvest for that segment.

For each inflorescence in Harvests 2, 3 and 4, all unopened buds above the topmost open flower were removed at harvest and discarded. Although this differed from the method for Harvest 1, these buds were removed on the assumption and prior observation that they would not set capsules without a pollen vector. These buds and flowers might divert assimilates away from the maturing capsules that were already on the now-detached inflorescence.

To allow for the analysis of the seed yield for each capsule at harvest maturity, each detached inflorescence was tied to a bamboo stake and secured vertically in a temperature-controlled cabinet set at 20°C, with 16 h of light and 8 h of darkness in each 24 h cycle. Relative humidity was not directly controlled. When each capsule reached harvest maturity, it was removed from the inflorescence by hand and seeds were screened into three size categories: (i) large (L) seeds retained by a 2.0 mm screen; (ii) medium (M) seeds passed through a 2.0 mm screen but not through a 1.5 mm screen; and (iii) small (S) seeds passed through a 1.5 mm screen but retained by a 1.2 mm screen. Particles that passed through the 1.2 mm screen were inspected and discarded as none contained a differentiated embryo.

Seeds harvested from the proximal, central and distal segments of each inflorescence were counted and stored separately in paper envelopes.

Germination tests

All seeds in each size category from each harvest were retained in separate paper envelopes in open storage (mean 3 pm RH 47–65%) for 8 months, before being tested for germination. As some reduction in seed size had occurred during storage, all seeds were sieved again into two sizes: (i) large (L), those retained by a 1.7 mm sieve; and (ii) small (S), those passed through a 1.7 mm sieve but retained by a 1.2 mm sieve. Less than 0.2% of seeds passed through a 1.2 mm sieve. These seeds were excluded from the germination test. Therefore, the seeds from each of four harvests were in six categories: the proximal, central or distal segment of the inflorescence and two sizes, large and small, a total of 24 treatments. Each treatment had two replicates and each replicate contained 25 seeds.

Seeds were germinated on filter paper, moistened with deionized water, in covered plastic Petri dishes with one replicate per dish. The Petri dishes were placed in a randomized block design in a growth cabinet at 20°C and illuminated for 8 h a day. Germination was recorded every 2 days for 6 weeks and was deemed to have occurred when the radicle had emerged from the testa by 1 mm.

Statistical analysis

Harvest treatment responses were analysed for significance using one-way analysis of variance (ANOVA) and the general linear model. Means were separated by Fisher's least significant difference (LSD) procedure (MINITAB Release 10.51). ANOVA of germination percentages was based on arcsine-transformed data. Germination percentages presented in Table 4.4 represent non-transformed values.

Results

Fruit set and duration of harvest

At harvest, each inflorescence comprised persistent pedicels of aborted flowers or capsules, capsules at a range of developmental stages, open flowers and unopened buds. On average, capsules comprised 46% of potential reproductive sites in the inflorescence. Harvesting inflorescences obtained from Harvest 1 (harvesting from intact plants) spanned an average of 33 days from the harvest of the first to the final capsule. This was more than twice the mean harvest duration of inflorescences from Harvest 2 (14 days), Harvest 3 (16 days) and Harvest 4 (13 days).

Harvest yield

The mean number of seeds per capsule did not differ significantly between the different harvests ($P > 0.05$) (Table 4.1A). However, there were differences in mean seeds per capsule within Harvests 2, 3 and 4, depending on the position of the capsule within the inflorescence, with the distal segment yielding significantly fewer seeds per capsule than the proximal and central segments ($P < 0.05$).

For each harvest treatment, the number of seeds shed before the commencement of harvest increased in a predictable manner. Average seed losses per harvested inflorescence increased from no seeds for Harvest 1 to 36.5 seeds for Harvest 4 (Table 4.1B).

Because of variability in capsule numbers between inflorescences, the total number of capsules per treatment was not uniform. To take proper account of the impact of harvest treatments on potential yield (including the initial discarding of capsules

Table 4.1. Capsule and seed numbers, and seeds per capsule for inflorescence segments and total inflorescences of *Bulbine bulbosa* for four harvest treatments, each of ten inflorescences: (A) total yield and the distribution of yield between proximal, central and distal inflorescence segments and (B) capsules and seeds that were lost prior to the start of each harvest as a direct consequence of the harvest treatment.

A

Harvest treatment	Designated segment	Number of capsules	Number of seeds	Seeds per capsule
1 (mature)	Proximal	105	653	6.2
	Central	103	707	6.9
	Distal	101	646	6.4
	Post	41	170	4.1
	Total	350	2176	6.2[a]
2 (early)	Proximal	93	671	7.2
	Central	91	665	7.3
	Distal	88	472	5.4
	Total	272	1808	6.6[a]
3 (intermediate)	Proximal	107	729	6.8
	Central	104	740	7.1
	Distal	100	435	4.4
	Total	311	1904	6.1[a]
4 (late)	Proximal	98	549	5.6
	Central	96	542	5.6
	Distal	93	451	4.8
	Total	287	1542	5.4[a]

[a]NS = no significant difference between means ($P > 0.05$).

B

	Potential yield loss prior to harvest		
Harvest treatment	Number of capsules per inflorescence	Number of seeds per capsule	Number of seeds per inflorescence
1 (mature)	NA	NA	NA
2 (early)	1	6.6	6.6
3 (intermediate)	3	7.1	21.2
4 (late)	6	6.4	36.5

Post = capsules set after the beginning of harvest; NA = not applicable.

from Harvests 2, 3 and 4) total seed yield from each harvest treatment was analysed from a common starting point. All capsules on each inflorescence at the beginning of harvest were included as the covariate. Capsules that were discarded from Harvests 2, 3 and 4 (but not their seed) were included. Capsules from Harvest 1 that set after the beginning of harvest were excluded (but their harvested seeds were included). When compared on this basis, there were significant differences in yield between harvest treatments ($P < 0.05$), with highest seed yield in Harvest 1 (Table 4.2).

Table 4.2. The analysis of seed yield per inflorescence of *Bulbine bulbosa* from four harvest treatments. To take account of variations in the number of capsules harvested per inflorescence, total capsule number per inflorescence at the time harvest commenced was included as the covariate (including capsules discarded prior to harvest). Values followed by the same letter do not differ significantly ($P > 0.05$).

Seed yield per inflorescence			
Harvest 1 (capsules matured on intact plants)	Harvest 2 (one capsule discarded)	Harvest 3 (three capsules discarded)	Harvest 4 (six capsules discarded)
222.5 a	200.8 ab	179.2 bc	140.5 c

Seed size

At harvest, seeds were sieved into three size categories: large, medium and small. For each inflorescence segment (proximal, central and distal) of each harvest, the number of seeds in each size category was expressed as a proportion of the total number of seeds harvested from that segment (Table 4.3).

Harvest 1 (64%) and Harvest 2 (52%) had the highest proportions of large seeds and the smallest proportions of small seeds. The proportion of small seeds increased in Harvests 3 and 4, particularly in the distal segment of the inflorescences.

Seed germination

All harvest treatments yielded germinable seeds. The time course for germination typically showed a lag phase of 5–6 days before germination was observed, followed by consistent germination for the following 14 days. For most treatments there was little additional germination before the completion of the experiment, 36 days after imbibition. The cumulative percentage of germination at 21 days after imbibition was analysed (using arcsine-transformed values) by harvest, inflorescence segment and post-storage seed size (Table 4.4). There were significant differences in germination after 21 days between harvests ($P < 0.05$), with Harvest 1 having the highest mean germination (89%). In each of the four harvests, seeds from capsules in the central segments of the inflorescences gave the highest mean germination values after 21 days. However, for the pooled values for segments, germination was significantly lower ($P < 0.05$) only for seeds from the distal segments (Table 4.4). Although, in the majority of germination tests, large seeds recorded higher germination values than small seeds, this trend was not significant at 21 days after imbibition, and this difference had narrowed further after 36 days.

From individual tests of inflorescence segment and seed-size combinations, the highest and narrowest range of values for germination were for seeds that had matured while still attached to the plant (Harvest 1). Similar germination values were recorded for some seeds harvested from detached inflorescences, but the range of values recorded was much higher. Small seeds harvested from the distal segment of detached inflorescences consistently gave significantly lower cumulative

Table 4.3. The total and relative proportions of freshly harvested *Bulbine bulbosa* seeds of three size categories from three inflorescence segments (i.e. proximal, central and distal).

Harvest treatment	Inflorescence segment	Seed number per size category per inflorescence segment			Percentage of seed per size category per inflorescence segment		
		L	M	S	L	M	S
1 (mature)	Proximal	339	297	17	52	45	3
	Central	462	236	9	65	33	1
	Distal	460	181	5	71	28	1
	Post	137	28	5	63	36	2
	Total	1398	742	36	64	34	2
2 (early)	Discard	39	26	1	59	39	2
	Proximal	405	256	1	61	39	0
	Central	357	283	15	55	43	2
	Distal	162	268	42	33	58	9
	Total	963	833	59	52	45	3
3 (intermediate)	Discard	72	108	32	34	51	15
	Proximal	277	378	74	38	52	10
	Central	296	375	69	40	51	9
	Distal	122	222	91	28	51	21
	Total	695	975	234	37	51	12
4 (late)	Discard	185	179	21	48	46	5
	Proximal	241	284	24	44	52	4
	Central	253	272	17	47	50	3
	Distal	129	277	45	29	61	10
	Total	623	833	86	40	54	6
Total (all harvests)		3647	3383	415	49	45	6

Seed size categories: L = large (>2.0 mm); M = medium (<2.0 mm and >1.5 mm); S = small (<1.5 mm and >1.2 mm); Post = seeds harvested from capsules that set after harvest commenced; Total = sum of all harvested seed (excludes discarded seed); Discard = seed collected from capsules that were discarded at harvest.

germination values after 21 days ($P < 0.05$) than other combinations of seed size and inflorescence segment (data not shown).

Discussion

This experiment showed that it is possible to recover germinable *B. bulbosa* seed from fruiting inflorescences harvested at a range of capsule maturities. The finding is significant because of its implications for the design of large-scale systems for harvesting this species and, possibly, related species from the same or similar plant communities (e.g. *Arthropodium*, *Burchardia umbellata* R. Br., *Caesia*, *Chamaescilla corymbosa* (R. Br.) Benth., *Hypoxis* and *Wurmbea dioica* (R. Br.) F. Muell.). Harvest 1 (intact inflorescences remained on the growing plant) gave the highest seed yields, in part because the racemose inflorescences of *B. bulbosa* continued to flower and produce capsules well after the earliest capsules had been harvested. Harvest 1

Table 4.4. Mean values for pooled data for germination percentage of *B. bulbosa* seeds 21 days after imbibition. Values followed by the same letter do not differ significantly ($P > 0.05$).

Pooled values	Germination (%)
Harvest	
1 (mature)	89 a
2 (early)	74 b
3 (intermediate)	81 ab
4 (late)	73 b
Inflorescence segment	
Proximal	83 a
Central	88 a
Distal	67 b
Seed size	
Large	84 a
Small	74 a

produced a greater proportion of larger seeds and there was less tendency for seed size to reduce in the distal segment, when compared with the detached inflorescences. After 8 months of dry storage, a higher percentage of seeds of Harvest 1 germinated within 21 days of imbibition. However, hand-harvesting was a very intensive activity that required careful handling of individual plants so as not to dislodge seeds from opening capsules. Daily harvesting was needed to recover all seeds for the purposes of this experiment.

In contrast, detaching and drying inflorescences offers the potential for a much less intensive harvest system. Such a system would allow *B. bulbosa* (and similar species) to be harvested less frequently than the hand-harvesting in the present study, at a predetermined inflorescence maturity. The yield, seed size and germination capacity of seeds harvested from the detached inflorescences (particularly from earlier harvests) compared favourably with seeds from hand-harvested plants. There was no advantage in harvesting inflorescences at a later stage of maturity (e.g. Harvest 4). Much potentially high-quality seed was lost from the discarded early capsules, and the yields from the upper segments of the inflorescences in Harvest 4 were lower, with lower germination percentage. The capsules on the detached inflorescences opened fully at maturity and released their seeds freely, suggesting that it would be a straightforward task to design an efficient seed collection and processing system.

The mean number of seeds per capsule was lower in the central and distal portions of the detached stems, whereas there was no significant difference in this measure in hand-harvested inflorescences. This suggests that, in the detached inflorescences, some immature seeds are aborted and available assimilates stored in the scape are directed to those that remain.

These results should still be treated with some caution because the seeds harvested from this experiment were neither evaluated for germination performance under nursery and field conditions nor was subsequent seedling growth assessed.

When detached from the parent plant, scapes in Harvests 2, 3 and 4 held some capsules at early stages of development. Very few capsules were aborted and 68% of them produced more than five seeds. However, only one set of drying conditions for inflorescences was used. This study did not investigate the optimum conditions for maturation of detached inflorescences.

The sequential removal of fruiting inflorescences, as a harvest strategy, may have other benefits for the standing crop. The removal of maturing inflorescences may improve the movement of assimilates to remaining inflorescences and possibly increase the induction of new inflorescences and, likewise, seed yield.

Whether these findings for *B. bulbosa* can be extrapolated to other indigenous liliaceous forbs is also a subject for further research. For instance, the capsules of *Arthropodium strictum* R. Br. tend not to open fully if harvested before reaching full maturity and *Caesia calliantha* RJF Hend. capsules fail to open if they are detached from the scape before dehiscence. In both cases, the capsule is a very strong structure from which it is difficult to remove seeds. However, it is possible that both these, and other species, would also mature and release their seeds freely if the capsules remained on a detached scape during gradual drying.

Conclusions

This experiment demonstrated that when inflorescences of *B. bulbosa*, at a range of maturities, are harvested and dried slowly at 20°C, immature capsules can continue to grow, mature and produce seeds of similar number, size and germination capacity to capsules that mature on the intact plant. The harvesting and drying of inflorescences at a predetermined stage of maturity could be of great benefit to large-scale production of seeds from this and, perhaps, other related undomesticated species with an extended period of flowering and seed production.

Further studies are needed to compare seeds harvested from detached inflorescences with those from intact inflorescences with regard to their storage potential, germination capacity and seedling establishment under nursery and field conditions.

Acknowledgement

I acknowledge the invaluable guidance and support provided by Dr Gerald Halloran in the completion of this study.

References

Bowring, J.D.C., Evans, A.W. and Sneddon, J.L. (1980) Objects and methods of seed production. In: Hebblethwaite, P.D. (ed.) *Seed Production.* Butterworths, London, pp. 3–14.

Conran, J.G. and Walsh, N.G. (1994) Bulbine. In: Walsh, N.G. and Entwistle, T.J. (eds) *Flora of Victoria*, Vol. 2. *Ferns and Allied Plants, Conifers and Monocotyledons.* Inkata Press, Melbourne, Australia, pp. 667–670.

Copeland, L.O. and McDonald, M.B. (1995) *Principles of Seed Science and Technology*, 3rd edn. Macmillan, New York.

Delpratt, C.J. (1997) A model for the cultivation of indigenous herbaceous wildflowers for seed production. In: Bellairs, S.M. and Osborne, J.M. (eds) *Proceedings of the Second Australian Native Seed Biology for Revegetation Workshop*. Australian Centre for Minesite Rehabilitation Research, Kenmore, Australia, pp. 107–111.

Desai, B.B., Kotecha, P.M. and Salunkhe, D.K. (1997) *Seeds Handbook*. Marcel Dekker, New York.

Thomson, J.R. (1979) *An Introduction to Seed Technology*. Leonard Hill, London.

5 Alterations in Gene Expression During Loss and Re-establishment of Desiccation Tolerance in Germinating and Germinated *Medicago truncatula* Seeds

J.M.R. Faria,[1,2] M. Lammers,[1] W. Ligterink,[1] J. Kodde[1] and H.W.M. Hilhorst[1]

[1]*Laboratory of Plant Physiology, Wageningen University, Arboretumlaan 4, 6703 BD Wageningen, The Netherlands;* [2]*Departamento de Ciências Florestais, Universidade Federal de Lavras, CP 3037, CEP 37200-000, Lavras, Minas Gerais, Brazil*

Abstract

This study investigated the expression of various genes during loss and re-establishment of desiccation tolerance (DT) in germinating and germinated *Medicago truncatula* Gaertn. 'Jemalong' A17 seeds. Osmotic treatment with polyethylene glycol (PEG) substantially re-established DT in germinated seeds with a protruded radicle up to 2 mm long, just before the resumption of the cell cycle. Real-time polymerase chain reaction (PCR) with gene-specific primers for *M. truncatula* showed that gene expression at the transcriptional level was clearly affected by germination, and responded to osmotic treatment and dehydration. Stress-related genes (i.e. *EM6*, *Per1* and *sHSP*) were downregulated during germination and upregulated by osmotic treatment, suggesting their involvement in the loss and reacquisition of DT in seeds. Cell cycle-associated genes (i.e. *CDC2a*) and cytoskeleton-associated genes (i.e. *ACT* and *TUB*) were upregulated during germination and were affected in various ways by PEG treatment. Upon dehydration, the mRNA levels of these genes decreased. No clear correlation was found between the expression of developmental genes (i.e. *ABI3*, *LEC1* and *CTS*) and DT.

Introduction

Orthodox seeds acquire DT during their development and are shed in a metabolically quiescent state. DT is a complex feature, and its acquisition has been related to the expression of various developmental genes, such as *ABSCISIC ACID-INSENSITIVE 3*

 Seeds: Biology, Development and Ecology (eds S. Adkins, S. Ashmore and S.C. Navie)

(*ABI3*) and *LEAFY COTYLEDON 1* (*LEC1*) (Koornneef *et al.*, 1984; Parcy *et al.*, 1997), as well as stress-related genes, such as *Early methionine* (*Em*), ABA-responsive protein kinases (*PKABA*), peroxiredoxins (*Per*) and heat-shock proteins (*HSP*) (Raghavan, 1997; Haslekas *et al.*, 1998; Wehmeyer and Vierling, 2000; Xiong and Zhu, 2001). Upon imbibition of non-dormant orthodox seeds, metabolism is rapidly resumed and thousands of genes are turned on or off (Bewley and Black, 1994; Bradford *et al.*, 2000). The *COMATOSE* (*CTS*) gene affects the metabolism of stored lipids and increases the sensitivity of the seed to gibberellins and pre-chilling, enhancing the germination potential (Russell *et al.*, 2000; Holdsworth *et al.*, 2001). Cell cycle, which is regulated by the activity of cyclin-dependent kinases (CDKs) (Huntley and Murray, 1999), is also resumed during germination or early seedling growth when DNA synthesis and cell division are required. *CDC2a* is an A-type CDK and is a key regulator in the G1-to-S and G2-to-M transitions in the cell cycle (Mironov *et al.*, 1999; Vazquez-Ramos and de la Paz Sanchez, 2003). Expression of *CDC2a* in growing radicles is likely confined to the root apical meristem and vascular tissues (Burssens *et al.*, 2000). During germination, microtubules and actin filaments, the two key components of the cytoskeleton, are also active (Jing *et al.*, 1999; Kost *et al.*, 2002).

DT is lost as germination progresses (Reisdorph and Koster, 1999), and several cellular and molecular changes are associated with this transition from a tolerant to a sensitive state, such as the resumption of DNA synthesis (Sargent *et al.*, 1981) and cell division (Berrie and Drennan, 1971), cell elongation (Dasgupta *et al.*, 1982) and decrease in the content of sucrose and larger oligosaccharides (Koster and Leopold, 1988). Loss of DT on or after germination can be re-established by mild stress, such as osmotic treatment through incubation in PEG (Bruggink and van der Toorn, 1995; Buitink *et al.*, 2003). This study aimed to investigate the changes in gene expression during loss and reacquisition of DT in germinated seeds of *M. truncatula.*

Materials and Methods

Seed germination and assessment of DT

M. truncatula Gaertn. 'Jemalong' A17 seeds were chemically scarified in concentrated H_2SO_4 for 5 min and were cold-imbibed (4°C) for 36 h before they were left to germinate in the dark at 20°C (modified from Sieberer *et al.*, 2002). To characterize the loss of DT after germination, germinated seeds with various radicle lengths were dehydrated over a saturated solution of K_2CO_3 (43% RH) in a closed box for 3 days by circulating air at 23°C. After dehydration, germinated seeds were pre-humidified in humid air (100% RH) for 24 h at 20°C to avoid imbibitional damage and were then rehydrated in petri dishes over moistened filter papers at 20°C in the dark. Germinated seeds that resumed radicle growth and developed into normal seedlings were considered to be desiccation-tolerant.

Re-establishment of DT in germinated seeds

The technique to re-establish DT in germinated seeds is described by Faria *et al.* (2005). Germinated seeds with various radicle lengths were incubated for 3 days in a −1.8 MPa solution of PEG 6000 at 5°C in the dark. After incubation, the germinated seeds were rinsed thoroughly in distilled water and then dehydrated, pre-humidified and rehydrated. Germinated seeds that resumed radicle growth and developed into normal seedlings were considered to be desiccation-tolerant.

Assessment of radicle moisture content

According to the International Seed Testing Association (ISTA, 1996), moisture content (MC) was assessed in four replications of ten radicles, by oven drying at 103°C for 17 h. MC is expressed on a dry weight basis (i.e. in g H_2O/g dw or simply in g/g).

RNA isolation and real-time PCR

The radicle is the most desiccation-sensitive part in germinated seeds of *M. truncatula* (Faria *et al.*, 2005). Thus, to avoid mixing seed parts with different levels of DT, RNA was isolated only from radicles. Radicles were excised from the seeds before, during and after germination (i.e. radicle 1, 2 and 3 mm long), after PEG treatment and after dehydration of PEG-treated and untreated germinated seeds. Total RNA was extracted using the RNeasy Plant Mini Kit (Qiagen Benelux B.V., Venlo, The Netherlands), according to the manufacturer's directions, and treated with DNase. cDNA synthesis was conducted using the iScript cDNA synthesis kit (Bio-Rad Laboratories, Hercules, California, USA) according to the manufacturer's protocol, in a thermal cycler (iCycler; Bio-Rad). Real-time PCR was performed using the same equipment, with gene-specific primers for *M. truncatula* (Table 5.1), cDNA and iQ SYBR green supermix (Bio-Rad). A specific primer for the *18S* gene was used as an internal control to normalize the other products. Fold changes in gene expression in relation to the control (i.e. dry seed) were calculated using the $2^{-\Delta\Delta Ct}$ method.

Table 5.1. Specific primers for *Medicago truncatula* used for real-time polymerase chain reaction (PCR).

Name	Forward	Reverse
sHSP18.2	CACGTGTGGACTGGAAAGAA	TCTCTCAACACGATGCCACT
EM6	GGCAAAGCAAGGAGAGACTG	ACCTTCCTCTTCAGCACGTT
PER1	GCACATACTCCAGGTGCAAA	ACCACCCTCAGCACTTCATC
PKABA1	GGCGCTTATCCTTTTGAAGA	CAGCTTCCACCTTCCCATTA
ABI3	ACCGGTGATTTTGTGAAAGC	ATGGCGTGCCATTATTATCC
LEC1	TGGGTTTTGGAAAATGGAAG	CAGCTCCGAATGAAAAGACC
CTS	GGTCTTCCACATGGCAAGTAA	AAACCCAAGGCCAAGTAACA
α-5-TUB	GGATAACGAGGCGATCTACG	CGAGGATACGGCACAAGATT
ACT-11	TCCATCATGAAGTGCGATGT	AACCTCCGATCCAGACACTG
CDC2a	ACCCCAGTTGATGTTTGGTC	CCACGGTTGCTAGGTCCTTA
18S	TGACGGAGAATTAGGGTTCG	CCTCCAATGGATCCTCGTTA

Results and Discussion

Changes in moisture content caused by PEG treatment and dehydration

The MC in the radicles of germinated seeds decreased from 3.74 to 2.24g/g after 3 days of PEG incubation. Dehydration of both PEG-treated and untreated germinated seeds caused a sharp decrease in the MC of radicles (0.64g/g) within the first hour, followed by a slower loss of water over the remaining period of dehydration. After 3 days of dehydration, the MC of radicles was 0.20 and 0.15g/g in PEG-treated and untreated germinated seeds, respectively, which is close to that of the radicles of dry seeds (0.19g/g) (Faria *et al.*, 2005).

Loss and re-establishment of DT in germinated seeds

As soon as visible germination occurred, with a protruded radicle length of 1mm, only 12% of the germinated seeds were still desiccation-tolerant (i.e. able to survive dehydration). Seeds with a radicle length of 2mm or longer lost DT completely. By incubating the germinated seeds in PEG before dehydration, DT could be substantially reinduced (84%) in seeds with a protruded radicle of up to 2mm, just before the resumption of the cell cycle (Faria *et al.*, 2005). In 2–3mm long radicles there was an abrupt drop in DT, decreasing its value to 33%. DT decreased even further in radicles that were 3–4mm long to near zero (Faria *et al.*, 2005).

Gene expression

Stress-related genes (sHSP, Em6, Per1 and PKABA)

The transcript levels of small *HSP* (*sHSP*) and *Em6* decreased during imbibition and, more intensively, after germination, with radicle growth (Table 5.2). The expression of *Per1* was also downregulated after germination, but not during imbibition. Protection against damaging reactive oxygen species (ROS), in which peroxiredoxins act, is also needed during imbibition, when respiration is resumed (Haslekas *et al.*, 2003). In the three radicle lengths analysed (i.e. 1, 2 and 3mm), PEG treatment was able to upregulate the expression of *sHSP*, *Em6* and *Per1*, bringing it back to levels similar to that found in the dry seed (Table 5.2). In general, dehydration of PEG-treated seeds had little effect on the expression of these genes. Since PEG treatment was able to upregulate the expression of these genes even in 3mm long radicles that showed low survival rates after dehydration (33%), it seems that these genes are involved in the loss and reacquisition of DT in seeds and their upregulation is not enough per se to re-establish DT. The expression of *PKABA1* was affected differently than the other genes, with a slight upregulation during germination, and then decreased afterwards. PEG treatment increased the abundance of *PKABA1* mRNAs that did not change significantly with subsequent dehydration (Table 5.2). There was no correlation between the expression of *PKABA1* and the re-establishment of DT.

Table 5.2. Changes in transcript levels of ten genes in radicles of *Medicago truncatula* seeds during and after germination; after germination and PEG treatment; and after germination, PEG treatment and dehydration. Changes are in relation to levels found in radicles of dry seeds.

	During imbibition (before radicle protrusion)		After germination			After germination and PEG treatment			After germination, PEG treatment and dehydration		
			Radicle length (mm)			Radicle length (mm)			Radicle length (mm)		
Gene	36 h at 4°C	36 h at 4°C + 6 h at 20°C	1	2	3	1	2	3	1	2	3
sHSP18.2	––	––	––	––	––	+	0	0	–	–	–
Em6	–	–	––	––	––	–	–	––	–	–	––
Per1	0	+	–	––	––	+	0	–	+	++	+
PKABA1	+	+	+	0	0	++	++	++	+	++	++
ABI3	++	++	+	–	––	+	0	0	+	+	0
LEC1	0	++	+	+	+	+	+	+	+	++	++
CTS	+	+	+	+	0	+	+	+	0	+	+
α-5-TUB	++	++	++	++	++	++	++	++	++	++	++
ACT-11	+	++	++	++	++	++	++	++	+	+	++
CDC2a	++	++	++	++	++	++	++	++	++	++	++

0 = less than twofold; – or + = twofold to tenfold; –– or ++ = more than tenfold; + and – signs refer to upregulation and downregulation, respectively.

Developmental genes (ABI3, LEC1 and CTS)

Abundance of *ABI3* mRNAs increased during imbibition and then progressively decreased as the radicle grew (Table 5.2). PEG treatment increased the *ABI3* levels, which hardly changed with subsequent dehydration. *LEC1* levels did not change during cold imbibition, but increased after the seeds were transferred to 20°C. The levels remained stable after germination and after PEG treatment. Subsequent dehydration led to upregulation in the expression of *LEC1*, in 2–3 mm long radicles. *CTS* levels increased during imbibition, as shown by Footitt *et al.* (2002) in *Arabidopsis thaliana* (L.) Heynh., but did not change after germination. Neither PEG treatment nor dehydration significantly changed the *CTS* mRNA levels. In general, there was no correlation between the expression of the developmental genes studied and the loss and re-establishment of DT.

Cytoskeleton (TUB and ACT) and cell cycle (CDC2a) genes

α-5-Tubulin (*TUB*) transcript levels increased steeply during imbibition and remained high after germination (Table 5.2). Germinating seeds require tubulin for building up microtubules, which are needed for cell expansion and later for cell division. PEG treatment and dehydration of PEG-treated germinated seeds did not change the *TUB* levels. The same pattern was observed for the *cell division cycle 2* (*CDC2a*) transcripts. *Actin-11* (*ACT*) transcript levels increased on imbibition and remained at the same level after germination. Like tubulin, actin is part of the cytoskeleton, and is required to be functional, not only in dividing or elongating cells but also in metabolically active cells in general, acting in intracellular motility and cell-to-cell communication (Kost *et al.*, 1999). PEG treatment of germinated seeds had no effect on *ACT* expression, but the expression was lowered by subsequent dehydration of 1 and 2 mm long radicles.

Acknowledgements

We thank CNPq (National Council for Scientific and Technological Development, from the Ministry of Science and Technology, Brazil) for its financial support for the studies of J.M.R. Faria.

References

Berrie, A.M.M. and Drennan, D.S.H. (1971) The effect of hydration–dehydration on seed germination. *New Phytologist* 70, 135–142.

Bewley, J.D. and Black, M. (1994) *Seeds: Physiology of Development and Germination*, 2nd edn. Plenum Press, New York.

Bradford, K.J., Chen, F., Cooley, M.B., Dahal, P., Downie, B., Fukunaga, K.K., Gee, O.H., Gurusinghe, S., Mella, R.A., Nonogaki, H., Wu, C.T. and Yim, K.O. (2000) Gene expression prior to radicle emergence in imbibed tomato seeds. In: Black, M., Bradford, K.J. and Vazquez-Ramos, J. (eds) *Seed Biology: Advances and Applications*. CAB International, Wallingford, UK, pp. 231–251.

Bruggink, T. and van der Toorn, P. (1995) Induction of desiccation tolerance in germinated seeds. *Seed Science Research* 5, 1–4.

Buitink, J., Vu, B.L., Satour, P. and Leprince, O. (2003) The re-establishment of desiccation tolerance in germinated radicles of *Medicago truncatula* Gaertn. seeds. *Seed Science Research* 13, 273–286.

Burssens, S., Himanen, K., van de Cotte, B., Beeckman, T., Van Montagu, M., Inzé, D. and Verbruggen, N. (2000) Expression of cell cycle regulatory genes and morphological alterations in response to salt stress in *Arabidopsis thaliana*. *Planta* 211, 632–640.

Dasgupta, J., Bewley, J.D. and Yeung, E.C. (1982) Desiccation-tolerant and desiccation-intolerant stages during the development and germination of *Phaseolus vulgaris* seeds. *Journal of Experimental Botany* 33, 1045–1057.

Faria, J.M.R., Buitink, J., van Lammeren, A.A.M. and Hilhorst, H.W.M. (2005) Changes in DNA and microtubules during loss and re-establishment of desiccation-tolerance in germinating *Medicago truncatula* seeds. *Journal of Experimental Botany* 56, 2119–2130.

Footitt, S., Slocombe, S.P., Larner, V., Kurup, S., Wu, Y., Larson, T., Graham, I., Baker, A. and Holdsworth, M. (2002) Control of germination and lipid mobilization by *COMATOSE*, the *Arabidopsis* homologue of human ALDP. *The EMBO Journal* 21, 2912–2922.

Haslekas, C., Stacy, R.A.P., Nygaard, V., Culianez-Macia, F.A. and Aalen, R.B. (1998) The expression of a peroxiredoxin antioxidant gene, *AtPer1*, in *Arabidopsis thaliana* is seed specific and related to dormancy. *Plant Molecular Biology* 36, 833–845.

Haslekas, C., Viken, M.K., Grini, P.E., Nygaard, V., Nordgard, S.H., Meza, T.J. and Aalen, R.B. (2003) Seed 1-cysteine peroxiredoxin antioxidants are not involved in dormancy, but contribute to inhibition of germination during stress. *Plant Physiology* 133, 1148–1157.

Holdsworth, M., Lenton, J., Flintham, J., Gale, M., Kurup, S., McKibbin, R., Bailey, P., Larner, V. and Russell, L. (2001) Genetic control mechanisms regulating the initiation of germination. *Journal of Plant Physiology* 158, 439–445.

Huntley, R.P. and Murray, J.A.H. (1999) The plant cell cycle. *Current Opinion in Plant Biology* 2, 440–446.

ISTA (1996) International rules for seed testing. *Seed Science and Technology* 24 (Suppl.).

Jing, H.C., van Lammeren, A.A.M., de Castro, R.D., Bino, R.J., Hilhorst, H.W.M. and Groot, S.P.C. (1999) β-tubulin accumulation and DNA synthesis are sequentially resumed in embryo organs of cucumber (*Cucumis sativus* L.) seeds during germination. *Protoplasma* 208, 230–239.

Koornneef, M., Reuling, G. and Karssen, C.M. (1984) The isolation and characterization of abscisic acid-insensitive mutants of *Arabidopsis thaliana. Physiologia Plantarum* 61, 377–383.

Kost, B., Mathur, J. and Chua, N.H. (1999) Cytoskeleton in plant development. *Current Opinion in Plant Biology* 2, 462–470.

Kost, B., Bao, Y.Q. and Chua, N.H. (2002) Cytoskeleton and plant organogenesis. *Philosophical Transactions of the Royal Society of London: Series B* 357, 777–789.

Koster, K.L. and Leopold, A.C. (1988) Sugars and desiccation tolerance in seeds. *Plant Physiology* 88, 829–832.

Mironov, V., de Veylder, L., van Montagu, M. and Inzé, D. (1999) Cyclin-dependent kinases and cell division in plants – the nexus. *The Plant Cell* 11, 509–522.

Parcy, F., Valon, C., Kohara, A., Misera, S. and Giraudat, J. (1997) The *ABSCISIC ACID-INSENSITIVE3, FUSCA3*, and *LEAFY COTYLEDON1* loci act in concert to control multiple aspects of Arabidopsis seed development. *The Plant Cell* 9, 1265–1277.

Raghavan, V. (1997) *Molecular Embryology of Flowering Plants.* Cambridge University Press, New York.

Reisdorph, N.A. and Koster, K.L. (1999) Progressive loss of desiccation tolerance in germinating pea (*Pisum sativum*) seeds. *Physiologia Plantarum* 105, 266–271.

Russell, L., Larner, V., Kurup, S., Bougourd, S. and Holdsworth, M. (2000) The *Arabidopsis COMATOSE* locus regulates germination potential. *Development* 127, 3759–3767.

Sargent, J.A., Mandi, S.S. and Osborne, D.J. (1981) The loss of desiccation tolerance during germination: an ultrastructural and biochemical approach. *Protoplasma* 105, 225–239.

Sieberer, B.J., Timmers, A.C.J., Lhuissier, F.G.P. and Emons, A.M.C. (2002) Endoplasmic microtubules configure the subapical cytoplasm and are required for fast growth of *Medicago truncatula* root hairs. *Plant Physiology* 130, 977–988.

Vazquez-Ramos, J.M. and de la Paz Sanchez, M. (2003) The cell cycle and seed germination. *Seed Science Research* 13, 113–130.

Wehmeyer, N. and Vierling, E. (2000) The expression of small heat shock proteins in seeds responds to discrete developmental signals and suggests a general protective role in desiccation tolerance. *Plant Physiology* 122, 1099–1108.

Xiong, L. and Zhu, J.K. (2001) Abiotic stress signal transduction in plants: molecular and genetic perspectives. *Physiologia Plantarum* 112, 152–166.

6 ASP53, a 53 kDa Cupin-containing Protein with a Dual Role: Storage Protein and Thermal Protectant

L. Mtwisha,[1] J.M. Farrant,[1] W. Brandt,[1] C. Hlongwane[2] and G.G. Lindsey[1]

[1]*Department of Molecular and Cell Biology, University of Cape Town, Rondebosch 7700, South Africa;* [2]*CSIR Bio/Chemtek, Pretoria 0001, South Africa*

Abstract

Acacia erioloba E. Mey. grows in the semiarid regions of southern Africa and the shed seeds are often exposed to soil temperatures higher than 70°C before summer rains facilitate germination. We have purified a 53 kDa heat-soluble protein, termed *Acacia* seed protein of molecular size 53 kDa (ASP53), from mature seeds of *A. erioloba.* The gene sequence was determined and found to contain two cupin motifs, with homology to seed storage proteins. Immunocytochemistry localized the protein not only to the vacuoles in cotyledons, confirming a seed storage role for the protein, but also to the cell walls of the axes and cotyledons in mature seeds. Circular dichroism (CD) spectroscopy demonstrated that the protein displayed defined secondary structure, which was maintained even at high temperature. ASP53 was found to completely inhibit the temperature-dependent aggregation of egg white protein, suggesting a protective role against heat stress.

Introduction

The Camel Thorn tree (*Acacia erioloba* E. Mey.) grows in the semiarid regions of southern Africa. The seed pods are shed in midwinter and are an important source of nutrition for many animals, especially the gemsbok (*Oryx gazella* L.). Each pod contains approximately 20 seeds, which are indigestible and are excreted in the surrounding area. The seeds only germinate when the rains arrive in mid to late summer, before which they are exposed to ground temperatures regularly exceeding 70°C. Survival under these conditions might suggest that these seeds be classified as extremophiles, and also that there is the possible presence of unusual late embryogenesis abundant (LEA)-like proteins in the mature seeds of this tree.

It is well documented that most species produce seeds that, when mature, are desiccation-tolerant (Roberts, 1973) and thus enable species survival during periods of unsuitable environmental conditions. Acquisition of tolerance occurs in the mid to final stages of reserve accumulation, before or at the same time as the onset of maturation drying, depending on the species (Vertucci and Farrant, 1995), and is coincident with the synthesis of non-reducing sugars and compatible solutes (Blackman *et al.*, 1992; Sun and Leopold, 1993; Horbowicz and Obendorf, 1994), antioxidants (Hendry *et al.*, 1992; Smirnoff, 1993; Leprince *et al.*, 1994) and stress-induced proteins (Baker *et al.*, 1988; Close *et al.*, 1989; Dure *et al.*, 1989; Blackman *et al.*, 1991; Roberts *et al.*, 1993; Bray, 1997), all believed to play a protective role in the tolerance of severe water deficit. These molecules are rapidly degraded during germination to provide a source of amino acids for protein synthesis and metabolic energy in the developing seedlings, a process that has been shown to coincide with the loss of desiccation tolerance (DT) (Blackman *et al.*, 1991; Leprince *et al.*, 1993; Vertucci and Farrant, 1995). The presence of seed storage proteins has been well reported (Galili *et al.*, 1993; Melo *et al.*, 1994; Bharali and Chrungoo, 2003; Makri *et al.*, 2005) but, to date, there have been no reports of such proteins having a protective role in desiccation stress.

Stress-induced proteins synthesized during seed maturation include the heat shock proteins (HSPs) and LEA proteins. The LEA proteins constitute up to 4% of total cellular proteins (Roberts *et al.*, 1993). Although the exact role of LEA proteins in plants has yet to be determined, these proteins are characterized by having a high hydrophilic amino acid content (Baker *et al.*, 1988; Close *et al.*, 1989; Dure *et al.*, 1989; Goyal *et al.*, 2005), a feature that has been proposed to allow them to act as water replacement molecules thereby stabilizing macromolecules against desiccation stress (Garay-Arroyo *et al.*, 2000). Alternative hypotheses are that LEA proteins sequester ions accumulated during dehydration and interact with the exposed hydrophobic surfaces of partially denatured proteins to prevent aggregation (Campbell and Close, 1997; Close, 1997). Strong circumstantial evidence exists that LEA proteins play a role in stress tolerance with specific roles suggested for some LEA and LEA-like proteins. The LEA-like protein HSP12p in the yeast *Saccharomyces cerevisiae* has been proposed to act as a plasticizer in the cell wall (Motshwene *et al.*, 2004; Karreman *et al.*, 2005) and the *Typha latifolia* pollen D-7 protein was found to stabilize sugar glasses in an *in vitro* system (Wolkers *et al.*, 2001). In the nematode *Aphelenchus avenae* Bastian, a group III LEA-like protein inhibited desiccation-induced aggregation of citrate synthase (Goyal *et al.*, 2005).

This chapter reports the identification of a cupin-containing protein present in the cell walls and the vacuoles of the mature seeds of *A. erioloba*, and outlines its possible role.

Materials and Methods

Plant material

Newly shed *A. erioloba* seed pods were collected from Auob in southern Namibia and stored at −20°C until used.

Protein purification

The seeds were homogenized with a Kinematica CH-6010 Ultra-turrex homogenizer in ice-cold extraction buffer consisting of 10 mM Tris-(hydroxymethyl)-aminomethane-hydrochloride (Tris-HCl), 50 mM NaCl, 5 mM $MgCl_2$ and 1 mM phenylmethanesulfonylfluoride (PMSF) at pH 7.4. The homogenate was centrifuged at 27,000 *g* using a Beckman JA rotor at 4°C for 10 min, after which the supernatant was incubated at 80°C for 30 min. Heat-coagulated protein was removed by similar centrifugation. Heat-soluble protein was fractionated on a Whatman CM 52 cation exchange column previously equilibrated with 50 mM acetate and 50 mM NaCl at pH 5.0. Proteins were eluted with a gradient of 50–500 mM NaCl in this buffer. Further purification was performed by gel filtration with a Sephadex G-100 column using 20 mM HCl as the buffer. Final purification was performed by high-performance liquid chromatography (HPLC) using a Jupiter C18 reverse-phase column equilibrated in 0.1% heptafluorobutyric acid (HFBA). Proteins were eluted using a linear 0–70% gradient of acetonitrile in 0.1% HFBA. Protein used for secondary structure analysis was not heat-treated, and the gel filtration step was performed using 50 mM phosphate buffer (PB) at pH 7.0. Proteins were eluted with this buffer, pooled, concentrated by Amicon ultrafiltration and reapplied to the same column. No HPLC purification was applied to this protein preparation.

Peptide identification and sequencing

A gel piece of 1 mm diameter containing ASP53 was excised from a Coomassie-stained gel and in-gel digested with trypsin (Wilm *et al.*, 1996). Peptides were mixed with an equal volume of α-cyano-4-hydroxy-cinnamic acid and analysed by matrix-assisted laser desorption/ionization time of flight (MALDI-TOF) mass spectrometry. The rest was guanidinated with *O*-methylisourea-hydrogen sulphate and sulphonated with the chemically assisted fragmentation (CAF) reagent (Ettan) before being analysed by MALDI-TOF mass spectrometry.

Polymerase chain reaction and 5' rapid amplification of cDNA ends

Total RNA was extracted from mature seeds of *A. erioloba* using TRI REAGENT® (Molecular Research Center, Inc.). cDNA was synthesized from DNaseI-treated total RNA (2 μg) using Moloney murine leukaemia virus reverse transcriptase (M-MLV RT; Promega) and an anchored oligo dT primer (5'-GGGATCCT18(CAG)I-3'). cDNA together with the oligo dT primer and a degenerate primer, gene-specific primer (GSP 1) (i.e. 5'-AA(CT)CA(AG)TA(CT)GA(CT)GCI(CTA)TI(CA)G-3'), designed from the amino acid sequence of the 1577.7 Da peptide obtained by post-source decay (PSD) MALDI-TOF mass spectroscopy were used to amplify the *ASP53* gene. The sample was denatured at 94°C for 4 min before being subjected to 35 cycles of denaturation at 94°C for 1 min, primer annealing at 55°C for 1 min and extension at 72°C for 2 min, followed by a final extension at 72°C for 7 min in a GeneAmp® PCR System 9700, Perkin Elmer (Applied Biosystems). The purified polymerase chain reaction (PCR) product was cloned into pGEM®-T Easy vector and sequenced (Westermeier and Naven, 2002).

5' rapid amplification of cDNA ends (RACE) was performed to determine the sequence of the 5' end of the *ASP53* gene. cDNA synthesis was carried out using an Expand Reverse Transcriptase kit (Roche) in the presence of SMART IV oligonucleotide (Creator™ SMART™ cDNA Library Construction kit, Clontech) and an anchored oligo dT primer. To amplify the 5' end of the gene, a gene-specific antisense primer, GSP 2, designed from the DNA sequence of the reverse transcriptase polymerase chain reaction (RT-PCR) product, was used in conjunction with the 5' PCR primer (complementary to the SMART IV oligonucleotide that was ligated to the 3' end of the cDNA). PCR was performed using the Expand High Fidelity^{PLUS} PCR System (Roche) in a GeneAmp® PCR System 9700, Perkin Elmer (Applied Biosystems) using 30 cycles of sample denaturation at 94°C for 30 s, primer annealing at 61°C for 45 s and extension at 72°C for 2 min, followed by a final extension at 72°C for 7 min. The purified 5' RACE product was cloned into pGEM®-T Easy vector and sequenced.

Circular dichroism and thermal denaturation

The circular dichroism (CD) spectra of samples in 100 mM K_2HPO_4 at pH 7.4 were determined at temperatures between 20°C and 80°C. Spectra were obtained over the range of 184–250 nm with a Jasco J-810 spectropolarimeter using a quartz cell of 1 cm path length at a protein concentration of 0.5 mg/ml in 100 mM K_2HPO_4 at pH 7.4. Predictions of the relative secondary structure were made using CDNN software (http://bioinformatik.biochemtech.uni-halle.de/cdnn).

Protein aggregation was monitored by measuring the light scattering of the solution in 100 mM K_2HPO_4 at pH 7.4 at 350 nm using a Pye-Unicam SP1800 spectrophotometer with a custom-made heating block interfaced to an IBM PC through an Oasis A/D converter. Absorbance values were recorded at every 0.2°C between 20°C and 90°C at a heating rate of 1°C/min.

Immunocytochemistry

Tissue processing for transmission electron microscopy

More than 20 cubes of 5 mm^3 of cotyledons and axes from at least five different mature seeds were processed for transmission electron microscopy (TEM) (Sherwin and Farrant, 1996). Epoxy resin-infiltrated specimens (Spurr, 1969) were sectioned at approximately 75 nm using a Reichert Ultracut-S microtome and collected on nickel grids.

Antibody preparation and purification

The antibody was prepared by injecting HPLC-purified protein in Freund's complete adjuvant into rabbits (*Oryctolagus cuniculus* L.). Titres and specificity were determined by enzyme-linked immunosorbent assay (ELISA) as described by Harlow and Lane (1988). Immunocytochemistry was performed by using this antibody (Motshwene *et al.*, 2004).

Immunogold labelling

Sections were sequentially incubated on drops of 1% (w/v) bovine serum albumin (BSA) in 0.1 M PB at pH 7.2 to prevent non-specific antibody binding, and on 0.02 M glycine to block unreactive aldehyde groups, prior to incubation with the primary antibody diluted at 1:1000 in 1% (w/v) BSA/PB at room temperature for 16 h. After washing in 1% (w/v) BSA/PB, the grids were floated on a drop of goat anti-rabbit immunoglobin, to which 10 nm diameter gold particles were attached, diluted at 1:50 in 1% (w/v) BSA/PB. Incubation was carried out for a minimum of 60 min at room temperature, after which the grids were rinsed with 1% (w/v) BSA/PB. Sections were fixed by floating the grids on 1% (v/v) glutaraldehyde for 10 min. After thoroughly rinsing the grids in water, samples were stained using 3% (w/v) uranyl acetate and 1% (w/v) lead citrate. Sections were examined for gold labelling using a Zeiss 200 SX (Germany) TEM. Control sections were probed with pre-immune serum in place of the antisera.

Results

The most abundant protein (constituting ~47% of total soluble proteins) from the seeds of *A. erioloba* was a 53 kDa heat-stable protein, termed ASP53 (Fig. 6.1). Once purified to homogeneity it was used to immunize rabbits for antibody production. Immunocytochemistry using this antibody showed label to occur only in the cell wall of the axis (Fig. 6.2a) and the cell walls and protein storage vacuoles in cotyledons (Fig. 6.2b) of mature seeds. No gold particles, indicative of the presence of ASP53, were detected when similar sections were probed with the pre-immune serum (Fig. 6.2c). These locations suggest a dual role for ASP53, as a storage protein in the vacuoles and a possible stress-associated function in the cell wall.

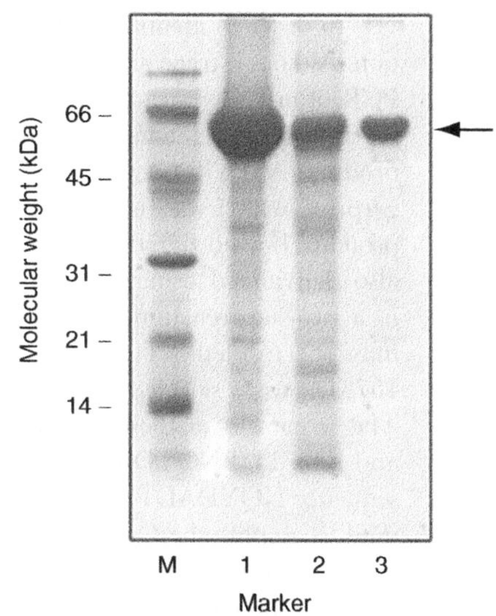

Fig. 6.1. SDS-PAGE of total proteins extracted from mature seeds of *Acacia erioloba* (column 1), soluble proteins after incubation at 80°C for 30 min (column 2) and HPLC-purified protein (column 3). The molecular weights (in kDa) of markers (M) of known size are denoted on the left-hand side of the figure. The arrow shows the electrophoretic migration of the protein selected for further study.

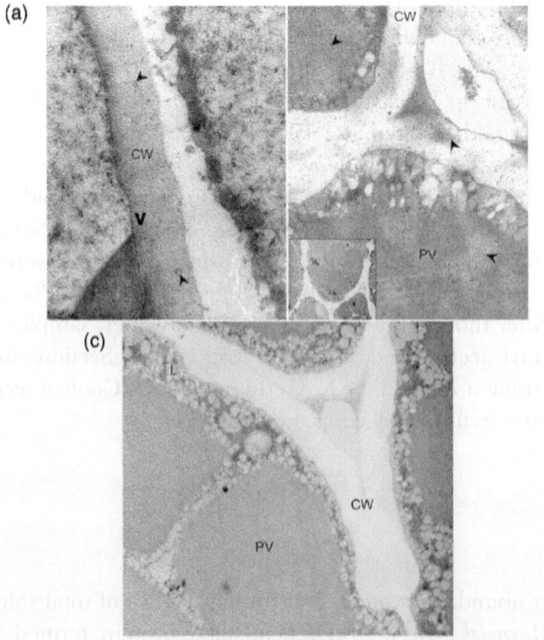

Fig. 6.2. Immunolocalization of ASP53 in mature *Acacia erioloba* seeds using the anti-ASP53 antibody. Sections of (a) axes and (b) cotyledons from mature seeds were examined. Arrows indicate gold label in the cell walls (CW) and protein vacuoles (PV). (c) Sections of cotyledons from mature seeds were also probed with the pre-immune serum as a control. The inset in (b) is a lower magnification view of the cotyledonary tissue from mature seeds, included to show the area examined in the higher magnification view. Magnifications used were: (a) ×45,000; (b) ×17,500; inset ×1588; and (c) ×9420.

In order to identify ASP53, purified protein was subjected to Edman degradation. Although trace amounts of the sequence GSEQQRQDEPT were found, RT-PCR of mRNA extracted from *A. erioloba* seeds using a degenerate primer directed to this sequence, together with an oligo dT primer, failed to produce a PCR product. ASP53 was further digested with trypsin to obtain a peptide mass fingerprint, which was used to determine whether ASP53 was homologous to known proteins, but no significant homology was found (data not shown). The digest was also derivatized using *O*-methylisourea-hydrogen sulphate and the CAF reagent in a two-step reaction, and the derivatized digest was analysed by MALDI-TOF mass spectrometry. Two sulphonated peptides with molecular masses of 1357 and 1578 Da were selected for sequencing by PSD-MALDI-TOF mass spectrometry. The sequences of these peptides were determined to be HL(I)QF(Mo)L(I)QDYR and TFVTL(I)NQYDAL(I)R (Fig. 6.3), respectively. A degenerate primer to the sequence NQYDAL(I)R of the latter peptide together with an oligo dT primer were used to amplify the C-terminal part of the *ASP53* gene. A primer to this sequence was fortuitously chosen because it was less degenerate than primers directed to the other peptide. The complete gene sequence was then determined by using 5′

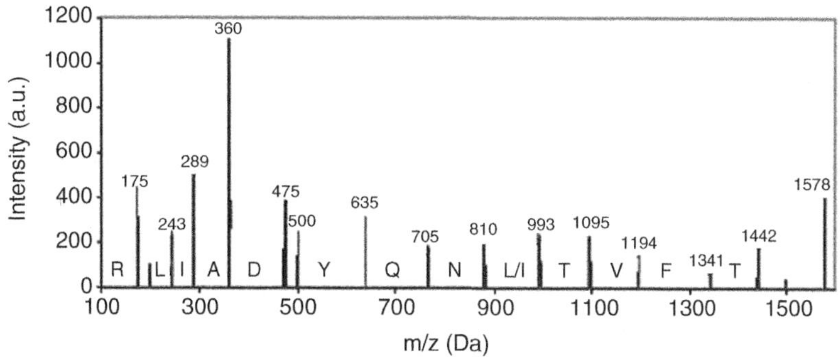

Fig. 6.3. PSD-MALDI-TOF mass spectrometric analysis of one of the sulphonated peptides derivatized using *O*-methylisourea-hydrogen sulphate and the CAF reagent. The sequence was determined from the difference in the masses of adjacent peaks to be TFVTL(I)NQYDAL(I)R. The intensity of the signal representing each peptide is given in arbitrary units (a.u.).

RACE with internal GSP (Fig. 6.4). This sequence showed an open reading frame (ORF) of 1358 bp between positions 81 and 1434 encoding a protein of 453 amino acids with a molecular weight of 51 kDa. The ORF contained a start codon and a stop codon, indicating that the full-length gene sequence had been obtained.

The presence of the sequence of the second peptide, identified by PSD-MALDI-TOF mass spectrometry, confirmed that the correct product had indeed been obtained. Comparison of the sequence obtained with that of known proteins in the databases showed that ASP53 was related to the known storage proteins β-conglutin, β-conglycinin, canavalin and vicilin with amino acid homologies of 57%, 55%, 55% and 53%, respectively. The members of this group of storage proteins all have a cupin motif. ASP53 was proposed to have two such cupin domains, between residues 52 and 169, and between 253 and 414, suggesting that ASP53 should be classified as a member of the cupin superfamily. Prediction of the structure of ASP53 using fold recognition (McGuffin and Jones, 2003) suggested that ASP53 had close structural similarity to a number of cupin motif-containing proteins. These proteins (shown in decreasing order of structural similarity), like *Glycine max* (L.) Merr. glycinin precursor, *Bacillus subtilis* (Ehr.) Cohn oxalate decarboxylase, *Canavalia ensiformis* (L.) DC. canavalin, *Aspergillus japonicus* Saito quercetin 2,3-dioxygenase, *Hordeum vulgare* L. oxalate oxidase and *Homo sapiens* L. homogentisate 1,2-dioxygenase, were from different phyla within the bacterial and eukaryotic kingdoms. The predicted structure of ASP53 was fitted to the structures of these proteins with net scores ranging between 0.959 for the *G. max* glycinin precursor and 0.847 for *H. sapiens* homogentisate 1,2-dioxygenase (maximum score = 1), and the confidence level of all scores was proposed to be 'certain'. Regions of maximum homology were found to be located in regions of β-sheet, with regions containing sequence insertions and deletions proposed to be present in the loop regions between segments of β-sheet and outside of the cupin barrel structure (Fig. 6.5). The N-terminals of ASP53 and of the proteins to which ASP53 displayed similarity are of variable length. As a result, the N-terminal region of ASP53 was not fitted to the core structure of these proteins, and the structure of this region was therefore predicted using neural

1 AAGCAGTGGTATCAACGCAGAGTGGCCATTACGGCCGGGGACCATATACATCTTCATCG
1

61 ATTCATTTGAGTGATCAACC***ATG***AAAGCCAGGCTTGCCCTTCTCTGGTTTCTGGGTTCTG
20 M K A R L A L L W F L G S

121 TTTTCCTGGCGGCGGCCTCCGTTGGCCTTGCCCGCTCGGAACAACAAGGACAAGATGAGC
40 V F L A A A S V G L A R S E Q Q G Q D E

181 GAACACGGGTTAAACATAACCCGTATCACATACACTCGGATAGGTTCCAAACCCGATTCG
60 R T R V K H N P Y H I H S D R F Q T R F

241 AGAACAAAAATGGTCAGATCCGGGTCCTCGAACAGTTCGACAAGCTGTCCCGACACCTTC
80 E N K N G Q I R V L E Q F D K L S R **H L**

301 AGTTTCTCCAAGACTACCGGATCGTCCAATATAAAGCCGAACCCAAACATAGTCCTGC
100 **Q F L Q D Y R** I V Q Y K A R T Q T I V L

361 CCCACTATTCCGATGCTGAGTTCCTCCTCATCATCACCAGAGGAAGCGCCTTCGCTGCTG
120 P H Y S D A E F L L I I T R G S A F A A

421 TTTCGTTTGGTCCGAATGTGGCCAACAGAACATTCGTCACCCTCAATCAATACGACGCGC
140 V S F G P N V A N R **T F V T L N Q Y D A**

481 TGAGGGTTCCGGCCGGAGCCGTATGTTAATACCATTAACAATGGTGGATCCAGTGGTGATG
160 **L R** V P A G A V C Y T I N N G G S S G D

541 AAGATCTTGAGTTCATCAAACTCGCACTACCCTCTAACCTTCCTGGTCAATTTGAGGACT
180 E D L E F I K L A L P S N L P G Q F E D

601 TGTACCCCACGGGCAAACGCTTGCCAACTTCCCTCTACAAAGTCTTCGATAAGAACACAC
200 L Y P T G K R L P T S L Y K V F D K N T

661 TCCAGGCCGCATTCAACGCCCCGTACGAGGAGATTCAAGACGTGTTGTGGGGATCAAATC
220 L Q A A F N A P Y E E I Q D V L W G S N

721 AGTGGCAAGAAGAGGCGATAGTGGCATTGCCGGGGAGTGAATTAAGATCACTTATCAGAC
240 Q W Q E E A I V A L P G S E L R S L I R

781 ATGCCCAATCGAGGTCAGAGGGACATGATGTACCAGCAACAAAATTATTAGGTCCTATCA
260 H A Q S R S E G H D V P A T K L L G P I

841 GATTGAGACGCATTAAACCTCGCTACTCCAACGACCATGGCTCCCTTCATGAGGCTTGGC
280 R L R R I K P R Y S N D H G S L H E A W

901 CGTTAGAAAAATACCCAGCTCTTGATGACCTAGACATCACTGCCTCCTACCTAAGACTCA
300 P L E K Y P A L D D L D I T A S Y L R L

961 ACAAGGGATCACTGTTTCTGCCCCACTACAATTCAAAAGCGATAGTAGTGTCCCTTTGTTG
320 N K G S L F L P H Y N S K A I V V S F V

1021 CTAACGGAAAAGGACAGAGTGAACTAGGGAGCCCATATCNTGTGAACAGACAGCGCGGCAGC
340 A N G K G Q S E L G S P Y X V N R Q R Q

1081 GGCAGCAACAACAACCACAGGAGGCAGAAGAAGGATATATTGAGAGCTTGCCAGCGAATC
360 R Q Q Q Q P Q E A E E G Y I E S L P A N

1141 TAAACTGCAATGATCTTTACATAGTCCCGGCAGGATATCCAACGGCCCTCAGTGCCTCAG
380 L N C N D L Y I V P A G Y P T A L S A S

1201 AGAACAACAACTTGGAGGTGGTCCAGTTCATCCTTAATGCCCGCAACAACTCCAGAATGT
400 E N N N L E V V Q F I L N A R N N S R M

1261 TCCTCACAGGTGCTGGAGACAATGTGGTGAAGCATATACCCCGAGAGTTGCTTTCAAGGT
420 F L T G A G D N V V K H I P R E L L S R

```
1321   CCCCTGAACTAGTGAAGAAGCTATTGGGACAACAACAGCAGTCATTCTTTGTGAACCGTA
440    S  P  E  L  V  K  K  L  L  G  Q  Q  Q  Q  S  F  F  V  N  R

1381   AGGATAAACCGAAGTCCCCCAGAAAGGGAGATTGGGTTGCTTGGTCTTCATCCTAGTTTG
460    K  D  K  P  K  S  P  R  K  G  D  W  V  A  W  S  S  S  *

1441   AAGTCTATCATGTGAGTGATGAAGTGAGCGAGCGAGTGGTGTCCACTACTATGTAATAAA
480                                                     *     *

1501   TAACGAATAAATAAATATATCAAAGTCCTGTTGGCCAATAAGGCAACTTACGTGACGGCT
500           *                                *

1561   TATCCCCATGTGAGGAAGGAGTACTATGTATCTATCTATATCAATGTTATGAATAAATGA
520                                                          *

1621   GGAGGTAGCTTCTGTCTTCCAACTACCTCCTCCTTTTTTCTCCTAAAAAAAAAAAAAAAA
540        *

1681   AAAGGATCCC
560
```

Fig. 6.4. The full-length nucleotide and translated amino acid sequence of ASP53 together with 79 bp of 5' flanking sequence and the 3' untranslated region. The sequences of the gene-specific antisense primer (GSP 2) and the 5' PCR primer used to amplify the 5' end of the *ASP53* gene are underlined. The sequences of the two peptides sequenced by PSD-MALDI-TOF mass spectroscopy are shown in bold type. The start and stop codons at positions 81 and 1434 are shown in bold type italics. No translation of the 5' and 3' untranslated regions is given except that stop codons are indicated (*).

networks and hidden Markov models (http://www.cbs.dtu.dk/services/SignalP/). These predicted that there was 100% probability of there being a signal peptide, that the helical region of this peptide was between residues L5 and A19, and that there was a 65% probability that this was removed by cleavage after R25. This would generate the sequence SEQQGQDERT, which is similar to the sequence GSEQQRQDEPT determined earlier, suggesting that a limited amount of ASP53 stored in the *A. erioloba* seed was processed for export from the cells.

The exposure of *A. erioloba* seeds to high daytime temperatures throughout the summer months prompted us to investigate the effect of temperature on the secondary structure of ASP53. CD spectroscopic measurements showed that considerable secondary structure was associated with ASP53, the conformation of which was largely unchanged up to 80°C (Fig. 6.6). Deconvolution of the spectrum suggested

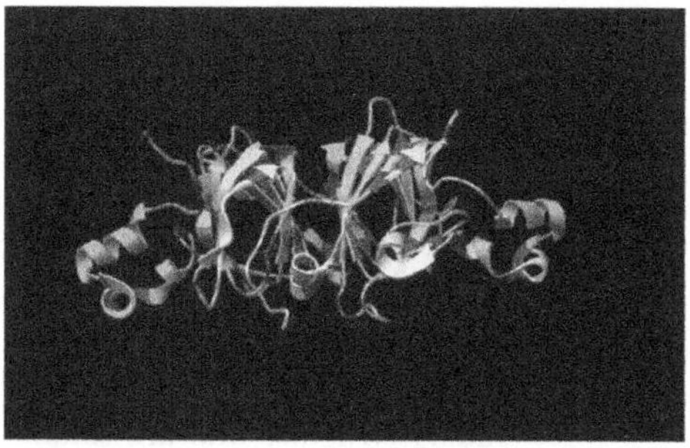

Fig. 6.5. Predicted structure of ASP53 using fold recognition theory (McGuffin and Jones, 2003) modelled on the structure of the *Glycine max* glycine precursor.

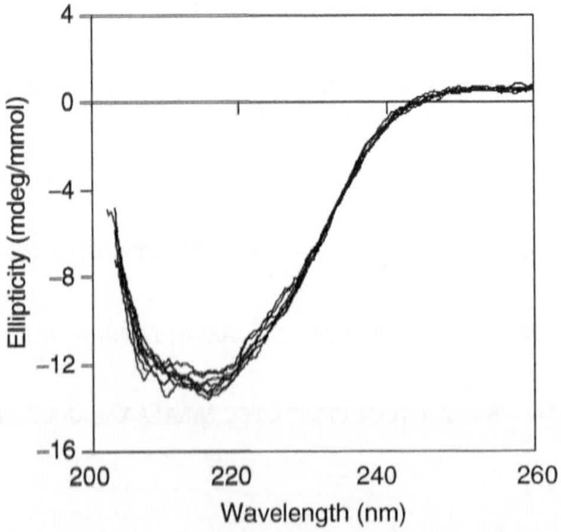

Fig. 6.6. Effect of temperature on the secondary structure of ASP53 as determined by circular dichroism (CD) spectroscopy of ASP53 in 100 mM K_2HPO_4 at pH 7.4. The CD spectra of ASP53 were determined at 10°C intervals between 20°C and 80°C. The graph depicts the variation in ellipticity as a function of wavelength.

that the protein consisted mainly of antiparallel β-pleated sheets (41%) with a further 33% of the protein in a random coil and 19% in a β-turn conformation. Only small parts of the protein were in an α-helical (7%) or in a parallel β-pleated sheet conformation (5%).

LEA proteins, with which ASP53 displays a number of common features including solubility at high temperature and which are present in substantial quantities, have been proposed to protect biological systems against desiccation and thermal stress. Thus, it has recently been demonstrated that a nematode group III LEA protein and group I Em wheat LEA protein inhibited the desiccation-induced aggregation of citrate synthase (Goyal *et al.*, 2005). We therefore investigated whether ASP53 could prevent the aggregation of a group of unfolded proteins. The temperature-dependent aggregation of a diluted solution of total chicken egg white protein was used as a model system. Increasing the temperature of this solution resulted in no turbidity, determined from the 350 nm absorbance, up to 70°C, after which it increased markedly, reaching a plateau at ~80°C (Fig. 6.7). Inclusion of ASP53 in the egg white protein solution at a molar ratio of 0.6:1 resulted in a marked decrease in the turbidity. A further increase in the concentration of ASP53 to 2.3:1 completely abolished protein aggregation (Fig. 6.7).

Discussion

ASP53 was identified as the most abundant protein present in the mature seeds of *A. erioloba.* This is the first study that reports a non-chaperone protein that can protect other proteins against thermal denaturation. This protein was mainly located in the cell wall of embryonic axis cells and in the cell walls and protein-containing storage vacuoles of cotyledons from mature seeds. This location suggests a possible dual function of this protein – as a protectant against (thermal) stress and as a storage reserve for metabolism upon germination. ASP53, like LEA proteins, was found to be soluble at high temperatures. Although LEA proteins have little secondary structure and are soluble at high temperature on account of their high

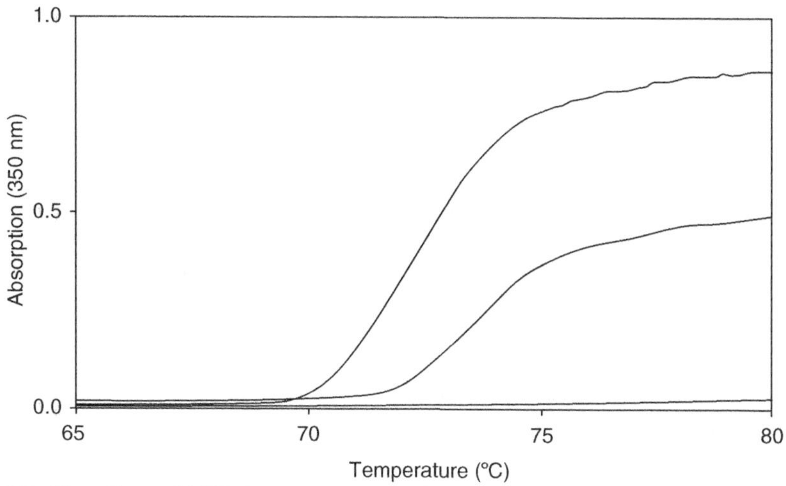

Fig. 6.7. The effect of ASP53 on temperature-induced aggregation of total chicken egg white proteins. The turbidity of the protein solution, determined from the 350 nm absorption, was determined alone (top), in the presence of 0.38 mg/ml ASP53 (middle) or in the presence of 1.5 mg/ml ASP53 (bottom).

hydrophilicities (Russouw *et al.*, 1997), ASP53 displayed considerable secondary structure, which was maintained independent of the temperature. ASP53 was found to be related to a number of globulin seed storage proteins, namely β-conglycinin, vicilin and canavalin, all of which also contain two cupin domains. Although the sequence of ASP53 showed relatively poor homology (<60%) to globulin seed storage proteins, which is consistent with previous observations (Dunwell, 1998), fitting the protein sequence to the three-dimensional structures of these proteins resulted in more convincing homology with a 'certain' fit. Regions of maximum homology were located in regions of β-sheet, with insertions and deletions in the ASP53 sequence proposed to be present in the loop regions between these segments of β-sheet and outside of the cupin barrel structure. It is therefore proposed that ASP53 is a member of the globulin seed storage protein family.

The cupin domain consists of six β-strands forming a β-barrel structure, and it has been suggested that the widespread occurrence of this domain in a variety of proteins is on account of its thermal stability and resistance to proteolysis (Dunwell *et al.*, 2004). Cupin domains have been reported to be present in proteins from a wide variety of phyla within the bacterial and eukaryotic kingdoms and include not only seed storage proteins but also dioxygenases, isomerases, epimerases, decarboxylases, auxin-binding proteins and transcription factors. A number of these proteins, e.g. anthocyanidin synthase (Wilmouth *et al.*, 2002), quercetin dioxygenase (Fuzetti *et al.*, 2002) and phosphomannose isomerase (Wilmouth *et al.*, 2002), have biochemical activities associated with the cell wall. In addition, germin-like proteins have been shown to have enzymatic activities including oxalate oxidase and superoxide dismutase (Woo *et al.*, 2000) as well as serine protease inhibitory activity (Segarra *et al.*, 2003). The demonstration that ASP53 can protect other proteins against thermal denaturation and aggregation is a novel activity. This activity is mediated by ASP53 acting as a water replacement molecule by forming hydrogen bonds with target macromolecules. The presence of the thermally stable cupin fold is presumably a requirement for ASP53 to maintain its hydrophilic exterior at high temperature. Although we have not investigated whether ASP53 has associated enzymatic or inhibitory activity, we have recently proposed that one function of the yeast *S. cerevisiae* hydrophilic stress response protein HSP12 is to act as a plasticizer in the cell wall (Motshwene *et al.*, 2004; Karreman *et al.*, 2005). ASP53 might have a similar function as it would inhibit hydrogen-bonded interactions between adjacent polysaccharide chains in the cell wall, thereby leading to decreased flexibility. Maintenance of the correct cell wall flexibility might be important in the desiccated state so as to retain the cell wall integrity upon water ingress during germination. This process would be analogous to maintenance of membranes in the liquid crystalline phase during desiccation, which is brought about by trehalose (Mansure *et al.*, 1994) and HSP12 (Sales *et al.*, 2000) in yeast.

Acknowledgements

G.G. Lindsey and J.M. Farrant would like to acknowledge the support of the South African National Research Foundation (NRF) and the University of Cape Town (UCT) Research Fund.

References

Baker, J.C., Steele, C. and Dure, L. III. (1988) Sequence and characterization of 6 LEA proteins and their genes from cotton. *Plant Molecular Biology* 11, 277–291.

Bharali, S. and Chrungoo, N.K. (2003) Amino acid sequence of the 26 kDa subunit of legumin-type seed storage protein of common buckwheat (*Fagopyrum esculentum* Moench): molecular characterization and phylogenetic analysis. *Phytochemistry* 63, 1–5.

Blackman, S.A., Wettlaufer, S.F., Obendorf, R.L. and Leopold, A.C. (1991) Maturation proteins associated with desiccation tolerance in soybean. *Plant Physiology* 96, 868–874.

Blackman, S.A., Obendorf, R.L. and Leopold, A.C. (1992) Maturation proteins and sugars in desiccation tolerance of developing soybean seeds. *Plant Physiology* 100, 225–230.

Bray, E.A. (1997) Plant responses to water deficit. *Trends in Plant Science* 2, 48–54.

Campbell, S.A. and Close, T.J. (1997) Dehydrins: genes, proteins, and associations with phenotypic traits. *New Phytologist* 137, 61–74.

Close, T.J. (1997) Dehydrins: a commonality in the response of plants to dehydration and low temperature. *Physiologia Plantarum* 100, 291–296.

Close, T.J., Kort, A.A. and Chandler, P.M. (1989) A cDNA-based comparison of dehydration-induced proteins (dehydrins) in barley and corn. *Plant Molecular Biology* 13, 95–108.

Dunwell, J.M. (1998) Cupins: a new superfamily of functionally diverse proteins that include germins and plant seeds storage proteins. *Biotechnology and Genetic Engineering Reviews* 15, 1–32.

Dunwell, J.M., Purvis, A. and Khuri, S. (2004) Cupins: the most functionally diverse protein superfamily? *Phytochemistry* 65, 7–17.

Dure, L. III, Crouch, M., Harada, J., Ho, T.H.D., Mundy, J., Quatrano, R. and Sung, Z.R. (1989) Common amino acid domains among LEA proteins of higher plants. *Plant Molecular Biology* 12, 475–486.

Fuzetti, F., Schroter, K.H., Steiner, R.A., van Noort, P.I., Pijning, T., Rozeboom, H.J., Kalk, K.H., Egmond, M.R. and Dijkstra. B.W. (2002) Crystal structure of the copper-containing quercetin 2,3-dioxygenase from *Aspergillus japonicus*. *Structure* 10, 259–268.

Galili, G., Altschuler, Y. and Levanony, H. (1993) Assembly and transport of seed storage proteins. *Trends in Cell Biology* 3, 437–442.

Garay-Arroyo, A., Colmenero-Flores, J.M., Garciarrubio, A. and Covarrubias, A.A. (2000) Highly hydrophilic proteins in prokaryotes and eukaryotes are common during conditions of water deficit. *Journal of Biological Chemistry* 275, 5668–5674.

Goyal, K., Walton, L.J. and Tunnacliffe, A. (2005) LEA proteins prevent protein aggregation due to water stress. *Biochemical Journal* 388, 151–157.

Harlow, E. and Lane, D. (1988) *Antibodies – A Laboratory Manual.* Cold Spring Harbour Laboratory, Cold Spring Harbour, New York.

Hendry, G.A.F., Finch-Savage, W.E., Thorpe, P.C., Atherton, N.M., Buckland, S.M., Nilsson, K.A. and Seel, W.E. (1992) Free-radical processes and loss of seed viability during desiccation in the recalcitrant species *Quercus robur* L. *New Phytologist* 122, 273–279.

Horbowicz, M. and Obendorf, R.L. (1994) Seed desiccation tolerance and storability: dependance on flatulence-producing oligosaccharides and cyclitols – review and survey. *Seed Science Research* 4, 385–406.

Karreman, R.J., Brandt, W.F. and Lindsey, G.G. (2005) The yeast *Saccharomyces cerevisiae* stress response protein HSP 12 decreases the gel strength of agarose used as a model system for the β-glucan layer of the cell wall. *Carbohydrate Polymers* 60, 193–198.

Leprince, O., Hendry, G.A.F. and McKersie, B.D. (1993) The mechanisms of desiccation tolerance in developing seeds. *Seed Science Research* 3, 231–246.

Leprince, O., Atherthon, N.M., Deltour, R. and Hendry, G.A. (1994) The involvement of respiration in free radical processes during loss of desiccation tolerance in germinating *Zea mays* L. (an electron paramagnetic resonance study). *Plant Physiology* 104, 1333–1339.

Makri, E., Papalamprou, E. and Doxastakis, G. (2005) Study of functional properties of seed storage proteins from indigenous European legume crops (lupin, pea, broad bean) in admixture with polysaccharides. *Food Hydrocolloids* 19, 583–594.

Mansure, J.J., Panek, A.D., Crowe, L.M. and Crowe, J.H. (1994) Trehalose inhibits ethanol effects on intact yeast cells and liposomes. *Biochimica et Biophysica Acta* 1191, 309–316.

McGuffin, L.J. and Jones, D.T. (2003) Improvement of the GenTHREADER method for genomic fold recognition. *Bioinformatics* 19, 874–881.

Melo, T.S., Ferreira, R.B. and Teixeira, A.N. (1994) The seed storage proteins from *Lupinus albus*. *Phytochemistry* 37, 641–648.

Motshwene, P., Karreman, R., Kgari, G., Brandt, W. and Lindsey, G.G. (2004) The LEA-like protein Hsp12 is present in the cell wall and enhances the barotolerance of the yeast *Saccharomyces cerevisiae*. *Biochemistry Journal* 377, 769–774.

Roberts, E.H. (1973) Predicting the storage life of seeds. *Seed Science Technology* 1, 499–514.

Roberts, J.K., DeSimone, N.A., Lingle, W.L. and Dure, L. III. (1993) Cellular concentrations and uniformity of cell-type accumulation of two Lea proteins in cotton embryos. *The Plant Cell* 5, 769–780.

Russouw, P.S., Farrant, J.M., Brandt, W.F. and Lindsey G.G. (1997) The most prevalent protein in a heat-treated extract of pea (*Pisum sativum*) embryos is a LEA group I protein; its conformation is not affected by exposure to high temperature. *Seed Science Research* 7, 117–123.

Sales, K., Brandt, W.F., Rumbak, E. and Lindsey G.G. (2000) The LEA-like protein HSP 12 in *Saccharomyces cerevisiae* has a plasma membrane location and protects membranes against desiccation and ethanol induced stress. *Biochimica et Biophysica Acta* 1463, 267–278.

Segarra, C.I., Casalongue, C.A., Pinedo, M.L., Ronchi, V.P. and Conde, R.D. (2003) A germin-like protein of wheat leaf apoplast inhibits serine proteases. *Journal of Experimental Botany* 54, 1335–1341.

Sherwin, H.W. and Farrant, J.M. (1996) Differences in rehydration of three desiccation-tolerant angiosperm species. *Annals of Botany* 78, 703–710.

Smirnoff, N. (1993) The role of active oxygen species in the response of plants to water deficit and desiccation. *New Phytologist* 125, 27–58.

Spurr, A.R. (1969) A low viscosity epoxy resin embedding medium for electron microscopy. *Journal of Ultrastructure Research* 26, 31–43.

Sun, W.Q. and Leopold, A.C. (1993) The glassy state and accelerated aging of soybeans. *Physiologia Plantarum* 89, 767–774.

Vertucci, C.W. and Farrant, J.M. (1995) Acquisition and loss of desiccation tolerance. In: Kigel, J. and Galilli, G. (eds) *Seed Development and Germination.* Marcel Dekker, New York, pp. 237–271.

Westermeier, R. and Naven, T. (2002) *Proteomics in Practise: A Laboratory Manual of Proteome Analysis.* Wiley-VCH, Weinheim, Germany.

Wilm, M., Shevchenko, A., Houthaeve, T., Breit, S., Schweigerer, L., Fotsis, T. and Mann, M. (1996) Femtomole sequencing of proteins from polyacrylamide gels by nano-electrospray mass spectrometry. *Nature* 379, 466–469.

Wilmouth, R.C., Turnbull, J.J., Welford, R.W., Clifton, I.J., Prescott, A.G. and Schofield, C.J. (2002) Structure and mechanism of anthocyanidin synthase from *Arabidopsis thaliana. Structure* 10, 93–103.

Wolkers, W.F., McCready, S., Brandt, W., Lindsey, G.G. and Hoekstra, F.A. (2001) Isolation and characterisation of a D-7 LEA protein from pollen that stabilises glasses *in vitro. Biochimica et Biophysica Acta* 1544, 196–206.

Woo, E.J., Dunwell, J.M., Goodenough, P.W., Marvier, A.C. and Pickersgill, R.W. (2000) Germin is a manganese-containing homohexamer with oxalate oxidase and superoxide disputase activities. *Nature Structural Biology* 7, 1036–1040.

7 Possible Involvement of Programmed Cell Death Events During Accelerated Ageing of *Glycine max* Axes

S.Q. Song,[1,3] X. Tian[2,3] and J.R. Fu[3]

[1]*Xishuangbanna Tropical Botanical Garden, The Chinese Academy of Sciences, Menglun, Mengla 666303, Yunnan, China;* [2]*School of Life Sciences, Jishou University, Jishou 416000, China;* [3]*School of Life Sciences, Sun Yat-Sen University, Guangzhou 510275, China*

Abstract

Soybean (*Glycine max* (L.) Merr. 'Aijiaozao') axes that had undergone accelerated ageing (40°C, 100% relative humidity (RH)) for different periods of time were examined to determine the possible involvement of programmed cell death (PCD) events. During accelerated ageing the moisture content (MC) of seeds increased, whereas the germination percentage and the rate of germination as well as subsequent seedling growth decreased. Changes in the respiratory rates of the seeds were similar to changes in the production of superoxide radicals and hydrogen peroxide (H_2O_2). All these changes reached a peak after 10 days of accelerated ageing, and were then followed by a slight decrease. Over this period of time, DNA became degraded but a typical DNA ladder appearance was not observed. The number of nucleosomes determined by enzyme-linked immunosorbent assay (ELISA) also increased rapidly. Terminal deoxynucleotidyl transferase-mediated dUTP nick end labelling (TUNEL) showed that some of the procambial and peripheral cell nuclei were stained brown. Although total nuclease activity increased, activity of caspase-3-like enzyme was not detected. These results indicate that PCD events occur during accelerated ageing of soybean axes.

Introduction

Programmed cell death (PCD) is a genetically encoded active process that results in the death of individual cells, tissues and whole organs. PCD is one of the many mechanisms used by plants for normal developmental elaboration of their life cycle (Gunawardena *et al.*, 2004). Activation of PCD in plants takes place during a number of processes, including the differentiation of tracheary elements (Fukuda, 2000) and female gametophytes (Wu and Cheun, 2000), the development of cereal endosperm and aleurone cells (Young and Gallie, 2000; Fath *et al.*, 2002), responses to external stimuli such as attacks by pathogens (Beers and McDowell, 2001) and severe abiotic stresses (Rao and Davis, 2001).

At present, however, the signalling pathways that lead plant cells towards death via apoptosis, and how this death occurs, are far from being elucidated. PCD in mammals and plants shares several morphological and biochemical features, including cytoplasm shrinkage, nuclear condensation, DNA laddering, expression of caspase-like proteolytic activity and the release of cytochrome *c* from mitochondria (Kim *et al.*, 2003). However, not all PCD events involve all of these changes (Pennell and Lamb, 1997).

Differences in seed vigour are important in determining crop establishment, and can have considerable economic implications for crop production. Loss of seed vigour leads to a reduction in seed quality, performance and seedling establishment. Tobacco (*Nicotiana tabacum* L. bright yellow-2) cells undergoing PCD showed an immediate burst in H_2O_2 and superoxide anion ($\cdot O_2^-$) production. Consistently, death was prevented by antioxidants such as ascorbate (ASC) and superoxide dismutase (SOD). Although it was reported that one of the main causes leading to seed deterioration is the production of reactive oxygen species (ROS), and subsequent lipid peroxidation (McDonald, 1999), nothing is known about PCD cues during seed ageing.

In this research, soybean (*Glycine max* (L.) Merr.) seeds that had undergone accelerated ageing were used as experimental material to determine the possible involvement of PCD events. Germination percentage was used to express changes in seed vigour during accelerated ageing. Viability staining of cells, light microscopy and transmission electron microscopy of soybean axes were used to characterize PCD morphological features. Total DNA electrophoresis, nucleosome ELISA, TUNEL and activity of nuclease were used to assess PCD biochemical features.

Materials and Methods

Plant material

Harvested soybean seeds (purchased from the Institute of Oil Crops Research, Chinese Academy of Agricultural Sciences, Wuhan, China) were placed in a nylon mesh bag, suspended in a closed desiccator (22 cm diameter) and subjected to accelerated ageing (40°C and 100% RH) for 0, 5, 10, 15 and 20 days.

Determination of moisture content

The MC of seeds was determined gravimetrically (103°C for 18 h) by sampling 30 seeds for each determination. MC is expressed on a dry weight basis (i.e. in g H_2O/g dw or g/g).

Germination test

Batches of 50 seeds were germinated on two filter papers with 15 ml deionized water in petri dishes (12 cm diameter) at 20°C in the dark for 5 days. Seeds showing radicle emergence of 2 mm or more were scored as germinated. The fresh weight of seedlings produced by germinating seeds did not include cotyledons.

Viability stain

The transverse and near-median longitudinal sections of soybean axes (~3 mm thick) were stained in 0.1% (w/v) Evans blue for 2 min following the method of Levine *et al.* (1994). Stained sections were rinsed in deionized water for 30 min and photographed using Kodak MAX 400 film under an Olympus stereomicroscope.

Assay of light microscopy and transmission electron microscopy

For the light microscopy assay, soybean radicles from the seeds subjected to accelerated ageing for different periods of time were fixed in formalin, acetic acid and alcohol solution (FAA), (i.e. 50% ethanol/glacial acetic acid/formaldehyde at 18:1:1 v/v/v). After dehydration in different concentrations of ethanol, the radicles were embedded in paraffin and sectioned on a Leica 2016 microtome. These sections were then stained with toluidine blue, observed and photographed under an Olympus microscope.

For the transmission electron microscopy assay, soybean radicles from seeds subjected to accelerated ageing for different periods of time were pre-fixed in 2.5% glutaraldehyde in 50 mM sodium cacodylate buffer for 24 h. After rinsing three times, the radicles were fixed in 0.5% aqueous osmium tetroxide for 24 h, followed by routine dehydration and embedding. Radicles were then sectioned on a Leica LKB 2088V microtome and the sections were stained with uranyl acetate and lead citrate. They were then viewed and photographed with a JEM-1230 (JEOL Ltd., Japan) transmission electron microscope.

DNA extraction and electrophoresis

DNA extraction was based on the method of Young and Gallie (2000), which was modified as follows. Thirty axes from seeds subjected to accelerated ageing for different periods of time were homogenized to a fine powder with a mortar and pestle under liquid nitrogen. DNA was subsequently extracted by grinding the powder in 200 μl of an extraction buffer composed of 10 mM tris-(hydroxymethyl)-aminomethane-hydrochloride (tris-HCl) at pH 8.0, 10 mM NaCl, 10 mM ethylene-diaminetetraacetic acid (EDTA) and 1% (w/v) sodium dodecyl sulphate (SDS). Following this, 50 μl (1.0 mg/ml) protease K was added to the homogenate and incubated at 37°C for 30 min. After incubation, 200 μl 2% (w/v) cetyltrimethylammonium bromide (CTAB), 400 μl chloroform:isopentanol (24:1) and 400 μl trisphenol were added to the mixture to remove proteins. The mixture was centrifuged at 12,000*g* for 5 min and then 100 µl isopropanol was added to the supernatant. The supernatant was centrifuged at 12,000*g* for 5 min once again, and the resulting DNA was washed in 70% (v/v) alcohol, air-dried and resolved in 100 µl tris-EDTA buffer. Then 1 µl (1 mg/ml) RNase A was added to the DNA solution to remove RNA.

DNA electrophoresis was carried out according to Young and Gallie (2000). DNA fragments (20 µg DNA/lane) were separated on a 2% (w/v) agarose gel, followed by visualization by staining with ethidium bromide.

Nucleosome enzyme-linked immunosorbent assay

Thirty axes from seeds subjected to accelerated ageing for different periods of time were homogenized to a fine powder with a mortar and pestle under liquid nitrogen. According to the manufacture's instructions, nucleosome contents were determined using a Nucleosome ELISA kit (Calbiochem-Novabiochem, International, Inc.). The sample provided by the Nucleosome ELISA kit was used as a standard, and nucleosome contents were expressed as unit/mg protein.

Terminal deoxynucleotidyl transferase-mediated dUTP nick end labelling (TUNEL) assay

Paraffin-embedded tissue sections of soybean radicles from seeds subjected to accelerated ageing for different periods of time were prepared as described in the light microscopy assay. The TUNEL assay was performed using a TdT-FragEL™ DNA fragmentation detection kit (CN Biosciences, Inc., Cambridge, MA), according to the manufacture's instructions. After counterstaining with methyl green, cells that had insoluble chromagen substrates (i.e. stained brown) in the nuclei were regarded to have undergone PCD.

Total nuclease activity

Total nuclease activity was determined as described by Young and Gallie (2000), which was modified as follows. Thirty axes from seeds subjected to accelerated ageing for different periods of time were homogenized to a fine powder under liquid nitrogen. The soluble proteins were then extracted by grinding the powder in an extraction buffer composed of 150 mM tris-HCl at pH 6.8, 1 mM dithiothreitol (DTT), 0.5 mM phenylmethanesulfonyl fluoride and 20 µM leucin. The homogenate was centrifuged at 15,000*g* for 10 min, after which 50 µl of the supernatant was added to 1.5 ml reaction medium including 50 mM tris-HCl at pH 6.8, 1 mM $CaCl_2$, 1 mM $MgCl_2$ and 1 mM $ZnCl_2$. Then 50 µl denatured (100 µg/ml) salmon sperm DNA was added to the reaction mixture, incubated at 37°C for 15 min and the absorbance of the mixture at 260 nm was determined. In this assay one unit of nuclease is defined as the amount required to decrease the absorbance of DNA at 260 nm by 0.01 AU. The specific activity of the nuclease was expressed as unit/mg protein/min.

Activity of caspase-3

The activity of caspase-3 was assayed using a caspase-3 cellular activity assay kit (CalBiochem, Inc.), according to the manufacture's instructions. The specific activity of caspase-3 was expressed as pmol pNA/mg protein/min.

Results

Changes in moisture content and germination of seeds

The MC of dried soybean seeds was originally $0.102 \pm 0.002\,g/g$. During the early stages of accelerated ageing it increased rapidly (40°C, 100% RH), then very slowly until after 15 days and then again rapidly (Fig. 7.1a).

The final germination percentage of soybean seeds decreased with accelerated ageing and reached zero after 20 days (Fig. 7.1a). The period of accelerated ageing at which final seed germination was decreased to 50% (T_{50}) was ~10.5 days. Seed vigour, measured by determining the fresh weight of the seedling produced after 5 days of imbibition (Fig. 7.1a) and the time course of germination of seeds (Fig. 7.1b), also decreased with the period of accelerated ageing.

Viability stain

The viability of soybean radicle cells during accelerated ageing was examined by staining fresh handmade sections in Evans blue. Very little Evans blue staining of cells was observed in the transverse and near-median longitudinal sections of non-aged soybean radicles (Fig. 7.2a and b). The proportion of radicle cells, especially meristematic cells, stained by Evans blue gradually increased with the period of accelerated ageing (Fig. 7.2c–j), indicating loss of cell viability over time.

Changes in cellular structure and ultrastructure

In transverse sections of non-aged soybean radicles the cambium cells appear integrated, do not show any plasmolysis and their nuclei are spheroidal and normally stained by toluidine blue (Fig. 7.3a). The separation of plasma membranes from their cell walls, cytoplasm shrinkage, vacuolation and nuclei abnormality (i.e. nuclei becoming elongated and flattened) gradually increased with the period of accelerated ageing (Fig. 7.3b–e).

Transmission electron microscopy indicated that cambium cells from non-aged soybean radicles had integrated nuclei, the nuclei had normal electron densities and were located in the centre of cells (i.e. they had not been marginalized) (Fig. 7.4a). The nuclei of cambium cells of radicles from seeds subjected to accelerated ageing for 10 days became more elongated than non-aged ones, and their electron-density increased. These nuclei were often located in the centre of cells, but appeared to be ovoid in shape, while some were occasionally marginalized (Fig. 7.4b). With an

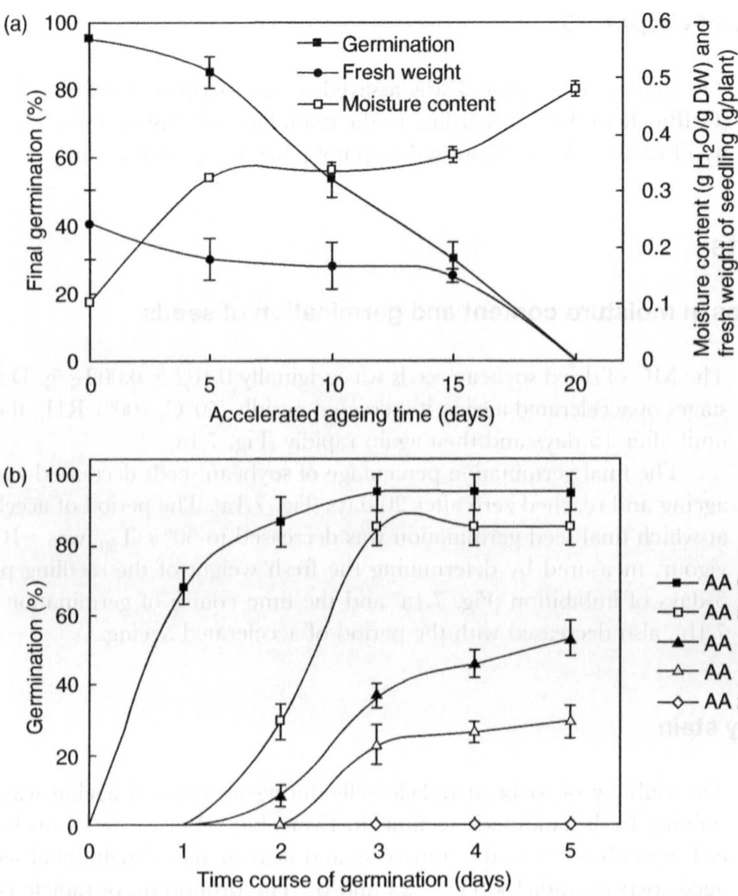

Fig. 7.1. Changes in (a) moisture content, final germination, seedling fresh weight and (b) time course of germination during the accelerated ageing of soybean seeds. After seeds were treated with accelerated ageing (40°C and 100% RH) for the indicated period of time, the moisture content of seeds was determined, and they were germinated at 20°C in the dark for the indicated period of time. Seeds showing radicle emergence of 2 mm were scored as germinated. The fresh weight of seedlings produced by germinating seeds does not include cotyledons. All values are means ± standard deviation of three replicates of 50 seeds. AA = accelerated ageing.

increasing period of accelerated ageing, nuclear membranes gradually degraded and appeared to become hollow so that nuclei and other organelles could no longer be recognized (Fig. 7.4c–f).

Total DNA electrophoresis

Agarose gel electrophoresis showed that DNA fragmentation (extensive smearing) increased with an increasing period of accelerated ageing of soybean axes (Fig. 7.5), and the typical DNA ladder could no longer be observed.

(a) (b) (c) (d) (e) (f) (g) (h) (i) (j)

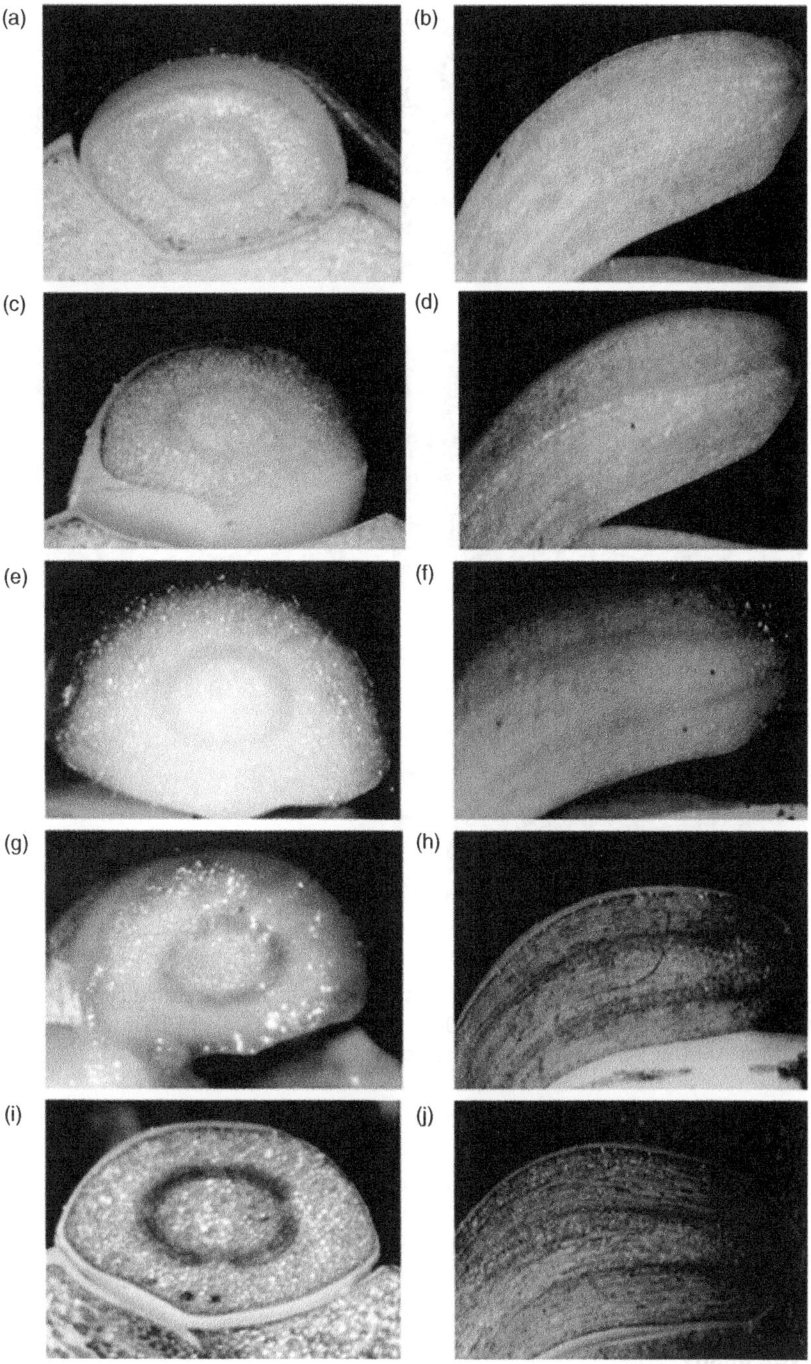

Fig. 7.2. Progress of radicle cell death during the accelerated ageing of soybean seeds, as indicated by Evans blue staining of transverse and longitudinal sections of soybean radicles. Accelerated ageing times shown are (a, b) 0 days; (c, d) 5 days; (e, f) 10 days; (g, h) 15 days; (i, j) 20 days. (a), (c), (e), (g) and (i) are transverse sections, while (b), (d), (f), (h) and (j) are longitudinal sections.

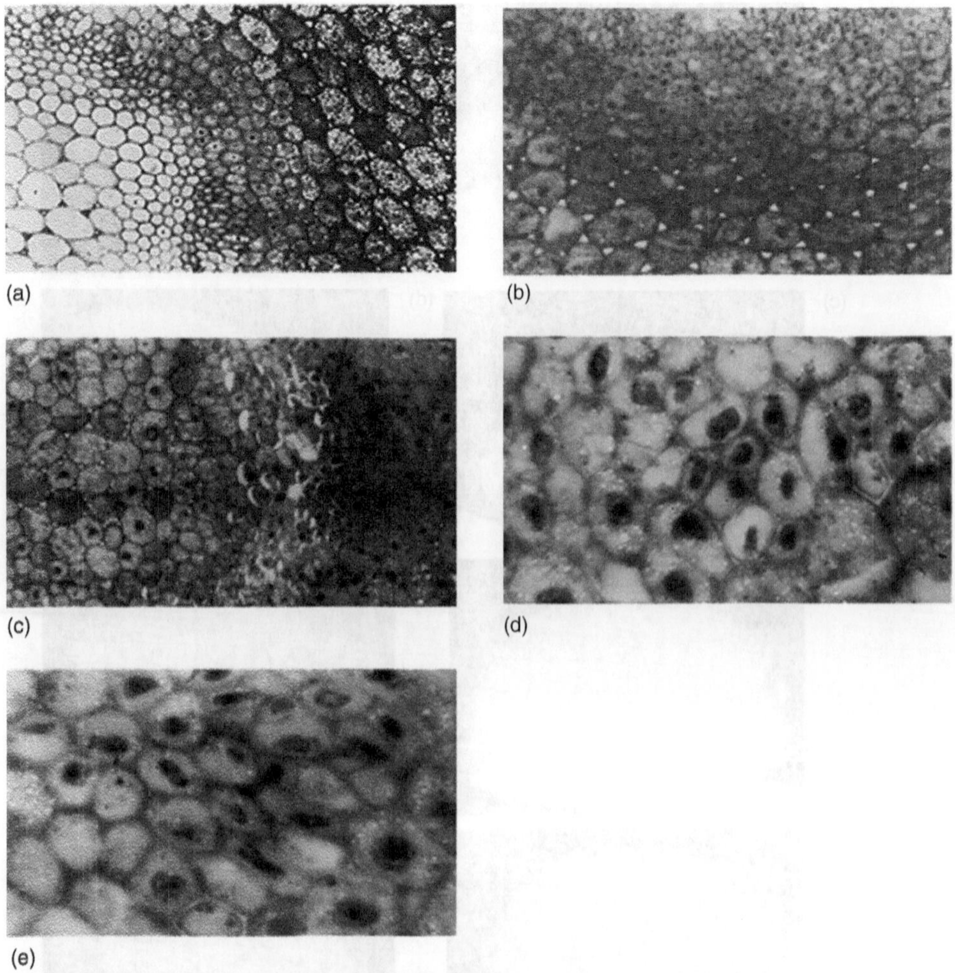

Fig. 7.3. Light micrographs of transverse sections of soybean radicles from seeds subjected to accelerated ageing for (a) 0 days (100×); (b) 10 days (100×); (c) 15 days (100×) and (d and e) 15 days (1000×). All samples were stained with toluidine blue.

Nucleosome content determination

Nucleosome ELISA showed that the nucleosome content of axes increased slowly during the early stages of accelerated ageing, until 10 days of exposure, and then rapidly after this period of time (Fig. 7.6a). For example, the nucleosome content of axes subjected to accelerated ageing for 20 days was 23% higher than that of non-aged axes (Fig. 7.6a).

TUNEL assay

A dark brown diaminobenzidine (DAB) signal indicated that 3'-OH DNA fragments, produced by PCD, were present in the cambium cells of soybean radicles that were subjected to accelerated ageing, and these dark brown DAB signals increased with an increasing period of accelerated ageing (data not shown).

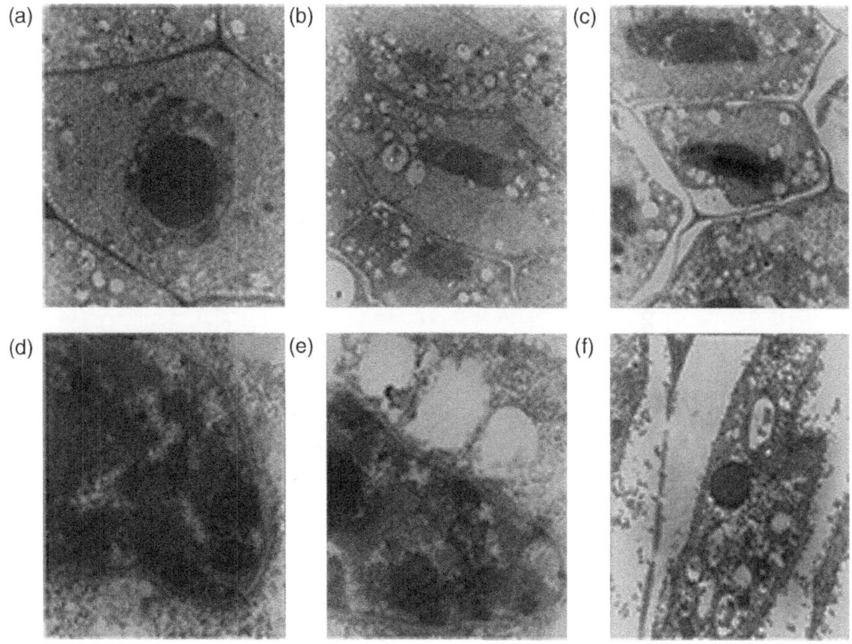

Fig. 7.4. Transmission electron micrographs of the nuclei of soybean radicles from seeds subjected to accelerated ageing for (a) 0 days (900×); (b) 10 days (900×); (c) 15 days (900×); (d and e) 15 days (10,000×); and (f) 20 days (2000×).

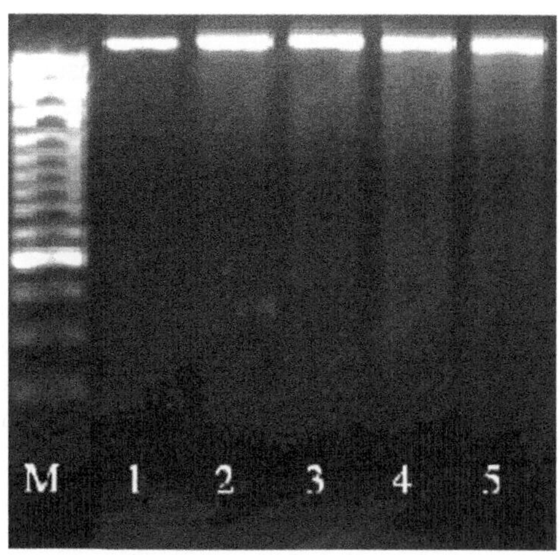

Fig. 7.5. Electrophoresis of total DNA from the soybean axes of seeds subjected to accelerated ageing for different periods of time. Column M represents molecular mass markers (100 bp); columns 1–5 indicate total DNA from soybean axes of seeds subjected to accelerated ageing for 0, 5, 10, 15 and 20 days, respectively.

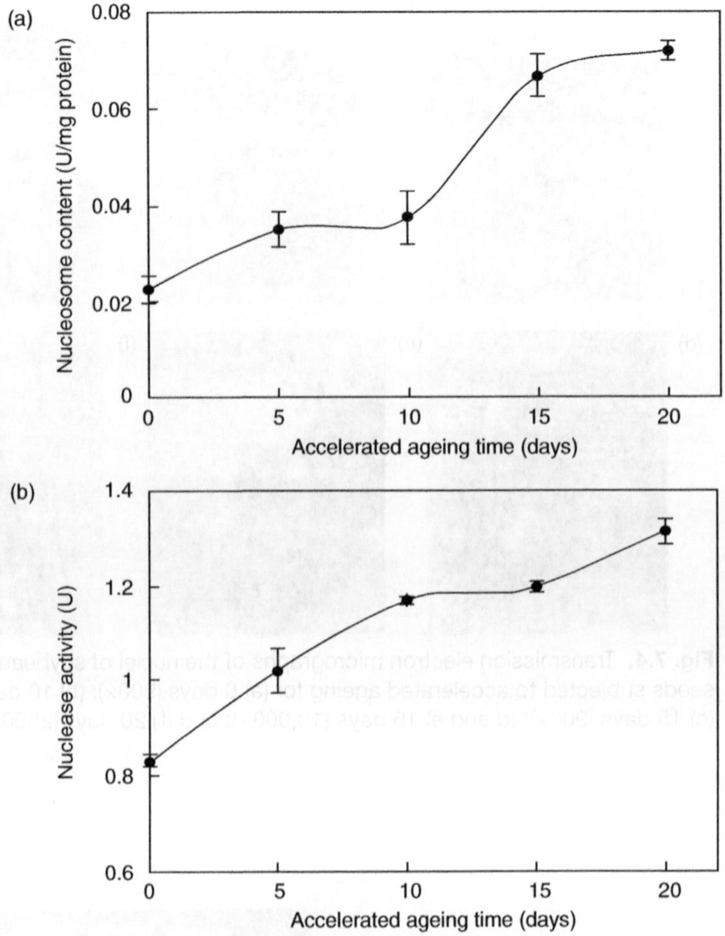

Fig. 7.6. Changes in (a) nucleosome content and (b) nuclease activity of soybean axes from seeds subjected to accelerated ageing.

Activity of total nuclease and caspase-3

Activity of total nuclease in axes gradually increased with an increasing period of accelerated ageing (Fig. 7.6b). For example, the activity of total nuclease increased by ~58% in aged axes for 20 days compared with non-aged ones. Caspase-3 activity, however, was not detected in non-aged or aged soybean axes.

Discussion

Under accelerated ageing (40°C, 100% RH), soybean seeds exhibit three phases of water uptake (i.e. the MC of seeds increase rapidly in the early stages of accelerated ageing, then very slowly until after 15 days of accelerated ageing and rapidly again after this period). This process was similar to that observed in the water uptake of dry non-aged seeds.

The final germination percentage, time course of seed germination and fresh weight of the seedlings produced by germinating seeds all decreased with an increasing period of accelerated ageing, as was previously found for *Arachis hypogaea* L. (Song *et al.*, 1992) and *Beta vulgaris* L. (Song *et al.*, 2001) seeds. The symptoms observed during accelerated ageing can be used to characterize the degree of ageing that is negatively correlated with storability. Stability against accelerated ageing has been recognized as a useful vigour test for some species (Priestley, 1986). The physiological and biochemical changes during rapid deterioration of seeds have been increasingly used as indices of ageing (Priestley, 1986).

Viability staining using Evans blue showed that the death of radicle cells, especially meristematic cells, gradually increased with accelerated ageing time. Previous studies using tetrazolium staining have shown that deterioration begins in the root and moves through the embryo in both naturally and artificially aged *T. aestivum* seeds (Das and Sen-Mandi, 1992), and that *A. hypogaea* axes were the most sensitive part of the seed to deteriorate (Fu *et al.*, 1988).

PCD has also been associated with several other processes in plants, including senescence (Schmid *et al.*, 2001), stress (Solomon *et al.*, 1999), development (Groover and Jones, 1999) and the hypersensitive response (HR) to pathogens (Mackey *et al.*, 2002; Abramovitch *et al.*, 2003). At present, very little is known about the fundamental processes that control and regulate PCD in plants. The hallmark features of PCD include some morphological and physiological changes. Cells undergoing apoptosis display morphological changes, including cell shrinkage, chromatin condensation and apoptotic body formation. Biochemically, apoptotic cells exhibit DNA fragmentation (also referred to as DNA laddering) and activation of a family of cysteine proteases called caspases (cysteine aspartases) (Vaux and Korsmeyer, 1999; Hengartner, 2000).

With accelerated ageing, cytoplasm shrinkage occurs in the cambium cells of soybean radicles, and vacuolation and nuclei abnormality (i.e. nuclei become long and flat) gradually increase. Nuclei are usually small, typically round, granular bodies. Transmission electron microscopy also indicates that hollows in the nuclear membrane increased gradually, nuclear membranes degraded progressively, and nuclei and other organelles could no longer be recognized after accelerated ageing. Previous morphological observations of dead and dying aleurone cells showed that the cytoplasm and nuclei could condense and shrink (Kuo *et al.*, 1996; Wang *et al.*, 1996). In apoptosis-like cell death, the nucleus is the first target of degradation, and dying cells exhibit characteristic features such as chromatin condensation, nuclear shrinkage and fragmentation or DNA laddering (Fukuda, 2000).

Cleavage of genomic DNA into smaller fragments is also a hallmark of PCD. In animal apoptosis, DNA is first cleaved into large fragments of 50–300 kB. Subsequently, DNA is cleaved at internucleosomal linker regions, producing fragments that are multimers of ~180 bp, which can be identified by a ladder-like pattern when separated electrophoretically. Finally, the fragmented DNA is digested completely by specific endonucleases (Gunawardena *et al.*, 2004). Electrophoresis of DNA from the dying barley (*Hordeum vulgare* L.) aleurone reveals oligonucleosome-sized DNA fragments that are characteristic of apoptotic cell death (Wang *et al.*, 1996). In the current study, agarose gel electrophoresis showed that DNA fragmentation (i.e. extensive smearing) increased with accelerated ageing of seeds, but the typical DNA laddering was not observed. These results were in accordance with those of Gunawardena *et al.* (2004), who could not find DNA laddering at any stage of perforation formation in lace plant (*Aponogeton madagascariensis* (Mirb.) H. Bruggen) leaves. The extensive smearing indicated that DNA fragments were of a continuous size range. Nucleosome contents and TUNEL-positive nuclei increased with accelerated ageing time, showing that DNA fragmentation had occurred.

Wang *et al.* (1996) considered that although electrophoresis can provide evidence for this specific pattern of DNA fragmentation, it cannot demonstrate tissue or cell specificity in plants. Furthermore, DNA laddering is difficult to demonstrate electrophoretically in extracts from tissues where only a fraction of the cells are undergoing the fragmentation process. Therefore, a combination of electrophoresis of isolated DNA and the TUNEL assay might be very useful to identify DNA fragmentation.

The activity of total nuclease in axes increased gradually with accelerated ageing, which was related to DNA fragmentation and increasing nucleosome content.

It was reported that some proteases are associated with several types of plant PCD, including PCD that is related to senescence (Schmid *et al.*, 2001; Eason *et al.*, 2002), oxidative stress (Solomon *et al.*, 1999), seed development (Wan *et al.*, 2002), tracheary element development (Groover and Jones, 1999) and the HR associated with pathogen attacks (Krüger *et al.*, 2002). However, characterization of these proteases reveals that they are degradative and not processive enzymes. Therefore, they are unlike caspases or other proteases involved in signalling pathways (Coffeen and Wolpert, 2004). Caspase-3 activity, however, has not been detected in non-aged and aged soybean axes.

Conclusions

The results of the study demonstrate that PCD does occur during accelerated ageing of soybean axes. In addition, the $\cdot O_2^-$ production rate and H_2O_2 content of axes increased; the activities of SOD, ASC peroxidase, catalase and glutathion reductase in axes decreased; and the malondialdehyde content of axes markedly increased during accelerated ageing (data not shown), showing that production and scavenging of ROS are closely related to PCD.

Acknowledgements

We are grateful to the Knowledge Innovation Program (KIP) Pilot Project (grant KSCX2-SW-117), the Hundred Talents Program of The Chinese Academy of Sciences, the Yunnan Natural Science Foundation (grant 2003C0068M), and the Ministry of Sciences and Technology of China (grant 2004DKA30430) for their support.

References

Abramovitch, R.B., Kim, Y.J., Chen, S., Dickman, M.B. and Martin, G.B. (2003) Pseudomonas type III effector AvrPtoB induces plant disease susceptibility by inhibition of host programmed cell death. *The EMBO Journal* 22, 60–69.

Beers, E.P. and McDowell, J.M. (2001) Regulation and execution of programmed cell death in response to pathogens, stress and developmental cues. *Current Opinion in Plant Biology* 4, 561–567.

Coffeen, W.C. and Wolpert, T.J. (2004) Purification and characterization of serine proteases that exhibit caspase-like activity and are associated with programmed cell death in *Avena sativa*. *The Plant Cell* 16, 857–873.

Das, G. and Sen-Mandi, S. (1992) Triphenyltetrazolium chloride staining pattern of differentially aged wheat seed embryos. *Seed Science and Technology* 20, 367–373.

Eason, J.R., Ryan, D.J., Pinkney, T.T. and O'Donoghue, E.M. (2002) Programmed cell death during flower senescence: isolation and characterization of cysteine proteinases from *Sandersonia aurantiaca*. *Functional Plant Biology* 29, 1055–1064.

Fath, A., Bethke, P., Beligni, V. and Jones, R. (2002) Active oxygen and cell death in cereal aleurone cells. *Journal of Experimental Botany* 53, 1273–1282.

Fu, J.R., Lu, X.H., Chen, R.Z., Zhang, B.Z., Liu, Z.S., Ki, Z.S. and Cai, C.Y. (1988) Osmoconditioning of peanut (*Arachis hypogaea* L.) seeds with PEG to improve vigor and some biochemical activities. *Seed Science and Technology* 16, 197–212.

Fukuda, H. (2000) Programmed cell death of tracheary elements as a paradigm in plants. *Plant Molecular Biology* 44, 245–253.

Groover, A. and Jones, A.M. (1999) Tracheary element differentiation uses a novel mechanism coordinating programmed cell death and secondary cell wall synthesis. *Plant Physiology* 119, 375–384.

Gunawardena, A.H.L.A.N., Greenwood, J.S. and Dengler, N.G. (2004) Programmed cell death remodels lace plant leaf shape during development. *The Plant Cell* 16, 60–73.

Hengartner, M.O. (2000) The biochemistry of apoptosis. *Nature* 407, 770–776.

Kim, M., Ahn, J.W., Jin, U.H., Chai, D., Poek, K.H. and Pai, H.S. (2003) Activation of the programmed cell death pathway by inhibition of proteasome function in plant. *Journal of Biological Chemistry* 278, 19406–19415.

Krüger, J., Thomas, C.M., Golstein, C., Dixon, M.S., Smoker, M., Tang, S., Mulder, L., and Jones, J.D.G. (2002) A tomato cysteine protease required for Cf-2-dependent disease resistance and suppression of autonecrosis. *Science* 296, 744–747.

Kuo, A., Cappelluti, S., Cervantes-Cervantes, M., Rodriguez, M. and Bush, D.S. (1996) Okadaic acid, a protein phosphatase inhibitor, blocks calcium changes, gene expression and cell death induced by gibberellin in wheat aleurone cells. *The Plant Cell* 8, 259–269.

Levine, A., Tenhaken, R., Alvarez, M., Palmer, R. and Lam, C.J. (1994) H_2O_2 from the oxidative burst orchestrates the plant hypersensitive disease resistance response. *Cell* 79, 583–593.

Mackey, D., Holt, B.F. III, Wiig, A. and Dangl, J.L. (2002) RIN4 interacts with pseudomonas syringae type III effector molecules and is required for RPM1-mediated resistance in *Arabidopsis*. *Cell* 108, 743–754.

McDonald, M.B. (1999) Seed deterioration: physiology, repair and assessment. *Seed Science and Technology* 27, 177–237.

Pennell, R.I. and Lamb, C. (1997) Programmed cell death in plants. *The Plant Cell* 9, 1157–1168.

Priestley, D.A. (1986) *Seed Aging*. Comstock Publishing Associates, Ithaca, New York.

Rao, M.V. and Davis, K.R. (2001) The physiology of ozone induced cell death. *Planta* 213, 682–690.

Schmid, M., Simpson, D., Sarioglu, H., Lottspeich, F. and Gietl, C. (2001) The ricinosomes of senescing plant tissue bud from the endoplasmic reticulum. *Proceedings of the National Academy of Sciences of the United States of America* 98 (9), 5353–5358.

Solomon, M., Belenghi, B., Delledonne, M., Menachem, E. and Levine, A. (1999) The involvement of cysteine proteases and protease inhibitor genes in the regulation of programmed cell death in plants. *The Plant Cell* 11, 431–443.

Song, S.Q., Fu, J.R. and Xia, W. (1992) Studies on accelerated aging and peroxidation of membrane lipids in peanut (*Arachis hypogaea* L.) seeds. *Oil Crops of China* 3, 31–33.

Song, S.Q., Fredlund, K.M. and Møller, I.M. (2001) Changes in low molecular weigh heat shock protein 22 of mitochondria during high temperature accelerated ageing of *Beta vulgaris* L. seeds. *Acta Phytophysiology Sinica* 27, 73–80.

Vaux, D.L. and Korsmeyer, S.J. (1999) Cell death in development. *Cell* 96, 245–254.

Wan, L., Xia, Q., Qiu, X. and Selvaraj, G. (2002) Early stages of seed development in *Brassica napus*: a seed coat-specific cysteine proteinase associated with programmed cell death of the inner integument. *The Plant Journal* 30, 1–10.

Wang, M., Oppedijk, B.J., Lu, X., Duijin, B.A. and Schilperoort, R.A. (1996) Apoptosis in barley aleurone during germination and its inhibition by abscisic acid. *Plant Molecular Biology* 32, 1125–1134.

Wu, H.M. and Cheun, A.Y. (2000) Programmed cell death in plant reproduction. *Plant Molecular Biology* 44, 267–281.

Young, T.E. and Gallie, D.R. (2000) Programmed cell death during endosperm development. *Plant Molecular Biology* 44, 283–301.

8 Storage and Germination Response of Recalcitrant Seeds Subjected to Mild Dehydration

S. EGGERS, D. ERDEY, N.W. PAMMENTER AND P. BERJAK

School of Biological and Conservation Sciences, University of KwaZulu-Natal, Durban 4041, South Africa

Abstract

A problem associated with the storage of fully hydrated recalcitrant seeds is germination in storage, and it has been suggested that this problem could be overcome by partial dehydration, which is sufficient to prevent germination but high enough to avoid desiccation damage (i.e. 'sub-imbibed' storage). However, partial drying (pd) is shown to stimulate germination, and this process could reduce storage lifespan. Data are presented on recalcitrant seeds from a number of species, demonstrating the enhancement of germination by rapid mild dehydration, and the adverse effects of this mild dehydration on subsequent storage at a range of temperatures.

Introduction

Recalcitrant seeds, by definition, are desiccation-sensitive and hence cannot be stored by the conventional methods employed for orthodox seeds. Not only does this make the long-term conservation of their genetic resources difficult, but it also places limitations on normal seed-handling procedures. To date the only 'successful' way of storing recalcitrant seeds is in the hydrated condition, at their shedding water content, but storage lifespan varies from several months, at best, to a week or two, depending on the species and the physiological condition of the seeds (King and Roberts, 1980). There are two main problems that are associated with hydrated storage: (i) seeds will often germinate in storage (King and Roberts, 1980); and (ii) the effects of fungal contamination can be severe, as the conditions (i.e. high humidity and temperature) necessary for hydrated storage also favour fungal proliferation (Mycock and Berjak, 1990).

Recalcitrant seeds are metabolically active (Berjak *et al.*, 1984; Farrant *et al.*, 1997), undergo continued development after shedding that grades into germination-associated changes, and often germinate more rapidly after a short period of hydrated storage than when initially shed (Pammenter *et al.*, 1984; Farrant *et al.*, 1986). It was suggested that the ultimate loss of viability of recalcitrant seeds in hydrated storage is a consequence of this germinative metabolism, and that the longevity of stored seeds is inversely related to the rate at which germination-associated events occur (Berjak *et al.*, 1989; Pammenter *et al.*, 1994).

Attempts to increase storage lifespan of recalcitrant seeds have centred on reducing the rate of metabolism and germination-associated events, and hence the extent of germination in storage. For chilling-tolerant seeds this can be achieved by reducing storage temperature (King and Roberts, 1980; Pritchard *et al.*, 1995), although this must be above 0°C to prevent freezing damage to hydrated tissue. However, the recalcitrant seeds of a variety of species, particularly those of tropical origin, are chilling-sensitive (King and Roberts, 1980; Corbineau and Côme, 1988), placing limitations on storage temperatures. Attention has also been paid to manipulating the composition of the storage atmosphere. Tompsett (1983) showed that oxygen was necessary (presumably for respiration) for retention of viability in *Araucaria hunsteinii* K. Schum. seeds, suggesting that anaerobic storage to reduce metabolic rate is not a viable option. In contrast, Sowa *et al.* (1991) showed that treatment of *Litchi chinensis* Sonn. and *Dimocarpus longan* Lour. seeds with nitrous oxide, which reduced metabolic rate, increased their storage lifespan.

Lowering seed water content slightly is another possible approach that has been suggested to curtail the extent of germination in storage (King and Roberts, 1980). Reduction of water content below the fully hydrated state, but to the extent that viability is not compromised (a process termed 'sub-imbibed' storage), may lead to an extension of storage lifespan. However, Pammenter *et al.* (1994) suggested that the ongoing metabolism in recalcitrant seeds in storage, in the absence of exogenous water, actually imposed a mild but prolonged water stress, and it is possible that partial drying before storage could exacerbate this problem. This is in keeping with the findings of Corbineau and Côme (1986, 1988) that partial drying shortened the longevity of the recalcitrant seeds of four tropical species. Similarly, Drew *et al.* (2000) showed that partial drying of *Trichilia dregeana* Sond. seeds before storage severely reduced their longevity, but the seeds were in poor physiological condition initially, which could have confounded the results. Despite these findings, and the warning by King and Roberts (1980) that the rate of degeneration could increase with a decrease in seed water content, the suggestion of sub-imbibed storage providing an extension of seed lifespan has been made periodically since 1980, and recently by Hong and Ellis (1996). Thus, this possibility needs to be systematically investigated further.

The present study investigated the effect of partial drying on storage and germination of the recalcitrant seeds of four dicotyledonous tree species and one gymnosperm: *T. dregeana*, *T. emetica* Vahl. and *Ekebergia capensis* Sparrm. (all of the family Meliaceae), which are widespread in eastern and southern Africa; *Syzygium cumini* (L.) Skeels (*Myrtaceae*), which is native to India and tropical Asia, but has been introduced to southern and eastern Africa; and *Podocarpus henkelii* (Stapf.) (*Podocarpaceae*), which is a gymnosperm widely distributed in southern Africa.

Materials and Methods

Mature fruit of *T. dregeana*, *T. emetica* and *E. capensis* were harvested from trees locally, in and around Durban, South Africa. Seeds were manually removed from the fruit and surface-sterilized by soaking them in a fungicidal solution for a short period of time. Mature fruit of *S. cumini* were hand-harvested from trees in Tanzania, and seeds were extracted from the fruit and despatched to Durban. Fallen seeds of *P. henkelii* were collected from the ground below the trees in Pietermaritzburg, and transported 80 km to Durban within a few days of collection, where they were surface-sterilized.

For storage experiments, seeds were partially dried to a non-lethal water content (Table 8.1) by burying them in activated silica gel in sealed plastic bags. After partial drying, seeds were placed in pre-sterilized buckets, which were then filled with vermiculite. The contents were mixed, the buckets sealed and placed in storage at 6°C, 16°C or 25°C. Fully hydrated seeds were similarly treated. At appropriate intervals, samples were withdrawn for germination assessment. Germinating seeds were assessed daily to enable calculation of mean time to germinate (MTG):

$$\mathrm{MTG} = \sum (t \cdot n)/\sum n$$

where t is the time in days from the start of the germination trial and n is the number of seeds completing germination on day t (Bewley and Black, 1994).

Water content was measured gravimetrically. The data presented here are for embryonic axes (whole embryos in the case of *P. henkelii*) and are expressed on a dry mass basis.

Results

Storage response

The initial water content (iwc) of the embryonic axes, the axis water content after partial drying to the levels at which the seeds were stored and the storage period for the five species are shown in Table 8.1. The extent of dehydration averaged a removal of 25% of the axis water initially present, but ranged from being fairly severe in *T. emetica* (60% loss) to very mild in *S. cumini* (2% loss). Irrespective of the extent of dehydration, all species were fully germinable following partial drying.

The effect of storage for 11 weeks at the initial water content, and storage after initial partial drying, on seeds of *P. henkelii* is shown in Fig. 8.1. Full germinability was retained in seeds stored at their initial water content at 6°C and 16°C (no data are available for seeds stored at 25°C because of severe fungal proliferation), but seeds stored after partial drying showed a loss of viability. This treatment appears to have imposed some chilling sensitivity on seeds of *P. henkelii* because the extent of viability loss increases with decreasing storage temperature. The results of the storage experiment for all five species are summarized in Table 8.2. The seeds of all species except *P. henkelii* showed some chilling sensitivity, and seeds stored at 6°C either did not survive the storage period or showed very poor germination. Partial drying before storage reduced viability in all species except *T. dregeana*. The extent

Table 8.1. The water content (± one standard deviation) of the embryonic axes ($n = 10$) of seeds at collection and after partial drying of the seeds for storage, and the time for which the seeds were stored.

Species	Initial water content (g/g)	Drying time (h)	Water content after partial drying (g/g)	Storage time (weeks)
Podocarpus henkelii	1.82 ± 0.17	163	1.60 ± 0.14	11
Trichilia emetica	2.13 ± 0.13	18	0.80 ± 0.18	3
T. dregeana	2.92 ± 0.16	3.5	2.16 ± 0.30	22
Ekebergia capensis	1.28 ± 0.40	6	0.94 ± 0.21	8
Syzygium cumini	1.66 ± 0.31	23	1.64 ± 0.36	4

of viability loss caused by partial drying was generally worse after storage at 25°C than at 16°C. *T. emetica* seeds were more susceptible to partial drying before storage than the other species. It might be suggested that this was a consequence of the fairly severe drying before storage (60% axis water removal). However, seeds of *S. cumini*, which had had only 2% of axis water removed, suffered similar viability loss, and even showed some loss when stored at the initial water content. Although *T. dregeana* seeds did not show viability loss during storage after partial drying, there were indications of loss of vigour (i.e. MTG increased from 10.1 days in fresh seeds to 12.6 and 14 days in seeds stored partially dry at 16°C and 25°C, respectively).

Response of germination rate

The influence of drying *T. emetica* seeds, to different extents, on the germination time course is illustrated in Fig. 8.2. Short-term drying (i.e. 2.5 h, sufficient to remove 15% of axis water in this instance) actually enhanced the rate of germination. The rate of germination can be conveniently expressed as the MTG, and the responses of the five species to short-term dehydration are summarized in Table 8.3. In all cases, except *T. dregeana*, there was a reduction in the MTG (i.e.

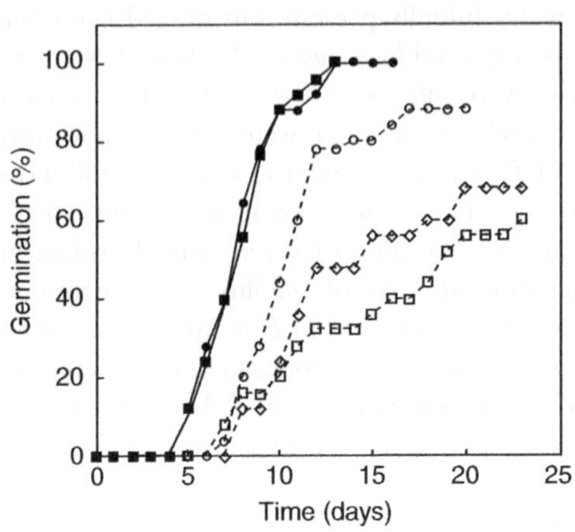

Fig. 8.1. Germination time course of seeds ($n = 25$) of *Podocarpus henkelii* stored at different temperatures at their initial water content (iwc) or after partial drying (pd) as described in Table 8.1. Solid lines and filled symbols = stored at iwc; broken lines and open symbols = stored after pd. Seeds were stored at 6°C (squares), 16°C (diamonds) or 25°C (circles).

Table 8.2. Final germination percentage of seeds ($n = 25$) stored at their initial water content (iwc), and after partial drying (pd) and storage as indicated in Table 8.1. Those species for which no data are presented for storage at 6°C had chilling-sensitive seeds that did not survive storage at that temperature. No data are available for *Podocarpus henkelii* seeds stored at their iwc at 25°C because of severe fungal infection.

		Final germination (%)					
		6°C		16°C		25°C	
Species	Fresh germination (%)	iwc	pd	iwc	pd	iwc	pd
Podocarpus henkelii	100	100	60	100	68		88
Trichilia emetica	100			100	20	100	0
T. dregeana	100	12	12	100	100	100	100
Ekebergia capensis	100			90	60	75	40
Syzygium cumini	100			60	15	100	25

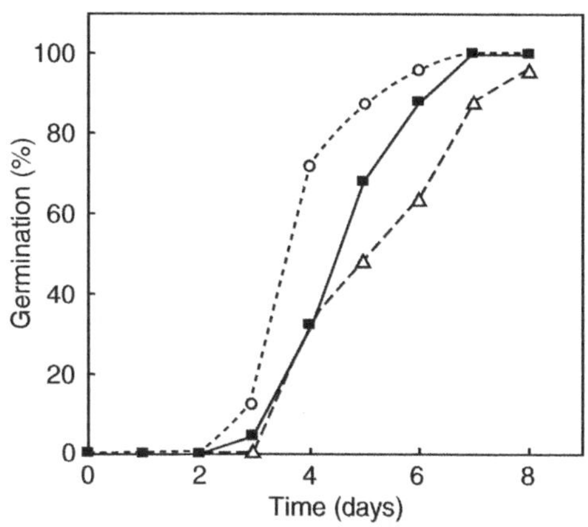

Fig. 8.2. The effect of drying time on the germination time course of seeds ($n = 25$) of *Trichilia emetica*. Solid line and filled squares = fresh undried seeds; broken lines = seeds partially dried for 2.5 h (open circles) or 18 h (open triangles). Corresponding water contents are given in Table 8.3.

Table 8.3. Axis water content (wc; g/g ± one standard deviation) and mean time to germinate (MTG; in days) of fresh seeds, seeds after mild dehydration and seeds after more severe partial dehydration. Drying times (dt; in hours) are indicated. For species where data for partial dehydration are not shown, drying beyond the level described as mild dehydration led to loss of germination capacity.

	Fresh		Mild dehydration			Partial drying		
Species	wc[a]	MTG	dt	wc	MTG	dt	wc	MTG
Podocarpus henkelii	1.82	11.5	22	1.80 ± 0.13	9.0	163	1.61 ± 0.13	14.6
Trichilia emetica	2.13	5.1	2.5	1.77 ± 0.23	4.3	18	0.80 ± 0.18	5.4
T. dregeana	2.92	8.9	3.5	2.16 ± 0.30	10.8			
Ekebergia capensis	1.28	3.3	6	0.94 ± 0.21	5.9			
Syzygium cumini	1.66	6.2	23	1.64 ± 0.36	5.9			

[a]Standard deviations for axes from fresh seeds are as in Table 8.1.

an increase in the germination rate) after mild dehydration. Further drying led to either an increase in the MTG (*P. henkelii* and *T. emetica*) or a loss of germinability, under which conditions the MTG becomes less meaningful.

Discussion

Recalcitrant seeds are metabolically active (Farrant *et al.*, 1997), initiate germination-associated developmental events (Berjak *et al.*, 1984; Farrant *et al.*, 1986, 1989) and often germinate in storage. Successful extension of medium-term storage would require a reduction in the rate of this metabolism and development. One possible approach is to reduce the water content to levels inhibiting germination (King and Roberts, 1980; Hong and Ellis, 1996). However, previous work has indicated that this is not a successful approach (Corbineau and Côme, 1986, 1988; Drew *et al.*, 2000). The data presented in this chapter confirm this. In every case, partial drying before storage led to a decline in storage lifespan, or a loss of vigour in the case of *T. dregeana*, relative to seeds stored at their initial water contents. The effect of temperature on the seeds was complex: partial drying before storage induced a chilling sensitivity in *P. henkelii*; *S. cumini* seeds survived marginally better at 25°C; but partially dried seeds of other species survived better at 16°C than at 25°C (Table 8.2). Although the greatest loss of viability occurred in the species that had been most severely dehydrated (i.e. *T. emetica*), the species that was least dehydrated before storage (i.e. *S. cumini*) showed the next greatest loss of viability. Generally, there were no apparent relationships between the extent to which axis water content was reduced by pre-storage partial drying, the rate of germination of fresh seeds and the performance in storage. Seeds were not stored at their initial water content until viability was lost, so the suggested relationship between storage lifespan and germination rate (Berjak *et al.*, 1989; Pammenter *et al.*, 1994) could not be assessed.

When the effect of drying on immediate post-dehydration germination was studied, it was observed that mild dehydration actually enhanced the rate of germination (Table 8.3). This observation has been reported before (Fu *et al.*, 1994; Pammenter *et al.*, 1998; Kioko *et al.*, 1999; Rodríguez *et al.*, 2000), although we suspect that it is an observation more often made than reported. Thus, seeds that are partially dried before storage could lose viability through one of the two-related causes: overdrying could lead to the accumulation of desiccation damage; or mild drying could stimulate germination events, which would likely lead to loss of viability in the absence of exogenous water.

The underlying processes leading to the enhancement of germination by mild dehydration are not yet understood. However, this phenomenon may have ecological implications. Partial dehydration is a risk to which most recalcitrant seeds on the soil surface are likely to be exposed. If this enhances germination, it could lead to the extending root accessing water deeper in the soil, thereby increasing the probability of survival of seeds exposed to a dry atmosphere at the soil surface. Many recalcitrant seeds are large (Daws *et al.*, 2005), providing them with the resources necessary for rapid germination.

References

Berjak, P., Dini, M. and Pammenter, N.W. (1984) Possible mechanisms underlying the differing dehydration responses in recalcitrant and orthodox seeds: desiccation associated subcellular changes in propagules of *Avicennia marina. Seed Science and Technology* 12, 365–384.

Berjak, P., Farrant, J.M. and Pammenter, N.W. (1989) The basis of recalcitrant seed behaviour. In: Taylorson, R.B. (ed.) *Recent Advances in the Development and Germination of Seeds*. Plenum Press, New York, pp. 89–108.

Bewley, J.D. and Black, M. (1994) *Seeds: Physiology of Development and Germination*, 2nd edn. Plenum Press, New York.

Corbineau, F. and Côme, D. (1986) Experiments on the storage of seeds and seedlings of *Symphonia globulifera* L.F. (Guttiferae). *Seed Science and Technology* 14, 585–591.

Corbineau, F. and Côme, D. (1988) Storage of recalcitrant seeds of four tropical species. *Seed Science and Technology* 16, 97–103.

Daws, M.I., Garwood, N.C. and Pritchard, H.W. (2005) Traits of recalcitrant seeds in a semi-deciduous tropical forest in Panama: some ecological implications. *Functional Ecology* 19, 874–885.

Drew, P.J., Pammenter, N.W. and Berjak, P. (2000) 'Sub-imbibed' storage is not an option for extending longevity of recalcitrant seeds of the tropical species *Trichilia dregeana* Sond. *Seed Science Research* 10, 355–363.

Farrant, J.M., Pammenter, N.W. and Berjak, P. (1986) The increasing desiccation-sensitivity of recalcitrant *Avicennia marina* seeds with increasing storage time. *Physiologia Plantarum* 67, 291–298.

Farrant, J., Pammenter, N.W. and Berjak, P. (1989) Germination associated events and the desiccation sensitivity of recalcitrant seeds: a study on three unrelated species. *Planta* 178, 189–198.

Farrant, J.M., Pammenter, N.W., Berjak, P. and Walters, C. (1997) Subcellular organization and metabolic activity during the development of seeds that attain different levels of desiccation tolerance. *Seed Science Research* 7, 135–144.

Fu, J.R., Jin, J.P., Peng, Y.F. and Xia, Q.H. (1994) Desiccation tolerance in two species with recalcitrant seeds: *Clausena lansium* (Lour.) and *Litchi chinensis* (Sonn.). *Seed Science Research* 4, 257–261.

Hong, T.D. and Ellis, R.M. (1996). A protocol to determine seed storage behaviour. In: Engels, J.M.M. and Toll, J. (eds) *IPGRI Technical Bulletin No. 1*. International Plant Genetic Resources Institute, Rome, Italy.

King, M.W. and Roberts, E.H. (1980) Maintenance of recalcitrant seeds in storage. In: Chin, H.F. and Roberts, E.H. (eds) *Recalcitrant Crop Seeds*. Tropical Press SDN.BDH, Kuala Lumpur, Malaysia, pp. 53–89.

Kioko, J., Berjak, P., Pritchard, H. and Daws, M. (1999) Studies of post-shedding behaviour and cryopreservation of seeds of *Warburgia salutaris*, a highly endangered medicinal plant indigenous to tropical Africa. In: Marzalina, M., Khoo, K.C., Jayanthi, N., Tsan, F.Y. and Krishnapillay, B. (eds) *IUFRO Seed Symposium 1998 Recalcitrant Seeds*. Forestry Research Institute Malaysia, Kuala Lumpur, Malaysia, pp. 365–371.

Mycock, D.J. and Berjak, P. (1990) Fungal contaminants associated with several homoiohydrous (recalcitrant) seed species. *Phytophylactica* 22, 473–418.

Pammenter, N.W., Farrant, J.M. and Berjak, P. (1984) Recalcitrant seeds: short-term storage effects in *Avicennia marina* (Forsk.) Vierh. may be germination associated. *Annals of Botany* 54, 843–846.

Pammenter, N.W., Berjak, P., Farrant, J.M., Smith, M.T. and Ross, G. (1994) Why do stored recalcitrant seeds die? *Seed Science Research* 4, 187–191.

Pammenter, N.W., Greggains, V., Kioko, J., Wesley-Smith, J., Berjak, P. and Finch-Savage, W.E. (1998) Effects of differential drying rates on viability retention of recalcitrant seeds of *Ekebergia capensis*. *Seed Science Research* 8, 463–471.

Pritchard, H.W., Tompsett, P.B., Manger, K. and Smidt, W.J. (1995) The effect of moisture content on low temperature responses of *Araucaria hunsteinii* seed and embryos. *Annals of Botany* 76, 79–88.

Rodríguez, M.D.C., Orozco-Segovia, A., Sánchez-Coronado, M.E. and Vázquez-Yanes, C. (2000) Seed germination of six mature neotropical rain forest species in response to dehydration. *Tree Physiology* 20, 693–699.

Sowa, S., Roos, E.E. and Zee, F. (1991) Anesthetic storage of recalcitrant seeds: nitrous oxide prolongs longevity of lychee and longan. *HortScience* 26, 597–599.

Tompsett, P.B. (1983) The influence of gaseous environment on the storage life of *Araucaria hunsteinii* seed. *Annals of Botany* 52, 229–237.

9 Immunocytochemical Localization of β-1,3-Glucanase in Wet-stored Recalcitrant Seeds of *Avicennia marina* Infected by *Fusarium moniliforme*

V. Merhar,[1] C. Calistru[2] and P. Berjak[2]

[1]*Electron Microscope Unit, University of KwaZulu-Natal, Durban 4041, South Africa;* [2]*School of Biological and Conservation Sciences, University of KwaZulu-Natal, Durban 4041, South Africa*

Abstract

Antiserum raised against a purified wheat β-1,3-glucanase was used to study the subcellular localization of this enzyme in *Fusarium moniliforme*-infected, hydrated-stored recalcitrant *Avicennia marina* (Forssk.) Vierh. seeds by transmission electron microscopy (TEM) of immunogold-labelled sections. In newly harvested seeds, the enzyme accumulated predominantly in the vacuoles. When these seeds were inoculated with *F. moniliforme* Sheld. and subsequently hydrated-stored for 8 days, an extensive network of fibrils (probably the result of fungal activity) was observed in the intercellular spaces. However, accumulation of β-1,3-glucanase in this vicinity was not found. When the seeds were subjected to 4 days' clean, hydrated storage prior to inoculation with the fungus, some fungal proliferation occurred, coincident with an increase of the β-1,3-glucanase label in the vicinity of the penetrating fungal hyphae. The results suggest that although newly shed *A. marina* seeds contain β-1,3-glucanase, this enzyme is confined to the vacuoles and is unable to reach the invading pathogen when the seeds are subsequently inoculated with *F. moniliforme*. During short-term storage, the enzyme appears in the cell walls and intercellular spaces where it is likely to come into contact with the invading pathogen, thus delaying the progress of infection.

Introduction

The seeds of *Avicennia marina* (Forssk.) Vierh. are highly recalcitrant (Farrant *et al.*, 1993), as they are very sensitive to desiccation and chilling. The only way to conserve these seeds in the short term is to store them in a saturated atmosphere and at ambient temperatures. This approach, termed wet or hydrated storage is, however, conducive

(eds S. Adkins, S. Ashmore and S.C. Navie)

to fungal proliferation that severely curtails seed postharvest lifespan (Calistru *et al.*, 2000). Although newly harvested *A. marina* seeds do harbour a variety of fungi (e.g. *Aspergillus, Penicillium* and *Fusarium*), the composition of the seed-associated mycoflora narrows during storage, and in time *Fusarium* species become dominant (Mycock and Berjak, 1990).

The resilience of *A. marina* seeds to *F. moniliforme* Sheld. (i.e. the most deleterious of the local recalcitrant seed-associated fungal species) increases after a few days of hydrated storage, but the effect is transient (Calistru *et al.*, 2000). The question arises as to whether *A. marina* seeds produce antifungal compounds that might suppress fungal proliferation during hydrated storage. Previously, the presence of two pathogenesis-related (PR) proteins (β-1,3-glucanase and chitinase) was reported in wet-stored *A. marina* seeds (Anguelova-Merhar *et al.*, 2003), and several flavonoid components, possibly phytoalexins, were isolated from these seeds (Calistru, 2004). Despite the fact that these antifungal compounds might play a role in curtailing fungal proliferation in wet-stored *A. marina* seeds, ultimately the seeds die. This is partly due to a mild, but prolonged, water stress (Pammenter *et al.*, 1994; Motete *et al.*, 1997). However, fungal depredation also contributes, implying that some vital defence factors are missing and/or the ability for defence in these seeds becomes ineffective during storage (Calistru *et al.*, 2000).

The present study is a part of ongoing research on the subcellular localization of antifungal enzymes in recalcitrant seeds of *A. marina* and focuses on β-1,3-glucanase, a (PR-2) protein with well-known antifungal properties (Van Loon, 1997; Fritig *et al.*, 1998; Leubner-Metzger and Meins, 1999), but also normally implicated in diverse physiological and developmental processes including germination (Leubner-Metzger, 2003). The subcellular localization of this enzyme has been studied in vegetative tissues of various plant species infected with pathogenic fungi, including *Fusarium oxysporum* Schltdl. (Benhamou *et al.*, 1989) and *Fusarium culmorum* (WG Smith) Sacc. (Kang and Buchenauer, 2002), but there is no information about the *in situ* localization of β-1,3-glucanase in recalcitrant seeds. The aim of the present study was to localize and quantify the production of β-1,3-glucanase in wet-stored *A. marina* seeds before and after inoculation with *F. moniliforme* and to ascertain whether this enzyme is already present or is induced during storage and/or pathogen infection.

Materials and Methods

Seed material, inoculation and sampling

Mature *A. marina* seeds were collected at the Beachwood Mangroves Nature Reserve, Durban, South Africa. After pericarp removal, seeds were surface-sterilized with 1% sodium hypochlorite for 20 min, after which a 0.25% solution of Previcur N® (AgrEvo, Pietermaritzburg, South Africa) fungicide was applied as an aerosol. A subsample of seeds was immediately inoculated with an identified strain of *F. moniliforme* and stored for 8 days, while the rest were placed in hydrated clean storage (Calistru *et al.*, 2000) for 4–10 days. After this period, the seeds were inoculated with *F. moniliforme* and wet-stored for a further 8 days. The following groups of seeds were used for the TEM studies: (i) fresh seeds; (ii) fresh seeds stored infected for 8 days;

(iii) seeds hydrated-stored for 4 days; (iv) seeds hydrated-stored clean for 10 days; (v) seeds hydrated-stored for 4 days, infected and re-stored for 8 days; and (vi) seeds hydrated-stored for 10 days, infected and re-stored for 8 days. The viability of each group of seeds was tested by their ability to germinate and establish seedlings (20 seeds per treatment).

Tissue processing for transmission electron microscopy

The mature propagule of *A. marina* consists of two large, folded cotyledons and an unusually well-developed embryonic axis (Farrant *et al.*, 1993). The hypocotyl tip has five (or, rarely, more) meristematic root primordia that are enclosed by a thin layer of hypocotyl tissue, from which a thick mass of bristle-like hairs protrudes. The present study was undertaken on the root primordia and cotyledonary surfaces, as these are the regions that first show fungal proliferation.

Five specimens per treatment (i.e. infected and non-infected tissues) were processed for TEM as described by Calistru *et al.* (2000). Ultra-thin sections were cut on a Reichert–Jung ultramicrotome (Wien, Austria) and collected on 300 mesh nickel grids for immunocytochemical processing.

Post-embedding immunogold labelling

The procedure described by Hu and Rijkenberg (1998) was used at room temperature unless otherwise stated. Grids were transferred in a drop of 0.01 M phosphate-buffered saline (PBS, pH 7.4), containing 0.2% (w/v) polyethylene glycol (PEG) 20,000 (Fluka, Buchs, Switzerland) for 5 min. Thereafter, the mounted sections were incubated for 30 min in a drop of blocking solution made up of 0.01 M PBS (pH 7.4), 10% (v/v) foetal bovine serum (FBS) (Delta Bioproducts, Kempton Park, South Africa), 1% (w/v) bovine serum albumin (BSA) (Sigma, St. Louis, Missouri, USA), 0.05% (v/v) Tween 20 (Sigma) and 0.2% (w/v) sodium azide (Fluka)). They were then incubated in 40 μl of primary antibody (rabbit antiserum raised against wheat β-1,3-glucanase) in 0.01 M PBS (pH 7.4) (1:500) in a moist chamber, overnight at 4°C. The primary antibody was supplied by Professor A.J. van der Westhuizen (see acknowledgements). Grids were then washed (3 × 5 min) in 0.01 M PBS (pH 7.4), containing 1% BSA and 0.05% Tween 20, and immersed for 1 h in a drop of 10 nm colloidal gold-conjugated goat antiserum to rabbit immunoglobulin (GAR-gold antibody [Sigma]) diluted at 1:20 in washing solution. Finally, the sections were fixed with 1% glutaraldehyde for 2 min, rinsed with sterile distilled water, counterstained with uranyl acetate and lead citrate and examined with a JEOL 100 CX TEM (Tokyo, Japan) at 80 kV. Specificity of labelling was assessed by examination of control sections in which either the primary or secondary antibody was omitted.

Quantification of labelling

The density of labelling over sections from different treatments was compared by determining the number of gold particles per square micrometre plus standard deviation

(SD). Densities were determined by counting the number of gold particles over each subcellular structure on five micrographs from different specimens and expressed as the number of gold particles per unit surface area (square micrometre).

Results and Discussion

Figure 9.1 shows the ultrastructure of root primordia cells of newly harvested *A. marina* seeds, which were 100% germinable. Although newly shed *A. marina* seeds have a broad spectrum of associated fungi (Mycock and Berjak, 1990, 1995), the meristematic root primordium cells did not show any evidence of fungal infection or degradation (Fig. 9.1a and inset). This is in accordance with previous observations that the fungi are generally associated with only the pericarp of newly shed *A. marina* seeds (Mycock and Berjak, 1990). Despite the lack of internal fungal mycelia, some gold particles were localized to the cell walls (0.95 gold particles/μm^2) and the vacuoles (2.65 particles/μm^2), indicating the presence of β-1,3-glucanase (Fig. 9.1a and b). The β-1,3-glucanases exist mainly in two forms: extracellular and vacuolar (Stintzi *et al.*, 1993). It is likely that this (and other) extracellularly located enzyme is capable of destroying fungal hyphae growing intercellularly, thus constituting a primary defence mechanism. Vacuolar enzymes, on the other hand, would come into contact with the pathogen only after decompartmentalization of the cells, or intracellular heterophagy of hyphae, thus constituting a secondary line of plant defence (Stintzi *et al.*, 1993; Van Loon, 1997). A considerable amount of β-1,3-glucanase (4.55 particles/μm^2) was localized in the meristem cell plastids of newly harvested seeds (Fig. 9.1b), suggesting that besides its defensive role, this enzyme might be involved in specific physiological processes connected to germination, as has been shown for some orthodox seeds (Petruzzelli *et al.*, 1999; Wu *et al.*, 2001; Leubner-Metzger, 2003). In the cotyledons of newly harvested seeds, β-1,3-glucanase labelling (not illustrated) was localized to the cell walls, the intercellular space and the vacuoles, although there were no signs of fungal infection associated with the cells. In general, the cotyledonary cells were characterized by a large central vacuole significantly compressing the cell contents against the wall.

When *A. marina* seeds were inoculated with *F. moniliforme* immediately after shedding and then subjected to storage for 8 days, fibrillar material (which may have been the consequence of the fungal infection) was observed in the intercellular spaces of the root primordium cells (Fig. 9.2 and inset). However, the β-1,3-glucanase labelling was practically absent from the spaces (Fig. 9.2). None of these seeds germinated although the infection was primarily localized in the cotyledons where the fungal penetration and proliferation was apparent (Fig. 9.3a and b). This was frequently accompanied by host cell wall distortion in the proximity of the penetrating intercellular hypha (Fig. 9.3a and inset). The label indicating the presence of β-1,3-glucanase, however, was associated with the fungal cell wall, but was absent from the host cell wall (Fig. 9.3a). The density of the gold label was ~6.7 particles/μm^2, much lower than that observed in highly incompatible plant–pathogen interactions (Hu and Rijkenberg, 1998; Kang and Buchenauer, 2002), probably accounting for the observation that the fungal hyphae actually penetrated cotyledon cells (Fig. 9.3b). Despite the β-1,3-glucanase association with

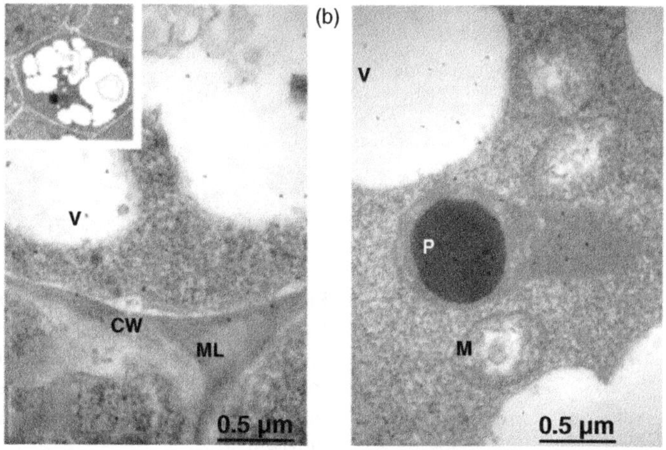

Fig. 9.1. Newly harvested *Avicennia marina* seeds. (a) Meristematic root primordium cells had a regular shape with no signs of fungal invasion (inset – ×10,000). Higher magnification (×50,000) revealed the presence of some β-1,3-glucanase label in the vacuoles while the cell wall was almost free of label. (b) A relatively high density of β-1,3-glucanase label was found associated with the plastids. Abbreviations for this and subsequent figures: CW = cell wall; FC = fungal cell; FCW = fungal cell wall; H = hypha; IS = intercellular space; M = mitochondrion; ML = middle lamella; P = plastid; and V = vacuole.

the hyphal wall (extracellularly and intracellularly) (Fig. 9.3b), the host cells were invaded by the fungus.

After 4 days of hydrated storage, 100% viability of *A. marina* seeds was retained, as indicated by their ability for seedling establishment, and no signs of fungal infection could be discerned internally, or on the cotyledonary surfaces of these seeds (Fig. 9.4). The meristematic root primordium cells of *A. marina* seeds stored for 4 days showed enhanced metabolic activity, as has been described previously (Farrant *et al.*, 1993; Calistru *et al.*, 2000), but there was no evidence of vacuolar localization of β-1,3-glucanase as seen in newly harvested material. Considerable β-1,3-glucanase labelling (17 particles/μm^2), however, was associated with the cell walls, especially surrounding the intercellular spaces (Fig. 9.4). The plastids were substantially labelled (not illustrated), as was the case for the fresh material. Cotyledonary ultrastructure was also well preserved in these clean-stored seeds and no traces of fungal infection were observed. There was substantial localization of β-1,3-glucanase in the wall and intercellular spaces as indicated by the density of the gold label associated with these areas (~40 particles/μm^2) (Fig. 9.5), indicating increased occurrence of the enzyme compared with the cotyledonary tissue of newly harvested seeds.

When *A. marina* seeds were clean-stored for 4 days prior to infection with *F. moniliforme*, then infected and subjected to a further storage for 8 days, some electron-dense material (probably the result of fungal activity) was observed in the intercellular spaces of the root primordium cells, and β-1,3-glucanase was localized to these areas, as well as to the contiguous cell walls (Fig. 9.6). Some fungal structures were observed in the epidermal cells of the cotyledons of 4-day-stored, 8-day-infected seeds (Fig. 9.7 and inset). Considerable β-1,3-glucanase label was associated with the fungal (Fig. 9.7)

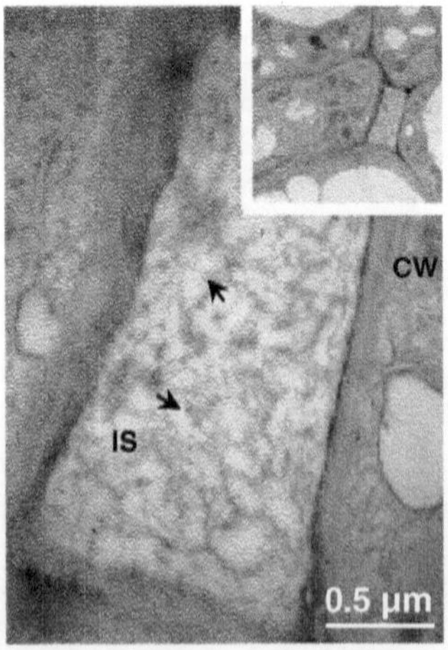

Fig. 9.2. *Avicennia marina* seeds stored, infected for 8 days. Meristematic root primordium cells showed a considerable accumulation of fibrillar material in the intercellular spaces (inset – ×10,000), but only a low density of β-1,3-glucanase label (arrows).

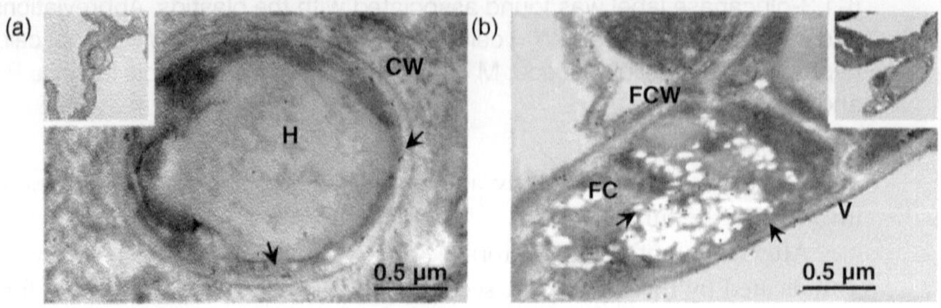

Fig. 9.3. *Avicennia marina* seeds stored, infected for 8 days. (a) Advanced deterioration was observed in the cotyledonary cells of these seeds as indicated by the irregularity of the cell walls in close proximity to the penetrating hypha (inset – ×12,000) and loss of intracellular organization (main micrograph). A scattering of gold particles could be seen associated with the fungal cell wall (arrows in the main micrograph), but the host cell wall was free of labelling. (b) In some cells, intracellular fungal penetration and proliferation was observed (inset – ×12,000), with a relatively large number of gold particles associated with the penetrating fungal hypha (arrows in the main micrograph).

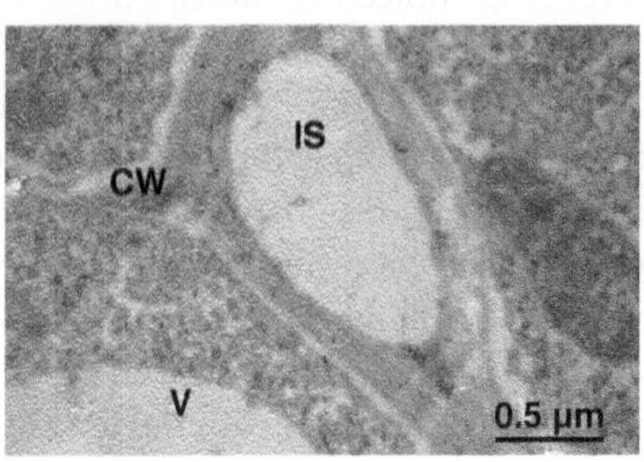

Fig. 9.4. *Avicennia marina* seeds stored clean for 4 days. β-1,3-glucanase label was absent from the vacuoles, but strong labelling was found in the cell walls and the intercellular spaces of these cells.

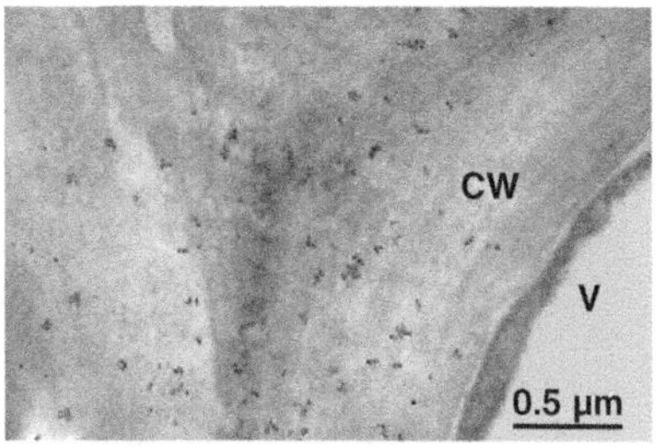

Fig. 9.5. *Avicennia marina* seeds stored clean for 4 days. Large numbers of gold particles were associated with the walls of the cotyledonary cells.

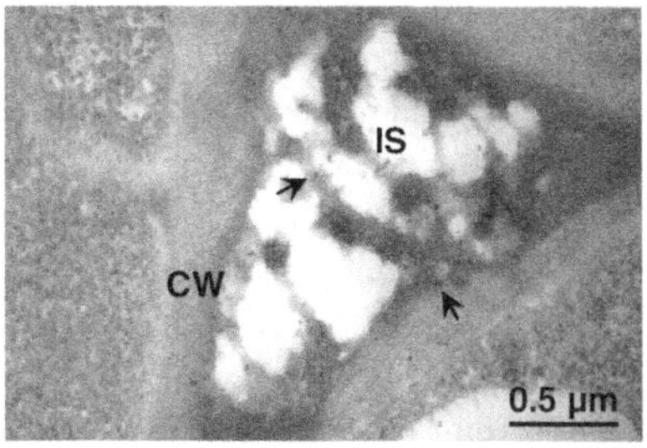

Fig. 9.6. *Avicennia marina* seeds stored clean hydrated for 4 days, infected and re-stored for a further 8 days. Gold label was confined to the cell walls and dense material that was observed in some of the intercellular spaces (arrows) of the meristematic root primordium cells.

and host cell walls. The occurrence of this enzyme in the cells, which the pathogen first invades, suggests that β-1,3-glucanase might play a role in the defence response of wet-stored *A. marina* seeds. The enzyme alone, or more likely in combination with other PR proteins (e.g. chitinase), has the potential to digest the invading *F. moniliforme* hyphae as has previously been shown for other *Fusarium* species (Sela-Buurlage *et al.*, 1993). This suggestion is supported by the observation that the spread of the pathogen into the internal tissue of infected-stored *A. marina* seeds was confined to only occasional hyphae. However, the density of the gold particles in infected-stored seeds did not differ significantly from that occurring in these seeds prior to infection with *F. moniliforme*, suggesting that production of enzyme(s) preceded inoculation, rather than being an induced effect. Similar results were obtained even when the seeds had been clean-stored for 10 days, infected and re-stored for another 8 days (results not shown), implying that events occurring early during wet storage probably play a role in the decreased susceptibility of *A. marina* seeds to fungal infection, compared with the newly harvested condition.

Conclusions

Results indicate that in newly shed *A. marina* seeds, β-1,3-glucanase is confined to the vacuoles and is unable to reach the invading pathogen when the seeds are subsequently inoculated with *F. moniliforme*. This might be one of the reasons for the high susceptibility of newly shed seeds to fungal infection. During short-term wet storage, β-1,3-glucanase becomes localized in the cell wall and intercellular spaces,

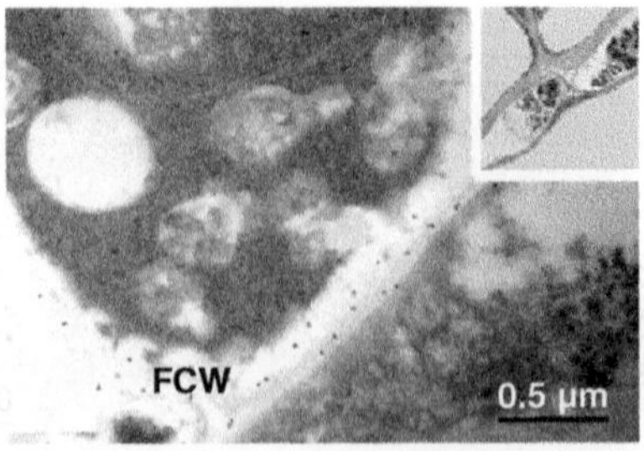

Fig. 9.7. *Avicennia marina* seeds stored clean hydrated for 4 days, infected and re-stored for a further 8 days. Fungal hyphae were observed in the epidermal layer of the cotyledonary tissue (inset – ×12,000). Gold particles were associated with the cell wall of the intracellular degenerating hyphae.

and is associated with germinative events in the hydrated seeds that occur in the first 4 days after shedding (Calistru, 2004). During infection, the penetrating fungal hypha is likely to come into contact with the extracellular (i.e. intercellular) enzymes, thus delaying degradation of the host tissue. Although hydrated-stored *A. marina* seeds do produce β-1,3-glucanase (and most probably other PR proteins), these propagules are far from being resistant, possibly because the quantity of enzyme(s) is fungistatic rather than fungicidal. Even the highest levels of β-1,3-glucanase in stored *A. marina* seeds were still below the effective level of the same enzyme found in resistant cultivars during some highly incompatible plant–pathogen interactions (Hu and Rijkenberg, 1998). Our results suggest that the timing of infection, and possibly the quantity of β-1,3-glucanase (and probably other PR proteins), may be a crucial factor for the successful defence response of hydrated-stored recalcitrant seeds.

Assessment of the causes of death of recalcitrant seeds in storage is not straightforward. These seeds will lose viability, even if the microbial contamination is controlled. This is probably the consequence of a mild, but prolonged water stress (Pammenter *et al.*, 1994, 1997). However, any step towards achieving long-term storage of such seeds is of major importance. The present contribution is the first report on the subcellular localization of PR protein in recalcitrant seeds and provides an appropriate framework for future studies on recalcitrant seeds' antimicrobial strategies.

Acknowledgements

We wish to thank Professor A.J. van der Westhuizen, University of the Orange Free State, South Africa, for supplying the antiserum against a purified wheat β-1,3-glucanase, as well as Priscilla Maartens, Gabrielle Turner and Visual Bharuth for their technical assistance. This work was supported by grant from the University of KwaZulu-Natal, South Africa.

References

Anguelova-Merhar, V.S., Calistru, C. and Berjak, P. (2003) A study of some biochemical and histopathological responses of wet-stored recalcitrant seeds of *Avicennia marina* infected by *Fusarium moniliforme*. *Annals of Botany* 92, 401–408.

Benhamou, N., Grenier, J., Asselin, A. and Legrand, M. (1989) Immunogold localization of β-1,3-glucanases in two plants infected by vascular wilt fungi. *The Plant Cell* 1, 1209–1221.

Calistru, C. (2004) Some implications of associated mycoflora during hydrated storage of recalcitrant seeds of *Avicennia marina* (Forssk.) Vierh. PhD thesis, University of KwaZulu-Natal, South Africa.

Calistru, C., McLean, M., Pammenter, N.W. and Berjak, P. (2000) The effects of mycofloral infection on the viability and ultrastructure of wet-stored recalcitrant seeds of *Avicennia marina* (Forssk.) Vierh. *Seed Science Research* 10, 341–353.

Farrant, J.M., Pammenter, N.W. and Berjak, P. (1993) Seed development in relation to desiccation tolerance: a comparison between desiccation-sensitive (recalcitrant) seeds of *Avicennia marina* and desiccation-tolerant types. *Seed Science Research* 3, 1–13.

Fritig, B., Thierry, H. and Legrand, M. (1998) Antimicrobial proteins in induced plant defense. *Current Opinion in Immunology* 10, 16–22.

Hu, G. and Rijkenberg, F.H.J. (1998) Subcellular localization of β-1,3-glucanase in *Puccinia recondita* f. sp. *tritici*-infected wheat leaves. *Planta* 204, 324–334.

Kang, Z. and Buchenauer, H. (2002) Immunocytochemical localization of β-1,3-glucanase and chitinase in *Fusarium culmorum*-infected wheat spikes. *Physiological and Molecular Plant Pathology* 60, 141–153.

Leubner-Metzger, G. (2003) Functions and regulation of β-1,3-glucanase during seed germination, dormancy release and after-ripening. *Seed Science Research* 13, 17–34.

Leubner-Metzger, G. and Meins, F. (1999) Functions and regulation of plant β-1,3-glucanases (PR-2). In: Datta, S.K. and Muthukrishnan, S. (eds) *Pathogenesis-related Proteins in Plants.* CRC Press, Boca Raton, Florida, pp. 49–76.

Motete, N., Pammenter, N.W., Berjak, P. and Frédéric, J.C. (1997) Response of the recalcitrant seeds of *Avicennia marina* to hydrated storage: events occurring at the root primordia. *Seed Science Research* 7, 169–178.

Mycock, D.J. and Berjak, P. (1990) Fungal contaminants associated with several homoiohydrous (recalcitrant) seed species. *Phytophylactica* 22, 413–418.

Mycock, D.J. and Berjak, P. (1995) The implication of seed-associated mycoflora during storage. In: Kigel, J. and Galili, G. (eds) *Seed Development and Germination.* Marcel Dekker, New York, pp. 747–766.

Pammenter, N.W., Berjak, P., Farrant, J., Smith, M.T. and Ross, G. (1994) Why do stored hydrated recalcitrant seeds die? *Seed Science Research* 4, 187–191.

Pammenter, N.W., Motete, N. and Berjak, P. (1997) The response of hydrated recalcitrant seeds to long-term storage. In: Ellis, R.H., Black, M., Murdoch, A.J. and Hong, T.D. (eds) *Basic and Applied Aspects of Seed Biology.* Kluwer Academic Publishers, Dordrecht, The Netherlands, pp. 673–687.

Petruzzelli, L., Kunz, C., Waldvogel, R., Meins, F. and Leubner-Metzger, G. (1999) Distinct ethylene- and tissue-specific regulation of β-1,3-glucanases and chitinases during pea seed germination. *Planta* 209, 195–201.

Sela-Buurlage, M.B., Ponstein, A.S., Bres-Vloemans, S.A., Melchers, L.S., van den Elzen, P.J.M. and Cornelissen, B.J. (1993) Only specific tobacco (*Nicotiana tabacum*) chitinases and β-1,3-glucanases exhibit antifungal activity. *Plant Physiology* 101, 857–863.

Stintzi, A., Heitz, T., Prasad, V., Wiedemann-Merdinoglu, S., Kauffmann, S., Geoffroy, P., Legrand, M. and Fritig, B. (1993) Plant 'pathogenesis-related' proteins and their role in defense against pathogens. *Biochimie* 75, 687–706.

Van Loon, L.C. (1997) Induced resistance in plants and the role of pathogenesis-related proteins. *European Journal of Plant Pathology* 103, 753–765.

Wu, C.-T., Leubner-Metzger, G., Meins, F. and Bradford, K.J. (2001) Class I β-1,3-glucanase and chitinase are expressed in the micropylar endosperm of tomato seeds prior to radicle emergence. *Plant Physiology* 126, 1299–1313.

10 Seed Development Transporting into the Post-genomic Era

W.M. Waterworth and C.M. Bray

Faculty of Life Sciences, University of Manchester, Oxford Road, Manchester, UK

Abstract

Seed development represents a critical stage in the plant life cycle and is of prime economic and agronomic importance, affecting both crop yield and nutritional quality. The recent availability of sequenced plant genomes and tagged or insertional mutant collections has been accompanied by the application of new technologies for high-throughput analysis of genes, proteins and metabolites and advances in methodologies such as the spatial imaging of metabolites. These approaches now allow us to dissect the complex networks of developmental and metabolic processes in model plant species, such as *Arabidopsis* (*Arabidopsis thaliana* (L.) Heynh.), and crop species. Consequently, our fundamental knowledge and understanding of the key developmental events and metabolic pathways that underlie seed development has significantly increased. In addition, we are now becoming aware of the importance of epigenetic control of seed development and nutrient regulation of seed storage reserve deposition. Recent progress and future perspectives relating to seed development will be reviewed in the context of these rapidly emerging methodologies using specific examples to highlight their usefulness and potential.

Introduction

The seed occupies a pivotal role in the plant life cycle, representing the culmination of one generation and the initiation of the next. Seeds have historically played an important role in the development of civilizations through their central role in agriculture and contributions to biotechnology. Seed development is a complex process that involves the interplay of a network of many developmental processes and metabolic pathways together with their interactions with the environment. Much progress has been made using biochemical and physiological approaches to study seed biology over the last century. The application of high-throughput post-genomic technologies is enabling rapid and exciting advances in seed research leading towards a detailed modelling of all aspects of seed development, germination and storage (i.e. conservation).

Seed development encompasses several phases: from embryogenesis, which is characterized by rapid cell division and histodifferentiation, to seed maturation, which correlates with the deposition of storage reserves and cell expansion. In orthodox seeds, there is a further stage of maturation drying, which is characterized by the onset of desiccation tolerance (DT) (Bewley and Black, 1994).

There is wide diversity in seed structure, developmental regulation and storage composition between species. However, in terms of dry weight, most orthodox seeds consist of more than 90% storage reserves. The nutrient reserves contain carbohydrates (often starch), lipids (usually triacylglycerides) and specialized storage proteins. These storage reserves are of critical importance during germination. Their mobilization supports the growth and metabolism of the embryo of a germinating seed until photosynthetic capacity is acquired by the developing seedling.

Seeds represent the staple food source of the human diet, either directly or indirectly through their use as livestock feed. The major food sources are represented principally by the seeds of cereals, including wheat (*Triticum aestivum* L.), rice (*Oryza sativa* L.), maize (*Zea mays* L.) and the protein-rich legumes. Thus, the study of nutrient reserve deposition in seed development is crucial to the improvement of agricultural production and the realization of the potential of biotechnology. The knowledge gained will be of fundamental importance to the improvement of crop yield, nutritional properties – in terms of overall protein, lipid and carbohydrate composition (i.e. quality and quantity) – and aesthetic considerations such as the stickiness of rice grains or the bread-making qualities of wheat. More recently, an understanding of the detailed metabolism of the developing seed has proved a key component in biotechnology by experiments designed to introduce novel compounds in seeds through genetic modification. The study of seed development is also important to the conservation of plant germplasm in terms of establishing the best point at which to harvest the seed of a particular wild species to retain maximum vigour and viability during short- and long-term storage.

This chapter describes how emerging post-genomic methodologies have facilitated recent advances in the field of seed development. Recent years have been very exciting for the plant sciences. Sequencing of the *Arabidopsis* and rice genomes in 2000 and 2005, respectively, has coincided with the availability of rapidly emerging technologies for the global high-throughput analyses of genes, proteins and metabolites (Arabidopsis Genome Initiative, 2000; Hirai *et al.*, 2005; Peck, 2005; Rensink and Buell, 2005). This has led to progress on a scale previously undreamed of. Databases arising out of extensive expressed sequence tag (EST) sequencing programmes for several other species are now available (e.g. wheat, barley (*Hordeum vulgare* L.), medicago (*Medicago truncatula* Gaertn.) and soybean (*Glycine max* (L.) Merr.). The widespread availability of *Arabidopsis* T-DNA insertion lines, in which specific known gene sequences have been disrupted, has enabled the use of mutant phenotypes to investigate gene function (Sessions *et al.*, 2002; Alonso *et al.*, 2003). This has facilitated advances in several areas including the regulation of early seed development and signalling. Knockout mutant phenotypes have contributed greatly to our understanding of processes in seed development and have frequently led to surprising results. For example, disruption of the *Arabidopsis* putative oligopeptide transporter (OPT3) (one of the multiple OPTs found in *Arabidopsis*) is embryo-lethal with development arrested at the octet stage (Stacey *et al.*, 2002).

Collectively, transcriptomics, proteomics and metabolomics represent the global profiling of gene expression (mRNA), proteins and low-molecular-weight metabolites, respectively. High-throughput techniques for analysis of mRNA transcripts, proteins and metabolites have now been developed. What impact have these technologies had on the study of seed development?

Transcriptomics

Transcriptomics is the global analysis of mRNA expression by hybridization of mRNA populations, ESTs, cDNAs or oligonucleotides immobilized on to DNA chips. Microarray technology has been widely available for some years, and has been utilized effectively in the study of seed development in several species over the last few years.

Girke *et al.* (2000) used microarray analyses to study seed development in *Arabidopsis*. They followed a more comprehensive study of seed filling in *Arabidopsis* (>35,000 genes), concentrating on analyses of the biosynthetic pathways for storage reserves (Ruuska *et al.*, 2002). These studies enabled the primary transcriptional networks that coordinate events in seed development to be visualized for the first time. Genes that perform similar functions or participate in the same pathways were often found to be expressed in a coordinated manner. For example, genes encoding major groups of enzymes associated with fatty acid, carbohydrate and storage protein biosynthesis demonstrated temporally distinct expression patterns. These studies on *Arabidopsis* were subsequently followed by microarray analyses of storage reserve deposition in other species. For example, Zhu *et al.* (2003) used gene chips comprising half the rice genome (21,000 genes) to study rice grain filling, and Sreenivasulu *et al.* (2004) used barley macroarrays to study barley grain development from fertilization to the early storage phase. The simultaneous expression patterns of thousands of genes could be studied for the first time, which in turn enabled clusters of genes and transcriptional networks primarily concerned with seed development to be delineated.

The major findings of these studies confirmed our existing knowledge of metabolic pathways. Significantly, cluster analysis of global changes in temporal expression patterns also enabled putative functions to be assigned to unknown genes, the prediction of metabolic pathways and the identification of genes whose involvement in seed development was previously unsuspected. An example of this was the discovery that expression of photosynthetic genes followed that of fatty acid biosynthesis in *Arabidopsis* seed development, which implicates a role in CO_2 fixation and cofactor supply (Ruuska *et al.*, 2002). Transcription factors responsible for the coordinated expression of specific genes or gene families can also be identified through promoter analysis (Zhu *et al.*, 2003).

Proteomics

Although transcriptomics allows us to visualize changes in gene expression, changes in mRNA levels do not necessarily correlate with protein levels and enzyme activities. For example, one gene can give rise to several proteins (e.g. through alternative mRNA splicing). In addition, covalent post-translational modifications, such as protein phosphorylation and glycosylation, can influence enzyme specificity, activity and/or protein–protein interactions essential for enzyme function. Proteomics methodologies overcome many of these limitations through the global analysis of proteins. This analysis is often performed using 2D electrophoretic separation of proteins followed by matrix-assisted laser desorption/ionization time of flight (MALDI-TOF) mass spectrometry (MS). The identification of proteins on the basis of peptide mass is reliant on genome sequence annotations (predicted protein sequences); so there are limitations for species whose genome has not been sequenced.

Over the last 3 years, studies have been initiated that aim to characterize global changes in protein profiles during seed development. One of the first proteomic analyses of seed development was the study of protein changes during barley grain filling and maturation (Finnie *et al.*, 2002). This study identified 19 proteins, with several potential roles in oxidative stress, and noted that α-amylase inhibitors accumulated in the seed throughout development.

Subsequently, Gallardo *et al.* (2003) studied changes in protein profiles throughout seed development, from embryogenesis to desiccation, in the model legume *M. truncatula*. A total of 120 proteins that displayed changes in expression were analysed using 2D electrophoresis and MALDI-TOF MS, and 84 proteins were identified on the basis of peptide mass. In particular, they characterized the kinetics of storage protein deposition and proteins associated with germination capacity. Hajduch *et al.* (2005) characterized seed filling in soybean and identified 422 proteins with changes in expression profile. The findings of these studies were largely consistent with our existing knowledge of seed development, but identified new proteins that may have a central role in seed developmental processes. An overriding feature of these studies was that proteins, which carry out similar functions, or participate in the same metabolic pathway, displayed similar patterns of expression during seed development.

The rapid developments in MS methodologies over recent years have impacted on the number of proteins identified through proteomic analyses. MS approaches, in combination with emerging microarray data, are now enabling the identification of coordinated processes at the genome-wide transcriptional and translational level.

Proteomics can also be used to answer questions concerning the subcellular compartmentalization of proteins. The use of proteomics to examine the protein profiles of subcellular organelles provides answers to questions such as which enzyme isoform is present in which subcellular compartment, and enables the identification of metabolic pathways and control processes localized to specific intracellular organelles (Peck, 2005).

Comparisons between the protein profiles of samples taken from different developmental stages are facilitated by recent advances in proteomic technologies such as isotope-coded affinity tag (ICAT) labelling, in which the relative abundance of proteins between two samples is compared using a density-labelled affinity tag approach. Protein samples to be compared are differentially labelled using heavy and light tags, and the samples are analysed by MS, circumventing the need for electrophoresis (reviewed by Turecek, 2002; Patton *et al.*, 2002).

Post-translational Regulation

There is widespread control of the expression and activity of many proteins at the post-translational level. Post-translational control by ubiquination, a process in which ubiquitin ligase enzymes tag unwanted proteins with a chain of ubiquitin molecules thereby committing these proteins for degradation by the 26S proteosome, is a common and universal mechanism (Hare *et al.*, 2003). In contrast to the transcriptome and the proteome, there are relatively few studies of post-translational control mechanisms in seed development. This can probably be attributed to a lack of widespread high-throughput methods for their study.

Phosphorylation has been demonstrated to be a ubiquitous post-translational protein modification through which control of cellular events is effected. Protein phosphorylation plays a central role in signalling networks that regulate seed development, such as in abscisic acid (ABA) signalling cascades (reviewed in Finkelstein and Gibson, 2002). The enzyme phosphoenol pyruvate carboxykinase (PEPCK), which acts at a key metabolic control point, is also known to be regulated by phosphorylation (reviewed in Izui *et al.*, 2004).

Recently, phosphorylation has been shown to play a key role in regulating nutrient transport in seed development. The barley small peptide transporter HvPTR1 is expressed in the scutellum during germination (West *et al.*, 1998) and in the embryo throughout development (Waterworth *et al.*, 2003). HvPTR1 peptide transport activity is downregulated by amino acids, and recent studies have shown that this regulation is mediated at the post-transcriptional level by phosphorylation (Waterworth *et al.*, 2005).

There is an increasing awareness that the functions of many proteins are mediated through their interactions with specific protein partners *in vivo* and that protein–protein interactions play an important role in many cellular processes. Bioimaging approaches such as fluorescence resonance energy transfer (FRET) and fluorescence recovery after photobleaching (FRAP) allow the elegant *in vivo* study and verification of protein–protein interactions. Riera *et al.* (2004) recently employed bioimaging techniques to study protein–protein interactions in signalling pathways during seed development. The maize Rab17 protein is a heavily phosphorylated late embryogenesis abundant (LEA) protein that interacts with the protein kinase CK2. Both proteins were tagged with green fluorescent protein (GFP) and transfected into onion (*Allium cepa* L.) epidermal cells, where they were transiently expressed and their cellular localization examined by confocal microscopy. A mutated form of Rab17, which cannot be phosphorylated, was sequestered into the nucleolus, but wild-type Rab17 and CK2 were spread throughout the cytoplasm, indicating that their intracellular localization is dependent on phosphorylation status.

Methodologies capable of genome-wide/global analyses of post-translational modifications to proteins are increasingly applied to plant systems and should provide exciting new insights into the control of seed development at the post-transcriptional level.

Metabolomics

Metabolomic profiling allows the quantitative measurement of all low-molecular-weight metabolites in a given cell or tissue. Complete metabolite analysis is achieved using a range of techniques including nuclear magnetic resonance (NMR) spectroscopy, gas chromatography (GC), ion mass spectrometry (IMS), liquid chromatography/mass spectroscopy (LC/MS) and high-performance liquid chromatography (HPLC). A single technique is not capable of resolving the large range of metabolites found in living cells. Advances in the development of high-throughput methods for the analysis of low-molecular-weight metabolites in *Arabidopsis* seeds during development have been described previously (Gibon *et al.*, 2002).

A key discovery is that there is substantial variation in the spatial concentration of metabolites in tissues at the subcellular level. *In vivo* techniques for the spatial analysis of metabolite distribution in the large-seeded legume *Vicia faba* L. have been used to characterize sucrose, ATP and O_2 levels in developing seeds (Borisjuk *et al.*, 2002, 2003). In *V. faba*, the cotyledons are the principal storage tissue of the seed, and the developing seed contains a developmental gradient of cells (reviewed in Weber *et al.*, 2005). Hexoses (glucose) accumulate in young cells undergoing mitotic cell division, while sucrose, acting as both a nutrient and a developmental signal, is associated with seed maturation and storage reserve deposition in terminally differentiated non-dividing cells. Metabolites in cryosections from *V. faba* cotyledons at different developmental stages were coupled by enzymic reaction to bioluminescence and single photon counting. This approach allowed non-invasive analysis of metabolites to near single cell resolution. In addition, metabolite concentrations can be compared with spatial expression patterns of specific mRNAs and proteins in identical sections by *in situ* hybridization and immunolocalization. For example, steep sucrose gradients were imaged in sections, in particular showing that high sucrose levels overlie the epidermal transfer layer of cells in which the sucrose transporter VfSUC1 was localized, reflecting its role in nutrient supply during seed development (Borisjuk *et al.*, 2002).

Advances in NMR spectroscopy will also facilitate spatial metabolite imaging, and in conjunction with stable isotope labelling techniques facilitate the measurement of metabolic fluxes through pathways. Schwender and Ohlrogge (2002) and Schwender *et al.* (2003) used radio-labelled sugars in combination with GC/MS to investigate metabolic flux *in vivo* in cultured *Brassica* embryos.

Metabolic profiling is important to obtain the complete picture of events in seed development. Unlike proteomics and transcriptomics, metabolomics is dependent on our knowledge of the vast array of plant metabolites and the approaches available to analyse these in a comprehensive and high-throughput way, rather than the availability of genome sequence information. This presents a formidable challenge for metabolomics in the immediate future.

Epigenetic Control of Seed Development

Recent studies have shown that early seed development is under maternal control mediated by the epigenetic mechanism of gene imprinting (i.e. expression is dependent on parent of origin). The developing endosperm, not the embryo, is subject to control by the epigenetic mechanism of gene imprinting in species including maize and *Arabidopsis* (reviewed by Autran *et al.*, 2005). In contrast to Mendelian genetics, imprinted genes are predominantly expressed from one allele in a parent-of-origin manner, probably mediated by changes in DNA methylation and chromatin structure. In early seed development, the paternal allele is silenced by DNA methylation. An example of a gene imprinted in the endosperm during seed development is the maternal-effect dominant embryonic arrest (MEDEA) polycomb group protein, which functions as a global repressor of gene activity (Choi *et al.*, 2002). This protein participates in protein complexes that modify histones to maintain a repressed state of gene expression, acting as a master regulator of endosperm development. DEMETER (DME), a protein that contains a DNA glycosylase domain, was identified as a gene necessary for maternal allele expression of the imprinted *MEDEA* gene (Choi *et al.*, 2002). DME is expressed in the central cell of the female gametophyte, the progenitor of the endosperm. DME was shown to have *in vivo* glycosylase activity and to nick the *MEA* promoter before fertilization, thereby activating maternal expression of the maternal *MEA* gene by reducing the level of methylation. DEMETER also regulates imprinting of a second gene, *FWA*, which is expressed in the central cell nucleus and developing endosperm (Kinoshita *et al.*, 2004). Allele-specific reverse transcription polymerase chain reaction (RT-PCR) assays of appropriate crosses showed that expression of *FWA* was dependent on the maintenance of methylation by the DNA methyltransferase MET1, and that activation occurred upon removal of direct repeats of the 5' region of the *FWA* gene promoter by DEMETER.

Conclusions

Genome sequencing and post-genomic technologies have led to rapid progress in plant sciences. Over the last 3 years, these technologies have been increasingly employed to bring a new dimension to the study of seed development. These, in conjunction with our previous knowledge base, are providing increasing insight into the developmental, signalling and metabolic networks involved in seed development, such that a picture of seed development at the genome-wide level is now emerging. However, much is still to be achieved as an understanding of gene regulation at all levels is necessary to construct seed metabolic and developmental control networks. It is also timely to remember that the study of genes, proteins and metabolites in isolation will not be enough. Integrated studies at all levels are necessary to formulate a comprehensive picture of seed development. While the establishment of developmental and metabolic pathways in model species is important, to see the 'complete' picture we must also ask to what extent these processes are conserved in diverse plant species. Transcriptomic and proteomic studies of plant species are currently difficult except in those species for which genome sequence information is available. Investigations into other interesting 'non-model' species are limited as identification of genes and proteins is restricted to those that display strong cross-species sequence homology.

There is an increasing awareness that it is the network of integrated biological structures and pathways in a cell, and not merely the individual molecular components, that control all aspects of growth and development. The large experimental data-sets being generated by high-throughput technologies present a new and exciting challenge. Can we now progress investigations on seed development through a study of the 'system' rather than the individual components? This emerging systems biology approach has been defined by Leroy Hood (http://www.systemsbiology.org/Scientistsandresearch) as 'the science of discovering, modelling, understanding and ultimately engineering at the molecular level the dynamic relationships between the biological molecules that define living organisms'. Such a systems biology approach can be applied to seed development and seed germination. It will require the construction of predictive mathematical models to describe the process and parallel experimental approaches to test the robustness of these models. With this preparation it should be possible to design better strategies to alter the behaviour of the system to meet required outcomes. We can now anticipate that rapid progress in the development and application of both existing and emerging methodologies will continue to drive significant and far-reaching advancements in our knowledge of seed development over the next few years.

Acknowledgements

We thank the UK Biotechnology and Biological Sciences Research Council (BBSRC) for continuing funding for W.M. Waterworth and for the support of ongoing research projects in C.M. Bray's laboratory.

References

Alonso, J.M., Stepanova, A.N., Leisse, T.J., Kim, C.J., Chen, H., Shinn, P., Stevenson, D.K., Zimmerman, J., Barajas, P., Cheuk, R., Gadrinab, C., Heller, C., Jeske, A., Koesema, E., Meyers, C.C., Parker, H., Prednis, L., Ansari, Y., Choy, N., Deen, H., Geralt, M., Hazari, N., Hom, E., Karnes, M., Mulholland, C., Ndubaku, R., Schmidt, I., Guzman, P., Aguilar-Henonin, L., Schmid, M., Weigel, D., Carter, D.E., Marchand, T., Risseeuw, E., Brogden, D., Zeko, A., Crosby, W.L., Berry, C.C. and Ecker, J.R. (2003) Genome-wide insertional mutagenesis of *Arabidopsis thaliana. Science* 301, 653–657.

Arabidopsis Genome Initiative (2000) Analysis of the genome sequence of the flowering plant *Arabidopsis thaliana. Nature* 408, 796–815.

Autran, D., Huanca-Mamani, W. and Vielle-Calzada, J.P. (2005) Genomic imprinting in plants: the epigenetic version of an Oedipus complex. *Current Opinion in Plant Biology* 8, 19–25.

Bewley, J.D. and Black, M. (1994) *Seeds: Physiology of Development and Germination*, 2nd edn. Plenum Press, New York.

Borisjuk, L., Walenta, S., Rolletschek, H., Mueller-Klieser, W., Wobus, U. and Weber, H. (2002) Spatial analysis of plant metabolism: sucrose imaging within *Vicia faba* cotyledons reveals specific developmental patterns. *The Plant Journal* 29, 521–530.

Borisjuk, L., Rolletschek, H., Walenta, S., Panitz, R., Wobus, U. and Weber, H. (2003) Energy status and its control on embryogenesis of legumes: ATP distribution within *Vicia faba* embryos is developmentally regulated and correlated with photosynthetic capacity. *The Plant Journal* 36, 318–329.

Choi, Y., Gehring, M., Johnson, L., Hannon, M., Harada, J.J., Goldberg, R.B., Jacobsen, S.E. and Fischer, R.L. (2002) DEMETER, a DNA glycosylase domain protein, is required for endosperm gene imprinting and seed viability in *Arabidopsis. Cell* 110, 33–42.

Finkelstein, R.R. and Gibson, S.I. (2002) ABA and sugar interactions regulating development: cross-talk or voices in a crowd? *Current Opinion in Plant Biology* 5, 26–32.

Finnie, C., Melchior, S., Roepstorff, P. and Svensson, B. (2002) Proteome analysis of grain filling and seed maturation in barley. *Plant Physiology* 129, 1308–1319.

Gallardo, K., Le Signor, C., Vandekerckhove, J., Thompson, R.D. and Burstin, J. (2003) Proteomics of *Medicago truncatula* seed development establishes the time frame of diverse metabolic processes related to reserve accumulation. *Plant Physiology* 133, 664–682.

Gibon, Y., Vigeolas, H., Tiessen, A., Geigenberger, P. and Stitt, M. (2002) Sensitive and high throughput metabolite assays for inorganic pyrophosphate, ADPGlc, nucleotide phosphates, and glycolytic intermediates based on a novel enzymic cycling system. *The Plant Journal* 30, 221–235.

Girke, T., Todd, J., Ruuska, S., White, J., Bening, C. and Ohlrogge, J. (2000) Microarray analysis of developing *Arabidopsis* seeds. *Plant Physiology* 124, 1570–1581.

Hajduch, M., Ganapathy, A., Stein, J.W. and Thelen, J.J. (2005) A systematic proteomic study of seed filling in soybean. Establishment of high-resolution two-dimensional reference maps, expression profiles, and an interactive proteome database. *Plant Physiology* 137, 1397–1419.

Hare, P.D., Seo, H.S., Yang, J.Y. and Chua, N.H. (2003) Modulation of sensitivity and selectivity in plant signaling by proteasomal destabilization. *Current Opinion in Plant Biology* 5, 453–462.

Hirai, M.Y., Klein, M., Fujikawa, Y., Yano, M., Goodenowe, D.B., Yamazaki, Y., Kanaya, S., Nakamura, Y., Kitayama, M., Suzuki, H., Sakurai, N., Shibata, D., Tokuhisa, J., Reichelt, M., Gershenzon, J., Papenbrock, J. and Saito, K. (2005) Elucidation of gene-to-gene and metabolite-to-gene networks in *Arabidopsis* by integration of metabolomics and transcriptomics. *Journal of Biological Chemistry* 280, 25590–25595.

Izui, K., Matsumura, H., Furumoto, T. and Kai, Y. (2004) Phosphoenolpyruvate carboxylase: a new era of structural biology. *Annual Review of Plant Biology* 55, 69–84.

Kinoshita, T., Miura, A., Choi, Y., Kinoshita, Y., Cao, X., Jacobsen, S.E., Fischer, R.L. and Kakutani, T. (2004) One-way control of FWA imprinting in *Arabidopsis* endosperm by DNA methylation. *Science* 303, 521–523.

Patton, W.F., Schulenberg, B. and Steinberg, T.H. (2002) Two-dimensional gel electrophoresis; better than a poke in the ICAT? *Current Opinion in Biotechnology* 13, 321–328.

Peck, S.C. (2005) Update on proteomics in *Arabidopsis*. Where do we go from here? *Plant Physiology* 138, 591–599.

Rensink, W.A. and Buell, C.R. (2005) Microarray expression profiling resources for plant genomics. *Trends in Plant Science* 10, 603–639.

Riera, M., Figueras, M., Lopez, C., Goday, A. and Pages, M. (2004) Protein kinase CK2 modulates developmental functions of the abscisic acid responsive protein Rab17 from maize. *Proceedings of the National Academy of Sciences of the United States of America* 29, 9879–9884.

Ruuska, S.A., Girke, T., Benning, C. and Ohlrogge, J.B. (2002) Contrapuntal networks of gene expression during *Arabidopsis* seed filling. *The Plant Cell* 14, 1191–1206.

Schwender, J. and Ohlrogge, J.B. (2002) Probing *in vivo* metabolism by stable isotope labeling of storage lipids and proteins in developing *Brassica napus* embryos. *Plant Physiology* 130, 347–361.

Schwender, J., Ohlrogge, J.B. and Shachar-Hill, Y. (2003) A flux model of glycolysis and the oxidative pentosephosphate pathway in developing *Brassica napus* embryos. *Journal of Biological Chemistry* 278, 29442–29453.

Sessions, A., Burke, E., Presting, G., Aux, G., McElver, J., Patton, D., Dietrich, B., Ho, P., Bacwaden, J., Ko, C., Clarke, J.D., Cotton, D., Bullis, D., Snell, J., Miguel, T., Hutchison, D., Kimmerly, B., Mitzel, T., Katagiri, F., Glazebrook, J., Law, M. and Goff, S.A. (2002) A high-throughput Arabidopsis reverse genetics system. *The Plant Cell* 14, 2985–2994.

Sreenivasulu, N., Altschmied, L., Radchuk, V., Gubatz, S., Wobus, U. and Weschke, W. (2004) Transcript profiles and deduced changes of metabolic pathways in maternal and filial tissues of developing barley grains. *The Plant Journal* 37, 539–553.

Stacey, M.G., Koh, S., Becke, J. and Stacey, G. (2002) AtOPT3, a member of the oligopeptide transporter family, is essential for embryo development in *Arabidopsis*. *The Plant Cell* 14, 2799–2811.

Turecek, F. (2002) Mass spectrometry in coupling with affinity capture-release and isotope-coded affinity tags for quantitative protein analysis. *Journal of Mass Spectrometry* 37, 1–14.

Waterworth, W.M., Ashley, M.K., West, C.E. and Bray, C.M. (2003) Peptide transport in the developing barley grain. In: Nicolas, G., Bradford, K.J., Come, D. and Pritchard, H.W. (eds) *The Biology of Seeds*. CAB International, Wallingford, UK, pp. 85–92.

Waterworth, W.M., Ashley, M.K., West, C.E., Sunderland, P.A. and Bray, C.M. (2005) A role for phosphorylation in the regulation of the barley scutellar peptide transporter HvPTR1 by amino acids. *Journal of Experimental Botany* 56, 1545–1552.

Weber, H., Borisjuk, L. and Wobus, U. (2005) Molecular physiology of legume seed development. *Annual Review of Plant Biology* 56, 253–279.

West, C.E., Waterworth, W.M., Stephens, S.M., Smith, C.P. and Bray, C.M. (1998) Cloning and functional expression of a peptide transporter expressed in the scutellum of barley grain during the early stages of germination. *The Plant Journal* 15, 221–229.

Zhu, T., Budworth, P., Chen, W., Provart, N., Chang, H.-S., Guimil, S., Su, W., Estes, B., Zou, G. and Wang, X. (2003) Transcriptional control of nutrient partitioning during rice grain filling. *Plant Biotechnological Journal* 1, 59–70.

11 Biogenesis of the Compound Seed Protein Storage Vacuole

Y.C. TSE, J. WANG, Y. MIAO AND L. JIANG

Department of Biology, The Chinese University of Hong Kong, Shatin, New Territories, Hong Kong, China

Abstract

Seeds accumulate and store proteins in protein storage vacuoles (PSVs) during their development and maturation. These stored proteins are then rapidly degraded to provide nutrients for the development of the seedlings during seed germination. PSVs in most seeds, including tobacco (*Nicotiana tabacum* L.), contain three morphologically distinct subcompartments: matrix, crystalloid and globoid. The first two are storage compartments, while the third represents a lytic compartment. The mechanisms by which these PSV subcompartments are formed and packed within a single membrane-bound PSV are unknown. Several marker proteins have been used to specifically define PSVs and their subcompartments. α- and δ-tonoplast intrinsic proteins (TIPs) are used as markers for the PSV tonoplast, while the dark intrinsic protein (DIP) and vacuolar pyrophosphatase (V-PPase) are used as markers for the crystalloid and globoid subcompartments, respectively. Recent studies have also demonstrated that the formation of the PSV and its subcompartments are regulated during seed development. PSV biogenesis during seed development is currently being studied using established markers as probes, and the tobacco seed as a model system. The long-term goal of this research is to elucidate the mechanism underlying PSV assembly during seed development and provide novel information on PSV biogenesis and function.

Introduction

All eukaryotic cells contain an endomembrane system that is comprised of several different distinct organelles involved in the secretory pathway. These organelles are often defined by their morphology and the specific proteins found on, or within, their membranes (Mellman and Warren, 2000; Jiang and Rogers, 2003). The endoplasmic reticulum (ER) and the Golgi apparatus are the main organelles that make up the endomembrane system. Soluble proteins, along with signal peptides, are first directed to the ER lumen where they are folded and given a three-dimensional (3D) conformation before they are exported to the Golgi apparatus. From this location, the proteins containing vacuolar sorting determinants are sorted by vacuolar sorting receptor (VSR) proteins prior to their delivery to vacuoles by prevacuolar compartments (PVCs) (Jiang and Rogers, 2003). In the default pathway, proteins lacking specific targeting information are packaged into poorly characterized vesicles by a presumably unselective process for their secretion to the cell exterior (Jiang and Rogers, 2003).

In contrast to mammalian and yeast cells, where only a single type of lysosome or vacuole exists, plant cells contain two functionally and biochemically distinct vacuoles: the protein storage vacuole (PSV) and the lytic vacuole (LV) (Okita and Rogers, 1996; Neuhaus and Rogers, 1998). LVs and lysosomes are common in eukaryotic cells, where they maintain an acidic pH and function as digestive compartments (Okita and Rogers, 1996). However, PSVs seem to be unique to plant cells, and in seeds the PSV is a compound organelle (Jiang *et al.*, 2000, 2001, 2002). Within the seed PSV, there are three distinct subcompartments: matrix, globoid and crystalloid. These subcompartments are clearly visible in PSVs from mature tobacco (*Nicotiana tabacum* L.) seeds (Fig. 11.1).

The three subcompartments are described as follows: (i) the matrix represents the 'proteinaceous material that is structurally amorphous'; (ii) the crystalloid represents the 'protein in a definite crystalline lattice arrangement'; and (iii) the globoid is an 'inclusion which is often spherical in shape' (Lott, 1980). The matrix and crystalloid subcompartments serve as storage compartments, while the globoid subcompartment contains salts of phytic acid (Weber and Neumann, 1980). Moreover, it has been assumed that these various components assembled themselves into the appropriate structures within the PSV (Lott, 1980). However, recent studies have demonstrated that the seed PSV is a compound organelle with the globoid subcompartment probably representing a membrane-bound lytic compartment within the PSV, while the crystalloid subcompartment also contains membrane proteins (Jiang *et al.*, 2002).

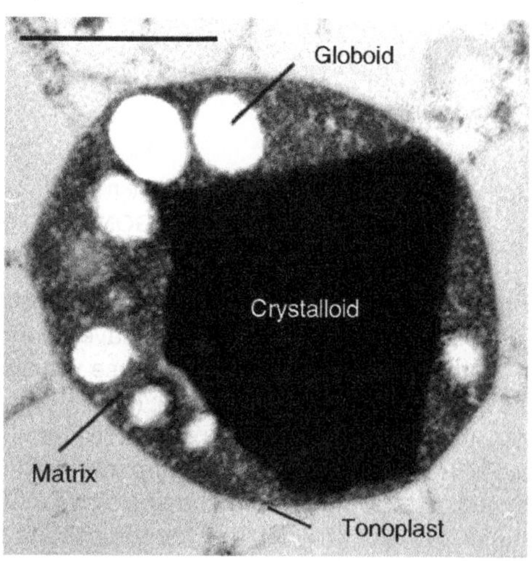

Fig. 11.1. The morphology of a protein storage vacuole (PSV) from a mature tobacco seed, with the tonoplast and three subcompartments indicated. Scale bar = 1 μm.

Vesicular Pathways for the Protein Storage Vacuole

Proteins reach the seed PSV and its subcompartments through distinct vesicular pathways (Fig. 11.2). In developing pea (*Pisum sativum* L.) cotyledons, the storage proteins vicilin and legumin are first transported from the Golgi apparatus to a putative PVC, which is 0.5 µm to several micrometres in diameter, by dense vesicles (DVs). DVs are transport vesicles ~100 nm in diameter, with a dense appearance and no visible cytoplasmic coat. The PVCs are then fused with the PSV for protein deposition (Fig. 11.2, route 1) (Robinson *et al.*, 1998). In developing pumpkin (*Cucurbita* cv Kurokawa Amakuri Nankin) cotyledons, storage proteins reach the PSVs through a Golgi apparatus-independent pathway mediated by ER-derived precursor-accumulating (PAC) vesicles that are 200–400 nm in diameter with an electron-dense core and electron-translucent outer layer (Fig. 11.2, route 2) (Hara-Nishimura *et al.*, 1998). Several other studies have also demonstrated that PAC vesicles receive proteins from the Golgi apparatus. Shimada *et al.* (1997, 2002) found that PV72 (a VSR protein) recruited pro2S albumin into PAC vesicles. Similarly, in developing tobacco seeds, cytosolic DIP organelles (0.5–1 µm in diameter) may serve as PVCs for both ER-derived and Golgi apparatus-derived proteins reaching PSVs (Jiang *et al.*, 2000). Hydrolytic enzymes may also reach PSVs through a

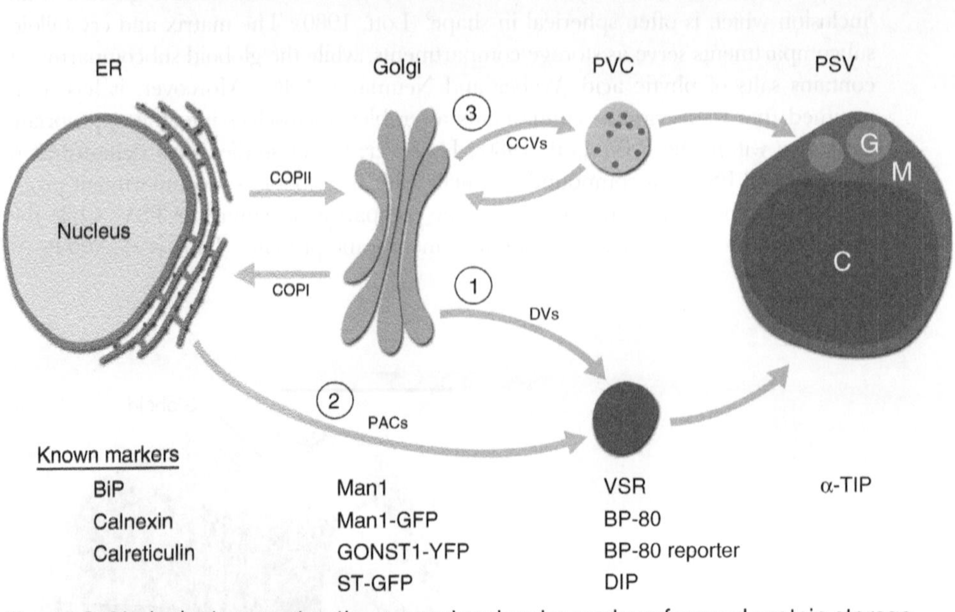

Fig. 11.2. Vesicular transport pathways and molecular markers for seed protein storage vacuoles (PSVs). Three pathways have been proposed for protein traffic into seed PSVs: (i) proteins reach PSVs through dense vesicles (DVs) from the Golgi apparatus to a prevacuolar compartment (PVC); (ii) precursor-accumulating (PAC) vesicles mediate protein transport directly from the endoplasmic reticulum (ER) to PSVs, and hence PACs may serve as PVCs; and (iii) an ER–Golgi apparatus–PVC–PSV globoid subcompartment pathway that is mediated by clathrin-coated vesicles (CCVs), which may represent a lytic pathway. Known markers and/or proteins that can be used to define individual organelles within the pathway are also included.

Golgi apparatus-dependent pathway mediated by clathrin-coated vesicles (CCVs) and VSR proteins (Fig. 11.2, route 3). Subcellular fractionation and immunocytochemical electron microscopy (ImmunoEM) studies have demonstrated that highly purified CCVs from pea cotyledons contain large amounts of BP-80, a VSR that transports proteases to LVs, but few storage proteins (Hohl *et al.*, 1996; Robinson *et al.*, 1998). These studies demonstrate that proteins are transported to PSVs and their subcompartments by multiple pathways, including lytic PVCs and the PVCs of PSVs. However, the molecular mechanisms by which PSVs and their subcompartments are formed during seed development remain unclear.

Markers for the Protein Storage Vacuole and Its Subcompartments

Several proteins have been used as markers in order to define the seed PSV and its subcompartments. In mature tobacco seeds, the PSV tonoplast is defined by the presence of both α- and δ-TIPs (Jiang *et al.*, 2000), while the globoid subcompartment membrane is marked by γ-TIP and V-PPase (Jiang *et al.*, 2000, 2001). In addition, DIP characterizes the crystalloid subcompartment, whereas the matrix subcompartment contains the storage proteins, chitinase and glucanase.

Two reporter proteins are localized to distinct compartments in transgenic tobacco seeds. The BP-80 reporter protein, which is transported from the ER to the Golgi apparatus and then to lytic PVCs (Jiang and Rogers, 1998), is co-localized with the γ-TIP-marked globoid subcompartment in mature tobacco seeds (Jiang *et al.*, 2001). In contrast, the α-TIP CT reporter protein, which is transported directly from the ER to a PSV-like compartment (Jiang and Rogers, 1998), is localized in the cytosolic DIP organelles in developing seeds, but was found in the crystalloid subcompartment of mature seeds (Jiang *et al.*, 2000). During the early stages of tobacco seed development, cytosolic PVCs are mainly localized outside the α-TIP-marked PSVs (Fig. 11.3a). However, in the later stages of seed development, these PVCs are taken up into the PSVs and assembled into the crystalloid subcompartment (Fig. 11.3b and c). Therefore, the cytosolic DIP organelle may represent the PVC of the PSV crystalloid subcompartment.

Current and Future Research

We are interested in understanding the molecular mechanisms by which the PSV subcompartments are assembled within a single membrane-bound PSV during seed development. Previous research indicates that the formation of the PSV and its subcompartments is developmentally regulated during seed development. However, the PSV and its subcompartments have yet to be defined morphologically and biochemically. Here we hypothesize that distinct cytosolic PVCs in developing seeds are taken up into, or are fused with, PSVs and assembled into PSV subcompartments upon seed maturation (Fig. 11.4a and b). We further hypothesize that the PVC for the lytic pathway, which is assembled into the PSV globoid subcompartment, is marked by both γ-TIP and the BP-80 reporter protein, while the PVC for

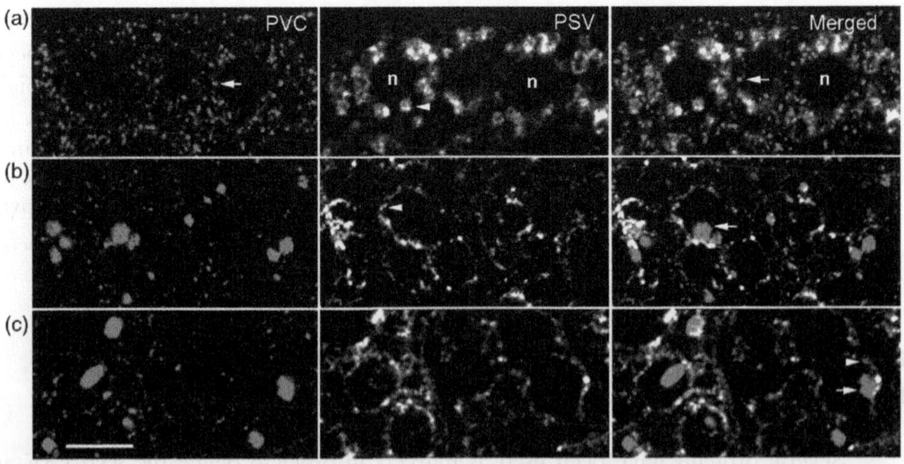

Fig. 11.3. The uptake of prevacuolar compartments (PVCs) into protein storage vacuoles (PSVs) during tobacco seed development. PVCs (indicated by arrows) are marked by DIP antibodies, while the PSV tonoplasts (indicated by arrowheads) are detected by anti-α-TIP antibodies. (a, b, c) Samples collected from different developmental stages of tobacco seeds. (a) During early stages, PVCs are mainly localized in the cytosol, outside the PSVs. (b, c) In the later stages, DIP-labelled contents are taken up into the PSVs for assembly into crystalloid subcompartments as aggregates. n = nucleus. Scale bar = 10 μm.

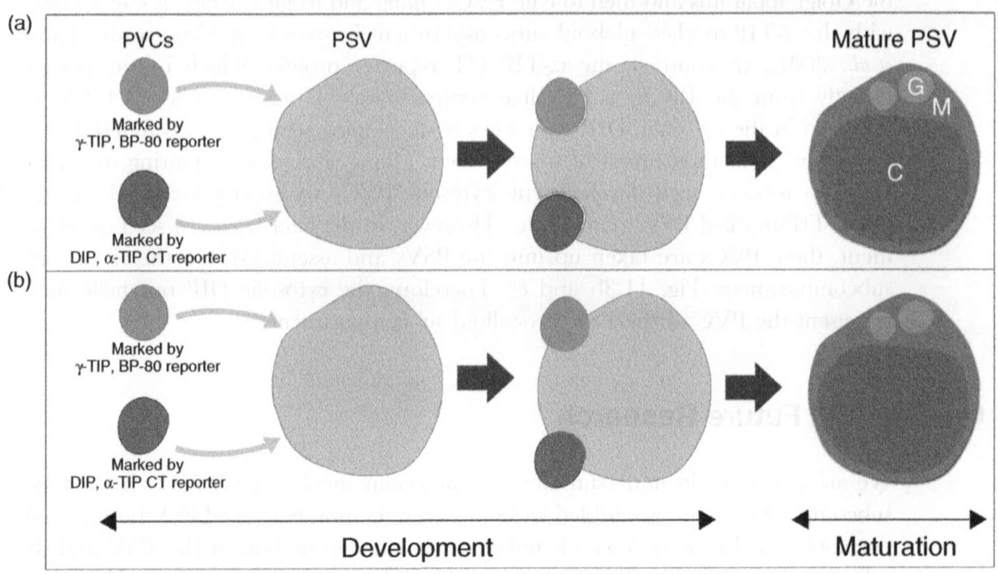

Fig. 11.4. Biogenesis of protein storage vacuoles (PSVs) and their subcompartments during seed development. Two working models are shown, which are currently being tested using transgenic tobacco plants expressing two reporter proteins. The BP-80 reporter protein marks the lytic prevacuolar compartments (PVCs) outside the PSVs of developing seeds, which will be taken up into the PSVs and assembled into globoid subcompartments in maturing seeds. The α-TIP reporter marks the PVCs outside the PSVs, which will either be (a) taken up into or (b) fused with the PSVs and assembled into crystalloid subcompartments upon seed maturation.

the PSV pathway, which is assembled into the PSV crystalloid subcompartment, is marked by both DIP and the α-TIP CT reporter protein (Fig. 11.4a and b).

As a first step to test these hypotheses, transgenic tobacco plants have recently been generated that express either the GFP-BP-80 reporter protein or the GFP-α-TIP CT reporter protein. These green fluorescent proteins (GFPs), which are fusion reporter proteins, will enable the uptake of PVCs into PSVs to be directly observed in the living cells of the transgenic plants. The morphological changes in the PSV and its subcompartments at different stages of seed development are also being studied using structural transmission electron microscopy (TEM). Putative PSVs and their subcompartments can be identified during the early stages of seed development (Fig. 11.5a and b), while typical PSVs with visible subcompartments can be readily identified in later stages of seed development (Fig. 11.5c and d).

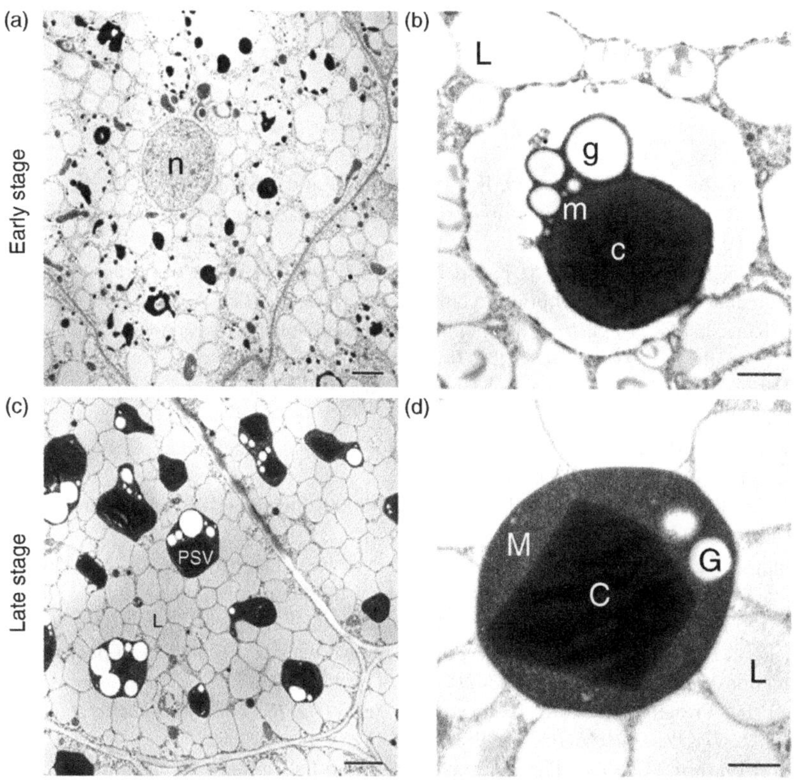

Fig. 11.5. Morphological changes in protein storage vacuoles (PSVs) during tobacco seed development. Tobacco seeds at early and later stages of development were collected and prepared for structural TEM study. (a, b) Images derived from seeds at an early stage of development, where the putative PSV and its subcompartments are indicated (g = globoid; c = crystalloid; m = matrix). (c, d) Sections from seeds at a later stage of development, where typical PSVs and their subcompartments are indicated (G = globoid; C = crystalloid; M = matrix). n = nucleus; L = lipid body. Scale bar = 2 μm in (a) and (c), and 500 nm in (b) and (d).

With these tools in hand, several questions will be raised using other experimental approaches such as ImmunoEM and confocal immunofluorescence: What are the membrane origins of PSVs and their subcompartments during the early development of seeds? Which vesicles are responsible for transporting proteins to PSVs and their subcompartments? How are the subcompartments assembled within the PSV?

This continuing research not only addresses the fundamental questions of cell biology and organelle biogenesis, but also provides a basis for future applications in plant biotechnology, and particularly the use of seeds as bioreactors for the production of pharmaceutical recombinant proteins (Jiang and Sun, 2002).

Acknowledgements

This work is supported by the Research Grants Council of Hong Kong (CUHK4156/01M, CUHK4260/02M, CUHK4307/03M and CUHK4580/05M) and a UGC Area of Excellence grant to L. Jiang.

References

Hara-Nishimura, I., Shimada, T., Hatano, K., Takeuchi, Y. and Nishimura, M. (1998) Transport of storage proteins to protein storage vacuoles is mediated by large precursor-accumulating vesicles. *The Plant Cell* 10, 825–836.

Hohl, I., Robinson, D.G., Chrispeels, M.J. and Hinz, G. (1996) Transport of storage proteins to the vacuole is mediated by vesicles without a clathrin coat. *Journal of Cell Science* 109, 2539–2550.

Jiang, L. and Rogers, J.C. (1998) Integral membrane protein sorting to vacuoles in plant cells: evidence for two pathways. *The Journal of Cell Biology* 143, 1183–1199.

Jiang, L. and Rogers, J.C. (2003) Sorting of lytic enzymes in the plant Golgi apparatus. *Annual Plant Review* 9, 114–140.

Jiang, L. and Sun, S.S.M. (2002) Membrane anchors for vacuolar targeting: application in plant bioreactors. *Trends in Biotechnology* 20, 99–102.

Jiang, L., Phillips, T.E., Rogers, S.W. and Rogers, J.C. (2000) Biogenesis of the protein storage vacuole crystalloid. *The Journal of Cell Biology* 150, 755–770.

Jiang, L., Phillips, T.E., Hamm, C.A., Drozdowicz, Y.M., Rea, P.A., Maeshima, M., Rogers, S.W. and Rogers, J.C. (2001) The protein storage vacuole: a unique compound organelle. *The Journal of Cell Biology* 155, 991–1002.

Jiang, L., Erickson, A. and Rogers, J. (2002) Multivesicular bodies: a mechanism to package lytic and storage functions in one organelle? *Trends in Cell Biology* 12, 362–367.

Lott, J.N.A. (1980) Protein bodies. In: Tolbert, N.E. (ed.) *The Biochemistry of Plants: A Comprehensive Treatise*, Vol. 1, *The Plant Cell.* Academic Press, New York, pp. 589–623.

Mellman, I. and Warren, G. (2000) The road taken: past and future foundations of membrane traffic. *Cell* 100, 99–112.

Neuhaus, J.M. and Rogers, J.C. (1998) Sorting of proteins to vacuoles in plant cells. *Plant Molecular Biology* 38, 127–144.

Okita, T.W. and Rogers, J.C. (1996) Compartmentation of proteins in the endomembrane system of plant cells. *Annual Review of Plant Physiology and Plant Molecular Biology* 47, 327–350.

Robinson, D.G., Hinz, G. and Holstein, S.E. (1998) The molecular characterization of transport vesicles. *Plant Molecular Biology* 38, 49–76.

Shimada, T., Kuroyanagi, M., Nishimura, M. and Hara-Nishimura, I. (1997) A pumpkin 72-kDa membrane protein of precursor-accumulating vesicles has characteristics of a vacuolar sorting receptor. *Plant Cell Physiology* 38, 1414–1420.

Shimada, T., Watanabe, E., Tamura, K., Hayashi, Y., Nishimura, M. and Hara-Nishimura, I. (2002) A vacuolar sorting receptor PV72 on the membrane of vesicles that accumulate precursors of seed storage proteins (PAC vesicles). *Plant Cell Physiology* 43, 1086–1095.

Weber, E. and Neumann, D. (1980) Protein bodies, storage organelles in plant seeds. *Biochemie und Physiologie der Pflanzen* 175, 279–306.

12 Embryo Development and Time of Cutting in Cool Temperate Carrot Seed Crops

A. Geard, C.J. Spurr and P.H. Brown

Tasmanian Institute of Agricultural Research, School of Agricultural Science, University of Tasmania, Private Bag 98 Hobart, Tasmania 7001, Australia

Abstract

A survey of carrot (*Daucus carota* L.) seed production in south-eastern Australia between 2002 and 2004 identified the occurrence of seeds with rudimentary embryos as a major cause of poor germination. Relationships among embryo size, cutting time, seed quality and yield were studied in Nantes, Amsterdam and Kuroda cultivars grown in Tasmania during the 2004 season. Under field conditions, maximum seed size was attained 40–60 days after full bloom, but embryo development continued up to 90 days after full bloom. In five commercial seed lines, >85% germination was recorded for seeds harvested between 55 and 87 days from full bloom, after the onset of yield losses through shattering in some lines. Seed chlorophyll content was a more reliable indicator of seed maturity than seed moisture content or thermal time from peak bloom. The data obtained in this study are consistent with the theory that the occurrence of rudimentary embryos in carrot seed lots is linked to the time of cutting and indicate that the current commercial practice of harvesting at the onset of seed shatter may be inappropriate for optimum seed quality in cool temperate climates. Hence, there appears to be potential to develop a reliable indicator of carrot seed maturity based on seed chlorophyll content.

Introduction

Poor seed germination is a common problem in plants belonging to the family *Apiaceae*, including carrot (*Daucus carota* L.) (Robinson, 1954; Dean *et al.*, 1989). In the USA and Australia, feeding of insects of the *Lygus* and *Nysius* genera on the embryos of developing carrot seeds is a significant cause of loss of viability (Flemion, 1949; Spurr, 2003). Implementation of control strategies for these insects has significantly improved seed quality, but some seed lots still fail to meet the industry standard of 85% germination under International Seed Testing Association (ISTA) testing conditions (ISTA, 1999). One explanation for these crop failures is the occurrence of seeds with small, underdeveloped (i.e. rudimentary) embryos (Borthwick, 1931a; Dean *et al.*, 1989; Spurr, 2003).

The causes of formation of rudimentary embryos in carrot seed are not completely understood. Evidence suggests that the underlying problem relates to the developmental pattern of carrot seed in which the endosperm grows first, with most embryo growth occurring later in seed maturation (Borthwick, 1931b; Gray *et al.*, 1984). Given the asynchronous flowering pattern of the carrot plant and once-over harvesting strategy employed by carrot seed producers (Rubatzky *et al.*, 1999), seed immaturity may be a factor that causes poor germination of commercial carrot seed. The germination percentage of carrot seed from higher-order (i.e. later-flowering) umbels is often less than that of the seed from lower-order (i.e. earlier-flowering) umbels when all seeds are harvested at the same time (Borthwick, 1931a; Gray, 1979; Jacobsohn and Globerson, 1980; Steiner *et al.*, 1990). Gray (1979) found that where differences in germination between umbel orders existed, there was a positive correlation between germination percentage and embryo size. Later cutting times have been shown to improve the germination percentage of seed from higher-order umbels in some studies (Gray and Steckel, 1982), but not in others (Gray, 1979; Sandin, 1980). Sandin (1980) and Hawthorn *et al.* (1961) found that even with late cutting, germination remained below 80%.

Determination of cutting time in Australian carrot seed crops is largely based on recommendations from northern hemisphere studies of time elapsed from peak bloom (50–79 days from first flowering depending on climate and cultivar) or appearance of the seed heads (Hawthorn *et al.*, 1961; Gray and Steckel, 1983; Gray *et al.*, 1984; Steckel *et al.*, 1989). Often these studies have been conducted on plants grown in polythene tunnels (Gray and Steckel, 1983; Gray *et al.*, 1984; Steckel *et al.*, 1989), or at field sites with hot summertime conditions (Hawthorn *et al.*, 1961). Much of the carrot seed production in Australia is field-based in cool temperate climates, so the recommendations of these earlier studies may not apply. This work was undertaken to establish whether the occurrence of rudimentary embryos in carrot seed is linked to time of cutting, and to identify appropriate indicators for determination of time of cutting of carrot seed crops grown under south-eastern Australian conditions.

Materials and Methods

Plant material

Carrot seed lines studied in this work were open-pollinated and hybrid Nantes, Amsterdam and Kuroda types. All seed lines were grown under field conditions during the 2004 season as commercial crops or trial plot isolations near Cambridge (42.5°S, 147.3°E), Bushy Park (42.4°S, 146.5°E) and Whitemore (41.3°S, 146.5°E) in Tasmania, Australia. Current management practices for seed to seed production were used. Temperature conditions during the study and long-term averages are shown for each site in Table 12.1.

Table 12.1. Season 2003/04 and long-term temperature data for the Bushy Park, Cambridge and Whitemore field sites during carrot flowering and seed maturation (i.e. December to March). Long-term averages are for a minimum of 47 years. (From Australian Bureau of Meteorology.)

		Mean maximum temperature (°C)		Mean minimum temperature (°C)	
Location	Month	2003/04	Long term	2003/04	Long term
Bushy Park	December	24.0	21.9	8.8	9.1
	January	22.5	23.7	9.1	10.0
	February	23.6	23.8	8.8	10.0
	March	21.4	21.7	7.2	8.5
Cambridge	December	22.4	20.5	11.9	10.6
	January	21.7	22.4	11.1	11.9
	February	21.9	22.3	11.5	12.0
	March	20.6	20.6	10.3	10.7
Whitemore	December	23.0	21.2	9.3	7.5
	January	21.7	23.2	9.0	8.2
	February	22.2	23.1	10.5	8.5
	March	21.0	21.0	7.0	7.2

Examination of the relationship between embryo size and germination

The relationship between embryo size and seed germination was examined in 21 trial-and-commercial seed lines produced in Tasmania and South Australia between 2002 and 2004, using the methodology of Gray and Steckel (1983). Fifty seeds from each line were soaked in formalin acetic alcohol (FAA) (50% ethanol, 6.5% formalin and 2.5% glacial acetic acid) for 24 h. After soaking, each seed was cut to remove the caruncle and the embryo extruded by applying pressure to the back of the seed with a scalpel blade. Individual embryo lengths were recorded under a dissecting microscope. Three replicates of 100 seeds from each line were tested for germination according to ISTA guidelines (ISTA, 1999).

Embryo growth patterns

Embryo growth patterns were studied in the hybrid crosses 30 (Nantes × Amsterdam) and 68,145 (Nantes × Imperator) grown near Cambridge. On 12 January 2004, 250 male sterile plants in peak bloom were tagged in each hybrid cross. At 3-day intervals from 24 days after peak bloom, the seeds of the primary and fourth secondary umbels of ten randomly selected plants were harvested. The embryos of a sample of 40 seeds from each umbel order were extracted and measured for length using the above method. Dry weight data were obtained for a sample of 100 seeds from each umbel order using the low constant temperature method (ISTA, 1999).

Time of harvest studies

The relationships between time of harvest and seed quality and yield were examined in four hybrid carrot seed lines, i.e. 30, 68,145 and 963 (Nantes) grown near Cambridge and line I (Kuroda) grown at Whitemore and an open-pollinated line, Nantes 2 (Nantes) grown at Bushy Park. Samples were collected from individual lines 3–6 times at an interval of 7–12 days. Sampling dates were chosen so that the commercial cutting occurred in the middle of the sampling period. At each site, the time of cutting treatments was arranged in a randomized complete block design with four replicates. Each replicate consisted of 20 consecutive plants from a row. The umbels of 18 plants from each plot were removed, separated into primary, secondary and higher-order classes and dried in a forced air cabinet at 25°C for 2 weeks. These were threshed in a modified garden mulcher (Stihl, Virginia Beach, USA). The seed was debearded using a laboratory thresher (Wintersteiger, Salt Lake City, USA) and cleaned using air-screen (Blount Agri-Industrial Corporation, Indiana, USA) and density separators (Seedburo, Chicago, USA). A total of 100 carrot seeds from each sample were germinated according to ISTA guidelines (ISTA, 1999). Seed yields and 100 seed weights were determined on a dry weight basis using the low constant temperature method (ISTA, 1999).

Evaluation of seed maturity indicators

The relationship between germination percentage and three seed maturity indicators (i.e. thermal time from peak bloom, seed moisture content and seed chlorophyll content) was examined using samples collected from seven carrot seed lines. These included the five lines described previously plus the open-pollinated lines, Amsterdam 2 (Amsterdam) and Nantes OP (Nantes). Temperature data collected at canopy level at each trial site were used in the calculation of thermal time from peak bloom for each sample date using the equation: thermal time (day degrees) $= \Sigma((T_{min}°C + T_{max}°C)/2) - 10$. Representative seed samples from the remaining two plants collected from each time of harvest plot were used to determine seed moisture content and seed chlorophyll content. Seed moisture content was measured using the low constant temperature method (ISTA, 1999). Seed samples weighing 2 g were immersed in 20 ml solutions of 90% methanol for 24 h in darkness. The methanol extract was decanted and absorbance (A) measured at 652 and 665 nm in a spectrophotometer (Shimadzu UV-160, Melbourne, Australia). Total chlorophyll concentration was calculated for each sample using the equation: total chlorophyll (µg/ml) $= 0.28A_{665} + 27.64A_{652}$ (Litchtenthaler, 1987).

Statistical analysis

Statistical analysis was performed with SPSS 11.5 (SPSS Inc., Chicago, USA). The relationship between embryo size and evaluation of maturity indicators was examined using regression analysis. The effects of harvest time on seed yield and quality were analysed using ANOVA for randomized complete block designs. Significantly different means were determined at the 5% level by Fisher's least significant difference (LSD).

Results and Discussion

Examination of the relationship between embryo size and germination

In 21 trial-and-commercial carrot seed lots from three seasons in south-eastern Australia, germination percentage under ISTA testing conditions (ISTA, 1999) was closely correlated ($P < 0.001$, $r^2 = 0.84$) to the proportion of seeds with embryos >0.7 mm (Fig. 12.1). For embryos <0.7 mm in length, the cotyledons and radicle were incompletely differentiated. The occurrence of rudimentary embryos may therefore be the main reason for poor germination of carrot seeds grown in Australia since the adoption of management strategies for *Nysius* insects (Spurr, 2003).

Embryo growth patterns

Embryo growth patterns were examined in seeds of the primary and secondary umbel orders, as these typically account for >90% of total yield (Hawthorn *et al.*, 1961; Hiller and Kelly, 1985). In lines 30 and 68,145, the growth of embryos from the primary and fourth secondary umbel orders continued until 84–90 days after peak bloom (Fig. 12.2). There was a 7- to 12-day lag in development of embryos from the secondary umbel seed compared with the embryos of primary umbel seed, but these ultimately reached the same maximum size (i.e. 2.0 mm in line 30 and 1.8 mm in line 68,145). Compared with previous reports of maximum embryo size at 60–75 days after first flowering (Gray *et al.*, 1984; Steckel *et al.*, 1989), the importance of later cutting dates for carrot seed crops grown in cool temperate conditions to avoid production of seeds with underdeveloped embryos was highlighted.

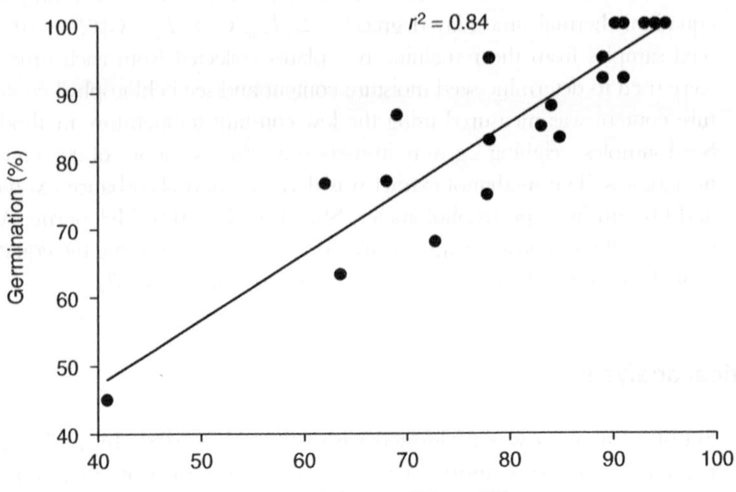

Fig. 12.1. The relationship between the percentage of seeds with embryos >0.7 mm in length and germination percentage of carrot seed lots grown in South Australia and Tasmania between 2002 and 2004. The relationship is significant ($P < 0.001$).

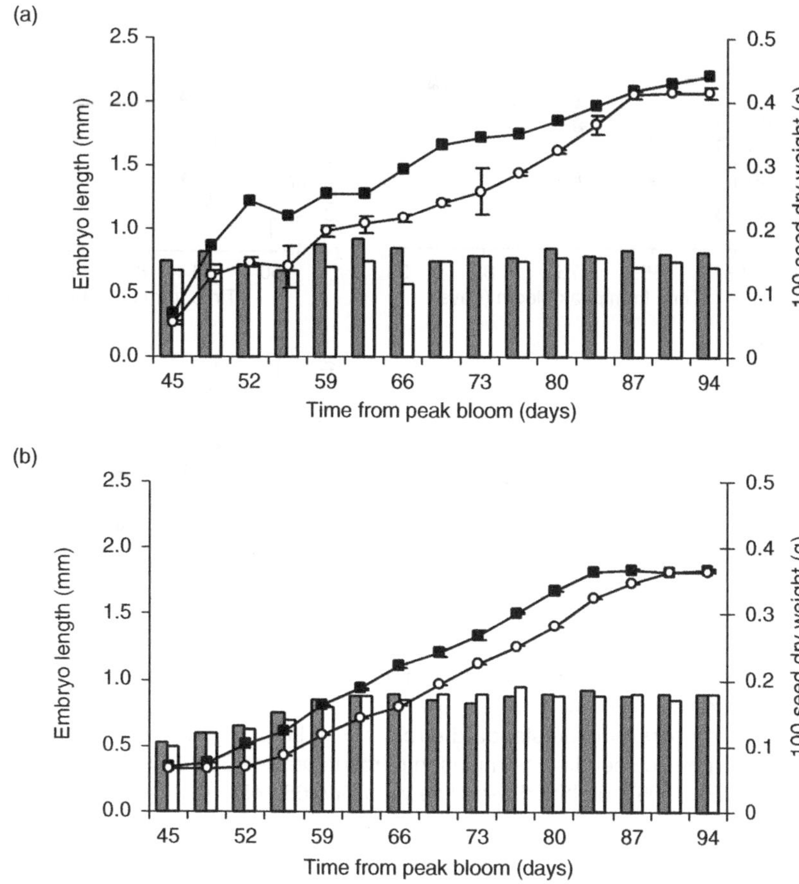

Fig. 12.2. The mean lengths of embryos from the primary (■) and fourth secondary (o) umbels during seed maturation for (a) line 68,145 and (b) line 30. Changes in dry weight of seeds from the primary (■) and fourth secondary (□) umbels are shown in the columns. Error bars indicate standard errors, $n = 4$.

Time of harvest studies

For each of the five lines studied, germination percentages exceeding 85% were recorded when crops were cut later than 55–87 days from peak bloom (Figs 12.3 and 12.4). Delaying harvest time to maximize seed quality must be balanced against potential yield losses due to shattering of mature umbels. In line 68,145 a significant ($P < 0.001$) yield loss of 21% was recorded prior to achieving 85% germination. Similar non-significant trends of yield loss were observed in other lines (Fig. 12.3). Under cool temperate conditions, yield losses may be inevitable in certain crops in order to achieve an acceptable germination percentage.

Evaluation of seed maturity indicators

While most carrot seed producers judge time of cutting on the basis of time from peak bloom or onset of seed shattering, this study indicates the need for a more reliable

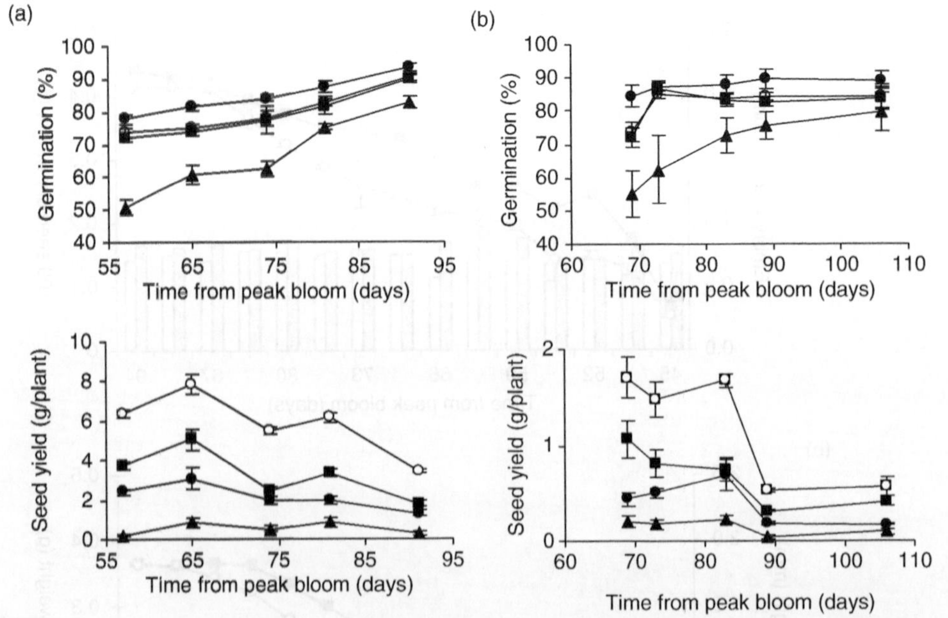

Fig. 12.3. The effect of time of cutting on mean germination percentage and seed yield from the primary (•), secondary (■) and tertiary umbel orders (▲) of (a) line 68,145 and (b) line 30. The mean germination percentage and yield of the combined seed of all umbel orders is also shown (o). Error bars indicate standard errors, $n = 4$. Least significant difference (LSD) values ($P < 0.05$) combined seed line germination percentage: line 68,145 = 1.2%; line 30 = 6.7%. LSD values ($P < 0.05$) combined seed line yield: line 68,145 = 0.11 g; line 30 = 0.48 g.

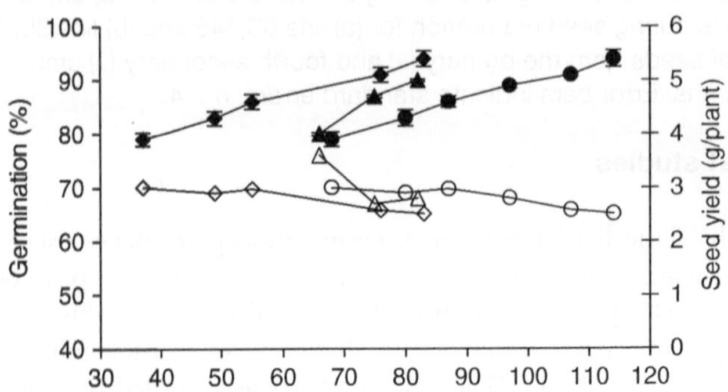

Fig. 12.4. The effect of time of cutting on mean seed germination (filled symbols) and yield (open symbols) on seed of lines 963 (♦), (•) and 'Nantes' 2 (▲). Error bars for germination data represent standard errors, $n = 4$. Least significant difference (LSD) ($P < 0.05$): line 963 = 6.1%; line I = 5.2%; 'Nantes' 2 = 4.8%. The effects of cutting time on seed yield were not significant in any of the lines.

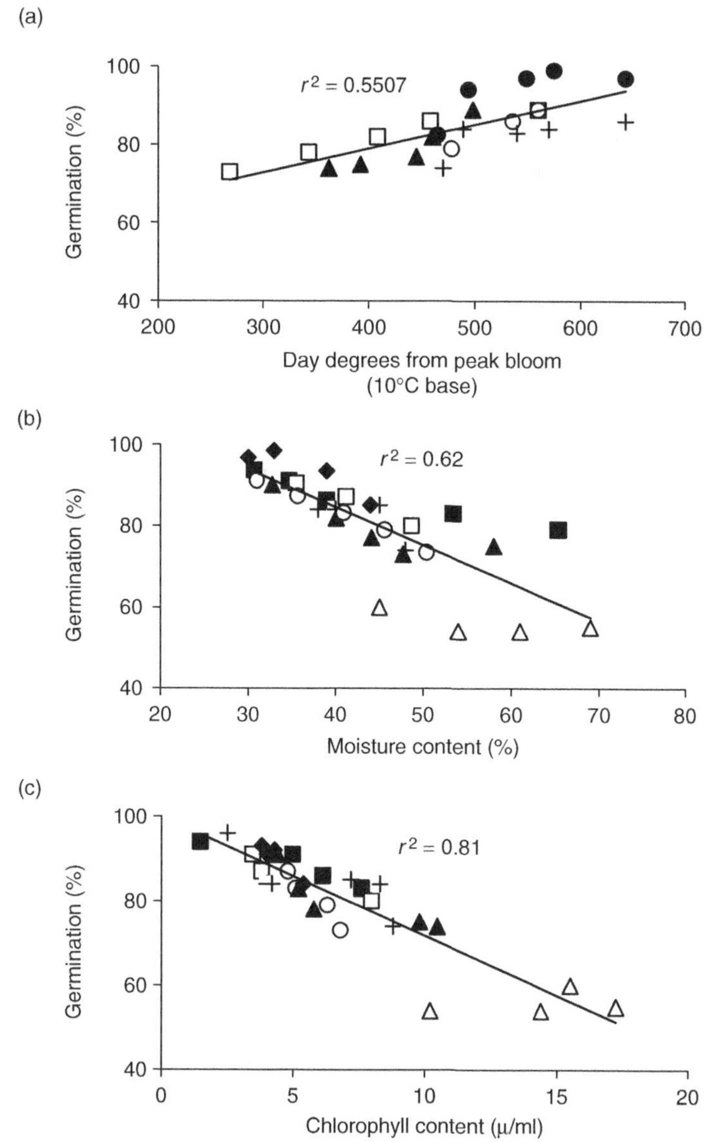

Fig. 12.5. The relationship between the mean germination percentage for carrot seed lots harvested in Tasmania in 2004 and the maturity indicators thermal time from (a) peak bloom (10°C base temperature), (b) seed moisture content and (c) seed chlorophyll content. The relationships between all maturity indicators and seed germination were significant ($P < 0.001$). The graphs show r^2 values. Samples of different maturity from the same line are denoted by a common symbol: + = line 30; ♦ = line 963; o = line 68,145; ▲ = Amsterdam 2; □ = Nantes 2; Δ = Nantes OP; and ■ = line I.

method. Of the indicators evaluated (Fig. 12.5), seed chlorophyll content provided a more accurate estimate of seed quality ($P < 0.001$, $r^2 = 0.81$) than either seed moisture content ($P < 0.001$, $r^2 = 0.62$) or thermal time from peak bloom ($P < 0.001$, $r^2 = 0.54$). Germination percentages >85% were obtained for samples cut when the seed had a chlorophyll content <7 µg/ml (Fig. 12.5). In a previous study, carrot seed lots with chlorophyll content <6 µg/ml had the highest germination percentage (Steckel *et al.*, 1989). The agreement of these data-sets suggests that a reliable and simple carrot seed maturity test based on chlorophyll content could be developed.

Acknowledgements

The authors gratefully acknowledge the financial and practical support provided by Horticulture Australia Limited, Rijk Zwaan, South Pacific Seeds Australia and Bejo Seeds.

References

Borthwick, H.A. (1931a) Carrot seed germination. *Proceedings of the American Society for Horticultural Science* 28, 310–314.

Borthwick, H.A. (1931b) Development of the macrogametophtye and embryo of *Daucus carota*. *Botanical Gazette* 92, 23–44.

Dean, B.B., Noland, T. and Maguire, J.D. (1989) Correlation of low seed quality with growing environment of carrot. *HortScience* 24, 247–249.

Flemion, F. (1949) *Lygus* bugs in relation to the occurrence of embryoless seeds in the Umbelliferae. *Science* 109, 364–365.

Gray, D. (1979) The germination response to temperature of carrot seeds from different umbels and times of harvest of the seed crop. *Seed Science and Technology* 7, 169–178.

Gray, D. and Steckel, J.R.A. (1982) The effects of ripening and drying of carrot seed crops on germination. *Annals of Applied Biology* 103, 327–334.

Gray, D. and Steckel, J.R.A. (1983) Seed quality in carrots: the effects of seed crop plant density, harvest date and seed grading on seed and seedling variability. *Journal of Horticultural Science* 58, 393–401.

Gray, D., Ward, J.A. and Steckel, J.R.A. (1984) Endosperm and embryo development in *Daucus carota* L. *Journal of Experimental Botany* 35, 459–465.

Hawthorn, L.R., Toole, E.H. and Toole, V.K. (1961) Yield and viability of carrot seeds as affected by position of umbel and time of harvest. *Journal of the American Society for Horticultural Science* 80, 401–407.

Hiller, L.K. and Kelley, W.C. (1985) *Daucus carota*. In: Halevy, A.H. (ed.) *CRC Handbook of Flowering*. CRC Press, Boca Raton, Florida, pp. 419–429.

ISTA (1999) International rules for seed testing. *Seed Science and Technology* 27, supplement.

Jacobsohn, R. and Globerson, D. (1980) *Daucus carota* seed quality: II. The importance of the primary umbel in carrot seed production. In: Hebbelthwaite, P.D. (ed.) *Seed Production*. Butterworths, Boston, pp. 643–646.

Litchtenthaler, H. (1987) Chlorophylls and carotenoids: pigments of photosynthetic biomembranes. *Methods in Enzymology* 148, 350–382.

Robinson, R. (1954) Seed germination problems in the Umbelliferae. *The Botanical Review* 9, 531–550.

Rubatzky, V.E., Quiros, C.F. and Simon, P.W. (1999) *Carrots and Related Vegetable Umbelliferae*. CAB International, Wallingford, UK.

Sandin, N.H. (1980) Optimum harvest time for *Daucus carota* crops in Sweden. In: Hebbelthwaite, P.D. (ed.) *Seed Production*. Butterworths, Boston, pp. 553–559.

Spurr, C.J. (2003) Factors limiting hybrid carrot seed production in Australia. PhD thesis, University of Tasmania, Australia.

Steckel, J.R.A., Gray, D. and Rowse, H.R. (1989) Relationship between indices of seed maturity and carrot seed quality. *Annals of Applied Biology* 114, 177–183.

Steiner, J.J., Hutchmatcher, R.B., Mantel, A.B., Ayres, J.E. and Vail, S.S. (1990) Response of seed carrot to various water regimes. II. Reproductive development, seed yield and seed quality. *Journal of the American Society of Horticultural Science* 115, 722–727.

13 Seed Biotechnology: Translating Promise into Practice

K.J. Bradford

Department of Plant Sciences, Seed Biotechnology Center, One Shields Avenue, University of California, Davis, CA 95616-8780, USA

Abstract

In recent years, biotechnology has come to be synonymous with the use of recombinant DNA techniques to introduce genes into plant or animal genomes. Biotechnology has been applied to modify the composition of seeds as foods or raw products to improve their nutritional, flavour, postharvest or processing characteristics. Genetic changes to improve the production of seeds or their performance as propagules can also be achieved using biotechnology. Proof of concept has been demonstrated for many applications of seed biotechnology. However, the only biotechnology products that have achieved wide commercialization are targeted towards insect and weed management, rather than towards enhancement of the biological or nutritional aspects of seeds. The application of biotechnology to enhance seed production, performance and composition has thus far been stymied in practice by the regulatory requirements and economic risk associated with introducing such products into the commercial marketplace.

Introduction

More than 81 million ha of biotechnology crops developed by utilizing recombinant DNA techniques were grown around the world in 2004 (James, 2004). All of these had herbicide resistance, insect resistance or both traits combined. These 'input traits' that target crop production methods are highly valuable to the grower, allowing reduced cultivation and pesticide use, but leave the product essentially unchanged. A few 'output traits', or those that alter the end product qualities of crops, have been marketed, including the slower softening Flavr Savr™ tomato (*Lycopersicon esculentum* Mill.), rape (*Brassica napus* L.) with modified oil properties and carnations (*Dianthus caryophyllus* L.) engineered to produce blue flowers, but these are no longer sold and/or have little market impact (Clark *et al.*, 2004). Output traits that improve the nutritional quality of seeds, such as 'Golden Rice'

(*Oryza sativa* L.) containing high beta-carotene content (Paine *et al.*, 2005), seeds with improved protein quality (Mandal and Mandal, 2000; Gibbon and Larkins, 2005), and soybean (*Glycine max* (L.) Merr.) and rape seeds accumulating omega-3-fatty acids (Ursin, 2003), are currently under development. Other modifications to alter seed starch (Morell and Myers, 2005) or cotton (*Gossypium hirsutum* L.) fibre properties (Wilkins and Arpat, 2005) are also being pursued.

Seed Biotechnology: the Promise

The targets of seed biotechnology utilize seeds as a delivery system for agronomic traits or as a food, raw product or fibre commodity. However, input traits associated directly with seed production or seed performance as propagules have also been identified. Transferable and controllable male sterility would be valuable for hybrid seed production. Prevention of seed shattering at maturity could improve both seed quality and yield in many species. Apomixis, or the production of seeds carrying only the maternal genotype, would allow the asexual propagation of elite genotypes through seeds. Seed dormancy, whether too low as in the case of preharvest sprouting or too high as in the case of stand failures due to lingering dormancy, can cause problems for seed producers and farmers. Higher vigour, greater stress tolerance and longer storage life are seed characteristics that both seed producers and farmers would desire. What is the situation with respect to enhancement of these seed-specific traits using biotechnology?

Traits affecting seed production

Male sterility

Although not a seed trait per se, the ability to control male fertility is important in the production of hybrid seeds. A system using an anther-specific promoter to control expression of a bacterial RNAase (barnase) to disrupt pollen development (Mariani *et al.*, 1990) is used commercially for hybrid seed production in maize (*Zea mays* L.) and rape. Fertility can be restored in the F1 generation by introducing an inhibitor of barnase (barstar) under the control of the same promoter through the male parent (Mariani *et al.*, 1992). Although effective, this genic male sterility has the disadvantage of segregating during maintenance of the female parent (Perez-Prat and van Lookeren Campagne, 2002). A recent report of engineering cytoplasmic male sterility by expression of β-ketothiolase in chloroplasts provides an alternative strategy for developing male sterile parents for hybrid seed production (Ruiz and Daniell, 2005).

Seed shattering

In the wild, most plants shed their seeds at maturity to promote seed dispersal. The loss of this 'seed-shattering' trait is closely associated with the domestication of many crops, including cereals, beans and sunflowers (*Helianthus annuus* L.), which retain their seeds on the mother plant until harvested. However, many crops that have been more recently domesticated, such as rape, or that are grown primarily for their vegetative organs, e.g. cabbage (*Brassica oleracea* L.), onions (*Allium cepa* L.) and carrots (*Daucus carota* L.), exhibit seed shattering at maturity. In addition, these species also exhibit prolonged or indeterminate flowering patterns, making it necessary to harvest the seed crop at a time when seeds are at a range of maturity stages (Still and Bradford, 1998). This results in heterogeneity in seed quality, as less mature seeds generally have lower vigour, less stress tolerance and shorter storage lives (Zanakis *et al.*, 1994). Genes have been identified in *Arabidopsis* (*Arabidopsis thaliana* (L.) Heynh.) that prevent seed shattering. Mutation of two *SHATTERPROOF* (*SHP*) genes or overexpression of the *FRUITFULL* (*FUL*) gene blocks differentiation of the abscission zone in the siliqua, preventing seed shedding (Ferrándiz *et al.*, 2000; Liljegren *et al.*, 2000). Even more subtle is the effect of mutation of the *ALCATRAZ* (*ALC*) gene, which only prevents separation of a single layer of lignified cells in the abscission zone of the siliqua (Rajani and Sundaresan, 2001).

Apomixis

Apomixis is the development of a seed from maternal tissue without fertilization, which occurs naturally in some species, including dandelion (*Taraxacum officinale* Weber), citrus and some grasses (Van Dijk and Van Damme, 2000). Thus, the genotype of the progeny is identical to that of the mother plant. If apomixis could be introduced and controlled in crop plants, it would allow the genetic fixation of heterosis and eliminate the need to recreate F1 hybrid seeds in each generation. Apomixis is being intensively studied and some genes influencing it are known, but current understanding is still insufficient to develop transgenic approaches (Bicknell and Koltunow, 2004).

Controlling trait expression

Apomixis illustrates a major limitation of current genetic engineering approaches. As it bypasses sexual reproduction, it also prevents genetic improvement through breeding. Thus, a successful system to induce apomixis will also require a method to control the generation in which the trait is expressed. For example, sexual crosses and breeding would be used for developing parent lines, and then apomixis would be induced only in the F1 generation to allow further asexual propagation of hybrid plants through seeds. A system to control trait expression in particular generations has been developed (Oliver *et al.*, 1998). The system uses an inducible bacterial recombinase to allow introduced genes to be silent until treated with a specific inducer. The recombinase then joins a promoter with the coding sequence of the target gene, allowing the gene to be expressed at the normal time that the promoter is active. In one application, the inducer would be given to seeds, which would then germinate and produce a normal crop, but expression of a lethal gene later in seed development would prevent further propagation of the patented seeds. This system was labelled 'terminator' by groups opposed to its use as a technology protection system.

Although the appropriateness of this application has been debated from various viewpoints, many applications of controllable gene expression system would be extremely valuable for eventual applications in seed biotechnology. An obvious example is the *shrunken2* (*sh2*) mutation in ADP-Glc pyrophosphorylase that greatly reduces the starch content (from 65% to 25%) and increases the sucrose content (from 1.4% to 16%) of maize grains (Hannah, 1997), which is highly desirable for sweet maize. However, the mutation results in shrunken grains that have low vigour and are highly susceptible to pathogens. If a normal *SH2* allele was expressed during seed production to result in normal grains, and then silenced during subsequent production of the sweet maize crop, then both high seed quality and desirable product quality could be obtained. A similar situation would occur if seeds were utilized to produce altered starches, oils or other products that might compromise the ability of the seed to perform as a propagule. Male sterility and fertility restoration could similarly be controlled by a system in which the expression of a trait could be restricted to a specific generation. Controllable gene expression/suppression systems will be important components for full utilization of seed biotechnology.

Traits affecting seed quality

Seed size

Seed size is a major yield component and therefore has received considerable attention from plant breeders. It is also associated, to some degree, with seed vigour, so increasing seed size would be likely to have beneficial effects in terms of seed quality. In most agronomic crops, there is a trade-off between seed size and number, making it difficult to increase overall yield by only increasing seed size (Egli, 1998). However, recent work in *Arabidopsis* has indicated that suppression of the *APETALA2* gene could significantly increase seed mass and that the effect was not entirely attributable to smaller seed number (Ohto *et al.*, 2005). If confirmed by similar work in crop plants, this approach might also result in higher seed vigour.

Seed dormancy

Progress is being made in identifying quantitative trait loci (QTLs) associated with seed dormancy in *Arabidopsis* (Alonso-Blanco *et al.*, 2003; Clerkx *et al.*, 2004b), cereals (Li *et al.*, 2004; Gu *et al.*, 2005), lettuce (*Lactuca sativa* L.) (Argyris *et al.*, 2005), sunflower (Gandhi *et al.*, 2005) and other species. In at least one case, a specific gene is isolated that is responsible for a dormancy locus, but its function remains unknown (Koornneef *et al.*, 2004). A number of abscisic acid (ABA)-related and gibberellin (GA)-related genes are known to affect seed dormancy and germination (Finkelstein *et al.*, 2002; Koornneef *et al.*, 2002; Gubler *et al.*, 2005). For example, dormancy breakage during moist chilling of *Arabidopsis* seeds is associated with increased expression of a GA-3 oxidase enzyme (AtGA3ox1) catalysing the formation of bioactive GA_4 and suppression of expression of a GA-2 oxidase (AtGA2ox2) that inactivates active GAs (Yamauchi *et al.*, 2004). In rice, a seed-specific GA-20 oxidase gene involved in GA synthesis was associated with a QTL controlling dormancy (Li *et al.*, 2004). Genes involved in catabolism of ABA can also be activated during dormancy breakage and germination (Ali-Rachedi *et al.*, 2004; Gonai *et al.*, 2004; Nambara and Marion-Poll, 2005). The expression of dozens to hundreds of genes can be altered by dormancy (Bove *et al.*, 2005; Pateman *et al.*, 2005; Toorop *et al.*, 2005), so finding key controlling genes, and particularly those that are specific to seeds, will be important in applying biotechnology to alter seed dormancy behaviour.

Seed vigour

Seed vigour is an imprecise term that encompasses a range of seed quality traits. In general, vigour relates to stress tolerance, germination speed, resistance to pathogens and related traits that allow the seed to successfully develop into a plant under a wide range of conditions. Genetic studies in several species have identified a number of QTLs that contribute to seed vigour (Foolad *et al.*, 1999; Bettey *et al.*, 2000; Cui *et al.*, 2002; Clerkx, 2004b; Finch-Savage *et al.*, 2005). As it is a complex quantitative trait, it may be difficult to find one or a few genes that will have a dramatic impact on seed vigour. However, development of molecular markers associated with seed vigour should allow breeders to select for improved seed quality along with other desirable traits.

Seed longevity

Like seed vigour, seed longevity is a complex trait strongly affected by environment. None the less, QTL analyses have identified loci associated with seed storability or resistance to controlled deterioration conditions (Bentsink *et al.*, 2000; Miura *et al.*, 2002; Clerkx *et al.*, 2004a,b). These studies have not supported expected associations of seed longevity with the accumulation of oligosaccharides or with mechanisms for detoxifying reactive oxygen species. A proteomic approach that allows the identification of oxidized (carbonylated) proteins in aged seeds may provide new leads on early targets of seed deterioration (Job *et al.*, 2005).

Seed Biotechnology: the Practice

It is evident from this brief review that there already are, or are likely to be, numerous input and output traits that could be introduced using genetic engineering that are valuable for seed production, quality, composition and yield. However, regulatory, marketing and consumer acceptance issues are delaying or preventing utilization of recombinant DNA techniques specifically for seed improvement. For crops that have already introduced biotechnology varieties into the marketplace, it will be relatively straightforward to add these traits as they are developed and validated. However, for crops in which no biotechnology traits have been marketed, it is unlikely that seed production or quality traits alone will justify commercialization. Current regulatory systems distinguish between traits developed using recombinant DNA and sexual crosses or other types of genetic modifications, with the former requiring extensive safety and environmental data and regulatory approvals from multiple agencies prior to commercial release (Nap *et al.*, 2003). The expense of these additional requirements makes it economical to utilize biotechnology approaches only in crops grown on a large scale, such as maize, soybean, cotton and rape (Alston, 2004; Redenbaugh and McHughen, 2004). The global marketing of new biotechnology products is further complicated by the diversity of international regulations governing biotechnology traits, often requiring segregation of biotechnology and non-biotechnology commodities, product testing and other expenses. With few exceptions, traits primarily targeted towards seed production or quality do not represent sufficient potential economic gain to justify the costs of development and regulatory approval or to risk consumer rejection of the commodity product. Thus, commercial deployment of most seed production or quality traits in additional species are likely to be delayed until other traits with greater market potential are released.

There continues to be strenuous international debate about the relative benefits and risks of genetically engineered crops, even as increasing global adoption demonstrates that growers are finding value in them (Qaim and Zilberman, 2003; James, 2004). Agreements such as the Cartagena Protocol on Biosafety to the Convention on Biological Diversity threaten the creation of considerable additional hurdles for international trade in biotechnology crops (De Greef, 2004). A proper risk–benefit analysis should compare biotechnology crops with those developed using 'traditional' breeding techniques, which have long included wide (interspecies and even intergeneric) crosses, mutagenesis, protoplast fusion, polyploidy and other manipulations of crop genomes (Conner *et al.*, 2003; Nap *et al.*, 2003). A sensible approach would be to regulate biotechnology traits in proportion to their potential risk and in light of experience with both traditional and biotechnology methods of genetic improvement (Bradford *et al.*, 2005). Unfortunately, this approach is not widely utilized and many hurdles remain for commercializing biotechnology crops. These challenges for translating promise into practice are even greater for applications of seed biotechnology that do not have direct consumer benefits, but that could have major implications for agriculture. The Public Research and Regulation Foundation has been established to provide an avenue for input from public researchers into the international biotechnology regulatory process (www.pubresreg.org). Seed biotechnology researchers who hope to see the results of their efforts translated into agricultural practice are encouraged to become engaged in this process so that sensible regulatory protocols can guide the safe application of seed biotechnology for agricultural and consumer benefit.

References

Ali-Rachedi, S., Bouinot, D., Wagner, M.-H., Bonnet, M., Sotta, B., Grappin, P. and Jullien, J. (2004) Changes in endogenous abscisic acid levels during dormancy release and maintenance of mature seeds: studies with the Cape Verde Islands ecotype, the dormant model of *Arabidopsis thaliana*. *Planta* 219, 479–488.

Alonso-Blanco, C., Bentsink, L., Hanhart, C.J., Blankestijn-de Vries, H. and Koornneef, M. (2003) Analysis of natural allelic variation at seed dormancy loci of *Arabidopsis thaliana*. *Genetics* 164, 711–729.

Alston, J.M. (2004) Horticultural biotechnology faces significant economic and market barriers. *California Agriculture* 58, 80–88.

Argyris, J., Truco, M.J., Ochoa, O., Knapp, S.J., Still, D.W., Lenssen, G.M., Schut, J.W., Michelmore, R.W. and Bradford, K.J. (2005) Quantitative trait loci associated with seed and seedling traits in *Lactuca*. *Theoretical and Applied Genetics* 111, 1365–1376.

Bentsink, L., Alonso-Blanco, C., Vreugdenhil, D., Tesnier, K., Groot, S.P.C. and Koornneef, M. (2000) Genetic analysis of seed-soluble oligosaccharides in relation to seed storability of *Arabidopsis*. *Plant Physiology* 124, 1595–1604.

Bettey, M., Finch-Savage, W.E., King, G.J. and Lynn, J.R. (2000) Quantitative genetic analysis of seed vigour and pre-emergence seedling growth traits in *Brassica oleracea*. *New Phytologist* 148, 277–286.

Bicknell, R.A. and Koltunow, A.M. (2004) Understanding apomixis: recent advances and remaining conundrums. *The Plant Cell* 16, S228–S245.

Bove, J., Lucas, P., Godin, B., Ogé, L., Jullien, M. and Grappin, P. (2005) Gene expression analysis by cDNA-AFLP highlights a set of new signaling networks and translational control during seed dormancy breaking in *Nicotiana plumbaginifolia*. *Plant Molecular Biology* 57, 593–612.

Bradford, K.J., Van Deynze, A., Gutterson, N., Parrott, W. and Strauss, S.H. (2005) Regulating transgenic crops sensibly: lessons from plant breeding, biotechnology and genomics. *Nature Biotechnology* 23, 439–444.

Clark, D., Klee, H. and Dandekar, A. (2004) Despite benefits, commercialization of transgenic horticultural crops lags. *California Agriculture* 58, 89–98.

Clerkx, E.J.M., Blankestijn-De Vries, H., Ruys, G.J., Groot, S.P.C. and Koornneef, M. (2004a) Genetic differences in seed longevity of various *Arabidopsis* mutants. *Physiologia Plantarum* 121, 448–461.

Clerkx, E.J.M., El-Lithy, M.E., Vierling, E., Gerda, R.J., Blankestijn-De Vries, H., Groot, S.P.C., Vreugdenhil, D. and Koornneef, M. (2004b) Analysis of natural allelic variation of *Arabidopsis* seed germination and seed longevity traits between the accessions Landsberg *erecta* and Shakdara, using a new recombinant inbred line population. *Plant Physiology* 135, 432–443.

Conner, A.J., Glare, T.R. and Nap, J.-P. (2003) The release of genetically modified crops into the environment. Part 2. Overview of ecological risk assessment. *The Plant Journal* 33, 19–46.

Cui, K.H., Peng, S.B., Xing, Y.Z., Xu, C.G., Yu, S.B. and Zhang, Q. (2002) Molecular dissection of seedling vigor and associated physiological traits in rice. *Theoretical and Applied Genetics* 105, 745–753.

De Greef, W. (2004) The Cartegena Protocol and the future of agbiotech. *Nature Biotechnology* 22, 811–812.

Egli, D.B. (1998) *Seed Biology and the Yield of Grain Crops.* CAB International, New York.

Ferrándiz, C., Liljegren, S.J. and Yanofsky, M.F. (2000) Negative regulation of the *SHATTERPROOF* genes by *FRUITFULL* during *Arabidopsis* fruit development. *Science* 289, 436–438.

Finch-Savage, W.E., Côme, D., Lynn, J.R. and Corbineau, F. (2005) Sensitivity of *Brassica oleracea* seed germination to hypoxia: a QTL analysis. *Plant Science* 169, 753–759.

Finkelstein, R.R., Gampala, S.S.L. and Rock, C.D. (2002) Abscisic acid signaling in seeds and seedlings. *The Plant Cell* 14, S15–S45.

Foolad, M.R., Lin, G.Y. and Chen, F.Q. (1999) Comparison of QTLs for seed germination under non-stress, cold stress and salt stress in tomato. *Plant Breeding* 118, 167–173.

Gandhi, S., Heesacker, A., Freeman, C., Argyris, J., Bradford, K.J. and Knapp, S.J. (2005) The self-incompatibility locus (*S*) and quantitative trait loci for self-pollination and seed dormancy in sunflower. *Theoretical and Applied Genetics* 111, 619–629.

Gibbon, B.C. and Larkins, B.A. (2005) Molecular genetic approaches to developing quality protein maize. *Trends in Genetics* 21, 227–233.

Gonai, T., Kawahara, S., Tougou, M., Satoh, S., Hashiba, T., Hirai, N., Kawaide, H., Kamiya, Y. and Yoshioka, T. (2004) Abscisic acid in the thermoinhibition of lettuce seed germination and enhancement of its catabolism by gibberellin. *Journal of Experimental Botany* 55, 111–118.

Gu, X.-Y., Kianian, S.F., Hareland, G.A., Hoffer, B.L. and Foley, M.E. (2005) Genetic analysis of adaptive syndromes interrelated with seed dormancy in weedy rice (*Oryza sativa*). *Theoretical and Applied Genetics* 110, 1108–1118.

Gubler, F., Millar, A.A. and Jacobsen, J.V. (2005) Dormancy release, ABA and pre-harvest sprouting. *Current Opinion in Plant Biology* 8, 183–187.

Hannah, L.C. (1997) Starch synthesis in the maize seed. In: Larkins, B.A. and Vasil, I.K. (eds) *Cellular and Molecular Biology of Plant Seed Development.* Kluwer Academic Publishers, Dordrecht, The Netherlands, pp. 375–405.

James, C. (2004) Preview: Global Status of Commercialized Biotech/GM Crops: 2004. ISAAA Briefs No. 32. International Service for the Acquisition of Agri-biotech Applications. Available at: www.isaaa.org

Job, C., Rajjou, M., Lovigny, Y., Belghazi, M. and Job, D. (2005) Patterns of protein oxidation in *Arabidopsis* seeds and during germination. *Plant Physiology* 138, 790–802.

Koornneef, M., Bentsink, L. and Hilhorst, H. (2002) Seed dormancy and germination. *Current Opinion in Plant Biology* 5, 33–36.

Koornneef, M., Bentsink, L., Clerkx, E. and Alonso-Blanco, C. (2004) The genetic analysis of seed dormancy in *Arabidopsis*. In: *Third International Symposium on Plant Dormancy*. Wageningen University, Wageningen, The Netherlands, p. 43.

Li, C., Ni, P., Francki, M., Hunter, A., Zhang, Y., Schibeci, D., Li, H., Tarr, A., Wang, J., Cakir, M., Yu, J., Bellgard, M., Lance, R. and Appels, R. (2004) Genes controlling seed dormancy and pre-harvest sprouting in a rice–wheat–barley comparison. *Functional and Integrative Genomics* 4, 84–93.

Liljegren, S.J., Ditta, G.S., Eshed, Y., Savidge, B., Bowman, J.L. and Yanofsky, M.F. (2000) *SHATTERPROOF* MADS-box genes control seed dispersal in *Arabidopsis*. *Nature* 404, 766–770.

Mandal, S. and Mandal, R.K. (2000) Seed storage proteins and approaches for improvement of their nutritional quality by genetic engineering. *Current Science* 79, 576–589.

Mariani, C., De Beuckeleer, M., Truettner, J., Leemans, J. and Goldberg, R.B. (1990) Induction of male sterility in plants by a chimaeric ribonuclease gene. *Nature* 347, 737–741.

Mariani, C., Gossele, V., De Beuckeleer, M., De Block, M., Goldberg, R.B., De Greef, W. and Leemans, J. (1992) A chimaeric ribonuclease-inhibitor gene restores fertility to male sterile plants. *Nature* 357, 384–387.

Miura, K., Lin, S.Y., Yano, M. and Nagamine, T. (2002) Mapping quantitative trait loci controlling seed longevity in rice (*Oryza sativa* L.). *Theoretical and Applied Genetics* 104, 981–986.

Morell, M.K. and Myers, A.M. (2005) Towards the rational design of cereal starches. *Current Opinion in Plant Biology* 8, 204–210.

Nambara, E. and Marion-Poll, A. (2005) Abscisic acid biosynthesis and catabolism. *Annual Review of Plant Biology* 56, 165–185.

Nap, J.-P., Metz, P.L.J., Escaler, M. and Conner, A.J. (2003) The release of genetically modified crops into the environment. Part 1. Overview of current status and regulations. *The Plant Journal* 33, 1–18.

Ohto, M., Fischer, R.L., Goldberg, R.B., Nakamura, K. and Harada, J.L. (2005) Control of seed mass by *APETALA2*. *Proceedings of the National Academy of Sciences of the United States of America* 102, 3123–3128.

Oliver, M.J., Quisenberry, J.E., Trolinder, N.L.G. and Keim, D.L. (1998) Control of plant gene expression. US Patent No. 5,723,765.

Paine, J.A., Shipton, C.A., Chaggar, S., Howells, R.M., Kennedy, M.J., Vernon, G., Wright, S.Y., Hinchliffe, E., Adams, J.L., Silverstone, A.L. and Drake, R. (2005) Improving the nutritional content of Golden Rice through increased pro-vitamin A content. *Nature Biotechnology* 23, 482–487.

Pateman, C., Toorop, P.E., Hilhorst, H.W.M. and Finch-Savage, W.E. (2005) A microarray analysis of seed dormancy in *Arabidopsis*. In: *Abstracts, Eighth International Seeds Workshop*. Brisbane, Australia, p. 35.

Perez-Prat, E. and van Lookeren Campagne, M.M. (2002) Hybrid seed production and the challenge of propagating male sterile plants. *Trends in Plant Science* 7, 199–203.

Qaim, M. and Zilberman, D. (2003) Yield effects of genetically modified crops in developing countries. *Science* 299, 900–902.

Rajani, S. and Sundaresan, V. (2001) The *Arabidopsis* myc/bHLH gene *ALCATRAZ* enables cell separation in fruit dehiscence. *Current Biology* 11, 1914–1922.

Redenbaugh, K. and McHughen, A. (2004) Regulatory challenges reduce opportunities for horticultural biotechnology. *California Agriculture* 58, 106–115.

Ruiz, O.N. and Daniell, H. (2005) Engineering cytoplasmic male sterility via the chloroplast genome by expression of β-ketothiolase. *Plant Physiology* 138, 1232–1246.

Still, D.W. and Bradford, K.J. (1998) Using hydrotime and ABA-time models to quantify seed quality of brassicas during development. *Journal of the American Society for Horticultural Science* 123, 692–699.

Toorop, P.E., Marroco, R.M., Engler, G., Groot, S.P.C. and Hilhorst, H.W.M. (2005) Differentially expressed genes associated with dormancy or germination of *Arabidopsis thaliana* seeds. *Planta* 221, 637–647.

Ursin, V.M. (2003) Modification of plant lipids for human health: development of functional land-based omega-3 fatty acids. *Journal of Nutrition* 133, 4271–4274.

Van Dijk, P. and Van Damme, J. (2000) Apomixis technology and the paradox of sex. *Trends in Plant Science*, 5, 81–84.

Wilkins, T.A. and Arpat, A.B. (2005) The cotton fiber transcriptome. *Physiologia Plantarum* 124, 295–300.

Yamauchi, Y., Ogawa, M., Kuwahara, A., Hanada, A., Kamiya, Y. and Yamaguchi, S. (2004) Activation of gibberellin biosynthesis and response pathways by low temperature during imbibition of *Arabidopsis thaliana* seeds. *The Plant Cell* 16, 367–378.

Zanakis, G.N., Ellis, R.H. and Summerfield, R.J. (1994) Seed quality in relation to seed development and maturation in three genotypes of soybean (*Glycine max*). *Experimental Agriculture* 30, 139–156.

14 Stress-inducible Gene Expression and its Impact on Seed and Plant Performance: a Microarray Approach

W. Ligterink,[1] J. Kodde,[1] M. Lammers,[1] H. Dassen,[1] A.H.M. van der Geest,[2] R.A. de Maagd[2] and H.W.M. Hilhorst[1]

[1]*Laboratory of Plant Physiology, Wageningen University and Research Centre, Arboretumlaan 4, 6703 BD Wageningen, The Netherlands;* [2]*Plant Research International, Bornsesteeg 65, 6708 PD Wageningen, The Netherlands*

Abstract

In addition to the improvement of germination characteristics, such as germination rate and uniformity, osmotic treatment of seeds has a positive effect on the stress tolerance. This research aimed to gain evidence that exposure of seeds to (abiotic) stresses may lead to cross-tolerance for these and other stresses that are experienced during germination and seedling growth. In order to identify the genes that are involved in this phenomenon, tomato (*Lycopersicon esculentum* Mill.) microarrays were used. The expression profiles of treated vs untreated dry seeds were compared. Furthermore, differences in expression profiles between treated and untreated seeds during germination were analysed. Following these results, genes with an expression pattern that could indicate a possible role in the imposition of cross-tolerance will be further analysed for their precise role. The practical implications of this study will be twofold: (i) development of new or modified methods for the improvement of seed quality; and (ii) identification of molecular markers that are specific for stress resistance.

Introduction

As sessile organisms, plants must cope with a number of adverse environmental conditions including abiotic stresses, such as extreme temperatures and drought, and biotic challenges from a wide array of plant pathogens, such as viruses, bacteria and fungi. In order to withstand these environmental challenges plants have developed several sensing mechanisms. The cellular responses that are activated by these sensing mechanisms are remarkably similar and very often resistance to one type

 Seeds: Biology, Development and Ecology (eds S. Adkins, S. Ashmore and S.C. Navie)

of stress is accompanied by resistance to other stresses. This phenomenon is called cross-tolerance (Bowler and Fluhr, 2000). For instance, the exposure of tobacco (*Nicotiana tabacum* L.) cells to aluminium enhances their tolerance not only to aluminium but also to hydrogen peroxide (H_2O_2) and Fe^{2+} and Cu^{2+} (Devi *et al.*, 2003). Furthermore, the upregulation of a disease resistance gene in *Arabidopsis* (*Arabidopsis thaliana* (L.) Heynh.) was shown not only to convey broad-spectrum disease resistance but also to confer drought tolerance (Chini *et al.*, 2004). It is obvious that this phenomenon of cross-tolerance has extremely important consequences for agriculture.

There is good evidence that cross-tolerance is also working in seeds, and in fact the principle is used in the enhancement of seed and seedling quality by the seed industry. This treatment is known as seed-priming. Priming-induced cross-tolerance has already been shown in several systems. For example, treatment of seeds with polyethylene glycol (PEG), to develop osmopriming, has been shown to increase the tolerance of seeds to artificial ageing (Dearman *et al.*, 1986), tolerance to elevated temperatures (Kretschmer, 1982) and tolerance to salt (Cayuela *et al.*, 1996).

The aim of this research was to investigate whether exposure of seeds to (abiotic) stresses may lead to cross-tolerance for these and other stresses that are experienced during germination and seedling growth. A further aim was to identify those genes that are upregulated during the exposure of seeds to stress and that confer stress tolerance to a broad array of biotic and abiotic stresses to the seed, seedling and beyond. The practical implications of this study are likely to be twofold: (i) the development of new or modified methods for the improvement of seed quality; and (ii) the identification of molecular markers that are specific for stress resistance.

Materials and Methods

Germination test

Germination tests were carried out using three or four replicates of 50 tomato (*Lycopersicon esculentum* Mill.) seeds (cultivar GT, a TMV-resistant cultivar, derived from Moneyberg). Seeds were considered germinated when radicles emerged. Parameters related to germination, such as germination rate (t_{50}) and uniformity (u_{7525}), were calculated with the software package SEED CALCULATOR 3.0 (Plant Research International BV, Wageningen, The Netherlands).

Seed treatments

Osmopriming was performed by placing 0.5 g of tomato seeds in a 50 ml tube with 15 ml of −1.0 MPa PEG 8000. The tubes were closed with air-permeable polyethylene (PE) foil and placed on a roller bench for 7 days. After this treatment the seeds were washed with running tap water for 1 min and dried for 3 days at 32% RH at 20°C.

Priming with the seed conditioner (SC) consists of two steps: an imbibition phase followed by 6 days incubation in the SC. For the imbibition phase, 0.5 g of tomato seeds was incubated, as during osmopriming, for 1 day. After this treatment the seeds were washed with running tap water for 1 min, excess water was removed by centrifuging at 450*g* for 4 min and the seeds were then placed in the SC for 6 days. After incubation in the SC the seeds were dried for 3 days at 32% RH at 20°C.

cDNA Labelling and Microarray Hybridizations

Total RNA was prepared using the hot phenol/LiCl method (Verwoerd *et al.*, 1989). For labelling, 40 mg of total RNA was fluorescently labelled by reverse transcription in the presence of 5-(3-aminoallyl)-dUTP (Sigma-Aldrich, St. Louis, USA) using oligo dT as a primer (Isogen Bioscience). After precipitation of the RNA/DNA hybrids and degradation of the RNA, either Cy3 or Cy5 monofunctional dyes (Amersham, Freiburg, Germany) were linked to the amino-modified nucleotides in a separate coupling step, creating fluorescently labelled cDNA. Cy5-labelled target was mixed with Cy3-labelled reference (or vice versa) prior to hybridization.

The Tom1 array, from the Centre for Gene Expression at the Boyce Thompson Institute (http://bti.cornell.edu/CGEP/CGEP.html, CGEP, Boyce Thompson Institute, Ithaca, New York), was used for the microarray hybridizations. This array contains 13,400 printed elements that correspond to ~8700 unigenes derived from different developmental stages and treatments (Alba *et al.*, 2004). The processed slide was prehybridized in a large volume of hybridization buffer (50% formamide, 5× Denhardt's reagent, 5× saline sodium citrate (SSC), 0.2% sodium dodecyl sulphate (SDS), 0.1 mg/ml denatured fish DNA) for 2 h at 42°C to reduce background. Subsequently, the slides were hybridized with the denatured probe mix in hybridization buffer for 16 h at 42°C. The slides were then washed in 1× SSC, 0.1% SDS for 5 min, 0.1× SSC, 0.1% SDS for 5 min, rinsed with 0.1× SSC and dried. Fluorescence levels for both Cy3 and Cy5 of the individual cDNA spots were determined using a ScanArray3000 laser scanner (Perkin-Elmer, Foster City, California). All microarray experiments were replicated four times. As a common reference, 24 h imbibed seeds were used for all hybridizations and swapped dye experiments were performed to estimate experimental variation. After subtraction of background, data were normalized using an analysis of variance (ANOVA) approach.

In order to monitor the changes in the transcriptome during priming, seeds were osmoprimed for 0, 1, 3 or 7 days in −1 MPa PEG 8000. To monitor the changes in the transcriptome during germination of non-treated and osmoprimed seeds, seeds were isolated every 2 h from the start of imbibition up to the moment of radicle protrusion. These seeds were pooled to obtain samples of five time points during this process, varying from dry seeds (time 0) to seeds about to exhibit radicle protrusion (time 4).

Results and Discussion

Osmopriming induces cross-tolerance

In order to test the hypothesis that priming changes the expression of genes that confer cross-tolerance to the seeds and subsequent seedlings, we first wanted to confirm

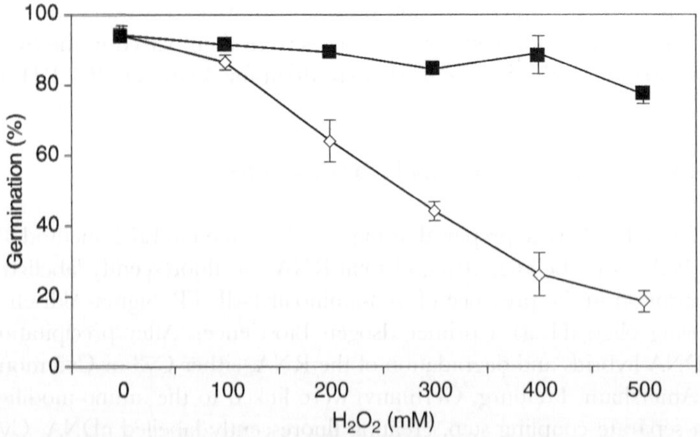

Fig. 14.1. Effect of hydrogen peroxide (H_2O_2) concentration on the mean germination percentages of non-treated (◊) and osmoprimed (■) tomato seeds. Germinations were performed with four replicates of 50 seeds. Bars represent ± standard error (SE).

that priming induces cross-tolerance. Therefore, we tested the effect of osmopriming on the cross-tolerance of tomato seeds. When germinated on filter paper with H_2O, osmoprimed seeds showed a typical reduction of t_{50} (the time taken for 50% of the seeds to germinate) with a factor 1.8 and an improved uniformity, measured by the u_{7525} (the time that elapsed between 25% and 75% of germination). During several stress treatments, the primed seeds performed better than the control. For example, oxidative stress caused a reduction in germination percentage with increasing concentrations of H_2O_2, whereas the germination percentage of primed seeds was hardly affected by concentrations of up to 500 mM H_2O_2 (Fig. 14.1).

A similar effect could be seen for germination at higher temperatures. Untreated seeds did not germinate at temperatures above 37°C (Fig. 14.2), whereas osmoprimed seeds still showed significant germination at temperatures up to 40°C.

Priming with the 'seed conditioner'

Several other priming methods are known besides osmopriming. Priming with the SC is known to improve seed germination characteristics (Boukens *et al.*, 1998, Seed Processing Holland, Enkhuizen, The Netherlands). We wanted to see if SC priming also resulted in cross-tolerance. The main difference between SC priming and standard osmopriming is that the moisture content (MC) of the seeds decreases slowly after the first day of imbibition in SC priming in contrast to a slight increase in the MC of seeds after the first day during standard osmopriming.

To test the effect of the SC treatment on seed cross-tolerance, the seeds were germinated in different concentrations of sodium chloride (NaCl). The positive effect of osmopriming on salt tolerance could only be seen in the t_{50} of the seeds, whereas SC-treated seeds showed a clear effect on the germination percentage, especially at high concentrations of NaCl. Germination of osmoprimed and non-treated seeds was reduced strongly at water potentials of −0.7 MPa and higher,

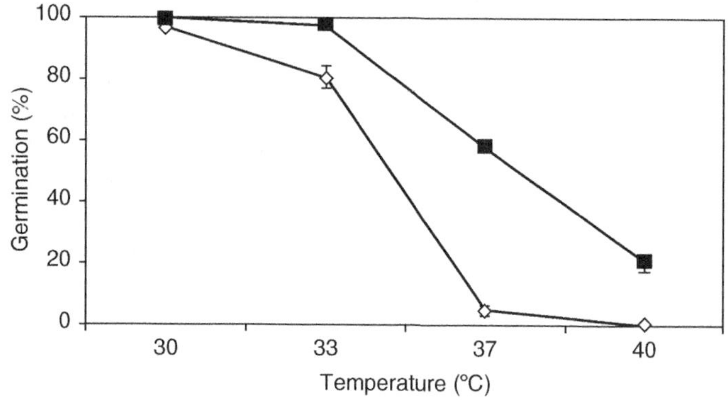

Fig. 14.2. Effect of temperature on the germination of untreated (◊) and osmoprimed (■) tomato seeds. Experiments were performed with three replicates of 50 seeds. Bars represent ± standard error (SE).

whereas the SC-treated seeds were still able to germinate at normal rates under these conditions (Fig. 14.3).

Priming-induced changes in gene expression

Since it has been clearly shown that different priming methods induce cross-tolerance, we were interested in the genes that are involved in this process. In order to identify the genes involved in priming-induced changes in the seed, we analysed the transcriptome with the aid of microarrays.

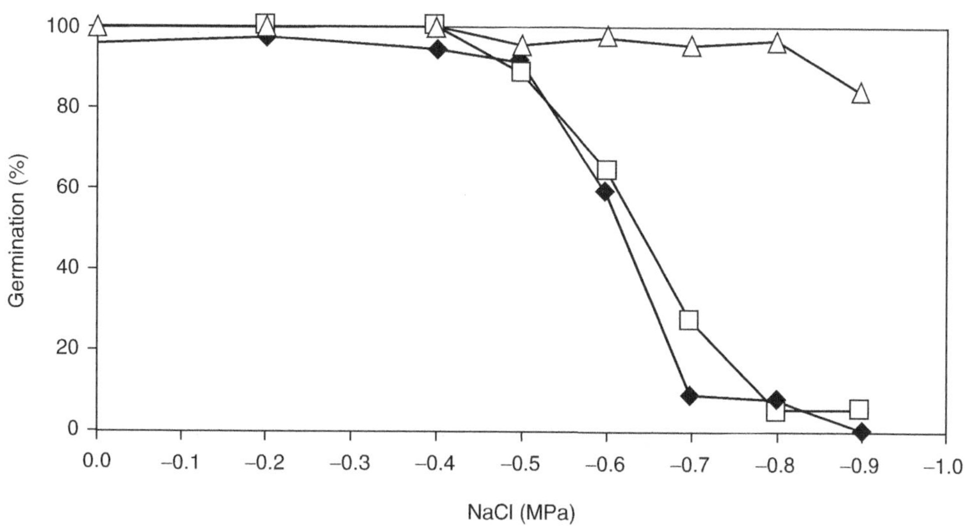

Fig. 14.3. Effect of different concentrations of sodium chloride (NaCl) on the germination of non-treated (□), osmoprimed (◆) and SC-treated (Δ) tomato seeds. Experiments were performed with 50 seeds.

To follow the changes in the transcriptome over time, seeds were osmoprimed for 0, 1, 3 or 7 days. Both the t_{50} and the u_{7525} of these seeds decreased with increasing priming time, with the greatest effect after 7 days (Fig. 14.4a). Furthermore, the seeds showed increased temperature tolerance at 40°C with increasing priming time (Fig. 14.4b).

Overall differences in the transcriptome between the different samples were small and also expression of individual genes did not show substantial differences. As a result of these small overall changes, large block effects were observed. The differences that were found (up to four times upregulated or downregulated) were reproducible. In total, ~20% of the genes showed a differential expression under the conditions tested.

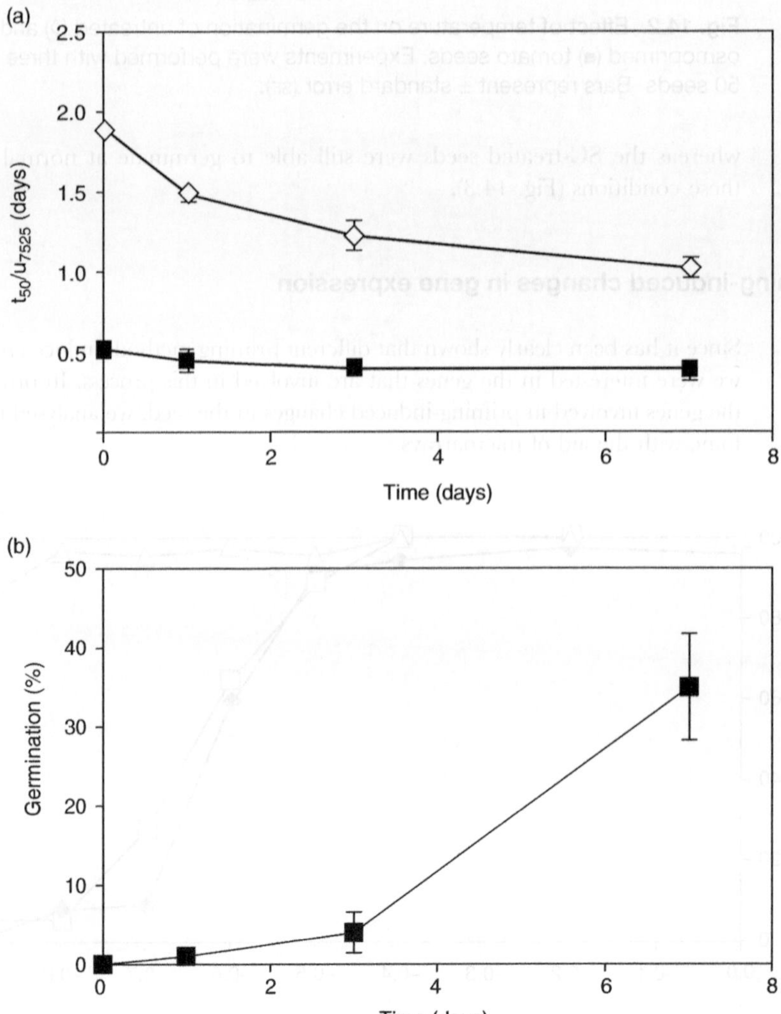

Fig. 14.4. Effect of the duration of osmopriming of tomato seeds on t_{50} (◊) and u_{7525} (■) of mean germination (a) in H_2O and (b) at 40°C. Experiments were performed with four replicates of 50 seeds. Bars represent ± standard error (SE).

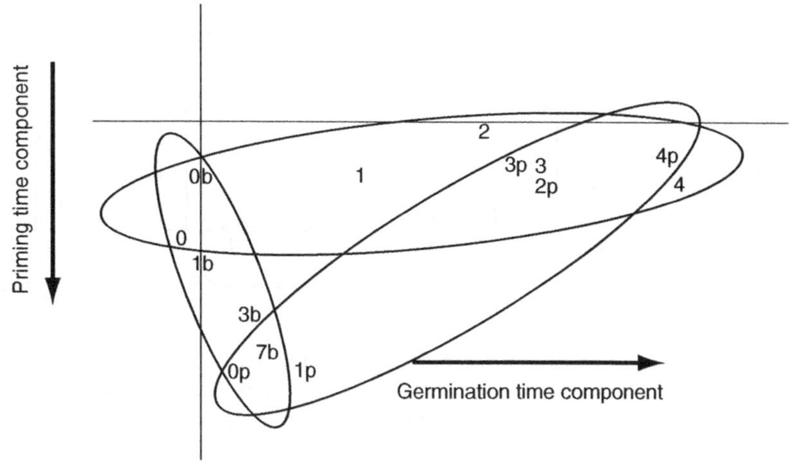

Fig. 14.5. Principal component analysis of the averages of three different experiments: where 0b, 1b, 3b and 7b are the four time points of the priming time experiment; 0, 1, 2, 3 and 4 are the different time points during germination of non-treated seeds; and 0p, 1p, 2p, 3p and 4p are the different time points during germination of osmoprimed seeds.

With this first microarray experiment, a list of genes was identified that were affected by priming. To find priming-specific gene regulation we also executed an experiment in which we monitored the changes in the transcriptome during germination of both non-treated and osmoprimed seeds. As with the priming time experiment, the differences in expression were not very high, but again a substantial portion of the genes showed a differential expression. Comparing these results with those from the priming time experiment resulted in 2600 genes that were differentially expressed in at least one of the experiments. When these 2600 genes were analysed using a principal component analysis, two main components could be seen: a priming time component on the *y*-axis and a germination time component on the *x*-axis (Fig. 14.5). It is interesting to see that during germination the overall expression profiles of non-treated seeds and primed seeds come together again (point 4 and 4p in Fig. 14.5). However, this does not mean that there are no individual genes with different expression levels at this stage.

According to the observed expression patterns during priming and germination, six major classes of genes could be listed: genes that were both upregulated and downregulated during priming and germination (Class 1 and 2, 52 and 35 genes, respectively); genes that were upregulated or downregulated during priming but not during germination, or that showed an opposite expression during germination (Class 3 and 4, 27 and 28 genes, respectively); and genes that were only upregulated or downregulated during germination (Class 5 and 6, 10 and 6 genes, respectively). Class 3 and 4 genes are especially interesting for our research, since they are priming-specific and thus likely to have a function in providing the seeds with the characteristics that are specific for primed seeds, including cross-tolerance. A typical expression pattern of such a gene is shown in Fig. 14.6. This gene was upregulated during the priming process, but no differential expression could be observed during

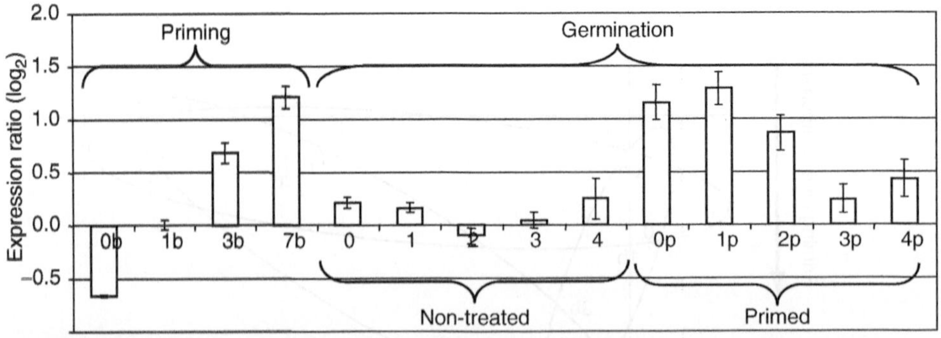

Fig. 14.6. Typical expression pattern of a priming-specific gene. The size of the bars represents the amount of expression at different time points in different treatments (see Fig. 14.5 for a description of the time points and treatments concerned). Bars represent ± standard error (SE).

the germination of non-treated seeds. During germination of primed seeds, this gene was downregulated, resulting in similar expression levels in both non-treated and primed seeds just before radicle protrusion.

The priming-specific genes were categorized with the Munich Information Centre for Protein Sequence (MIPS) annotation according to their cellular function (Mewes *et al.*, 2004). Significant subsets of the genes that were downregulated during priming had a function in signal transduction and in providing energy (Fig. 14.7a). For the genes that were upregulated, large subsets had a function in transcription and cell defence, but many of them had an unknown function (Fig. 14.7b).

Conclusions

It has been shown that priming of seeds can confer cross-tolerance. These effects exceed the normal enhancement of the seeds under non-stress conditions. This can be seen not only during germination under oxidative and temperature stress (Figs 14.1 and 14.2), but also during germination under high osmotic stress (data not shown). Furthermore, the cross-tolerance phenomenon can be observed after treatment of the seeds with different priming methods as observed after treatment of the seeds in the SC.

With the help of microarray hybridizations, we have been able to identify a set of genes that are differentially regulated during priming and germination. Future research will focus on the genes that show a priming-specific differential expression, since these are the best candidates to be responsible for the observed cross-tolerance phenotype. The upregulated genes that are annotated as cell defence genes are particularly interesting in this respect. However, the transcription-related genes are also significant, since they may be the key regulators of the observed differential expression. These genes will be further analysed for their exact role in cross-tolerance by studying their expression patterns under different conditions and in different seed batches and especially by the generation and phenotypical characterization of overexpression and gene-silencing mutants.

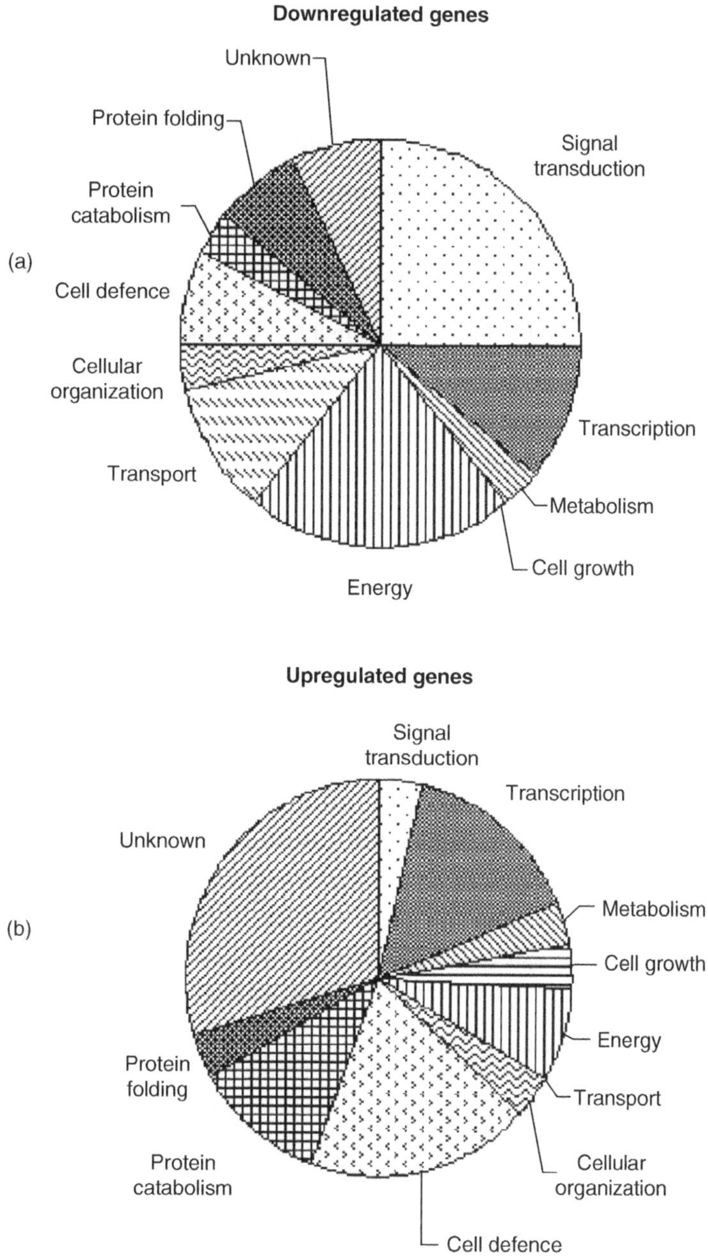

Fig. 14.7. Categorization of priming-specific genes that are either (a) downregulated or (b) upregulated. Annotations of the genes were made using the Munich Information Centre for Protein Sequence (MIPS) classification (Mewes *et al*., 2004).

In addition to a better understanding of the processes that lead to the positive effect of seed priming, the practical implications of this research will be the identification of possible marker genes that are specific for stress resistance which can be used to control and optimize the priming process, accurately predict the vigour of seedlings and aid in breeding programmes.

Acknowledgements

This work was supported by the Technology Foundation STW, the Applied Science Division of NWO and the Technology Program of the Ministry of Economic Affairs.

References

Alba, R., Fei, Z., Payton, P., Liu, Y., Moore, S.L., Debbie, P., Cohn, J., D'Ascenzo, M., Gordon, J.S., Rose, J.K.C., Martin, G., Tanksley, S.D., Bouzayen, M., Jahn, M.M. and Giovannoni, J. (2004) ESTs, cDNA microarrays, and gene expression profiling: tools for dissecting plant physiology and development. *The Plant Journal* 39, 697–714.

Boukens, M.A.N., De Boer, J. and Van Dok, I. (1998) Method and apparatus for priming seed. Patent WO 99/33331, Wittop Koning, The Hague, The Netherlands.

Bowler, C. and Fluhr, R. (2000) The role of calcium and activated oxygens as signals for controlling cross-tolerance. *Trends in Plant Science* 5, 241–246.

Cayuela, E., PerezAlfocea, F., Caro, M. and Bolarin, M.C. (1996) Priming of seeds with NaCl induces physiological changes in tomato plants grown under salt stress. *Physiologia Plantarum* 96, 231–236.

Chini, A., Grant, J.J., Seki, M., Shinozaki, K. and Loake, G.J. (2004) Drought tolerance established by enhanced expression of the *CC-NBS-LRR* gene, *ADR1*, requires salicylic acid, EDS1 and ABI1. *The Plant Journal* 38, 810–822.

Dearman, J., Brocklehurst, P.A. and Drew, R.L.K. (1986) Effects of osmotic priming and aging on onion seed germination. *Annals of Applied Biology* 108, 639–648.

Devi, S.R., Yamamoto, Y. and Matsumoto, H. (2003) An intracellular mechanism of aluminum tolerance associated with high antioxidant status in cultured tobacco cells. *Journal of Inorganic Biochemistry* 97, 59–68.

Kretschmer, M. (1982) Extension of the temperature tolerance of *Lactuca* achenes with PEG and red light. *Gartenbauwissenschaft* 47, 152–157.

Mewes, H.W., Amid, C., Arnold, R., Frishman, D., Guldener, U., Mannhaupt, G., Munsterkotter, M., Pagel, P., Strack, N., Stumpflen, V., Warfsmann, J. and Ruepp, A. (2004) MIPS: analysis and annotation of proteins from whole genomes. *Nucleic Acids Research* 32, D41–D44.

Verwoerd, T.C., Dekker, B.M.M. and Hoekema, A. (1989) A small-scale procedure for the rapid isolation of plant RNAs. *Nucleic Acids Research* 17, 2362.

15 The Use of Proteome and Transcriptome Profiling in the Understanding of Seed Germination and Identification of Intrinsic Markers Determining Seed Quality, Germination Efficiency and Early Seedling Vigour

L. Rajjou, L. Miché, R. Huguet, C. Job and D. Job

Centre National de la Recherche Scientifique (CNRS)–Bayer CropScience Joint Laboratory (UMR CNRS 2847), Bayer CropScience, 14/20 rue Pierre Baizet, BP 9163, F-69263 Lyon, Cedex 09, France

Abstract

In order to better understand seed germination and identify molecular markers of seed vigour, two global expression-profiling methods have been used (i.e. proteomics and transcriptomics) on the model plant *Arabidopsis* (*Arabidopsis thaliana* (L.) Heynh.). The requirements for germination, in terms of protein and RNA syntheses, were studied. Approximately 400 seed proteins were identified by mass spectrometry. In addition to the storage proteins, dry mature seeds contain a number of other proteins, including enzymes that are required for the resumption of metabolism during the early stages of germination. Moreover, the use of α-amanitin demonstrated that, even in the absence of transcription, radicle protrusion can still occur. This suggests that dry mature seeds possess all the stored mRNAs and proteins necessary and sufficient for germination. Transcriptomic experiments carried out using an oligo-based microarray, covering the entire genome of *Arabidopsis*, revealed that 6500 genes are differentially regulated during germination. Furthermore, a clustering analysis unravelled specific sets of genes that are expressed at different phases of the germination process. This combined approach allowed, for the first time, the comparison of proteomic and transcriptomic data for such a complex developmental process and the discovery of novel features of seed germination.

 Seeds: Biology, Development and Ecology (eds S. Adkins, S. Ashmore and S.C. Navie)

Introduction

Seed germination is a complex, multistage process that can be divided into three phases of imbibition, increased metabolic activity and initiation of growth, which loosely parallel the triphasic water uptake of dry mature seeds. Morphologically, initiation of growth corresponds to radicle emergence, and subsequent growth is generally defined as seedling growth. By definition, germination *sensu stricto* incorporates those events that start with the uptake of water by the non-dormant quiescent dry seed and terminate with the protrusion of the radicle and the elongation of the embryonic axis (Bewley and Black, 1994). Upon imbibition, the quiescent dry seed rapidly resumes metabolic activity. To explore genome-wide expression patterns of this complex developmental process, parallel protein and mRNA profiling studies of dry and imbibed seeds have been conducted, during the last few years, using the model plant *Arabidopsis* (*Arabidopsis thaliana* (L.) Heynh.). The widespread acceptance of *Arabidopsis* as a model plant is based on the genetic and genomic methods and resources that are available for it, which have facilitated the investigation of a range of biological problems (see Somerville and Koornneef, 2002 for a review). Our results show that new insights into this complex developmental pathway can be extracted from such global expression-profiling approaches.

Proteomic Analyses During Germination

Proteomic analyses during *Arabidopsis* seed germination were carried out by using non-dormant seeds from the *Landsberg erecta* (Ler) accession. In water, and under optimal conditions (i.e. 25°C), radicle protrusion started after 1.5 days of imbibition and it took almost 1.8 days for 50% of the seeds to reach this phase. Germination was completed 2 days after imbibition. Total soluble proteins were extracted from various seed samples (i.e. dry mature seeds, 1-day imbibed seeds and 2-day imbibed seeds; corresponding to germination *sensu stricto* and radicle protrusion, respectively), analysed by two-dimensional (2D) gel electrophoresis (Fig. 15.1) and characterized by mass spectrometry measurements as described by Job *et al.* (2005).

Approximately 400 seed proteins have been identified in this way that can be grouped into several functional categories using the ontological classification (Fig. 15.2).

These analyses revealed that the proteome of the dry mature seeds and of the germinated seeds is dominated by three major groups of proteins, corresponding to those involved in metabolism, protein metabolism and stress response.

Protein Oxidation in Dry Mature Seeds and During Germination

The above data indicated that a large number of proteins involved in stress response are present in dry mature seeds and in germinating seeds (Fig. 15.2). In this context, it is worth noting that increased cellular levels of reactive oxygen species (ROS) are known to occur during seed development and germination (reviewed by Bailly, 2004), but the consequences in terms of protein oxidation and damage

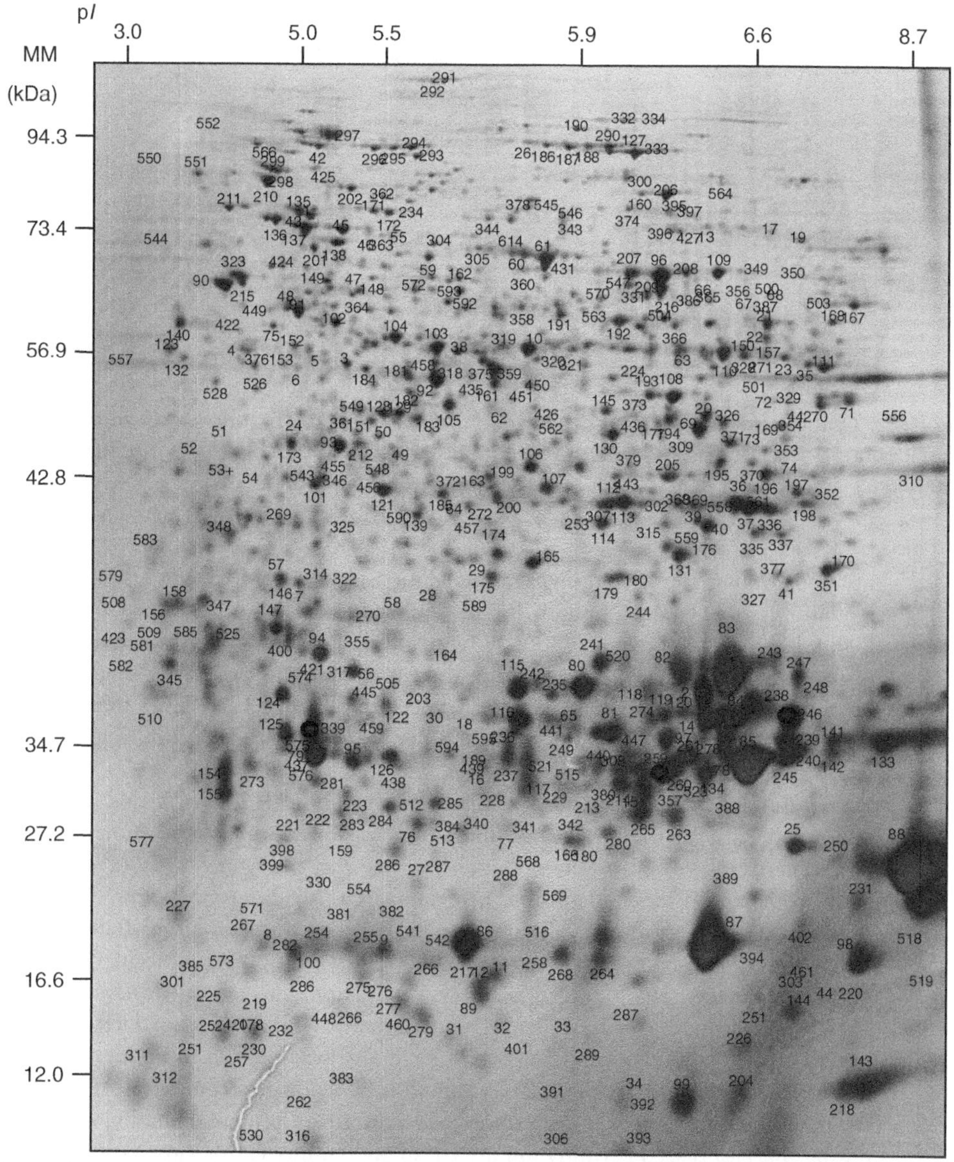

Fig. 15.1. A 2D gel of total soluble proteins from dry mature seeds of *Arabidopsis*. Proteins were first separated by electrophoresis according to charge. Isoelectrofocusing (IEF) was carried out with protein samples with an equivalent extract of ~100 seeds, corresponding to ~200 µg protein for all samples. Proteins were then separated according to size by sodium dodecyl sulfate polyacrylamide gel electrophoresis (SDS-PAGE) using ten polyacrylamide gels. Proteins were visualized by silver nitrate coloration (see Job *et al*., 2005 for details on protein extractions and proteomic analyses). Protein numbers refer to proteins identified by mass spectrometry (Gallardo *et al*., 2001, 2002a,b; Rajjou *et al*., 2004; Job *et al*., 2005; http://seed.proteome.free.fr).

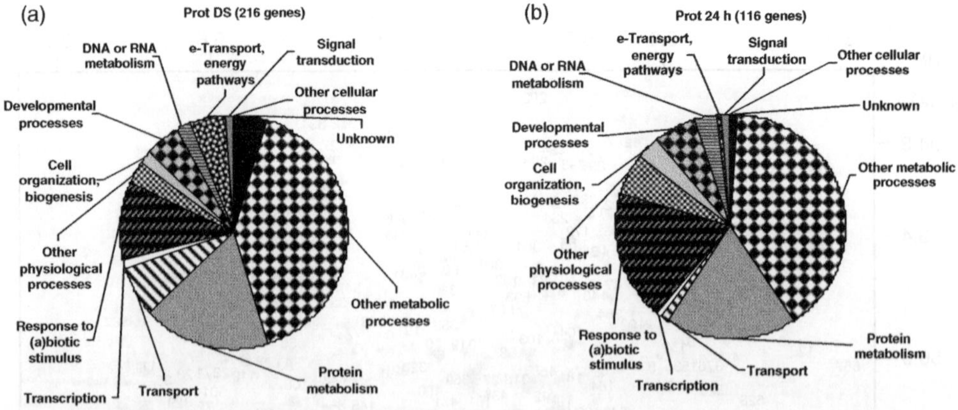

Fig. 15.2. Ontological classification of the seed proteins (a) in dry mature seeds and (b) in germinating (1-day imbibed) *Arabidopsis* seeds. Proteins were identified by proteomic analysis as described in Fig. 15.1.

are poorly characterized. Protein carbonylation is a widely used marker of protein oxidation and sensitive methods for its detection have been developed (Nyström, 2005). It occurs by direct oxidation of lysine, arginine, proline or threonine residues of proteins, thus inhibiting or altering their activities and increasing their susceptibility towards proteolytic attack (Nyström, 2005). Figure 15.3 shows that protein carbonylation can be detected in dry mature *Arabidopsis* seeds and during the first stages of germination (Job *et al.*, 2005).

In the dry mature seeds, the legumin-type globulins (12S cruciferins) were the major targets (Fig. 15.3a). During imbibition, various carbonylated proteins were accumulated (Fig. 15.3b). This oxidation damage was not evenly distributed among seed proteins and targeted specific proteins as glycolytic enzymes, mitochondrial ATP synthase, chloroplastic ribulose bisphosphate carboxylase large chain, aldose reductase, methionine synthase, translation factors and several molecular chaperones. Among the specific targets of protein oxidation detected in *Arabidopsis* seeds, Hsp70 chaperones, aconitase, translation initiation and elongation factors, ATP synthase β-subunit, actin and GAPDHc have also been shown to be highly sensitive to oxidation in bacterial, yeast or animal cells, leading to common features in a range of living organisms. Furthermore, the patterns of protein oxidation observed during *Arabidopsis* seed germination are strikingly reminiscent of those described for *Escherichia coli* and *Saccharomyces cerevisiae* cells exposed to oxidative stress (Job *et al.*, 2005), lending further support to the contention that oxidative stress normally accompanies seed germination (Bailly, 2004).

Although accumulation of carbonylated proteins is usually considered in the context of ageing in a variety of model systems (Nyström, 2005), this was not the case for the *Arabidopsis* seeds since they germinated at a high rate and yielded vigorous seedlings. The results indicate that the observed specific changes in protein carbonylation patterns are probably required for counteracting and/or utilizing the production of ROS caused by recovery of metabolic activity in the germinating seeds (Job *et al.*, 2005). In agreement with a previous proposal (Côme and

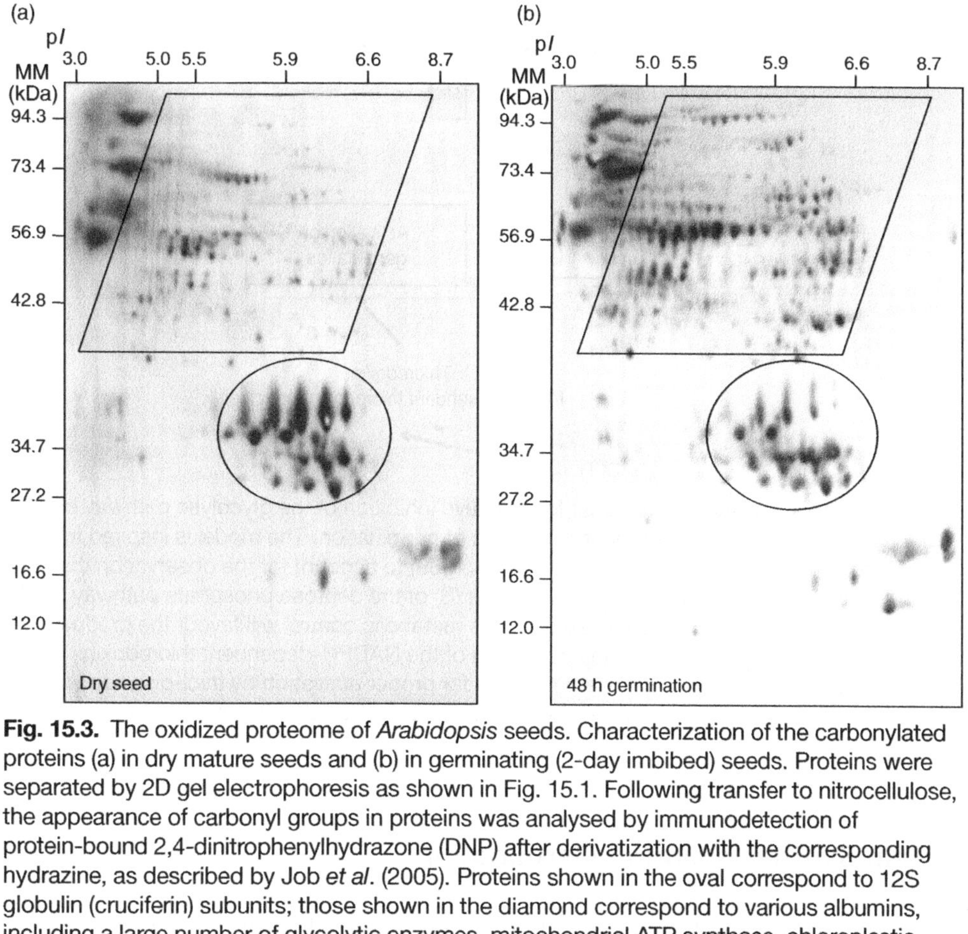

Fig. 15.3. The oxidized proteome of *Arabidopsis* seeds. Characterization of the carbonylated proteins (a) in dry mature seeds and (b) in germinating (2-day imbibed) seeds. Proteins were separated by 2D gel electrophoresis as shown in Fig. 15.1. Following transfer to nitrocellulose, the appearance of carbonyl groups in proteins was analysed by immunodetection of protein-bound 2,4-dinitrophenylhydrazone (DNP) after derivatization with the corresponding hydrazine, as described by Job *et al.* (2005). Proteins shown in the oval correspond to 12S globulin (cruciferin) subunits; those shown in the diamond correspond to various albumins, including a large number of glycolytic enzymes, mitochondrial ATP synthase, chloroplastic ribulose bisphosphate carboxylase large chain, aldose reductase, methionine synthase, translation factors and several molecular chaperones.

Corbineau, 1989), our results suggest that blocking glycolysis could be beneficial during conditions of oxidative stress since it would result in an increased flux of glucose equivalents through the pentose phosphate pathway (PPP), thus leading to the generation of NADPH (Fig. 15.4). This provides the reducing power for antioxidant enzymes, including the thioredoxin and GSH/glutaredoxin systems. The NADP/thioredoxin system, which is composed of NADPH, thioredoxin *h* and NADP–thioredoxin reductase, plays a crucial role in seed germination since it functions in the reduction of major storage proteins in seeds. It converts disulfide (S–S) bonds to the reduced SH state, thereby leading to increased solubility and mobilization (Wong *et al.*, 2004a).

Many of the carbonylated proteins detected in *Arabidopsis* seeds (Job *et al.*, 2005) have been identified as thioredoxin targets in wheat (*Triticum aestivum* L.) seeds (Wong *et al.*, 2004b). The results further suggest the existence of a link between ROS and redox regulatory events catalysed by thioredoxin in seeds (Wong *et al.*, 2004a).

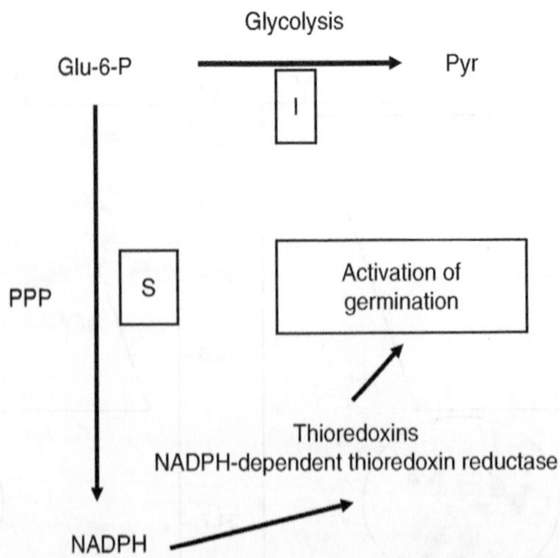

Fig. 15.4. A model to account for selective inhibition of the glycolytic pathway by protein oxidation in relation to activation of germination. The model is inspired from that proposed by Côme and Corbineau (1989) to account for the observation that inhibitors (I) of glycolysis and stimulators (S) of the pentose phosphate pathway (PPP) stimulate germination *sensu stricto*. This metabolic control will favour the production of NADPH necessary for the functioning of the NADPH-dependent thioredoxin reductase/thioredoxins system required for proper activation by thiol-disulfide exchanges of enzymes necessary for germination (see Wong *et al*., 2004a,b).

Protein Synthesis During Germination

Figure 15.2 suggests the importance of protein synthesis in the germination process. Presumably, protein synthesis can occur from the mRNA species stored in the dry mature seeds and also from those that are synthesized by *de novo* transcription during the early stages of seed germination (Raghavan, 2000). To address this question and to investigate the role of stored and neosynthesized mRNAs in seed germination, we examined the effect of α-amanitin, a transcriptional inhibitor targeting RNA polymerase II, on the germination of non-dormant *Arabidopsis* seeds (Rajjou *et al.*, 2004). We used *transparent testa* mutants, which have highly permeable seed coats (Debeaujon *et al.*, 2000), to better ascertain that the drug can reach the embryo during seed imbibition. Even with the most permeable mutant (*tt2-1*), germination (i.e. radicle protrusion) occurred in the absence of transcription, while subsequent seedling growth was blocked (Fig. 15.5). In contrast, seed germination proved to be totally inhibited in the presence of the translational inhibitor cycloheximide, indicating a requirement for protein synthesis in germinating embryos (Fig. 15.5). These combined results strongly suggest that germination-specific proteins are made from the long-lived stored mRNAs. These results highlight the role of stored proteins and stored mRNAs for germination in *Arabidopsis* and show that in this species the potential for germination is largely programmed during the seed maturation process.

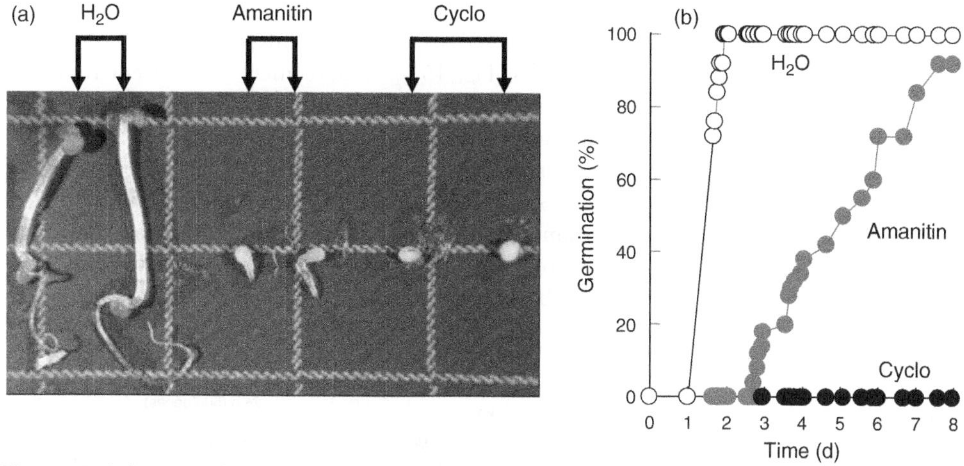

Fig. 15.5. Influence of α-amanitin and cycloheximide on seed germination and seedling establishment of the *tt2-1 Arabidopsis* transparent mutant (from Rajjou *et al.*, 2004). (a) Phenotypes observed 8 days after sowing the seeds in water, 500 μM α-amanitin (Ama) or 100 μM cycloheximide (Cyclo). Square: 3 × 3 mm. (b) Germination data under the conditions of (a).

The α-amanitin-resistant germination exhibited characteristic features:

1. This germination was strongly slowed, indicating that *de novo* transcription normally allows the synthesis of factor(s) activating the germination rate (Fig. 15.5b).
2. The sensitivity of germination to gibberellic acid was reduced 15-fold, confirming the role of this phytohormone in germination.
3. *De novo* synthesis of enzymes involved in reserve mobilization and resumption of metabolic activity was repressed, thus accounting for the failure in seedling establishment.
4. The proteome of seeds germinated on α-amanitin, characterized through the incorporation of [^{35}S]methionine into newly synthesized proteins, was markedly different from that of seeds germinated in water, as it was more reminiscent of a proteome characteristic of the seed maturation phase than of germination.

This indicates that germinating seeds can recapitulate at least part of the seed maturation programme as they are capable of using mRNAs stored during development. If such a use of stored mRNAs encoding maturation proteins also occurred during normal germination, particularly under unfavourable germination conditions, it might have physiological implications. As stressed by Lopez-Molina *et al.* (2002), it would allow germinating seeds to mount adaptive responses to environmental water stress. In summary, commitment to germination and plant growth requires transcription of genes allowing the imbibed seed to discriminate between mRNAs to be utilized in germination and those to be destroyed, thereby promoting the shift from a developmental to a germination programme.

Transcriptomic Analyses During Germination

Transcriptomic analyses during *Arabidopsis* seed germination were carried out by using a 70-mer oligonucleotide-based microarray that covers 26,090 unique and predicted genes of *Arabidopsis* (Qiagen/Operon; see Ma *et al.*, 2005). Among these 26,090 genes represented on the microarray, 11,025 mRNA species were quantitatively detected in mature seeds (Fig. 15.6), in agreement with recent results of Nakabayashi *et al.* (2005). This huge number of stored mRNA species corresponds to two classes of transcripts: those that were synthesized during late maturation and those that are required for early germination. However, a large majority of the stored mRNA species (70%) are only present in very small quantities (i.e. less than three times the background level) in dry mature seeds. In contrast, only very few transcripts (3% of the stored mRNA species) are present in high abundance (i.e. greater than 25 times the background level).

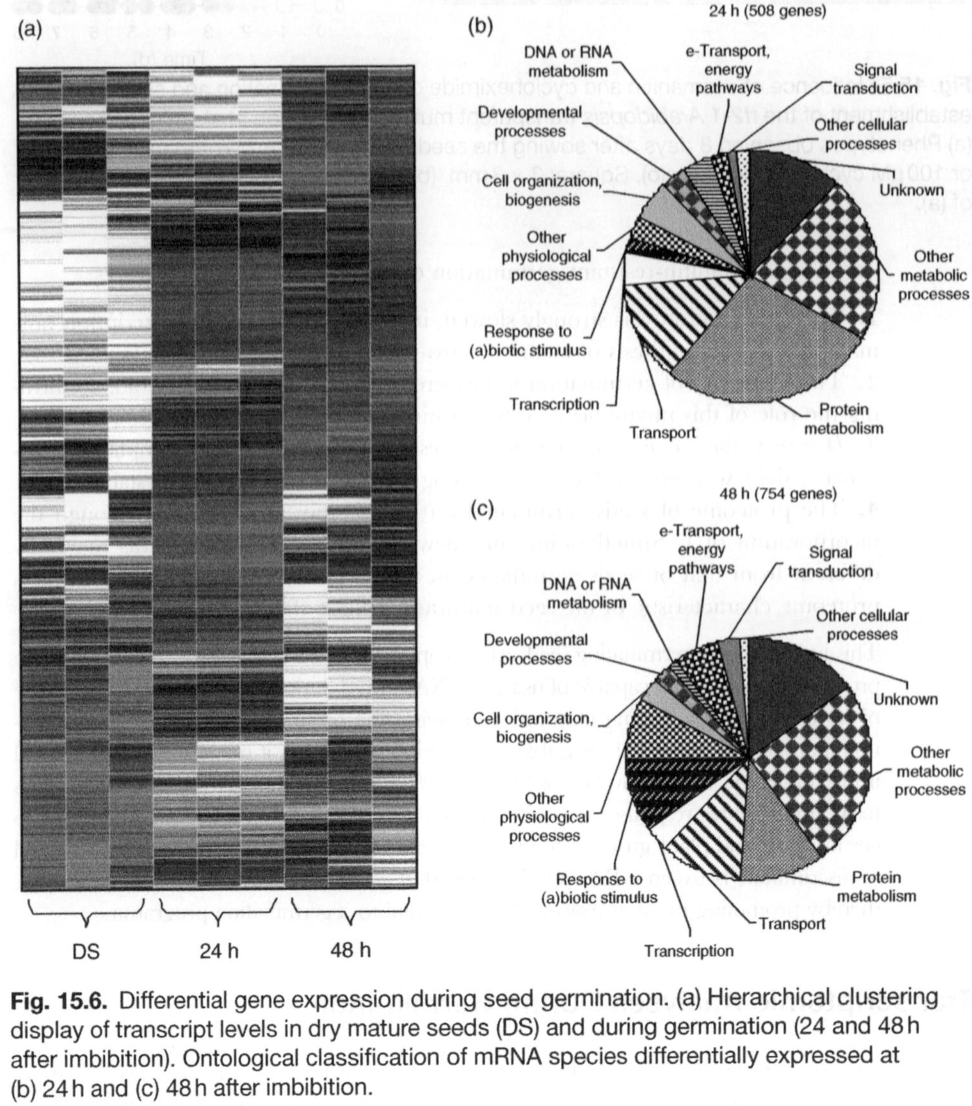

Fig. 15.6. Differential gene expression during seed germination. (a) Hierarchical clustering display of transcript levels in dry mature seeds (DS) and during germination (24 and 48 h after imbibition). Ontological classification of mRNA species differentially expressed at (b) 24 h and (c) 48 h after imbibition.

Expression profiles obtained from seeds at 1 day and 2 days after imbibition were also analysed, disclosing 6579 genes that were differentially expressed during the germination process (Fig. 15.6). In particular, the data in Fig. 15.6a document the occurrence of specific kinetic expression patterns, with some genes being specifically expressed in the dry mature seed, at the radicle protrusion step or at the time of seedling establishment. This specific feature is in agreement with recent data (Nakabayashi *et al.*, 2005). A recent study demonstrated that the germinating *Arabidopsis* seed has a unique genome expression profile, distinct from those found in other plant organs such as siliquae, flowers, hypocotyls, leaves and roots (Ma *et al.*, 2005). The differentially expressed genes can be grouped according to the ontological classification (Fig. 15.6b and c). These data confirmed the importance of protein metabolism in the early stages of seed germination, in total agreement with the proteomic data of Fig. 15.2 and with the observation that cycloheximide caused a strong block in seed germination (Fig. 15.5). They also highlight the fact that a large number of defence-related genes are synthesized during the early stages of seed germination, notably genes involved in oxidative stress, lending further support to the contention that oxidative stress normally accompanies seed germination (Bailly, 2004), a finding that is also in very good agreement with the proteomic data in Figs 15.2 and 15.3.

Conclusions and Future Research

This recent work has documented for the first time the proteome and transcriptome of dry mature *Arabidopsis* seeds at various stages during the germination process. Overall, there is good agreement between the proteomic and transcriptomic data. Protein synthesis from the stored mRNA species is of paramount importance for radicle protrusion, suggesting that the quality of the stored mRNAs is an important feature of seed vigour. These global expression-profiling studies also disclosed the occurrence of oxidative stress during germination. In view of the specificity of protein oxidation and the fact that the patterns of oxidized proteins closely resemble those of bacterial or yeast cells exposed to oxidative stress, it appears more likely that protein carbonylation does not simply reflect secondary epiphenomena, but is used as a means to adapt embryo metabolism to the oxidative conditions encountered during germination. In future work, it would be of interest to correlate seed vigour and the extent of protein oxidation in seeds of varying quality, and to understand the molecular basis of the specificity of protein carbonylation in developing and germinating seeds. This would allow the characterization of novel markers for seed vigour (see Rajjou *et al.*, Chapter 34, this volume). Similar approaches are being used to characterize protein markers of seed vigour in crops (see Catusse *et al.*, Chapter 17, this volume).

Acknowledgements

This work has been partly supported by a grant from the French Ministry of Research (ACI Physiologie Intégrative et Biologie du Développement). The PhD thesis of Loïc Rajjou is supported by Bayer CropScience and the French Ministry of Industry. Lucie Miché thanks Centre National de la Recherche Scientifique (CNRS) for a postdoctoral fellowship. We gratefully acknowledge Joël Vandekerckhove (University of Ghent, Belgium) and Maya Belghazi (INRA, Nouzilly, France) for the mass spectrometry determinations.

References

Bailly, C. (2004) Active oxygen species and antioxidants in seed biology. *Seed Science Research* 14, 93–107.

Bewley, J.D. and Black, M. (1994) *Seeds: Physiology of Development and Germination.* Plenum Press, New York.

Côme, D. and Corbineau, F. (1989) Some aspects of metabolic regulation of seed germination and dormancy. In: Taylorson, R.B. (ed.) *Recent Advances in the Development and Germination of Seeds.* Plenum Press, New York, pp. 165–179.

Debeaujon, I., Léon-Kloosterziel, K.M. and Koornneef, M. (2000) Influence of the testa on seed dormancy, germination, and longevity in *Arabidopsis. Plant Physiology* 122, 403–413.

Gallardo, K., Job, C., Groot, S.P.C., Puype, M., Demol, H., Vandekerckhove, J. and Job, D. (2001) Proteomic analysis of Arabidopsis seed germination and priming. *Plant Physiology* 126, 835–848.

Gallardo, K., Job, C., Groot, S.P.C., Puype, M., Demol, H., Vandekerckhove, J. and Job, D. (2002a) Proteomics analysis of *Arabidopsis* seed germination: a comparative study of wild-type and GA-deficient seeds. *Plant Physiology* 129, 823–837.

Gallardo, K., Job, C., Groot, S.P.C., Puype, M., Demol, H., Vandekerckhove, J. and Job, D. (2002b) Importance of methionine biosynthesis for *Arabidopsis* seed germination and seedling growth. *Physiologia Plantarum* 116, 238–247.

Job, C., Rajjou, L., Lovigny, Y., Belghazi, M. and Job, D. (2005) Patterns of protein oxidation in *Arabidopsis* seeds and during germination. *Plant Physiology* 138, 790–802.

Lopez-Molina, L., Mongrand, S., McLachlin, D.T., Chait, B.T. and Chua, N.-H. (2002) ABI5 acts downstream of ABI3 to execute an ABA-dependent growth arrest during germination. *The Plant Journal* 32, 317–328.

Ma, L., Sun, N., Liu, X., Jiao, Y., Zhao, H. and Deng, X.W. (2005) Organ-specific expression of *Arabidopsis* genome during development. *Plant Physiology* 138, 80–91.

Nakabayashi, K., Okamoto, M., Koshiba, T., Kamiya, Y. and Nambara, E. (2005) Genome-wide profiling of stored mRNA in *Arabidopsis thaliana* seed germination: epigenetic and genetic regulation of transcription in seed. *The Plant Journal* 41, 697–709.

Nyström, T. (2005) Role of oxidative carbonylation in protein quality control and senescence. *The EMBO Journal* 24, 1311–1317.

Raghavan, B. (2000) Seed germination. In: *Developmental Biology of Flowering Plants.* Springer, New York, pp. 7–24.

Rajjou, L., Gallardo, K., Debeaujon, I., Vandekerckhove, J., Job, C. and Job, D. (2004) The effect of α-amanitin on the *Arabidopsis* seed proteome highlights the distinct roles of stored and neosynthesized mRNAs during germination. *Plant Physiology* 134, 1598–1613.

Somerville, C. and Koornneef, M. (2002) A fortunate choice: the history of *Arabidopsis* as a model plant. *Nature Reviews Genetics* 3, 883–889.

Wong, J.H., Cai, N., Balmer, Y., Tanaka, C.K., Vensel, W.H., Hurkman, W.J. and Buchanan, B.B. (2004a) Thioredoxin targets of developing wheat seeds identified by complementary proteomic approaches. *Phytochemistry* 65, 1629–1640.

Wong, J.H., Cai, N., Tanaka, C.K., Vensel, W.H., Hurkman, W.J. and Buchanan, B.B. (2004b) Thioredoxin reduction alters the solubility of proteins of wheat starchy endosperm: an early event in cereal germination. *Plant and Cell Physiology* 45, 407–415.

16 A *Seed-GUS-Expression* Enhancer-trap Library for Germination Research

P.-P. Liu,[1] R.C. Martin,[2] J.R. Hewitt,[1] N. Koizuka,[1] T.M. Homrichhausen[3] and H. Nonogaki[1]

[1]*Department of Horticulture, Oregon State University, Corvallis, OR 97331, USA;* [2]*US Department of Agriculture–Agricultural Research Service (USDA-ARS), National Forage Seed Production Research Center (NFSPRC), 3450 SW Campus Way, Corvallis, OR 97331, USA;* [3]*Department of Chemical Engineering, Oregon State University, Corvallis, OR 97331, USA*

Abstract

Enhancer-trap lines are used to identify tissue- and stage-specific gene expression. An *Arabidopsis* (*Arabidopsis thaliana* (L.) Heynh.) enhancer-trap population from the Arabidopsis Biological Resource Center (ABRC), Ohio, USA, has been screened for β-glucuronidase (GUS) expression in germinating seeds. Application of these enhancer-trap lines for functional analysis of seed germination-associated genes has led to the successful identification of a GATA zinc finger transcription factor crucial for seed germination. A *Seed-GUS-Expression* enhancer-trap library was donated to the ABRC (stock no. CS24362–CS24480), and is now available to the international seed research community. This library can be used for seed germination research by combining northern blotting and reverse transcription polymerase chain reaction (RT-PCR) expression analysis with functional analysis using knockout plants. Here, an overview of the procedures for utilizing the enhancer-trap library for germination research including some potential pitfalls and hints for troubleshooting experiments is presented.

Introduction

A substantial amount of information on the biochemical and molecular mechanisms of seed germination has been obtained using tomato (*Lycopersicon esculentum* Mill.) as a model system (Bradford *et al.*, 2000). The tomato seed is an excellent system for seed germination research, because it has a distinct endosperm and an embryo at maturity, and therefore provides an excellent tool for analysing the physical and chemical interactions between these two tissues. Since the tomato seed

 Seeds: Biology, Development and Ecology (eds S. Adkins, S. Ashmore and S.C. Navie)

is larger than the seeds from many other plant species, such as tobacco (*Nicotiana tabacum* L.) and *Arabidopsis* (*Arabidopsis thaliana* (L.) Heynh.), it is feasible to dissect it. The molecular mechanisms of endosperm weakening and the generation of embryo growth potential are best understood in this model system. However, the number of genes identified in tomato seeds is still limited and the regulatory mechanisms of gene expression are not fully understood. Although multiple genes are expressed in tomato seeds in a tissue-specific manner, the mechanisms regulating tissue specificity are largely unknown. Therefore, characterization of the promoter regions of the germination-associated genes by identifying hormone responsive- and tissue-specific *cis*-elements and isolating the DNA-binding proteins specific to these elements is necessary. However, this is difficult in tomato due to limited availability of genomic resources.

Although *Arabidopsis* seeds are too small to analyse embryo–endosperm interactions in a physiological way, a large number of genetic tools are available for this model species. Considering the genomic resources available for *Arabidopsis*, and ongoing genome synteny between *Arabidopsis* and tomato (Ku *et al.*, 2000, 2001), the identification of seed germination-associated genes in *Arabidopsis* will significantly enhance characterization of their orthologues in tomato seeds. The transfer of information between *Arabidopsis* and tomato seeds has been applied successfully in studies of tomato testa (Downie *et al.*, 2003).

On the other hand, there is only limited information on *sensu stricto* germination-associated genes in *Arabidopsis*. Testa mutants of *Arabidopsis* seeds have advanced our knowledge of seed biology; however, the critical genes that determine the chemical and physical properties of the testa are expressed in the endothelium of developing seeds (Debeaujon *et al.*, 2001, 2003; Sagasser *et al.*, 2002). Gene expression in germinating *Arabidopsis* seeds needs to be characterized. There are emerging data from cDNA microarrays and proteomic studies concerning gene expression in germinating *Arabidopsis* seeds (Gallardo *et al.*, 2001, 2002; Ogawa *et al.*, 2003); however, the tissue specificity of the expressed genes is still unknown.

Enhancer-trap technology has been successfully applied to identify tissue- and stage-specific gene expression in *Drosophila* (Akimoto *et al.*, 2005), zebrafish (*Danio rerio* (Ham. Buc.); Balciunas *et al.*, 2004) and also in plants including *Arabidopsis* (Campisi *et al.*, 1999; He *et al.*, 2001; Vroemen *et al.*, 2003; Endo *et al.*, 2005; Engineer *et al.*, 2005), rice (*Oryza sativa* L.; Johnson *et al.*, 2005) and carrot (*Daucus carota* L.; Ko and Kamada, 2002). Enhancer-trap lines are generated by randomly integrating a cassette that contains a minimal promoter (MP) fused to a reporter gene into the genome. If the MP–reporter fusion integrates adjacent to an enhancer (or promoter proximal element), the expression of the reporter gene is detected in an organ-, tissue-, or cell-specific pattern as directed by the enhancer (Campisi *et al.*, 1999).

In a previous study, *Arabidopsis* enhancer-trap lines (Thomas Jack lines, 1130 pools of 10 lines, CS31086, Arabidopsis Biological Resource Center (ABRC) at Ohio State University) were screened for GUS expression in imbibed seeds and a library of *Seed-GUS-Expression* enhancer-trap lines was generated (Liu *et al.*, 2005a). The seeds have been donated to ABRC by the international research community to identify seed germination-associated genes. A GATA zinc finger transcription factor gene (termed *BME3-ZF*), whose expression is induced during cold stratification, was identified in this library. Seeds from two independent T-DNA knockout lines of *BME3-ZF* showed reduced germination, suggesting that this gene is a positive regulator of seed germination (Liu *et al.*, 2005b). This chapter is a proof of a concept that clearly demonstrates the utility of the *Seed-GUS-Expression* library. An overview of the seed enhancer-trap approach and the resources useful for conducting experiments is also presented.

Seed Order and Production

Seed-GUS-Expression enhancer-trap lines (CS24362–CS24480) are now available from ABRC (http://www.Arabidopsis.org/abrc). Detailed information about individual lines is also described at the NSF-funded Integrated Seed Biology web site at Oregon State University (http://www.science.oregonstate.edu/isb/). The general scheme of utilizing this library is summarized in Fig. 16.1. The initial step for using the enhancer-trap approach is to order seeds from ABRC. This can be done through The Arabidopsis Information Resource (TAIR) web site (http://www.arabidopsis.org/). Ordering multiple lines (5–10) is recommended since the identification of trapped genes may not always be successful, as discussed below. Each of the individual stock pools of *Seed-GUS-Expression* should express GUS in germinating or germinated seeds. However, the available seeds are a mixture of wild-type, heterozygous and homozygous enhancer-trap seeds; therefore, more than ten plants from each pool should be grown for seed production. Seeds need to be prechilled at 4°C for 3 days to break dormancy. When plants start to produce mature seeds, a few siliquae are collected and dried in a 25°C incubator for 5 days. Seeds are extracted from siliquae, given 3 days of prechilling treatment, incubated at 22°C for 22 h and tested for GUS activity. By conducting this preliminary GUS expression analysis, positive enhancer-trap individual plants are identified. In this way, seeds for future experiments should be harvested from the GUS-positive plants only. Several rosette leaves need to be collected, frozen and saved to extract gDNA for gene identification before terminating these positive individual plants.

Reporter Gene Analysis

After thoroughly drying the harvested seeds, ~50 seeds from each plant are placed on a small round filter paper (~7 mm in diameter), which has been moistened with water and placed on a metal stage for scanning electron microscopy (SEM) specimens (Fig. 16.2a). The metal stage is cleaned thoroughly after each specimen to avoid contamination. A large plastic Petri dish (15 cm in diameter) with two layers of filter paper (No. 2, Whatman Inc., Clifton, New Jersey, USA) can hold about 50 seeds from each of 100 lines (e.g. ten individual plants × ten ABRC stocks) (Fig. 16.2b). After seeds are imbibed at 4°C for 3 days and incubated at 22°C for 22 h, the small filter papers holding seeds are briefly blotted on dry filter papers to remove excessive water and thereafter transferred to GUS substrate solution. GUS staining of seeds is performed as previously described (Weigel and Glazebrook, 2002) using 100 mM sodium phosphate buffer (pH 7.0) containing

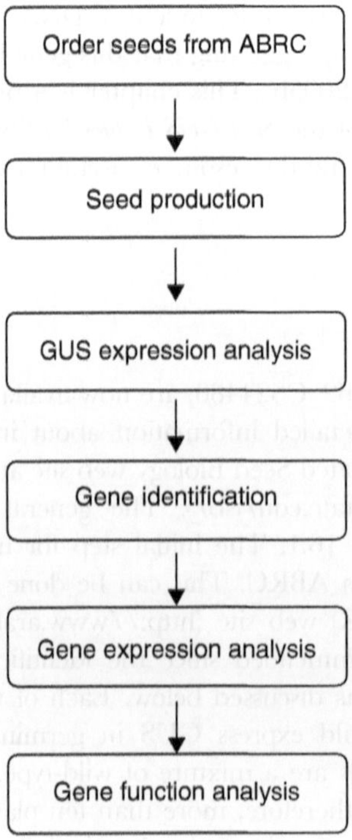

Fig. 16.1. A flow chart for identifying germination-associated genes using the *Seed-GUS-Expression* enhancer-trap library.

0.1% (v/v) Triton X-100 and 2 mM X-Gluc (RPI Co., Mount Prospect, Illinois, USA). Samples can be examined for GUS staining after an overnight incubation at room temperature.

The timing of GUS staining is critical for identifying germination-associated genes in the enhancer-trap lines. Imbibed *Arabidopsis* seeds initially exhibit only testa rupture, which is then followed by endosperm rupture indicating that germination is complete (Fig. 16.3a). If staining is done after endosperm rupture, the majority of GUS-positive samples may actually reflect only postgermination-associated gene expression. On the other hand, it seems difficult to penetrate the GUS substrate into the endosperm and the embryo before testa rupture, probably due to the impermeability of the testa to the GUS substrate. Therefore, GUS staining needs to be focused on the lag phase between testa and endosperm rupture (Liu *et al.*, 2005a). Initiation of GUS staining after 20 h of incubation at 22°C appears most appropriate to fit this narrow lag phase window of time (Fig. 16.3b).

Identification of Trapped Genes

The leaves collected from the GUS-positive individual plants are used for gDNA extraction. The QUICK-PREP method (http://www.biotech.wisc.edu/NewServicesandResearch/Arabidopsis/) works well for a scale of one to several leaves. The gDNA is run on a 1% (w/v) agarose gel with ethidium bromide staining to deter-

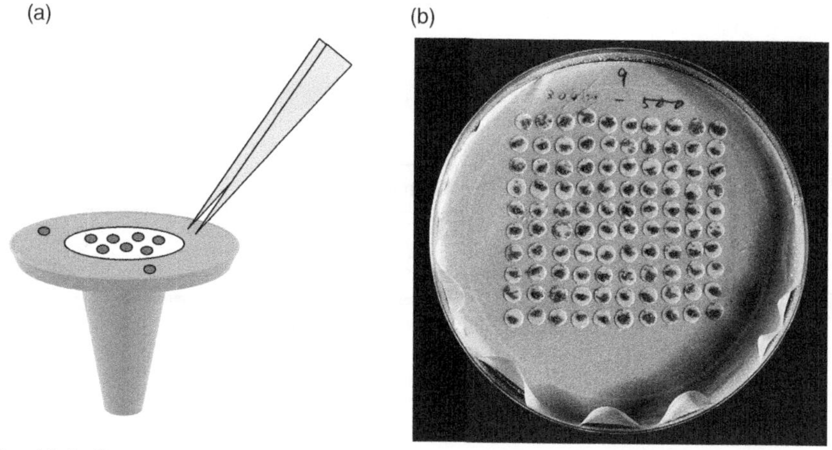

Fig. 16.2. Procedures for plating seeds to screen for GUS expression during germination. (a) Schematic representation of seed plating on the metal stage, which is generally used for SEM specimens. Small filter paper (~7 mm) moistened with water is placed on the stage and ~50 seeds are sown. (b) A 15 cm plastic Petri dish with two layers of filter paper moistened with 8 ml water, containing 50 seeds (placed on small filter paper) from each of 100 lines.

mine the quality of the DNA. Genome-walking PCR analysis is performed using a Genome Walker Kit (Clontech Laboratories, Inc., Palo Alto, California, USA), according to the manufacturer's manual. Briefly, gDNA extracted from the enhancer-trap lines is digested with *Dra*I. Adapter DNA provided with the kit is ligated to the *Dra*I-digested gDNA fragments. The gDNA region flanking the T-DNA insert is amplified using an adapter primer provided with the kit and a T-DNA right border-specific primer (RB1: 5'-TCTAGAGTCGACCTGCAGGCATGCAAGCTT-3'). Second-round PCR is performed using the nested adapter primer and the nested T-DNA right border-specific primer (RB2: 5'-TCCCAACAGTTGCGCACCTGAATGGCGAAT-3'). Three other enzymes (*Eco*RV, *Pvu*II and *Stu*I) contained in the kit can also be used for the same procedure, but *Dra*I gave the best results in our laboratory. Detailed methods can be found in Liu *et al.* (2005a). Figure 16.4a shows the examples of amplified DNA fragments from genome-walking PCR. The product is usually 0.1–2 kb long and can be directly sequenced. The DNA sequence flanking the T-DNA right border is analysed using the *SIGnAL T-DNA Express Arabidopsis Gene Mapping Tool* (http://signal.salk.edu/cgi-bin/tdnaexpress). An example showing the identification of a putative T-DNA insertion site is shown in Fig. 16.5. In this programme, the direction of the submitted sequence is displayed so that the direction of the T-DNA insertion relative to the trapped candidate gene is clear. The example shows a putative T-DNA insertion site in the vicinity of a cytochrome *P450* gene (At5g25180) with the right border facing the gene.

To verify the predicted T-DNA insertion site in the enhancer-trap line, a gene-specific primer (GSP) is designed in the region ~0.5 kb upstream of the putative insertion site. The primer is used together with the RB2 primer (described earlier) to perform PCR for both the trapped line and wild-type *Arabidopsis*. The predicted

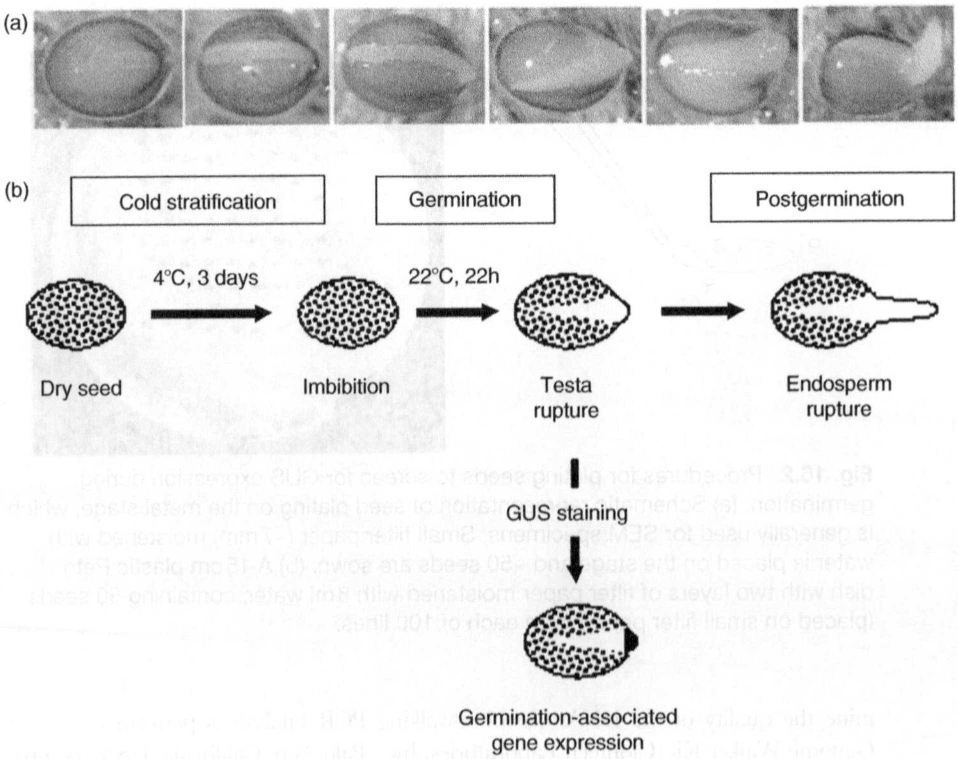

Fig. 16.3. (a) Morphological changes in imbibed *Arabidopsis* seeds (left to right). First visible change during germination is the rupture of testa; the radicle then penetrates a single cell layer of endosperm following some lag phase. (b) Schematic representation of the timing of GUS staining to identify germination-associated genes. Germinating seeds need to be transferred to GUS substrate solution immediately after testa rupture but before radicle emergence.

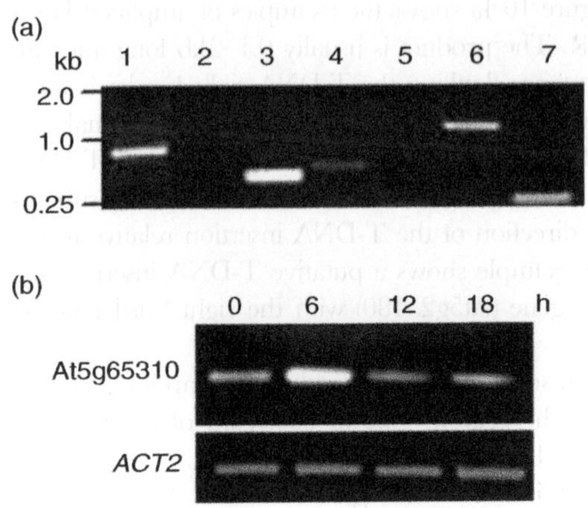

Fig. 16.4. (a) Example of PCR products obtained by genome-walking PCR. gDNA from seven independent enhancer-trap lines (1–7) are examined. No amplification was detected in lines 2 and 5. Successfully amplified products were sequenced to identify the putative T-DNA insertion sites. (b) Example of semi-quantitative RT-PCR to characterize stage-specific expression of the identified trapped genes (At5g65310 in this case). An actin gene *ACT2* (An *et al*., 1996) is used as a control.

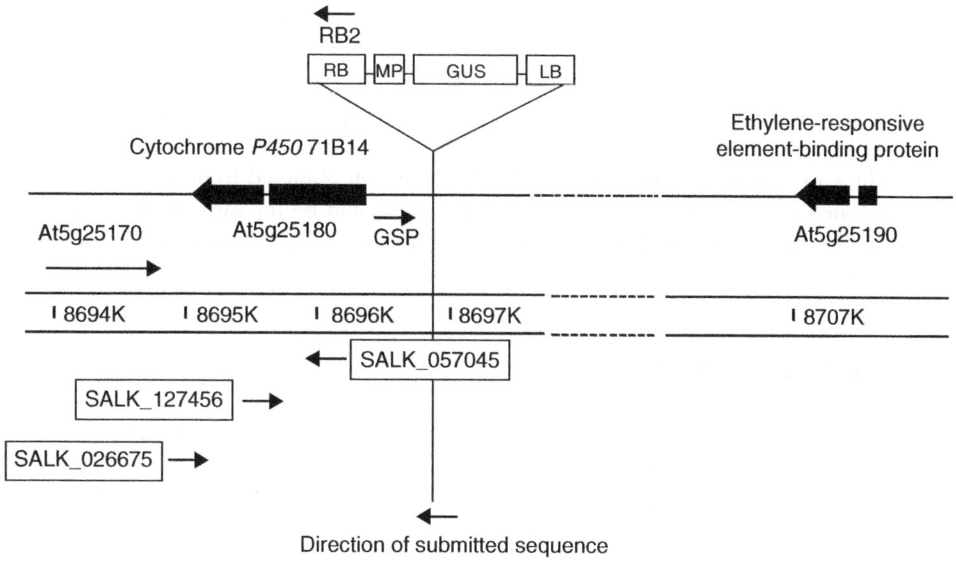

Fig. 16.5. Schematic representation of the process involved in identifying a putative T-DNA insertion site using *SIGnAL T-DNA Express Arabidopsis Gene Mapping Tool* (http://signal.salk.edu/cgi-bin/tdnaexpress). Simplified display of the web site is shown with an example of an enhancer-trap line that had a T-DNA insertion site in the vicinity of cytochrome *P450* gene. The sequence obtained from genome-walking PCR is copied, pasted and submitted to this web site to identify the putative insertion site. The programme on the web site indicates the location of the genome where the submitted sequence matches (vertical bar in the middle). It also shows the direction of the submitted sequence relative to the genome sequence (arrow at the bottom) so that the direction of T-DNA right border (RB), minimal promoter (MP), the glucuronidase-coding region (GUS) and left border (LB) relative to the insertion site can be deduced. The positions of T-DNA insertions in three putative knockout lines (SALK_026675, SALK_057045 and SALK_127456) are shown with arrows. RB: right border-specific primer. GSP: gene-specific primer.

size of DNA fragment should be amplified from the enhancer-trap line, but not from wild-type gDNA. The junction between the gDNA and T-DNA is verified by sequencing this amplified product.

Gene Expression Analysis

The presence of a T-DNA insertion immediately adjacent to the candidate gene does not necessarily prove that the GUS expression detected in the enhancer-trap line represents the expression patterns of this gene. One also needs to examine expression of the candidate gene during seed germination. Forward and reverse primers are designed in the coding region of the candidate gene and used for RT-PCR analysis to examine gene expression. A few time points (e.g. 12 and 24 h) during cold stratification at 4°C and incubation at 22°C can be tested for expression analysis. When expression is confirmed, detailed analysis is performed using semi-quantitative RT-PCR to characterize stage-specific expression (Fig. 16.4b).

It is possible to have multiple T-DNA insertion sites in an enhancer-trap line; therefore, it is important to characterize the 5' upstream regulatory sequence of the candidate gene using a promoter–reporter construct. If the candidate gene is the correct one, the tissue-specific GUS expression obtained from the plants transformed with the promoter–reporter construct should be the same as that observed in the enhancer-trap line. This approach has been used successfully in previous work (Liu *et al.*, 2005b).

Gene Function Analysis

The biological function of the genes identified from the enhancer-trap lines can be determined using knockout plants. For example, three independent knockout lines (SALK_026675, SALK_057045 and SALK_127456) for the cytochrome *P450* gene were found using the *SIGnAL T-DNA Express Arabidopsis Gene Mapping Tool* web site (Fig. 16.5). Knockout lines that have T-DNA insertions in the coding regions are the best candidates, although those carrying an insertion in the promoter region have also exhibited knockout phenotypes (Liu *et al.*, 2005b). Seeds of these putative knockout lines can be ordered through the TAIR web site. These seeds are used to grow plants from which the next generation of seeds will be harvested. To identify homozygous lines, segregation ratios can be determined by growing plants on media containing antibiotic (e.g. kanamycin). When putative homozygous lines showing 100% antibiotic resistance are obtained, target gene expression levels in the potential mutant seeds need to be compared with the level of gene expression in wild-type seeds. Knockout lines may not be available for some genes identified. In these cases, an RNA interference (RNAi) approach can be used to produce and examine loss-of-function mutants. It is possible that an apparent phenotype in seed germination or in other phases of plant development may not be observed due to redundancy in gene function. In this case, overexpression of the candidate genes using CaMV 35S or inducible promoters or examination of double mutants may be necessary. The knockout approach worked well for characterization of the GATA zinc finger protein expressed in germinating *Arabidopsis* seeds (Liu *et al.*, 2005b).

Perspectives

The enhancer-trap approach is well utilized to identify potentially important factors of seed germination. Multiple transcription factors have been identified as candidate trapped genes (Liu *et al.* 2005a). Use of *Arabidopsis* as a model plant is accelerating gene discoveries in seed biology. However, it is still important to identify and characterize homologues of these *Arabidopsis* genes in model species for germination research, such as tomato and lettuce (*Lactuca sativa* L.), and also in seeds of other agricultural crops. This approach is very important to elucidate universal mechanisms underlying seed dormancy and germination.

Acknowledgements

We are grateful to the Arabidopsis Biological Resource Center (ABRC), Ohio State University, USA, for propagating the *Seed-GUS-Expression* enhancer-trap lines for the international research community. This work was supported by the National Science Foundation grant IBN-0237562 to H. Nonogaki and the American Seed Research Foundation grant to H. Nonogaki and R.C. Martin.

References

Akimoto, A., Wada, H. and Hayashi, S. (2005) Enhancer trapping with a red fluorescent protein reporter in *Drosophila. Developmental Dynamics* 233, 993–997.

An, Y.Q., McDowell, J.M., Huang, S., McKinney, E.C., Chambliss, S. and Meagher, R.B. (1996) Strong, constitutive expression of the *Arabidopsis* ACT2/ACT8 actin subclass in vegetative tissues. *The Plant Journal* 10, 107–121.

Balciunas, D., Davidson, A.E., Sivasubbu, S., Hermanson, S.B., Welle, Z. and Ekker, S.C. (2004) Enhancer trapping in zebrafish using the Sleeping Beauty transposon. *BMC Genomics* 5, 62.

Bradford, K.J., Chen, F., Cooley, M.B., Dahal, P., Downie, B., Fukunaga, K.K., Gee, O.H., Gurusinghe, S., Mella, R.A., Nonogaki, H., Wu, C.-T. and Yim, K.-O. (2000) Gene expression prior to radicle emergence in imbibed tomato seeds. In: Black, M., Bradford, K.J. and Vazquez-Ramos, J. (eds) *Seed Biology: Advances and Applications.* CAB International, Wallingford, UK, pp. 231–251.

Campisi, L., Yang, Y., Yi, Y., Heilig, E., Herman, B., Cassista, A.J., Xiang, H. and Jack, T. (1999) Generation of enhancer trap lines in *Arabidopsis* and characterization of expression patterns in the inflorescence. *The Plant Journal* 17, 699–707.

Debeaujon, I., Peeters, A.J.M., Léon-Kloosterziel, K.M. and Koornneef, M. (2001) The TRANSPARENT TESTA12 gene of *Arabidopsis* encodes a multidrug secondary transporter-like protein required for flavonoid sequestration in vacuoles of the seed coat endothelium. *The Plant Cell* 13, 853–872.

Debeaujon, I., Nesi, N., Perez, P., Devic, M., Grandjean, O., Caboche, M. and Lepiniec, L. (2003) Proanthocyanidin-accumulating cells in *Arabidopsis* testa: regulation of differentiation and role in seed development. *The Plant Cell* 15, 2514–2531.

Downie, B.A., Zhang, D., Dirk, L.M.A., Thacker, R.R., Pfeiffer, J.A., Drake, J.L., Levy, A.A., Butterfield, D.A., Buxton, J.W. and Snyder, J.C. (2003) Communication between the maternal testa and the embryo and/or endosperm affect testa attributes in tomato. *Plant Physiology* 133, 145–160.

Endo, M., Nakamura, S., Araki, T., Mochizuki, N. and Nagatani, A. (2005) Phytochrome B in the mesophyll delays flowering by suppressing FLOWERING LOCUS T expression in *Arabidopsis* vascular bundles. *The Plant Cell* 17, 1941–1952.

Engineer, C.B., Fitzsimmons, K.C., Schmuke, J.J., Dotson, S.B. and Kranz, R.G. (2005) Development and evaluation of a Gal4-mediated LUC/GFP/GUS enhancer trap system in *Arabidopsis. BMC Plant Biology* 5, 9.

Gallardo, K., Job, C., Groot, S.P.C., Puype, M., Demol, H., Vandekerckhove, J. and Job, D. (2001) Proteomic analysis of *Arabidopsis* seed germination and priming. *Plant Physiology* 126, 835–848.

Gallardo, K., Job, C., Groot, S.P.C., Puype, M., Demol, H., Vandekerckhove, J. and Job, D. (2002) Proteomics of *Arabidopsis* seed germination: a comparative study of wild-type and gibberellin-deficient seeds. *Plant Physiology* 129, 823–837.

He, Y., Tang, W., Swain, J.D., Green, A.L., Jack, T.P. and Gan, S. (2001) Networking senescence-regulating pathways by using *Arabidopsis* enhancer trap lines. *Plant Physiology* 126, 707–716.

Johnson, A.A., Hibberd, J.M., Gay, C., Essah, P.A., Haseloff, J., Tester, M. and Guiderdoni, E. (2005) Spatial control of transgene expression in rice (*Oryza sativa* L.) using the GAL4 enhancer trapping system. *The Plant Journal* 41, 779–789.

Ko, S. and Kamada, H. (2002) Enhancer-trapping system for somatic embryogenesis in carrot. *Plant Molecular Biology Report* 20, 421a–421j.

Ku, H.M., Vision, T., Liu, J. and Tanksley, S.D. (2000) Comparing sequenced segments of the tomato and *Arabidopsis* genomes: large-scale duplication followed by selective gene loss creates a network of synteny. *Proceedings of the National Academy of Sciences of the United States of America* 97, 9121–9126.

Ku, H.M., Liu, J., Doganlar, S. and Tanksley, S.D. (2001) Exploitation of *Arabidopsis*-tomato synteny to construct a high-resolution map of the ovate-containing region in tomato chromosome 2. *Genome* 44, 470–475.

Liu, P.-P., Koizuka, N., Homrichhausen, T.M., Hewitt, J.R., Martin, R.C. and Nonogaki, H. (2005a) Large-scale screening of *Arabidopsis* enhancer-trap lines for seed germination-associated genes. *The Plant Journal* 41, 936–944.

Liu, P.-P., Koizuka, N., Martin, R.C. and Nonogaki, H. (2005b) The *BME3* (*Blue Micropylar End 3*) GATA zinc finger transcription factor is a positive regulator of *Arabidopsis* seed germination. *The Plant Journal* 44, 960–971.

Ogawa, M., Hanada, A., Yamauchi, Y., Kuwahara, A., Kamiya, Y. and Yamaguchi, S. (2003) Gibberellin biosynthesis and response during *Arabidopsis* seed germination. *The Plant Cell* 15, 1591–1604.

Sagasser, M., Lu, G.-H., Hahlbrock, K. and Weisshaar, B. (2002) *A. thaliana* TRANSPARENT TESTA 1 is involved in seed coat development and defines the WIP subfamily of plant zinc finger proteins. *Genes and Development* 16, 138–149.

Vroemen, C.W., Mordhorst, A.P., Albrecht, C., Kwaaitaal, M.A. and de Vries, S.C. (2003) The CUP-SHAPED COTYLEDON3 gene is required for boundary and shoot meristem formation in *Arabidopsis*. *The Plant Cell* 15, 1563–1577.

Weigel, D. and Glazebrook, J. (2002) *Arabidopsis: A Laboratory Manual.* Cold Spring Harbor Laboratory Press, Cold Spring Harbor, New York.

17 Betaomics: a Combined Proteome and Transcriptome Profiling Approach to Characterize Seed Germination and Vigour in Sugarbeet Seeds

J. Catusse,[1] E. Pestsova,[2] C. Bechetoille,[1] C. Job,[1] J. Meinhard,[3] A. Jagonak,[3] J.-M. Strub,[4] A. van Dorsselaer,[4] U. Fischer,[3] P. Westhoff[2] and D. Job[1]

[1]*Centre National de la Recherche Scientifique (CNRS)–Bayer CropScience Joint Laboratory, Bayer CropScience, 14-20 rue Pierre Baizet, 69263 Lyon, Cedex 9, France;* [2]*Institut für Entwicklungs und Molekularbiologie der Pflanzen, Heinrich-Heine-Universität, Universitätsstr, 1, 40225 Düsseldorf, Germany;* [3]*KWS SAAT AG, Grimsehlstr. 31, 37555 Einbeck, Germany;* [4]*Laboratoire de Spectrométrie de Masse Bio-Organique, ECPM, 25 rue Becquerel, 67087 Strasbourg, France*

Abstract

Producing high-quality seed lots, in terms of germinability, is a major concern of seed companies. The availability of molecular and/or biochemical markers of seed vigour would allow seed producers to better characterize their germplasm and to better design invigoration seed treatments such as priming. The aim of this work is to use global expression profiling methods (i.e. transcriptomics and proteomics) to find such markers for sugarbeet (*Beta vulgaris* L.) seeds. This combined approach allows, for the first time, the comparison of proteomic and transcriptomic data for sugarbeet seed, and the characterization of the molecular and biochemical mechanisms underlying seed vigour and germination.

Introduction

Sugarbeet (*Beta vulgaris* L.) is a biennial dicotyledonous plant of the *Chenopodiaceae* family. In the first (vegetative) growing season, sugarbeet plants develop a root that serves as a storage organ for sucrose. Plants not harvested after the first year for sugar production that are exposed to low temperature over an extended period of time (vernalization) produce a branched inflorescence in the second (generative) season under long-day conditions. Fruit develops after pollination, and contains the true seed within a pericarp structure. In the time course of germination, the covering structure of this pericarp is opened by the expanding cells of the seedling, enabling growth, emergence and the development of a new sugarbeet plant. The ability of the seed to germinate and emerge vigorously is determined not only by the genetic background of the seed but also by the environmental conditions during seed development and emergence (Dürr and Aubertot, 2000). Because most of the modern sugarbeet varieties are monogermic and seeds are drilled to the final stand, any failure in germination or emergence will directly influence the corresponding crop yield. In order to secure the best seed quality, it is therefore important to improve the knowledge of the physiological mechanisms that account for this property.

To this end, we started a characterization of the proteome of sugarbeet seeds by high-resolution 2D-gel electrophoresis (2D-PAGE) in combination with mass spectrometry (MS). In a parallel approach, we initiated mRNA profiling experiments during germination by macroarray analysis. We anticipate that the data gathered in this work will help in understanding the many processes that are important for the success of sugarbeet seed germination and to determine components involved in sugarbeet seed quality, particularly concerning the molecular mechanisms governing physiological functions such as seedling emergence and survival of mature seeds in the dried state. The usefulness of global expression profiling analyses for the study of sugarbeet seed vigour has recently been documented (de los Reyes and McGrath, 2003; de los Reyes *et al.*, 2003).

Osborne (1924) classified seed proteins into four categories according to their solubility in various solutions: (i) albumins, which are soluble in water at neutral or slightly acidic pH and correspond primarily to enzymes; (ii) globulins, which are insoluble in water but soluble in salt solutions (e.g. 0.5 M NaCl); (iii) prolamins, which are soluble in ethanol; and (iv) glutelins, which can be extracted with very acidic or very basic solutions. Globulins, prolamins and glutelins are seed storage protein reserves. These protein classes are not present systematically in the seeds of all plant species (Bewley and Black, 1983). For example, prolamins exist only in the herbaceous plant species and cereals as the major storage proteins, whereas only albumin, globulin and glutelin fractions are investigated in sugarbeet.

The literature concerning sugarbeet proteins is very limited. Thus, the first aim of this study was to document the various protein classes in dry mature seeds by a proteomic approach. The only publication on this subject goes back to 1990 (Lawrence *et al.*, 1990), dealing with an analysis of sugarbeet seed proteins by 1D-gel electrophoresis (1D-PAGE). A second objective is to characterize the proteome of seeds during germination.

It is well known that seed development and germination is accompanied by regulated gene expression. During the maturation phase mRNAs encoding proteins involved in the production of storage compounds are prevalent, whereas during the desiccation phase the levels of those mRNAs that encode proteins with protective function, like late embryogenesis abundant (LEA) and heat shock proteins

(HSPs), increase. Germination is accompanied by drastic changes in gene expression patterns, the genes involved in reinitiation of cell cycle, respiration, reserve mobilization and other processes necessary for seedling growth are activated. Gene expression programmes during seed maturation and germination have been studied in several plant species: *Arabidopsis thaliana* (L.) Heynh. (Girke *et al.*, 2000; White *et al.*, 2000; Clerkx *et al.*, 2004; Nakabayashi *et al.*, 2005), *Brassica oleracea* L. (Soeda *et al.*, 2005) and *Hordeum vulgare* L. (Potokina *et al.*, 2002; Sreenivasulu *et al.*, 2002). However, hardly anything is known about gene expression and regulation during seed development, maturation and germination in sugarbeet. Therefore, our third aim was to perform a transcriptome profiling of dry and germinating sugarbeet seeds through the generation of seed-specific cDNA libraries, construction of a unigene expressed sequence tag (EST) set and further analysis of gene expression by cDNA macroarray approach.

Materials and Methods

Plant material and germination experiments

Germination assays were carried out on four replicates of 100 sugarbeet seeds (KWS, Einbeck, Germany) and repeated in at least three independent experiments. Germination characteristics were evaluated in a pleated filter paper assay according to International Seed Testing Association (ISTA) regulations (ISTA, 1985) with 30 ml H_2O at 15°C. Germination was followed by daily counts, and final germination was determined after 14 days. A seed was considered germinated when the radicle protruded through the seed coat. Seed Calculator software (Plant Research International BV, Wageningen, The Netherlands) was used in curve-fitting analyses to estimate the germination parameters from the germination curves. The following values were calculated: T_1, T_{50} and T_{95}, corresponding to the completion of 1%, 50% and 95% of germination, respectively.

Proteome and transcriptome experiments

The three time points, T_1, T_{50} and T_{95}, were selected and used for further proteome and transcriptome analysis. The 1D-PAGE and proteomic experiments were carried out as described in Job *et al.* (2005).

Three cDNA libraries, each composed of 20,000 clones, have been developed from germinating (T_1, T_{50} and T_{95}) seeds using the pBluescript II XR cDNA library construction kit (Stratagene, California, USA) and one library of the same size was prepared out of dry seeds (T_0) using the SuperScript Plasmid System for cDNA Synthesis (Gibco-BRL, Eggenstein, Germany). Insert size has been found to vary from 0.5 kb to 3.5 kb with a mean value of 1.1 kb. Three sets of high-density filters containing 18,432 colonies/22.5 cm^2 were developed in cooperation with Max Planck Institute (MPIZ, Cologne, Germany). Colony hybridization was carried out using the DIG non-radioactive system (Roche, Mannheim, Germany).

Results and Discussion

Proteomics of the dry mature seed

In order to characterize the proteome of dry mature seeds, we first used the extraction protocols devised by Osborne (1924) and Lawrence *et al.* (1990) and analysed proteins extracted under these conditions using 1D-PAGE (Fig. 17.1).

A comparison of the different extraction protocols shows that the proteins of the albumin class resulting from the methods of Osborne (1924) (Fig. 17.1, track a) and Lawrence *et al.* (1990) (Fig. 17.1, track b) are very similar. However, the globulins of the 11S type, which are very abundant reserve proteins in sugarbeet seeds, are absent in the fractions prepared according to the methods of Osborne (1924) or Lawrence *et al.* (1990). The 11S globulins are the storage proteins of dicotyledons, which belong to the family of globulin-legumin (Gruis *et al.*, 2002). They exist in the form of hexameric structures, each subunit of the hexamers being made up of an acid chain (called the A chain of about 35 kDa) and a basic chain (called the B chain of about 25 kDa), both of which are derived from a single precursor. During the seed maturation process on the mother plant, the protein precursor undergoes a post-translational cleavage, resulting in the generation of two polypeptides: the acidic and basic chains that remain associated by a disulfide bridge (Adachi *et al.*, 2001, 2003). Using 1D sodium dodecyl sulfate polyacrylamide gel electrophoresis (1D SDS-PAGE) under reducing conditions, one expects to observe the two major bands of A and B subunits, with molecular masses of ~25 and 35 kDa, respectively. From the absence of such polypeptides in the globulin fraction (Fig. 17.1), we decided to develop a new protocol of extraction of sugarbeet proteins based on our previous work with *Arabidopsis* seeds (Gallardo *et al.* 2001, 2002; Rajjou *et al.*, 2004; Job *et al.*, 2005). In this new protocol, proteins are extracted in a sequential way

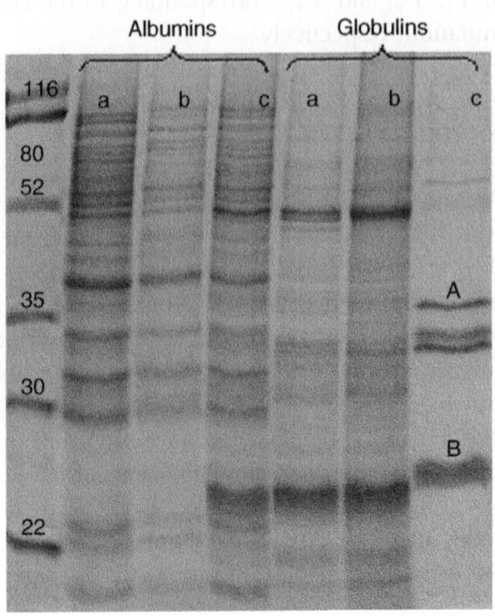

Fig. 17.1. 1D-gel electrophoresis of albumin and globulin proteins from 100 dry mature sugarbeet seeds according to the sequential extraction protocols of (a) Osborne (1924); (b) Lawrence *et al.* (1990); and (c) in the present work. A and B represent acidic and basic subunits of 11S globulins, respectively.

using the following series of solutions: an Hepes-EDTA buffer (Job *et al.*, 1997) of low ionic strength to solubilize the albumins; a buffer of higher ionic strength (Gallardo *et al.*, 2001) to release the globulins; sodium hydroxide (2% NaOH) to solubilize the glutelins and finally ethanol 70% to solubilize the prolamins. Figure 17.1 clearly shows that the globulin fraction obtained in this way contains all the three major bands previously identified as the 11–12S globulins of sugarbeet (Job *et al.*, 1997; Capron *et al.*, 2000). This result has been confirmed by 2D-PAGE (data not shown).

2D-PAGE allowed the visualization of about 800 albumins (Fig. 17.2) and 400 globulins (data not shown) on the 2D gels. Proteins were identified by MS, by nano LC MS/MS of peptide mixtures after digestion by trypsin. A total of 192 proteins (136 albumins and 56 globulins) have been identified on the reference protein maps for the dry mature sugarbeet seeds (Fig. 17.3).

Results of MS/MS analyses were correlated with the position of proteins on the 2D gels. The genome analysis of the plant biological system (GABI) beet database (http://gabi.rzpd.de/database/cgi-bin/Blast.pl.cgi) (Herwig *et al.*, 2002) was used to determine homologies with sugarbeet ESTs. A blast analysis was also

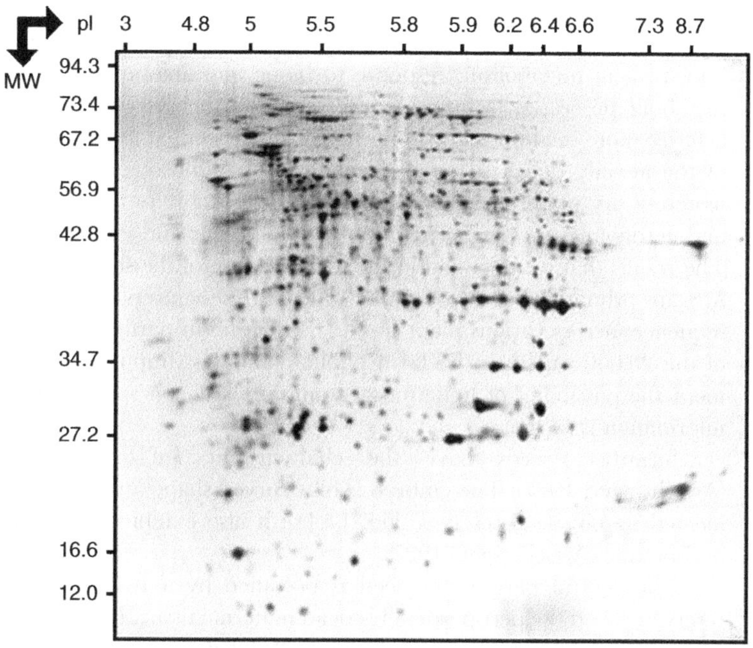

Fig. 17.2. 2D-gel electrophoresis (2D-PAGE) of the albumins extracted from dry mature seeds of sugarbeet. Proteins were first separated by electrophoresis according to charge. Isoelectrofocusing was carried out with protein samples with an extract equivalent to about 100 µg protein for all samples. Proteins were then separated according to size by sodium dodecyl sulfate polyacrylamide gel electrophoresis (SDS-PAGE) using 10% polyacrylamide gels. Proteins were visualized by silver nitrate coloration (see Job *et al.*, 2005 for details on protein extractions and proteomic analyses).

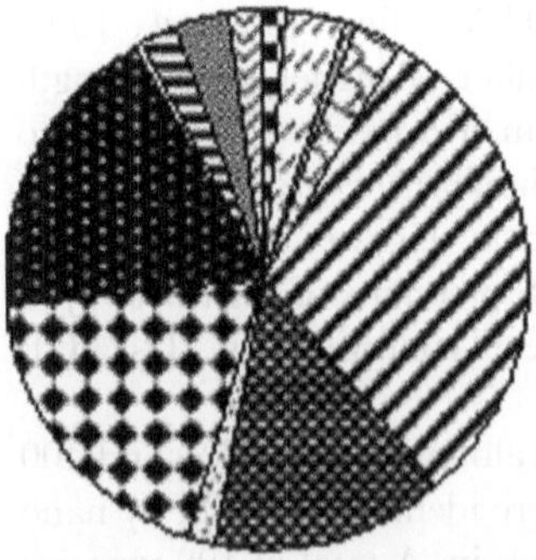

Fig. 17.3. Ontological classification of the seed proteins in dry mature sugarbeet seeds. Proteins were identified by proteomic analysis.

carried out using The Arabidopsis Information Resource (TAIR) database (http://www.arabidopsis.org/Blast/) to seek for orthologous proteins in *Arabidopsis*. One of our objectives is to compare the metabolic processes (Fig. 17.3) that are set up during germination in the model plant *Arabidopsis*, and in plants of agronomic interest such as *B. vulgaris*.

Four major classes of proteins dominate the actual proteome map: metabolism, protein metabolism, response to biotic and abiotic stimuli and cellular and metabolic processes. In future work, we will further accentuate our effort to characterize more seed proteins. In particular, it is essential to complete our knowledge of the already detected metabolic pathways. A thorough knowledge of the proteome of dry mature seeds will also make it possible to analyse various biological and agronomic questions (e.g. germination, vigour and environmental variability). The results that we obtained in the identification of the sugarbeet seed proteins by MS are primarily due to two mains factors: the conservation of the seed protein sequences across various plant species, *Arabidopsis* in particular; and the availability of the 20,000 sugarbeet ESTs in public databases. Importantly, our results document the possibility of initiating genomic research on species for which genomic information is lacking.

Sugarbeet seeds have the characteristics of campylotrope-type seeds (Artschwager, 1927). The embryo is of a curved shape and surrounds the centrally located starchy seed reserves (Fig. 17.4a); it also exhibits the peculiarity of being perispermic (Artschwager, 1927).

The central part of the seed is occupied by a tissue composed of starchy reserves, called the perisperm. This dead maternal tissue, derived from the nucellus, consists of a network of cells with fine walls filled with starch grains in the form of globules. In the mature seed, the endosperm is present only in the micropylar area where it consists of a tissue of two cellular layers surrounding the radicle–hypocotyl axis of the embryo. Sugarbeet seeds are shed with their pericarp intact.

We were also interested in characterizing the tissue specificity of the seed proteins. To this end, we have carried out dissections of dry seeds and set up a study of the tissue proteome. In spite of the small size of sugarbeet seed and the robustness of the pericarp, pure fractions of various seed tissues including root,

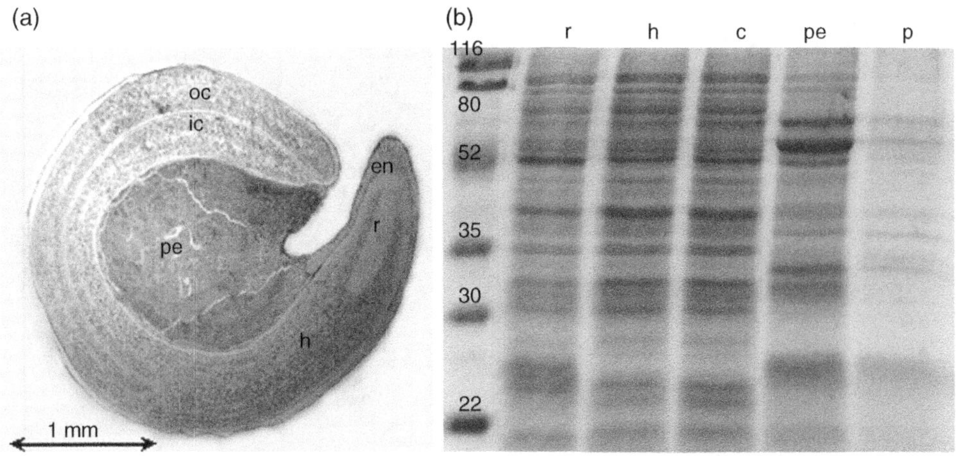

Fig. 17.4. (a) Transverse seed section of a dry mature sugarbeet seed and (b) 1D-gel electrophoresis (1D-PAGE) of albumin proteins from various tissues of mature dry sugarbeet seeds. oc = outer cotyledon; ic = inner cotyledon; en = endosperm; r = radicle; pe = perisperm; h = hypocotyl; c = cotyledon; and p = pericarp.

hypocotyl, cotyledons and perisperm can be prepared (Elamrani *et al.*, 1992, 1994). The sequential extraction protocol of sugarbeet seeds was applied to each of these tissues in order to evaluate, by a proteomic approach, whether specific proteins accumulate in these tissues. The proteins extracted in each tissue were first analysed by 1D-PAGE. The electrophoretic profiles of albumin proteins from the root, hypocotyl and cotyledons are very similar (Fig. 17.4b). Using 2D gel analysis, it is possible to visualize electrophoretic profiles of proteins with tissue specificity and further analysis of this is in progress.

Proteomics of the germinating seed

Optimum germination conditions were defined for this study. Three selected time points, T_1, T_{50} and T_{95}, correspond to physiological stages of germination: the end of the first rapid phase of imbibition, the phase characterized by the end of desiccation tolerance and the completion of germination and the beginning of seedling growth. A 2D gel analysis of proteins isolated at these stages was performed, and the characterization of proteins with variable levels of accumulation (Fig. 17.5) provided information about the metabolic processes that are set up in seeds during germination.

Development of a set of expressed sequence tags from dry and germinating seeds

The ESTs derived from single-pass partial sequences of cDNA clones represent invaluable resources for new gene discovery and the annotation of genomic sequences

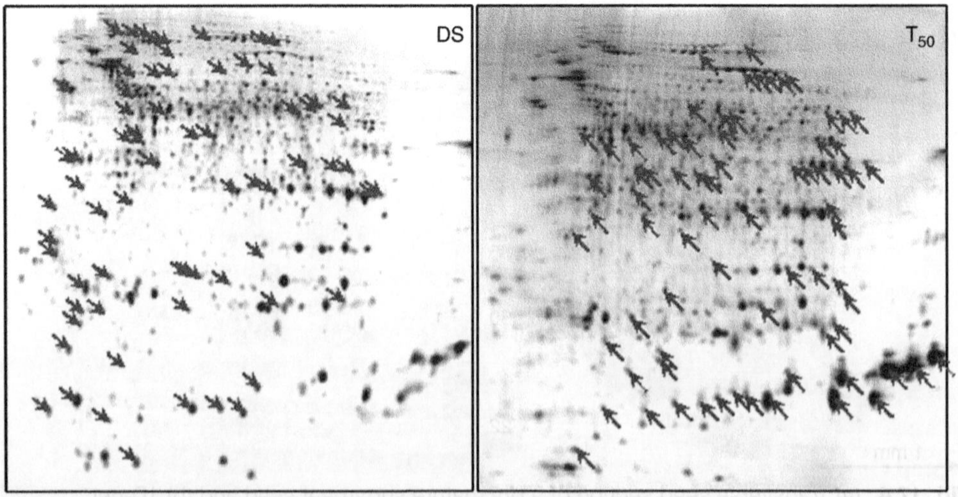

Fig. 17.5. 2D-gel electrophoresis (2D-PAGE) of the albumins extracted from dry mature sugarbeet seeds (DS) and seeds having completed 50% of germination (T_{50}). In DS, arrows indicate proteins whose spot volumes decrease at 50% of germination (the gel of dry mature seed proteins (T_0) is taken as a reference). In T_{50}, arrows indicate proteins whose spot volumes increase at 50% of germination.

expressed in specific plant tissues under defined growth conditions. Most of the commonly available sugarbeet ESTs originate from leaf-, root- or inflorescence-specific cDNA libraries (Bellin *et al.*, 2002; Herwig *et al.*, 2002). In order to gain insight into sugarbeet seed development and germination, we developed four cDNA libraries specific for dry mature seeds and for different germinating stages. A number of cDNA clones were sequenced non-randomly, and a special preselection technique applied in order to decrease the number of identical sequences originating from highly expressed genes and increase the percentage of unique ESTs. Sequencing of each of the 96 clones was followed by hybridization of the sequenced cDNAs with high-density filters spotted from the cDNA libraries. The clones giving a positive signal were eliminated from the set of the cDNAs intended to be sequenced. In total, 3200 sugarbeet cDNA clones were sequenced, and the estimated number of unique sequences comprise 2000 ESTs. Clustering and annotation analysis of the ESTs are in progress.

Our next goal is to use this set of unique, non-redundant cDNAs for macroarray development. A transcriptome map of sugarbeet germination will then be produced by comparison of gene expression profiles in dry and germinating seeds. Genes differentially expressed in seed samples varying in viability will be identified by comparing expression in seeds of poor and superior germination quality.

Conclusions

A reference map of the total proteome of dry mature sugarbeet seeds is in the course of establishment, with more than 192 proteins already identified. The analysis of the proteome of these seeds during the germination process will help to better understand how they are able to give a seedling when they are hydrated, and thus how to store them in the best conditions. There are no other examples in the development of eukaryotic organisms where interruption and conservation of the embryo's vigour occurs. Therefore, it is important to understand how seeds are preserved in a dry state and which biochemical mechanisms are set up when they start the process of returning to life, i.e. germination.

Acknowledgements

This work has been supported by a grant from Génoplante and GABI.

References

Adachi, M., Takenaka, Y., Gidamis, A.B., Mikami, B. and Utsumi, S. (2001) Crystal structure of soybean proglycinin A1aB1b homotrimer. *Journal of Molecular Biology* 305, 291–305.

Adachi, M., Kanamori, J., Masuda, T., Yagasaki, K., Kitamura, K., Mikami, B. and Utsumi, S. (2003) Crystal structure of soybean 11S globulin: Glycinin A3B4 homohexamer. *Proceedings of the National Academy of Sciences of the United States of America* 100, 7395–7400.

Artschwager, E. (1927) Development of flowers and seed in the sugar beet. *Journal of Agricultural Research* 34, 1–25.

Bellin, D., Werber, M., Theis, T., Schulz, B., Weisshaar, B. and Schneider, K. (2002) EST sequencing, annotation and macroarray transcriptome analysis identify preferentially root-expressed genes in sugar beet. *Plant Biology* 4, 700–710.

Bewley, J.D. and Black, M. (1983) *Physiology and Biochemistry of Seeds.* Springer, Berlin, Germany.

Capron, I., Corbineau, F., Dacher, F., Job, C., Côme, D. and Job, D. (2000) Sugar beet seed priming: effects of priming conditions on germination, solubilization of 11-S globulin and accumulation of LEA proteins. *Seed Science Research* 10, 243–254.

Clerkx, E.J.M., El-Lithy, M.E., Vierling, E., Ruys, G.J., Blankestijn-De Vries, H., Groot, S.P.C., Vreugdenhil, D. and Koornneef, M. (2004) Analysis of natural allelic variation of *Arabidopsis* seed germination and seed longevity traits between the accessions Landsberg *erecta* and Shakdara, using a new recombinant inbred line population. *Plant Physiology* 135, 432–443.

de los Reyes, B.G. and McGrath, J.M. (2003) Cultivar-specific seedling vigor and expression of a putative oxalate oxidase germin-like protein in sugar beet (*Beta vulgaris* L.). *Theoretical and Applied Genetics* 107, 54–61.

de los Reyes, B.G., Myers, S.J. and McGrath, J.M. (2003) Differential induction of glyoxylate cycle enzymes by stress as a marker for seedling vigor in sugar beet (*Beta vulgaris*). *Molecular Genetics and Genomics* 269, 692–698.

Dürr, C. and Aubertot, J.N. (2000) Emergence of seedlings of sugar beet (*Beta vulgaris* L.) as affected by the size, roughness and position of aggregates in the seedbed. *Plant and Soil* 219, 211–220.

Elamrani, A., Raymond, P. and Saglio, P. (1992) Nature and utilization of seed reserves during germination and heterotrophic growth of young sugar beet seedlings. *Seed Science Research* 2, 1–8.

Elamrani, A., Gaudillere, J.P., Saglio, P., Durr, C. and Raymond, P. (1994) Processus biochimiques au cours de la germination et de la croissance des plantules de betterave. In: Boiffin, J. and Choppin de Janvry, E. (eds) *L'implantation de la betterave industrielle, Les colloques de l'INRA 67.* INRA, Paris, pp. 29–38.

Gallardo, K., Job, C., Groot, S.P.C., Puype, M., Demol, H., Vandekerckhove, J. and Job, D. (2001) Proteomic analysis of *Arabidopsis* seed germination and priming. *Plant Physiology* 126, 835–848.

Gallardo, K., Job, C., Groot, S.P.C., Puype, M., Demol, H., Vandekerckhove, J. and Job, D. (2002) Proteomics of *Arabidopsis* seed germination: a comparative study of wild type and gibberellin deficient seeds. *Plant Physiology* 129, 823–837.

Girke, T., Todd, J., Ruuska, S., White, J., Benning, C. and Ohlrogge, J. (2000) Microarray analysis of developing *Arabidopsis* seeds. *Plant Physiology* 124, 1570–1581.

Gruis, D., Selinger, D.A., Curran, J.M. and Jung, R. (2002) Redundant proteolytic mechanisms process seed storage proteins in the absence of seed-type members of the vacuolar processing enzyme family of cysteine proteases. *The Plant Cell* 14, 2863–2882.

Herwig, R., Schulz, B., Weisshaar, B., Hennig, S., Steinfath, M. and Drungowski, M. (2002) Construction of a 'unigene' cDNA clone set by oligonucleotide fingerprinting allows access to 25000 potential sugar beet genes. *The Plant Journal* 32, 845–857.

ISTA (1985) International rules for seed testing. *Seed Science and Technology* 13, 322–447.

Job, C., Kersulec, M., Ravasio, L., Chareyre, C., Pépin, R. and Job, D. (1997) The solubilization of the basic subunit of sugar beet globulin during priming and early germination. *Seed Science Research* 7, 225–243.

Job, C., Rajjou, L., Lovigny, Y., Belghazi, M. and Job, D. (2005) Patterns of protein carbonylation in *Arabidopsis* seeds and during germination. *Plant Physiology* 138, 790–802.

Lawrence, D.M., Halmer, P. and Bowles, D.J. (1990) Mobilisation of storage reserves during germination and early seedling growth of sugar beet. *Physiologia Plantarum* 78, 421–429.

Nakabayashi, K., Okamoto, M., Koshiba, T., Kamiya, Y. and Nambara, E. (2005) Genome-wide profiling of stored mRNA in *Arabidopsis thaliana* seed germination: epigenetic and genetic regulation of transcription in seed. *The Plant Journal* 41, 697–709.

Osborne, T.B. (1924) *The Vegetable Proteins*. Logmans, Green and Co., London.

Potokina, E., Sreenivasulu, N., Altschmied, L., Michalek, W. and Graner, A. (2002) Differential gene expression during seed germination in barley (*Hordeum vulgare* L.). *Functional and Integrative Genomics* 2, 28–39.

Rajjou, L., Gallardo, K., Debeaujon, I., Vandekerckhove, J., Job, C. and Job, D. (2004) The effect of α-amanitin on the *Arabidopsis* seed proteome highlights the distinct roles of stored and neosynthesized mRNAs during germination. *Plant Physiology* 134, 1598–1613.

Soeda, Y., Konings, M.C., Vorst, O., van Houwelingen, A.M., Stoopen, G.M., Maliepaard, C.A., Kodde, J., Bino, R.J., Groot, S.P. and van der Geest, A.H.M. (2005) Gene expression programs during *Brassica oleracea* seed maturation, osmopriming, and germination are indicators of progression of the germination process and the stress tolerance level. *Plant Physiology* 137, 354–368.

Sreenivasulu, N., Altschmied, L., Panitz, R., Hähnel, U., Michalek, W., Weschke, W. and Wobus, U. (2002) Identification of genes specially expressed in maternal and filial tissues of barley caryopses: a cDNA array analysis. *Molecular Genetics and Genomics* 266, 758–767.

White, J.A., Todd, J., Newman, T., Focks, N., Girke, T., Martínez de Ilárduya, O., Jaworski, J.G., Ohlrogge, J.B. and Benning, C. (2000) A new set of *Arabidopsis* Expressed Sequence Tags from developing seeds: the metabolic pathway from carbohydrates to seed oil. *Plant Physiology* 124, 1582–1594.

18 Cell Cycle Activity, Membrane Integrity and Germination of Matriconditioned Lentil (*Lens culinaris* Medik.) Seeds

J. Sadowski and E. Sliwinska

Department of Genetics and Plant Breeding, University of Technology and Agriculture, Al. Kaliskiego 7, 85-789 Bydgoszcz, Poland

Abstract

Three samples of lentil (*Lens culinaris* Medik. cv. Anita) seeds, undeteriorated and deteriorated for 4 or 6 days, were matriconditioned by incubating them for 3 days at 20°C with humid vermiculite. Mean germination time (MGT) and germination capacity at 10°C, 15°C and 20°C were established. Cell cycle activity in radicle tips was studied using flow cytometry, and membrane integrity was determined using conductivity tests. Untreated seeds were then stored for 1 year at room temperature and all treatments and analyses were repeated. Matriconditioning decreased the MGT of all samples and improved the germination percentage of the deteriorated seeds at 10°C. The ratio of nuclei in the second growth phase of the cell cycle to those in the first growth phase (i.e. G_2/G_1 ratio) in the radicle tips increased after matriconditioning, which suggests that the treatment induced the onset of germination. The conductivity test showed that matriconditioning also improved membrane integrity. Matriconditioning is an efficient method for the enhancement of lentil seed vigour, and flow cytometry and measurements of leachate conductivity can be applied to determine the effectiveness of this treatment.

Introduction

Lentil (*Lens culinaris* Medik.) has been cultivated since ancient times. Its seeds have been found in Egyptian pyramids as well as in excavations in the Near East, Greece and Rome. At present, more than 70% of the world's lentils are grown in Asia. However, this species is also becoming more and more popular in Europe as an alternative crop, and is used extensively in vegetarian diets.

Lentil is a large-seeded legume plant and, typically for species from this group, its seeds are susceptible to imbibition injury. Therefore, hydropriming, the most popular method of seed enhancement of many species, does not seem to be suitable for quality improvement of lentil seeds. Another priming method that allows control of water uptake is matriconditioning. During this treatment, seeds, mixed with a solid insoluble matrix of particles and water, slowly imbibe to reach an equilibrium hydration level just below that required for radicle protrusion (Khan *et al.*, 1990; Halmer, 2004). Matriconditioning has proved successful in improving germination, the rate of emergence, stand size, cold hardiness and the prevention of germination thermoinhibition in small- and large-seeded crops (Khan *et al.*, 1990; McDonald, 2000). It also reduces electrolyte leakage, increases endo-β-mannase and dehydrogenase activity and stimulates hydrolysis of storage proteins in the embryo of some vegetable seeds (Habdas *et al.*, 2000; Kepczynska *et al.*, 2003). Dawidowicz-Grzegorzewska (1997) observed fine structural changes in the cells of matriconditioned carrot (*Daucus carota* L.) and cucumber (*Cucumis sativus* L.) seeds. However, there is no information on the cell cycle activity changes in seeds during matriconditioning. Since hydropriming and osmopriming usually commence DNA synthesis and increase the proportion of cells in the G_2 phase of the cell cycle (Lanteri *et al.*, 1994; Sliwinska and Babinska, 1999; Sliwinska and Jendrzejczak, 2002), similar processes can be expected during matriconditioning.

The aim of this study was to determine the effect of matriconditioning on cell cycle activity, membrane integrity and germination capacity of lentil seeds of differing quality (undeteriorated and deteriorated), especially at lower germination temperatures. The potential for use of flow cytometry and conductivity tests in predicting the effectiveness of the treatment was also studied.

Materials and Methods

Plant material

A commercial seed sample of lentil (*L. culinaris* Medik. cv. Anita), harvested in 2002, was provided by the breeding company 'Spójnia' Hodowla i Nasiennictwo (Nochowo, Poland). This cultivar is characterized by large yellowish-green seeds. The weight of 1000 seeds used in the experiments was 45 g, and the mean moisture content of these seeds was 11%.

Controlled deterioration and matriconditioning

Some of the seeds were exposed to controlled deterioration (CD) in sealed plastic bags for either 4 or 6 days at 45°C, at 20% moisture content (ISTA, 1995). After CD, the seeds were dried on a thin layer of filter paper for 48 h at room temperature. Both deteriorated and undeteriorated samples were then matriconditioned in vermiculite for 3 days at 20°C, using a seeds/vermiculite/water ratio of 1:0.7:0.8 (w/w/w), which was established in a preliminary experiment (data not shown). After matriconditioning, the seeds were again dried on filter paper for 48 h at room temperature.

Germination test

Laboratory germination tests were performed at 10°C, 15°C and 20°C, using four replicates of 100 seeds, according to ISTA rules (ISTA, 1993), with some minor modifications (i.e. pleated filter paper, 65% relative substrate moisture content and darkness). The seeds were scored daily for radicle emergence for 10 days. The mean germination time (MGT) was evaluated according to Nichols and Heydecker (1968). Germination percentage after 10 days was determined.

Flow cytometry

Embryo radicle tips were dissected from the seeds. Samples of three radicle tips were prepared as described previously (Sliwinska, 2003). For each sample, 7000–9000 nuclei were analysed using a Partec CCA (Münster, Germany) flow cytometer. Analyses were performed on ten replicates for each seed treatment. The computer program PARTEC DPAC v. 2.2 was used for histogram analysis. The number of nuclei at different nuclear DNA replication stages was established and the G_2/G_1 ratio (i.e. ratio of nuclei in the second growth phase of the cell cycle to nuclei in the first growth phase) calculated.

Conductivity test

The seeds (50 per sample, four replicates) were soaked in 250 ml of deionized water for 24 h at 20°C (ISTA, 1995). The conductivity of the leachate was measured using an N5711 conductometer (TEL-EKO SA, Poland). All treatments and analyses were performed in 2003 and repeated in 2004 after 12 months storage of the untreated seeds at room temperature (20°C ± 2°C), with no control of relative humidity (RH).

Statistical analysis

A single-factor analysis of variance (ANOVA) and a Duncan's test were performed on the results. The percentage data from the germination tests were subjected to ANOVA after arcsine transformation, although actual percentages are presented in Table 18.1.

Results

Germination

At optimal temperature (i.e. 20°C), untreated seeds showed a germination capacity of 96% in both years' analyses (i.e. 2003 and 2004). In 2003, germination decreased to about 80% and 70% when CD was applied for 4 and 6 days, respectively, and

Table 18.1. Effect of matriconditioning (MC) on mean percentage germination of lentil seeds that were either undeteriorated or exposed to controlled deterioration (CD) for 4 or 6 days. Germination studies were conducted for 10 days at either optimal (i.e. 20°C) or lower temperatures (i.e. 15°C and 10°C), duplicated in 2003 and 2004 and four replicates of 100 seeds used to obtain each data point. Values in columns followed by the same letter are not significantly different at $P = 0.05$ (Duncan's test).

	Mean germination (%)					
	20°C		15°C		10°C	
Seed treatment	2003	2004	2003	2004	2003	2004
No treatment	96.0 a	95.8 a	95.7 a	94.0 a	95.0 a	88.0 b
MC	94.7 a	94.2 a	95.0 a	94.8 a	93.7 a	94.2 a
4 days CD	82.3 b	67.5 b	74.7 b	61.0 b	61.7 c	46.2 de
4 days CD, then MC	78.7 bc	63.2 bc	71.7 c	59.2 b	72.3 b	57.8 c
6 days CD	70.0 cd	56.5 d	54.3 c	54.8 b	38.7 e	40.0 e
6 days CD, then MC	59.7 d	60.0 cd	55.3 c	56.5 b	49.7 d	48.2 d

an even greater decrease was observed in 2004 (Table 18.1). A similar tendency was observed when the seeds were germinated at lower temperatures. The lowest germination capacity (~40%) was detected at 10°C for the seeds exposed to CD for 6 days. Matriconditioning did not influence the germination percentage of any seed sample at 20°C and 15°C, but improved germination by about 10% for all seed samples at 10°C, except for the undeteriorated seeds in 2003. In 2004, the treatment even improved the germination capacity of untreated seeds at the lowest temperature to the level attained in 2003.

MGT of untreated seeds at 20°C was ~2 days and it increased at lower germination temperatures (Fig. 18.1). The seed samples exposed to CD in 2004 showed the longest MGT – ~5 days. A considerable decrease (i.e. 20–38%) in MGT of all matriconditioned seed samples was observed, regardless of the germination temperature. This was especially evident at the lowest germination temperature and for the seeds analysed 2 years after harvest (i.e. in 2004). Matriconditioning shortened MGT of these seeds by ~40% (i.e. 2 days).

Cell cycle activity

The majority of root tip nuclei isolated from the seeds of all samples was arrested in the G_0/G_1 phase of the cell cycle (Fig. 18.2). However, a considerable proportion of the nuclei was also arrested in the G_2 phase. This proportion was higher in matriconditioned than unmatriconditioned seeds. The G_2/G_1 ratio in the radicle tip of unmatriconditioned seeds was ~0.35–0.37 and it did not change upon CD (Table 18.2). Matriconditioning increased this ratio to >0.4 in all seed samples, and the highest G_2/G_1 ratio (~0.5) was detected in the radicle tips of undeteriorated seeds.

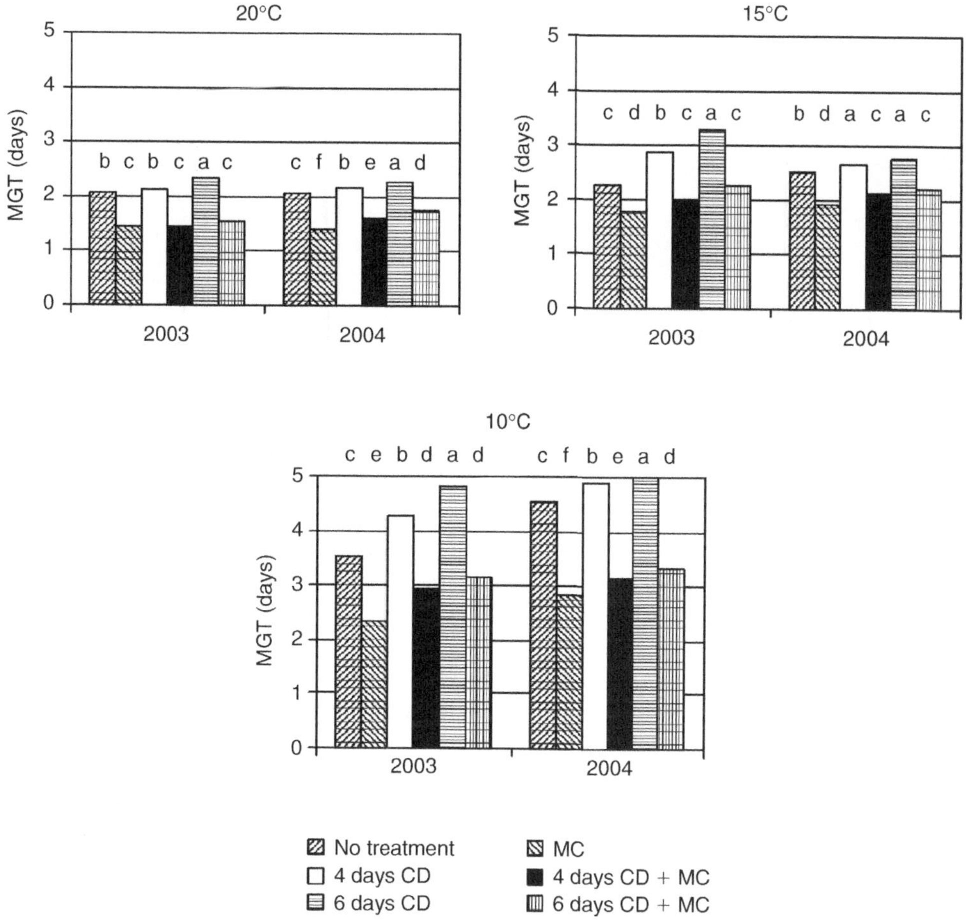

Fig. 18.1. Effect of matriconditioning (MC) on mean germination time (MGT) of lentil seeds that were either undeteriorated or exposed to controlled deterioration (CD) for 4 or 6 days. Germination studies involved four replicates of 100 seeds at 20°C, 15°C or 10°C, and were duplicated in 2003 and 2004. Values for each year within each temperature followed by the same letter are not significantly different at $P = 0.05$ (Duncan's test).

Conductivity

In the first year of analysis, the conductivity of the leachate of the untreated seeds was >20 μS/cm/g, and it increased to ~30 μS/cm/g after 12 months of seed storage (Table 18.2). CD strongly increased this parameter, especially when the longer deterioration period was applied. The highest conductivity, >70 μS/cm/g, was detected in the leachate of seeds exposed to 6 days CD in 2004. Matriconditioning did not change the conductivity of the leachates of the undeteriorated seeds.

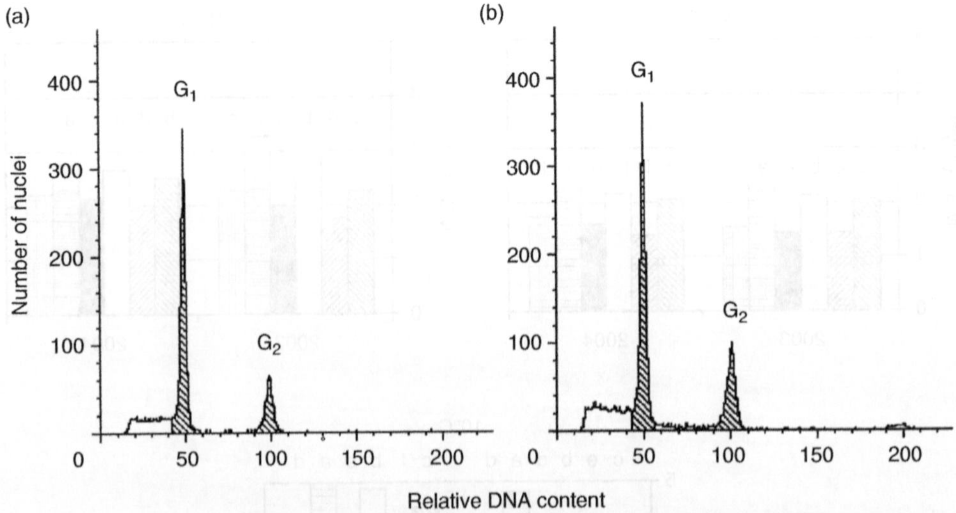

Fig. 18.2. Selected DNA histograms from nuclear preparations of radicle tips of lentil seeds (a) untreated and (b) matriconditioned in humid vermiculite for 3 days at 20°C.

Electrolyte leakage considerably decreased in the deteriorated seeds after treatment, but not to the level of the undeteriorated ones.

Discussion

Priming is a pre-sowing seed treatment, which is able to improve the vigour of seed lots. The cheapest, and, for some species, most effective priming method is to soak the seeds in water (i.e. hydropriming). For example, hydropriming was

Table 18.2. Effect of matriconditioning (MC) on the G_2/G_1 ratio and leachate conductivity of lentil seeds that were either undeteriorated or exposed to controlled deterioration (CD) for 4 or 6 days. Experiments were duplicated in 2003 and 2004. The G_2/G_1 ratio was determined using flow cytometry on 7000–9000 nuclei from three radicles, in ten replicates, while each of the conductivity tests involved four replicates of 50 seeds. Values in columns followed by the same letter are not significantly different at $P = 0.05$ (Duncan's test).

	G_2/G_1 ratio		Conductivity (μS/cm/g)	
Seed treatment	2003	2004	2003	2004
No treatment	0.370 c	0.352 dc	22.8 e	30.3 d
MC	0.483 a	0.449 a	20.8 e	28.4 d
4 days CD	0.376 c	0.367 c	55.5 b	63.0 b
4 days CD, then MC	0.433 b	0.422 ab	41.9 d	43.8 c
6 days CD	0.382 c	0.346 dc	67.6 a	71.3 a
6 days CD, then MC	0.426 b	0.403 bc	46.7 c	44.4 c

found to have a beneficial effect on sugarbeet (*Beta vulgaris* L.) seeds (Longden, 1973; Sliwinska and Jendrzejczak, 2002). The method, however, does not seem to be applicable to the large-seeded legume species, which are very sensitive to imbibitional injury (Powell, 1986; Taylor *et al.*, 1992). This was confirmed in a preliminary experiment when lentil seeds were soaked in water for different periods of time. Upon hydropriming, their germination capacity decreased to <80%. It is suggested that hydration killed a portion of the seeds, most probably as a result of damage to their membranes during the rapid water uptake. A treatment that better mimicked natural imbibition in the soil was presumed to be more suitable for this species. Matriconditioning, involving incubation of the seeds with solid carriers (e.g. Micro-Cel or vermiculite, which creates matric forces that hold water so that it is gradually absorbed by the seed), seems to fulfil this requirement. This priming treatment improves seed germination, in the laboratory as well as in the field, in legume species such as snap bean (*Phaseolus vulgaris* L.), soybean (*Glycine max* (L.) Merr.) (Khan *et al.*, 1990, 1995) and pea (*Pisum sativum* L.) (Harman *et al.*, 1989).

In this study, matriconditioning in humid vermiculite did not improve the germination capacity of lentil seeds at 15°C and 20°C. In some cases, even a slight decrease (not statistically proven) in germination was noticed at the optimal temperature, which was an effect of attack by seedborne fungi. Stimulation of fungal development during matriconditioning has been detected by Khan *et al.* (1990), who recommended that the water used in the conditioning mixture should be replaced by a thiram solution in lengthy treatments or field plantings. Habdas *et al.* (2000), when exposing cucumber seeds to matriconditioning, also found that the treatment resulted in seed damage caused by necrotic and pathogenic fungi, and that this was especially evident for seed lots with low viability. In the present experiment, this negative effect of priming was not observed when the seeds were germinated at 10°C, and at this suboptimal temperature there was a beneficial effect of the treatment. Similarly, Khan (1992) and Khan *et al.* (1995) noted that an improvement in germination percentage, and germination rate, caused by matriconditioning generally occurred at suboptimal temperatures. In this study, however, the germination rate (expressed as the MGT) was improved not only at suboptimal but also at optimal temperature, although to a lesser extent in the latter case. This is consistent with the results of other matriconditioning studies on grasses (Hardegree, 1994; Madakadze *et al.*, 2000), carrot, cucumber (Dawidowicz-Grzegorzewska, 1997) and onion (*Allium cepa* L.) (Kepczynska *et al.*, 2003) seeds. In all of these studies, higher germination rates were expressed at a range of temperatures after a matriconditioning treatment.

The induction of germination *sensu stricto* was confirmed by augmentation of the G_2/G_1 ratio in the radicle tips of the matriconditioned lentil seeds. Commencement of DNA synthesis also increased the proportion of G_2 cells in the radicle tips of various species during hydropriming and osmopriming (Lanteri *et al.*, 1994, 2000; Sliwinska and Babinska, 1999; Sliwinska and Jendrzejczak, 2002). The G_2/G_1 ratio has been suggested to be a molecular marker for the progression of germinative events preceding radicle protrusion, and the results presented here confirm this suggestion. Flow cytometric analysis showed that the G_2/G_1 ratio was not influenced by CD. In contrast to this observation in lentil, a decreased proportion of G_2 cells was detected after CD in the seeds of sugarbeet (Sliwinska, 2003).

Deterioration was suggested to have a negative effect on the viability of nuclei with a higher DNA content, because they are more sensitive to different kinds of stresses; however, this suggestion was not supported here.

CD did damage the membranes of lentil seeds. There was increased conductivity of leachates from deteriorated seeds when compared with the leachates from undeteriorated ones. As a consequence, CD seeds germinated poorly (i.e. they had a lower germination rate and percentage). Matriconditioning considerably improved the membrane integrity of the deteriorated seeds, but not of the undeteriorated ones. A similar beneficial effect of matriconditioning has been observed on the cytomembranes of artificially aged cucumber seeds (Habdas *et al.*, 2000). Khan *et al.* (1995) also noted a reduced electrolyte leakage from maize (*Zea mays* L.) seeds, following matriconditioning. The reduction was greatest in the seeds of a supersweet maize cultivar, which had a higher susceptibility to imbibitional chilling injury than a traditional maize cultivar. The relationship between the organization of biomembrane structure and susceptibility to imbibitional injury would also explain the beneficial effect of matriconditioning on low-quality lentil seed performance.

It is concluded that matriconditioning is an efficient method for improving the quality of lentil seed lots, especially those of lower vigour that are germinated at suboptimal temperatures. Thus, this treatment is recommended for enhancing the vigour of lentil seeds that are sown in regions with a cold spring, like Central Europe. The treatment induces the onset of germination and improves membrane integrity. These events can be recognized by flow cytometry and the measurement of leachate conductivity. Therefore, these methods can be applied to determine treatment effectiveness.

Acknowledgements

The authors wish to thank 'Spójnia' Hodowla i Nasiennictwo (Nochowo, Poland) for kindly providing the lentil seeds and Professor J.D. Bewley (University of Guelph, Canada) for his critical comments on the manuscript.

References

Dawidowicz-Grzegorzewska, A. (1997) Ultrastructure of solid matrix-primed endospermic and non-endospermic seeds. In: Ellis, R.H., Black, M. and Hong, T.D. (eds) *Basic and Applied Aspects of Seed Biology*. Kluwer Academic Publishers, Dordrecht, The Netherlands, pp. 479–487.

Habdas, H., Grzesik, M., Szafirowska, A. and Sokolowska, A. (2000) Cytological and physiological effect of matriconditioning on low viable cucumber seed germination. *Acta Horticulturae* 517, 113–120.

Halmer, P. (2004) Methods to improve seed performance in the field. In: Benech-Arnold, R.L. and Sanchez, R.A. (eds) *Handbook of Seed Physiology: Applications to Agriculture*. Haworth Press, New York, pp. 125–166.

Hardegree, S.P. (1994) Matric priming increases germination rate of great basin native perennial grasses. *Agronomy Journal* 86, 289–293.

Harman, G.E., Taylor, A.G. and Stasz, T.E. (1989) Combining effective strains of *Trichoderma harzianum* and solid matrix priming to improve biological seed treatments. *Plant Disease* 73, 631–637.

ISTA (1993) International rules for seed testing. *Seed Science and Technology* 21 (Suppl.), p. 288.

ISTA (1995) *Handbook of Vigour Test Methods.* The International Seed Testing Association, Zurich, Switzerland.

Kepczynska, E., Piekna-Grochala, J. and Kepczynski, J. (2003) Effects of matriconditioning on onion seed germination, seedling emergence and associated physical and metabolic events. *Plant Growth Regulation* 41, 269–278.

Khan, A.A. (1992) Preplant physiological seed conditioning. *Horticultural Reviews* 13, 131–181.

Khan, A.A., Miura, H., Prusinski, J. and Ilyas, S. (1990) Matriconditioning of seeds to improve emergence. *Proceedings of National Symposium on Stand Establishment for Horticultural Crops*, Minneapolis, Minnesota, pp. 19–40.

Khan, A.A., Ilyas, S. and Ptasznik, W. (1995) Integrating low water potential seed hydration with other treatments to improve cold tolerance. *Annals of Botany* 75, 13–19.

Lanteri, S., Saracco, F., Kraak, H.L. and Bino, R.J. (1994) The effects of priming on nuclear replication activity and germination of pepper (*Capsicum annuum*) and tomato (*Lycopersicon esculentum*) seeds. *Seed Science Research* 4, 81–87.

Lanteri, S., Portis, E., Bergervoet, J.H.W. and Groot, S.P.C. (2000) Molecular markers for the priming of pepper seeds (*Capsicum annuum* L.). *Journal of Horticultural Science and Biotechnology* 75, 607–611.

Longden, P.C. (1973) Washing sugar-beet seed. *Journal of International Institute for Sugar Beet Research* 6, 154–162.

Madakadze, I.C., Prithiviray, B., Madakadze, R.M., Stewart, K., Peterson, P., Coulman, B.E. and Smith, D.L. (2000) Effect of preplant seed conditioning treatment on the germination of switch grass (*Panicum virgatum* L.). *Seed Science and Technology* 28, 403–411.

McDonald, M.B. (2000) Seed priming. In: Black, M. and Bewley, J.D. (eds) *Seed Technology and Its Biological Basis.* Sheffield Academic Press/CRC Press, Sheffield, UK, pp. 287–325.

Nichols, M.A. and Heydecker, W. (1968) Two approaches to the study of germination data. *Proceedings of the International Seed Testing Association* 33, 531–540.

Powell, A.A. (1986) Cell membranes and seed leachate conductivity in relation to the quality of seed for sowing. *Journal of Seed Technology* 19, 81–99.

Sliwinska, E. (2003) Cell cycle and germination of fresh, dried and deteriorated sugarbeet seeds as indicators of optimal harvest time. *Seed Science Research* 13, 131–138.

Sliwinska, E. and Babinska, L. (1999) Effect of presowing hydration treatment on DNA replication activity in the embryo of sugar beet (*Beta vulgaris* L.). *Electronic Journal of Polish Agricultural Universities, Agronomy* 2(2). Available at: http://www.ejpau.media.pl/series/volume2/issue2/agronomy/art-04.html

Sliwinska, E. and Jendrzejczak, E. (2002) Sugar-beet seed quality and DNA synthesis in the embryo in relation to hydration–dehydration cycles. *Seed Science and Technology* 30, 597–608.

Taylor, A.G., Prusinski, J., Hill, H.J. and Dickson, M.D. (1992) Influence of seed hydration on seedling performance. *HortTechnology* 2, 336–344.

19 Are Dormant Seeds Lazy and Germinating Seeds Not?

H.W.M. Hilhorst

Laboratory of Plant Physiology, Wageningen University, Arboretumlaan 4, 6703 BD Wageningen, The Netherlands

Abstract

Seed dormancy and germination are generally ill-defined phenomena since their occurrence and completion is measured almost exclusively by protrusion of (usually) the radicle through the surrounding tissues. This has hampered significant progress in our understanding of mechanisms and regulation of dormancy and germination. Fortunately, modern molecular-genetic approaches, including genomics, have immensely accelerated the acquisition of relevant data to deepen our insights. However, the vast amount of data that is available threatens to blur our minds rather than present us with the clear picture that we pursue. To avoid this, it is necessary to retrieve the truly relevant questions, assess if the system under investigation is suitable to address these questions and, if so, whether these systems are properly used. This chapter provides a short survey of recently acquired data for *Arabidopsis* (*Arabidopsis thaliana* (L.) Heynh.) dormancy and germination, with an emphasis on the role of abscisic acid (ABA).

Introduction

Seed science is coming of age. The discipline is now at par with most other fields of modern plant science. Global expression studies, including transcriptomics and proteomics, of diverse seed aspects and functions are being published at a rapid rate (Girke *et al.*, 2000; Gallardo *et al.*, 2001, 2002; Rajjou *et al.*, 2004; Job *et al.*, 2005; Liu *et al.*, 2005; Nakabayashi *et al.*, 2005; Soeda *et al.*, 2005). The acquired data should add essential information to the previous as well as ongoing physiological and genetic studies. Thus, it is expected that a major leap forward can be made towards understanding the fundamentals of seed behaviour. However, in many instances the results of these studies are difficult to interpret and compare with other studies because of a lack of definition of the physiology of the experimental system. Standardized conditions for treatment of the seed material, in order to make comparisons possible, are non-existent.

Choice and Characterization of Seed Material for Physiological, Genetic and Global Expression Studies of Dormancy and Germination

Selection of most suitable species

There is some controversy among seed biologists as to how dormancy should be defined and which factors control the induction and breaking of dormancy (Vleeshouwers *et al.*, 1995). For example, the effects of light, nitrate and temperature on dormancy are difficult to distinguish from their effects on the process of germination itself, as dormancy can only be assessed post facto in a germination test. Thus, this apparent continuum between dormancy and germination is a major constraint to studying dormancy (Cohn, 1996). This is particularly true in crop species where dormancy is usually minimal and may occur only under stress conditions. This feature can make crop species unsuitable for molecular studies of dormancy responses as they cannot be distinguished from germination-related processes. For such studies, it is essential to have a well-characterized and unambiguous experimental system. It has been argued that annual soil seed-bank species are better subjects than domesticated crops and have particular features that make them suitable as a model species for physiological, genetic and molecular studies of seed dormancy (Cohn, 1996; Hilhorst, 1997). For example, the germination of *Sisymbrium officinale* (L.) Scop. seeds is absolutely dependent on such germination stimulants as light and nitrate, providing the tools to separate the state of non-dormancy from germination (reviewed by Hilhorst, 1997). *Arabidopsis* (*Arabidopsis thaliana* (L.) Heynh.) is another annual soil seed-bank species (Baskin and Baskin, 1972) with well-known advantages for molecular studies. Unfortunately, its seeds lose dormancy quite rapidly after ripening. However, accessions have recently been identified with deep dormancy that requires long periods of after-ripening to be relieved. One of these is the Cvi ecotype, originating from the Cape Verde Islands (Koorneef *et al.*, 2000), which has been used in physiological and genetic analysis of seed dormancy (Alonso-Blanco *et al.*, 2003; Ali-Rachedi *et al.*, 2004). Seeds of Cvi also have a dependency on germination stimulants like that of *S. officinale* and can cycle through dormant and non-dormant states unless specifically induced to germinate by light (Cadman *et al.*, 2006). Thus, this ecotype provides an ideal experimental system for physiological, genetic and molecular analysis of (non-deep, physiological) seed dormancy.

Physiological characterization of the experimental system

It is very useful to determine the class and type of dormancy of the selected species (Baskin and Baskin, 1998, 2004). This provides essential information required to assess whether a certain species is suitable for a particular study or not. Once the experimental system has been selected, serious attention should be given to the multiplication of the seeds:

- Grow all plants at the same time, under the same controlled conditions.
- Harvest seeds at the same (developmental) time with the same method.

- Dry seeds under controlled conditions (temperature, relative humidity (RH) and duration) to a standard moisture content.
- Store seeds under controlled conditions (sealed, constant temperature).

Variations in any of these conditions and manipulations may influence dormancy, germination and viability of the seeds. A germination time curve will give preliminary information about the level of dormancy under standard germination conditions. A viability test is required to assess the proportion of the non-germinating seeds that are dormant and dead. The next step is to monitor dormancy or germination under varying conditions and environmental factors. In this way the optimal conditions for the process under study can be identified (e.g. for breaking or inducing dormancy, or for initiation of germination). Evidently, germination conditions should be standardized for comparison. These include light and temperature, type of filter paper, use of water or agar as the germination medium and amount of water available per seed. The work by Cohn (1996) elaborates on these aspects and is mandatory reading for meaningful seed physiology.

Recent Advances in the Molecular Genetics of Seed Dormancy and Germination: Global Expression Studies

Over the last 20 years, the use of mutants, particularly those of *Arabidopsis*, has made a major contribution to our understanding of developmental processes in seeds, including dormancy and germination (Koornneef and Karssen, 1994; Bentsink and Koornneef, 2002; Koornneef *et al.*, 2002). A vast range of seed phenotypes have been described and loci and genes that are associated with dormancy and germination have been identified (listed in Bentsink and Koornneef, 2002). However, we are still far from a comprehensive view of the regulation and mechanisms of dormancy and germination. Despite the dissimilarities among used protocols and often erroneous methodology, outlines are gradually becoming visible. Most studies have focused on hormonal signalling involved in the regulation of these processes, particularly abscisic acid (ABA) and gibberellins (GAs), and most progress has been made in this field (reviewed by Finkelstein *et al.*, 2002; Olszewski *et al.*, 2002). Here, we focus on the action of ABA.

ABA signalling in dormancy and germination

ABA has long been implicated in the acquisition of dormancy during seed development. In many cases, the induction of primary dormancy is accompanied by a transient increase in ABA content, and ABA-deficient or ABA-insensitive mutants are characterized by the absence of dormancy. However, ABA is involved in the regulation of a number of developmental processes, including protein synthesis and the acquisition of desiccation tolerance. It has thus been suggested that dormancy is a developmental event and that ABA affects a multitude of developmental processes rather than directly inducing dormancy (Hilhorst, 1995). Genome-wide profiling of stored mRNA in *Arabidopsis* seeds has revealed that ~300 of the almost

500 most highly expressed genes in dry seeds (out of a total of 12,470 expressed genes) contained one or more ABA-responsive sequences (Nakabayashi *et al.*, 2005). Genome-wide gene expression profiling by massively parallel signature sequencing (MPSS) yielded 1357 genes that were upregulated or downregulated by ABA in *Arabidopsis* seedlings (Hoth *et al.*, 2002). In a gene-finding survey in existing databases, using the *cis*-regulatory ABA-responsive element (ABRE) and its coupling element (CE) to target at ABA-responsiveness, almost 2000 stress-inducible genes were found of diverse functional categories (Zhang *et al.*, 2005). These studies have made clear that hundreds of genes are potentially controlled by ABA. In addition, evidence is accumulating that extensive cross-talk exists among signalling pathways of ABA and other hormones. For example, the *etr1-2* mutant in *Arabidopsis*, which is impaired in downstream signalling from the ethylene receptor, displayed higher levels of ABA, GAs, cytokinins and auxins (Chiwocha *et al.*, 2005).

Dormancy in *Arabidopsis* Cvi seeds is clearly correlated with ABA content. Dormancy-breaking treatments reduced ABA content, whereas conditions that maintained dormancy induced an increase (Ali-Rachedi *et al.*, 2004). Thus, it is expected that expression of ABA-controlled genes will increase under conditions that are conducive to induction or maintenance of dormancy. This is indeed the case in seeds of *Arabidopsis* Cvi that were manipulated to go through several dormancy cycles by modulation of the temperature (Cadman *et al.*, 2006). This study showed that dormant seeds were as highly transcriptionally active as non-dormant seeds. Dormancy could be characterized by the expression of 442 genes which is at least twofold higher when compared with the non-dormant state. ABREs were overrepresented in the dormant gene set. Among the genes, many were identified as stress-related, including those encoding for small heat shock proteins (sHSPs), superoxide dismutase (SOD) and peroxiredoxin, as well as late embryo abundant (LEA) proteins and others related to seed development and maturation. This suggests that acquisition of primary dormancy may be related to expression of the same set of dormancy genes. Higher expression of these genes was also observed under non-stressed conditions, at 20°C in the dark when non-dormant seeds acquired secondary dormancy (Cadman *et al.*, 2006). Thus, it may be argued that any condition that induces an increase in ABA content will result in the expression of a similar set of ABA-controlled genes, apart from genes that may be specific for a certain response. In addition, it has been suggested that ABRE-mediated transcription may be affected by signals that do not alter ABA content (Nambara and Marion-Poll, 2003).

Signalling networks in the regulation of dormancy

What is the role of ABA in a plant cell when so many genes are potentially controlled by the hormone and how is the broad response of the transcriptome of a cell or tissue to ABA fine-tuned to a more specific response, such as dormancy? It is hard to assume that the increase in expression of a majority of genes is simply redundant (e.g. when their translation is blocked). Yet, inhibition of transcription in imbibing *Arabidopsis tt2-1* mutant seeds by α-amanitin did not inhibit germination *sensu stricto* but prevented further growth of the protruding radicle (Rajjou *et al.*,

2004). Thus, germination *sensu stricto* may only depend on preformed transcripts and not on *de novo* transcription. These results suggest that transcriptional activity during seed maturation may allow the subsequent stage of early germination, and transcriptional activity during germination may be in preparation for subsequent growth of the seedling.

Because of the large number of genes affected, it is clear that ABA signalling is not simply linear but consists of a complex signalling network. The *cis*-acting elements seem to play a crucial role and may be regarded as nodes within the network. The fast progress in transcriptome expression profiling has made it possible to identify various combinations of *cis*-acting elements involved in ABA-, stress- and other hormone responses. For example, among the dry seed transcriptome of *Arabidopsis* Cvi, the genes with the highest expression were more likely to contain multiple ABREs, a combination of ABRE with the CE, or a combination of ABRE with the seed-specific RY/Sph motif (RY: purine/pyrimidine repeat motif; Sph: a restriction enzyme site; Nakabayashi *et al.*, 2005). Now, it is also possible to distinguish between combinations that are affected by ABA and those that are not.

Superposed upon the organization of *cis*-acting elements are the networks of transcription factors that interact with them. Thus, the combination of certain transcription factors with certain combinations of *cis*-acting elements may then determine the promoter activity of a much reduced number of genes compared with the potential number of genes under ABA control (e.g. as described for osmotic and cold stress responses) (Yamaguchi-Shinozaki and Shinozaki, 2005). Furthermore, post-transcriptional and post-translational control mechanisms will be involved in further fine-tuning of the response.

Challenges for the Future

With the aid of '-omics' technology, rapid progress is now being made in the elucidation of signalling networks involved in the regulation of dormancy and germination. With complete genome sequences available for a growing number of species, gene functional analysis by reverse genetics is becoming a feasible option for an increasing number of seed scientists. Mutagenized (chemically or by fast neutron bombardment) populations of an increasing number of species are also available. Screening of pooled DNAs for deletions or point mutations (as in targeting-induced local lesions in genomes [TILLING]) in specific genes (or regions) may directly yield the gene responsible for the phenotype.

However, the occurrence of a certain (seed) phenotype depends not only on gene function but also (or even more) on the active gene product, i.e. the protein. Characterization of the proteome is a formidable task, which aims at defining the structures and functions of tens of thousands of proteins. Protein structure–function relationships, folding and stability, regulation through post-translational modifications, interaction with DNA, RNA, other proteins and small ligands may all contribute to the phenotype. Proteomics has already been applied successfully to the analysis of seed germination (Gallardo *et al.*, 2001). With chemical genetics methodology reverse genetics can also be used to elucidate protein function in a cell, tissue or organism (Stockwell, 2000).

A seed phenotype is also the result of the interaction between genome and environment. Forward genetics in *Arabidopsis* has yielded many loci that account for the natural variation in dormancy of several accessions (Alonso-Blanco *et al.*, 2003). Unfortunately, the subsequent identification of the genes involved is tedious and time-consuming. Recently, a reverse genetics method called Eco-TILLING has been proposed to identify polymorphisms in natural populations (Comai *et al.*, 2004). This method more rapidly identifies genes that contribute to natural variation (e.g. in dormancy). However, reverse genetics can only be powerful if efficient and reliable high-throughput phenotyping is available. Apart from these technologies, there is the need for input of expertise on plant and seed physiology and morphology. Functional genomics will undoubtedly boost our knowledge of signalling pathways and networks that are involved in the seed's 'decision' to germinate or remain dormant. However, only a combination of disciplines will enable us to address the fundamental questions in seed science.

References

Ali-Rachedi, S., Bouinot, D., Wagner, M.-H., Bonnet, M., Sotta, B., Grappin, P. and Jullien, M. (2004) Changes in endogenous abscisic acid levels during dormancy release and maintenance of mature seeds: studies with the Cape Verde Islands ecotype, the dormant model of *Arabidopsis thaliana. Planta* 219, 479–488.

Alonso-Blanco, C., Bentsink, L., Hanhart, C.J., Blankestijn-de Vries, H. and Koorneef, M. (2003) Analysis of natural allelic variation at dormancy loci of *Arabidopsis thaliana. Genetics* 164, 711–729.

Baskin, C.C. and Baskin, J.M. (1998) *Seeds: Ecology, Biogeography and Evolution of Dormancy and Germination.* Academic Press, San Diego, California.

Baskin, C.C. and Baskin, J.M. (2004) A classification system for seed dormancy. *Seed Science Research* 14, 1–16.

Baskin, J.M. and Baskin, C.C. (1972) Ecological life and physiological ecology of seed germination of *Arabidopsis thaliana. Canadian Journal of Botany* 50, 353–360.

Bentsink, L. and Koornneef, M. (2002) Seed dormancy and germination. In: Somerville, C.R. and Meyerowitz, E.M. (eds) *The Arabidopsis Book.* The American Society of Plant Biologists, Rockville, Maryland, pp. 1–18. Available at: http://www.aspb.org/publications/arabidopsis/

Cadman, C.S.C., Toorop, P.E., Hilhorst, H.W.M. and Finch-Savage, W.E. (2006) Gene expression profiles of *Arabidopsis* Cvi seeds during cycling through dormant and non-dormant states indicate a common underlying dormancy control mechanism. *The Plant Journal* 46, 805–822.

Chiwocha, S.D.S., Cutler, A.J., Abrams, S.R., Ambrose, S.J., Yang, J., Ross, A.R.S. and Kermode, A.R. (2005) The *etr1-2* mutation in *Arabidopsis thaliana* affects the abscisic acid, auxin, cytokinin and gibberellin metabolic pathways during maintenance of seed dormancy, moist-chilling and germination. *The Plant Journal* 42, 35–48.

Cohn, M.A. (1996) Operational and philosophical decisions in seed dormancy research. *Seed Science Research* 6, 147–153.

Comai, L., Young, K., Till, B.J., Reynolds, S.H., Greene, E.A., Codomo, C.A., Enns, L.C., Johnson, J.E., Burtner, C., Odden, A.R. and Henikoff, S. (2004) Efficient discovery of DNA polymorphisms in natural populations by Ecotilling. *The Plant Journal* 37, 778–786.

Finkelstein, R.R., Gampala, S.S.L. and Rock, C.D. (2002) Abscisic acid signalling in seeds and seedlings. *The Plant Cell* 2002 (Suppl.), S15–S45.

Gallardo, K., Job, C., Groot, S.P.C., Puype, M., Demol, H., Vandekerckhove, J. and Job, D. (2001) Proteomic analysis of Arabidopsis seed germination and priming. *Plant Physiology* 126, 835–848.

Gallardo, K., Job, C., Groot, S.P.C., Puype, M., Demol, H., Vandekerckhove, J. and Job, D. (2002) Proteomics of *Arabidopsis* seed germination: a comparative study of wild-type and gibberellin-deficient seeds. *Plant Physiology* 129, 823–837.

Girke, T., Todd, J., Ruuska, S., White, J., Benning, C. and Ohlrogge, J. (2000) Microarray analysis of developing *Arabidopsis* seeds. *Plant Physiology* 124, 1570–1581.

Hilhorst, H.W.M. (1995) A critical update on seed dormancy. 1. Primary dormancy. *Seed Science Research* 5, 61–73.

Hilhorst, H.W.M. (1997) Seed dormancy. *Seed Science Research* 7, 221–223.

Hoth, S., Morgante, M., Sanchez, J., Hanafey, M.K., Tingey, S.V. and Chua, N. (2002) Genome-wide gene expression profiling in *Arabidopsis thaliana* reveals new targets of abscisic acid and largely impaired gene regulation in the *abi1-1* mutant. *Journal of Cell Science* 115, 4891–4900.

Job, C., Rajjou, L., Lovigny, Y., Belghazi, M. and Job, D. (2005) Patterns of protein oxidation in *Arabidopsis* seeds and during germination. *Plant Physiology* 138, 790–802.

Koornneef, M. and Karssen, C.M. (1994) Seed dormancy and germination. In: Meyerowitz, E.M. and Somerville, C.R. (eds) *Arabidopsis.* Cold Spring Harbor Laboratory Press, Cold Spring Harbor, New York, pp. 313–334.

Koorneef, M., Alonso-Blanco, C., Bentsink, L., Blankestijn-de Vries, H., Debeajon, I., Hanhart, C.J., Léonkloosterziel, K.M., Peeters, T. and Raz, V. (2000) The genetics of seed dormancy in *Arabidopsis thaliana.* In: Viémont, J.-D. and Crabbé, J. (eds) *Dormancy in Plants.* CAB International, Wallingford, UK, pp. 365–373.

Koornneef, M., Bentsink, L. and Hilhorst, H.W.M. (2002) Seed Dormancy and Germination. *Current Opinion in Plant Biology* 5, 33–36.

Liu, P.P., Koizuka, N., Homrichhausen, T.M., Hewitt, J.R., Martin, R.C. and Nonogaki, H. (2005) Large-scale screening of *Arabidopsis* enhancer-trap lines for seed germination-associated genes. *The Plant Journal* 41, 936–944.

Nakabayashi, K., Okamoto, M., Koshiba, T., Kamiya, Y. and Nambara, E. (2005) Genome-wide profiling of stored mRNA in *Arabidopsis thaliana* seed germination: epigenetic and genetic regulation of transcription in seed. *The Plant Journal* 41, 697–709.

Nambara, E. and Marion-Poll, A. (2003) ABA action and interactions in seeds. *Trends in Plant Science* 8, 213–217.

Olszewski, N., Sun, T.-P. and Gubler, F. (2002) Gibberellin signalling: biosynthesis, catabolism, and response pathways. *The Plant Cell* 14, S61–S80.

Rajjou, L., Gallardo, K., Debeaujon, I., Vandekerckhove, J., Job, C. and Job, D. (2004) The effect of alpha-amanitin on the Arabidopsis seed proteome highlights the distinct roles of stored and neosynthesized mRNAs during germination. *Plant Physiology* 134, 1598–1613.

Soeda, Y., Konings, M.C.J.M., Vorst, O., van Houwelingen, A.M.M.L., Stoopen, G.M., Maliepaard, C.A., Kodde, J., Bino, R.J., Groot, S.P.C. and van der Geest, A.H.M. (2005) Gene expression programs during *Brassica oleracea* seed maturation, osmopriming, and germination are indicators of progression of the germination process and the stress tolerance level. *Plant Physiology* 137, 354–368.

Stockwell, B.R. (2000) Chemical genetics: ligand-based discovery of gene function. *Nature Reviews Genetics* 1, 116–125.

Vleeshouwers, L.M., Bouwmeester, H.J. and Karssen, C.M. (1995) Redefining seed dormancy: an attempt to integrate physiology and ecology. *Journal of Ecology* 83, 1031–1037.

Yamaguchi-Shinozaki, K. and Shinozaki, K. (2005) Organization of *cis*-acting regulatory elements in osmotic- and cold-stress responsive promoters. *Trends in Plant Science* 10, 88–94.

Zhang, W., Ruan, J., Ho, T.D., You, Y., Yu, T. and Quatrano, R.S. (2005) *Cis*-regulatory element based targeted gene finding: genome-wide identification of abscisic acid- and abiotic stress-responsive genes in *Arabidopsis thaliana. Bioinformatics* 21, 3074–3081.

20 Emerging and Established Model Systems for Endosperm Weakening

G. Leubner-Metzger, B. Kucera and K. Müller

Institute for Biology II, Botany/Plant Physiology, Albert-Ludwigs-University Freiburg, Schänzlestr. 1, D-79104 Freiburg i. Br., Germany

Abstract

Endosperm rupture is the main germination-limiting process in members of the *Asteraceae* (e.g. lettuce (*Lactuca sativa* L.)) and *Solanaceae* (e.g. tomato (*Lycopersicon esculentum* Mill.) and tobacco (*Nicotiana tabacum* L.)). About four decades ago a 'hatching enzyme' was proposed to cause endosperm weakening (i.e. a decline in the mechanical resistance of the micropylar endosperm), which is likely to be essential for seeds to complete germination. Although, there are established model systems among *Asteraceae* and *Solanaceae* for endosperm weakening, its molecular mechanism(s) still remain(s) a mystery. No single 'hatching enzyme' or universal molecular mechanism has been demonstrated explicitly. For the time being, the provisional conclusion is that endosperm weakening is likely to be achieved by the collaborative or successive action of several distinct molecular mechanisms. The knowledge gained from these established model systems will be compared and discussed. However, consideration of their severe experimental limitations shows that there is an urgent need for novel model systems. Such an optimal system has been recently found within the *Brassicaceae*. In this emerging model system for endosperm weakening, a complete study of the process is possible on each experimental level, from the direct measurement of the weakening by 'puncture force' to molecular investigations (e.g. proteome and transcriptome analyses).

Introduction

A major reason for the evolutionary success of the angiosperms is the 'invention' of seeds with double fertilization (Friedman, 1998; Judd *et al.*, 2002). In a typical angiosperm seed the diploid embryo is surrounded by two covering layers: (i) the triploid endosperm (i.e. nutritive tissue, living cells); and (ii) the diploid testa (i.e. the seed coat, maternal tissue, dead cells). Depending on the species, the endosperm is either maintained or obliterated during seed development. The evolutionary trend is towards cotyledon storage and endospermless seeds at maturity. Endosperm

 Seeds: Biology, Development and Ecology (eds S. Adkins, S. Ashmore and S.C. Navie)

development and its function as nutritive tissue have been studied thoroughly (Jacobsen *et al.*, 1995; Friedman, 1998; Baskin and Baskin, 2004). In contrast, little is known about the function of the endosperm as a constraint during endosperm-limited germination and coat-imposed dormancy (Bewley, 1997; Leubner-Metzger, 2003). Seed germination is a complex physiological process, water uptake by imbibition is followed by embryo growth, and radicle protrusion through all seed-covering layers is considered as the completion of germination. Environmental factors and plant hormones (e.g. gibberellins (GA), abscisic acid (ABA), brassinosteroids and ethylene) are regulators of germination and/or dormancy (Koornneef *et al.*, 2002; Kucera *et al.*, 2005).

The mature seeds of most angiosperms have a more or less abundant endosperm layer (Fig. 20.1). Mature angiosperm seeds differ in their 'embryo to seed' (E/S) ratios (Martin, 1946; Forbis *et al.*, 2002). Low E/S values, due to abundant endosperm tissue and tiny embryos, are typical for mature seeds of basal angiosperms (Fig. 20.1). High E/S values, due to obliterated endosperm tissue and the predominance of cotyledon storage, are evident in mature seeds of higher angiosperms of the rosid clade. Typical examples are the more or less endospermless *Brassicaceae* seeds with E/S values ~0.9 (Fig. 20.1). Although the mature seeds of *Brassica* species are completely endospermless, a single cell layer of endosperm is present in mature seeds of *Arabidopsis* (*Arabidopsis thaliana* (L.) Heynh.) (Liu *et al.*, 2005). E/S values between 0.4 and 0.5 are typical for mature seeds of higher angiosperms of the asterid clade (e.g. the *Solanaceae*), with abundant endosperm and additional embryo storage (Fig. 20.1; Bewley, 1997; Hilhorst *et al.*, 1998; Koornneef *et al.*, 2002; Leubner-Metzger, 2003).

The distribution of the different seed types and the E/S values in a modern phylogenetic tree (Fig. 20.1) support the following evolutionary seed trends (Martin, 1946; Forbis *et al.*, 2002; Baskin and Baskin, 2004): (i) in mature seeds of primitive angiosperms a small embryo is embedded in abundant endosperm tissue, and such seed types are prevalent among basal angiosperms; (ii) the general evolutionary trend within the higher angiosperms (i.e. core eudicots) is by the *Solanaceae*-like endospermic seed types of many asterids towards *Brassicaceae*-like more or less endospermless seed types of many rosids with storage cotyledons; and (iii) in addition to these general seed trends there are clade-specific differences. It has been proposed that endospermless seeds were 'invented' several times independently during evolution (Baskin and Baskin, 2004). Thus, our knowledge on endosperm abundance in mature seeds is quite complete.

In contrast, we do not know how endosperm-limited germination and/or endosperm-enhanced dormancy evolved during angiosperm seed phylogeny. Figure 20.1 shows (in bold) the few clades with at least some experimental evidence for seeds where the endosperm is important as a germination barrier (almost nothing is known for many rosid species). An exception is the work on perisperm weakening of *Cucurbitaceae* seeds (e.g. Welbaum *et al.*, 1995; Yim and Bradford, 1998). Established model systems for endosperm-limited germination (Fig. 20.1) are exclusively asterid species (Bewley, 1997; Hilhorst *et al.*, 1998; Leubner-Metzger, 2003; da Silva *et al.*, 2004, 2005): lettuce (*Lactuca sativa* L., *Asteraceae, Asterales*), tomato (*Lycopersicon esculentum* Mill., *Solanaceae, Solanales*), tobacco (*Nicotiana tabacum* L., *Solanaceae, Solanales*) and coffee (*Coffea arabica* L., *Rubiaceae, Gentianales*). Endosperm weakening, a decline

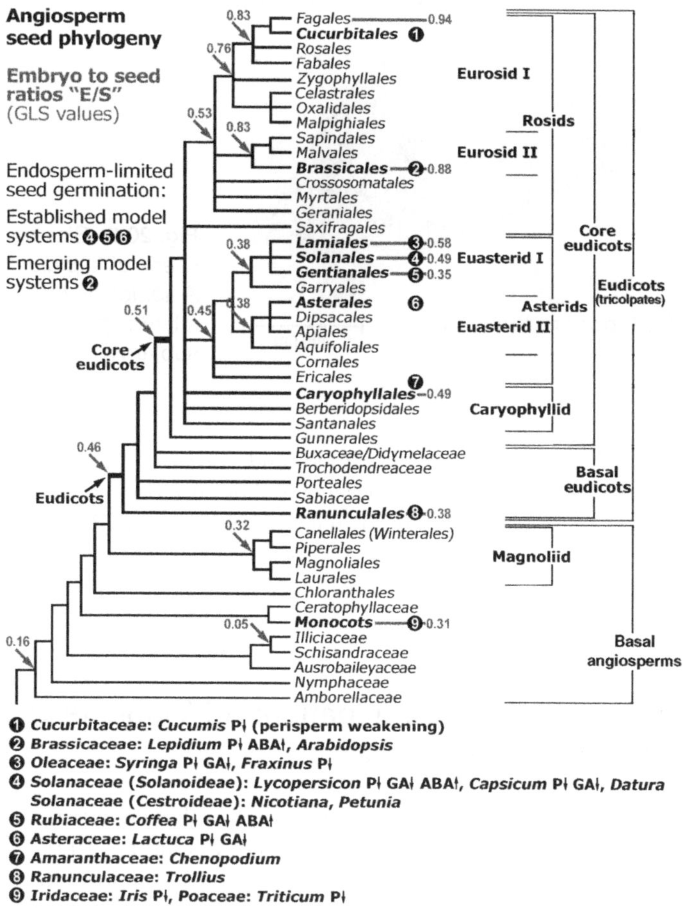

Fig. 20.1. Angiosperm seed phylogeny. Grey numbers in the phylogenetic tree represent 'embryo to seed' (E/S) ratios expressed as generalized least squares (GSL) values (Forbis *et al.*, 2002). Clades with experimental evidence for endosperm weakening are in bold and numbered; some of these provide model systems. P↓: endosperm weakening during germination was measured as decreasing puncture force; GA↓: endosperm weakening promoted by GA; and ABA↑: endosperm weakening inhibited by ABA.

in the mechanical resistance of the micropylar endosperm (i.e. endosperm covering the radicle tip), appears to be a prerequisite for the germination of these species. Endosperm weakening can be measured directly by the puncture-force method (see Müller *et al.*, Chapter 30, this volume). Puncture-force experiments are not possible with seeds as small as tobacco or *Arabidopsis* (Fig. 20.2), but have been performed for larger asterid seeds of the *Asterales* (Pavlista and Haber, 1970; Tao and Khan, 1979), *Gentianales* (da Silva *et al.*, 2004, 2005), *Solanales* (Watkins and Cantliffe, 1983;

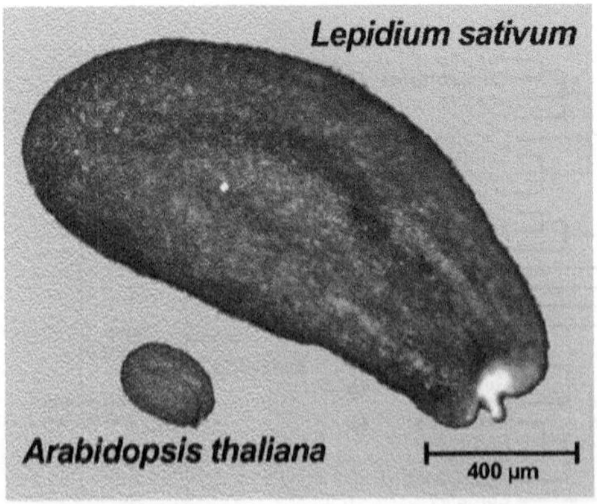

Fig. 20.2. Size comparison of dry seeds of *Lepidium sativum* and *Arabidopsis thaliana*, two closely related species of the *Brassicoideae* subfamily of the *Brassicaceae*.

Toorop *et al.*, 2000; Wu *et al.*, 2000) and *Laminales* (Junttila, 1973; Finch-Savage and Clay, 1997). All these experiments showed a decline in the puncture force of the micropylar endosperm prior to endosperm rupture (P↓ in Fig. 20.1). Promotion of endosperm weakening by GA appears to be a general phenomenon (GA↓ in Fig. 20.1). Thus, endosperm weakening prior to endosperm rupture seems to be widespread among asterid seeds, but has not been investigated in rosid seeds.

ABA Inhibition of Seed Germination: ABA Inhibits Endosperm Rupture, but not Testa Rupture

Another phenomenon of endospermic seeds is that endosperm rupture is inhibited by ABA (Karssen, 1976; Finch-Savage and McQuistan, 1991; Toorop *et al.*, 2000; Petruzzelli *et al.*, 2003). In the established model systems of the asterid clade it has been shown that ABA inhibits endosperm rupture, at least in part, by acting in an inhibitory manner on the micropylar endosperm (Ni and Bradford, 1993; Leubner-Metzger, 2003; da Silva *et al.*, 2004). The germination of intact tomato seeds is inhibited by 10–100 µM ABA, but surgical removal of the micropylar cap permits germination (i.e. initial embryo elongation) even in the presence of 1000 µM ABA (Liptay and Schopfer, 1983). The *Solanaceae* family can be divided into two large subgroups (Judd *et al.*, 2002; Petruzzelli *et al.*, 2003): (i) the *Solanoideae* (e.g. *Capsicum*, *Lycopersicon* and *Datura*); and (ii) the *Cestroideae* (e.g. *Nicotiana* and *Petunia*). In *Solanoideae*-type seeds, the micropylar covering layers, testa and endosperm form a cap-like structure (i.e. a micropylar cap) (Hilhorst *et al.*, 1998; Toorop *et al.*, 2000). A visible distinction between testa rupture and endosperm rupture is not possible for *Solanoideae*-type seeds. A typical feature of *Cestroideae*-type seeds like tobacco is a two-step germination with a visible distinction between testa rupture and endosperm rupture (Fig. 20.3; Leubner-Metzger, 2003; Petruzzelli *et al.*, 2003).

Two-step germination with separate testa and endosperm rupture is widespread over the entire phylogenetic tree and has been described for many species, including *Trollius* (*Ranunculaceae*, *Ranunculales*, basal eudicots) (Hepher and Roberts, 1985), *Chenopodium* (*Amaranthaceae*, *Caryophyllales*, caryophyllids) (Karssen, 1976) and

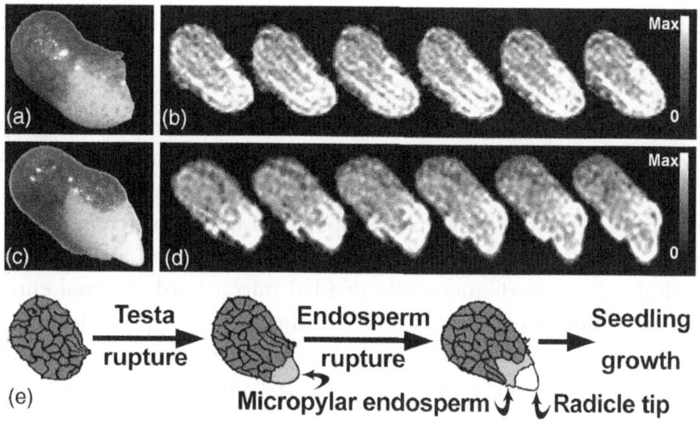

Fig. 20.3. Two-step germination and water uptake of tobacco seeds. (a, b and e) Testa rupture is followed by (c, d and e) endosperm rupture and subsequent seedling growth. (b and d) Nuclear magnetic resonance (NMR) images show that the micropylar endosperm and the radicle are major sites of water uptake. White: maximum water abundance; black: minimum water abundance (see NMR images in Manz *et al.*, 2005 for details).

Nicotiana (*Cestroideae, Solanaceae, Solanales,* asterids) (Leubner-Metzger, 2003). Recently, it has also been found in the *Brassicaceae* species, *Arabidopsis thaliana* L. (Heynh.) (Liu *et al.*, 2005) and *Lepidium* (*Lepidium sativum* L.) (see Müller *et al.*, Chapter 30, this volume). Separate testa and endosperm rupture is therefore a feature of endospermic seeds within the rosid clade. It was also found that ABA inhibits endosperm rupture, but not testa rupture, of *Arabidopsis* and *Lepidium* (see Müller *et al.*, Chapter 30, this volume). In agreement with this, ABA also does not inhibit testa rupture of endospermless *Brassica* seeds (Schopfer and Plachy, 1984). Separate testa rupture and endosperm rupture are an important experimental advantage of tobacco seeds when compared with tomato seeds. It has helped considerably to assign different enzymes, transcription factors and plant hormones to their target sites (Leubner-Metzger, 2003).

Although ABA inhibits the embryo growth potential and the transition to postgermination growth, it does not inhibit initial water uptake by imbibition or initial embryo elongation (Liptay and Schopfer, 1983; Schopfer and Plachy, 1984; Homrichhausen *et al.*, 2003; Manz *et al.*, 2005). We measured the regulation of water uptake of germinating tobacco seeds spatially and temporally by *in vivo* ^{1}H nuclear magnetic resonance (^{1}H-NMR) microimaging and ^{1}H-MAS NMR spectroscopy (Manz *et al.*, 2005). These non-destructive state-of-the-art methods show that water distribution in the water uptake phases II and III is inhomogeneous (Fig. 20.3). The micropylar seed end is the major entry point of water. The micropylar endosperm and the radicle show the highest hydration. ABA specifically inhibits endosperm rupture and phase III water uptake, but does not alter the spatial and temporal pattern of phase II water uptake. Taken together, these findings demonstrate that the micropylar endosperm is a main target for the ABA inhibition of endosperm rupture.

ABA, Endosperm Hydrolases and Endosperm Weakening in Established Asterid Model Systems

Direct measurements of the effect of ABA on endosperm weakening by puncture-force experiments have been published only for the asterid species like coffee and tomato (Toorop *et al.*, 2000; Wu *et al.*, 2000; da Silva *et al.*, 2004, 2005). The coffee embryo is enveloped by an endosperm tissue and surrounded by an endocarp (da Silva *et al.*, 2004). The endosperm is composed of a hard greenish tissue with polyhedral cells, is isodiametrically divided into a hard external endosperm and a soft internal endosperm, and belongs to the nuclear type. Endosperm weakening was measured in imbibed seeds with the endocarp mechanically removed. Endosperm weakening in coffee is biphasic. The first phase of endosperm weakening is ABA-insensitive, which is followed by the second phase that is inhibited by ABA. This second phase accounts for ~53% (420 mN) of the total difference in puncture force (da Silva *et al.*, 2004). Endosperm weakening of tomato is also biphasic, with a first phase that is ABA-insensitive. In this case the second phase, which is inhibited by ABA, has been reported to account for ~6% (30 mN; Wu *et al.*, 2000) or ~24% (80 mN; Toorop *et al.*, 2000) of the total tomato micropylar cap weakening. These results were obtained by incubating whole seeds in a medium with and without ABA, and dissecting the seeds when puncture-force measurements were performed (Groot and Karssen, 1987, 1992; Toorop *et al.*, 2000; Wu *et al.*, 2000). If micropylar caps are dissected from tomato seeds prior to the onset of endosperm weakening (3 h), a further 24 h incubation of isolated micropylar caps in medium with GA results in endosperm weakening (i.e. puncture force decreases by 170 mN), whereas incubation of isolated micropylar caps in a GA- plus ABA-containing medium inhibits endosperm weakening completely (Groot and Karssen, 1992). Although ABA clearly inhibits the second-phase endosperm weakening of coffee, the situation in tomato is less clear.

The micropylar cap of tomato seeds consists of endosperm and testa. No visible distinction between testa rupture and endosperm rupture is possible in tomato. Although the micropylar testa of tomato seeds accounts for only 20% of the initial puncture force and does not weaken during the first phase of the biphasic micropylar cap weakening, a significant decline in testa puncture force occurs during the second (i.e. ABA-sensitive) phase of tomato micropylar cap weakening just prior to radicle protrusion (Groot and Karssen, 1987). Tomato micropylar cap weakening is also highly dependent on the physiological seed stage. Endosperm weakening and endosperm rupture were delayed in freshly harvested tomato seeds when compared with after-ripened seeds (Groot and Karssen, 1992). ABA deficiency of the *sit*^{w} tomato mutant replaced this after-ripening effect, and micropylar cap weakening and endosperm rupture of freshly harvested and after-ripened *sit*^{w} seeds were equal. The ABA-deficient *sit*^{w} tomato mutant has a thinner testa (i.e. one cell layer thick) compared with wild-type seeds (i.e. 3–4 cell layers thick) and is therefore a testa mutant (Hilhorst and Downie, 1995). A species with separate testa rupture and endosperm rupture would provide a seed model system with experimental advantages for studying endosperm weakening (see Müller *et al.*, Chapter 30, this volume).

Little is known about the molecular mechanisms of endosperm weakening (Bewley, 1997; Toorop *et al.*, 2000; Leubner-Metzger, 2003; da Silva *et al.*, 2004,

2005). Ikuma and Thimann (1963) proposed the 'hatching hypothesis' of seed biology as 'the final step in the germination control process is the production of an enzyme whose action enables the tip of the radicle to penetrate through the coat'. Experiments to identify this 'hatching enzyme' have been conducted in a variety of species and have provided evidence for the contribution of various cell wall-modifying proteins (e.g. endo-β-1,4-mannanases, endo-β-1,3-glucanases and expansins) (Bewley, 1997; Hilhorst *et al.*, 1998; Koornneef *et al.*, 2002; Leubner-Metzger, 2003). Expression of endo-β-1,4-mannanase in the micropylar endosperm of coffee is associated with the second phase of endosperm weakening and is inhibited by ABA (da Silva *et al.*, 2004). In contrast, endo-β-1,4-mannanase in tomato seeds is associated with the first phase of endosperm weakening and is not inhibited by ABA (Nonogaki *et al.*, 2000; Toorop *et al.*, 2000). Expression of endo-β-1, 3-glucanase in the micropylar endosperm, its inhibition by ABA and the inhibition of endosperm rupture by ABA is widespread among the *Solanaceae* (Wu *et al.*, 2000; Leubner-Metzger, 2003; Petruzzelli *et al.*, 2003). The ABA inhibition of endosperm rupture is partially reverted in transgenic tobacco seeds that overexpress endo-β-1,3-glucanase in the seed-covering layers under the control of an ABA-inducible transgene promoter (Leubner-Metzger and Meins, 2000; Leubner-Metzger, 2002; Manz *et al.*, 2005). This directly proves that endo-β-1,3-glucanase is causally involved in promoting endosperm rupture, but it is not the sole 'hatching enzyme' (Leubner-Metzger, 2003). Conclusive evidence for a sole 'hatching enzyme' has not yet been found. In his review, Bewley (1997) stated that 'endosperm weakening is likely to be essential for some seeds to complete germination, how it is achieved remains a mystery'. Taken together, these findings support the view that germination control by the seed-covering layers is achieved by the collaborative or successive action of several cell wall-modifying proteins and various molecular mechanisms (Leubner-Metzger, 2003). The intriguing issue that arises is that there might be evolutionary conserved molecular mechanisms as well as species-specific adaptations for endosperm weakening.

ABA, Endosperm Weakening and Novel Molecular Mechanisms Studied in Emerging Rosid Model Systems

Within the rosid clade, there is prevalence for more or less endospermless seeds (Fig. 20.1). There are also several examples of endospermic rosid seeds (O'Brien and McCully, 1969), but not a single publication about rosid endosperm weakening. A new model system for endosperm-limited germination with molecular phylogenetic placement within the rosid clade is therefore a necessity (Mandoli and Olmstead, 2000; Soltis and Soltis, 2003). The genome of the rosid model plant *Arabidopsis* is completely sequenced. The placement of a novel rosid model system in close proximity with this 'molecular' model plant allows successful utilization of the *Arabidopsis* databases for modern molecular methods like transcriptomics and proteomics (Mandoli and Olmstead, 2000; Hall *et al.*, 2002).

Lepidium is closely related to *Arabidopsis*; both belong to the *Brassicoideae* subfamily of *Brassicaceae*. Both species have separate testa and endosperm rupture, and ABA inhibits endosperm rupture, but not testa rupture (Liu *et al.*, 2005; see Müller *et al.*,

Chapter 30, this volume). *Arabidopsis* seeds are too small to perform puncture-force measurements (Fig. 20.2). These measurements show that endosperm weakening occurs prior to *Lepidium* endosperm rupture (see Müller *et al.*, Chapter 30, this volume). This endosperm weakening is inhibited by ABA. In future, we want to exploit these experimental advantages to investigate the molecular mechanisms that regulate *Lepidium* endosperm weakening. *Lepidium* is an emerging model system for studying endosperm weakening. Reactive oxygen species are a novel molecular mechanism for endosperm weakening and embryo expansion during seed germination (see Müller *et al.*, Chapter 30, this volume).

Acknowledgements

We gratefully acknowledge travel support by the Deutsche Forschungsgemeinschaft (to G. Leubner-Metzger), the seed company Ernst Benary Samenzucht GmbH, Hann. Münden, Germany (to B. Kucera) and the Wissenschaftliche Gesellschaft Freiburg (to K. Müller) for attending the Eighth International Seed Workshop in Brisbane, Australia, May 2005. Our research is supported by grants from the Deutsche Forschungsgemeinschaft and the Wissenschaftliche Gesellschaft Freiburg.

References

Baskin, J.M. and Baskin, C.C. (2004) A classification system for seed dormancy. *Seed Science Research* 14, 1–16.

Bewley, J.D. (1997) Breaking down the walls: a role for endo-β-mannanase in release from seed dormancy? *Trends in Plant Science* 2, 464–469.

da Silva, E.A.A., Toorop, P.E., van Aelst, A.C. and Hilhorst, H.W.M. (2004) Abscisic acid controls embryo growth potential and endosperm cap weakening during coffee (*Coffea arabica* cv. Rubi) seed germination. *Planta* 220, 251–261.

da Silva, E.A.A., Toorop, P.E., Nijsse, J., Bewley, J.D. and Hilhorst, H.W.M. (2005) Exogenous gibberellins inhibit coffee (*Coffea arabica* cv. Rubi) seed germination and cause cell death in the embryo. *Journal of Experimental Botany* 413, 1029–1038.

Finch-Savage, W.E. and Clay, H.A. (1997) The influence of embryo restraint during dormancy loss and germination of *Fraxinus excelsior* seeds. In: Ellis, R.H., Black, M., Murdoch, A.J. and Hong, T.D. (eds) *Basic and Applied Aspects of Seed Biology*. Kluwer Academic Publishers, Dordrecht, The Netherlands, pp. 245–253.

Finch-Savage, W.E. and McQuistan, C.I. (1991) Abscisic acid: an agent to advance and synchronise germination for tomato (*Lycopersicon esculentum* Mill.) seeds. *Seed Science and Technology* 19, 537–544.

Forbis, T.A., Floyd, S.K. and deQueiroz, A. (2002) The evolution of embryo size in angiosperms and other seed plants: implications for the evolution of seed dormancy. *Evolution* 56, 2112–2125.

Friedman, W.E. (1998) The evolution of double fertilization and endosperm: an "historical" perspective. *Sexual Plant Reproduction* 11, 6–16.

Groot, S.P.C. and Karssen, C.M. (1987) Gibberellins regulate seed germination in tomato by endosperm weakening: a study with gibberellin-deficient mutants. *Planta* 171, 525–531.

Groot, S.P.C. and Karssen, C.M. (1992) Dormancy and germination of abscisic acid-deficient tomato seeds. *Plant Physiology* 99, 952–958.

Hall, A.E., Fiebig, A. and Preuss, D. (2002) Beyond the *Arabidopsis* genome: opportunities for comparative genomics. *Plant Physiology* 129, 1439–1447.

Hepher, A. and Roberts, J.A. (1985) The control of seed germination in *Trollius ledebouri*: the breaking of dormancy. *Planta* 166, 314–320.

Hilhorst, H.W.M. and Downie, B. (1995) Primary dormancy in tomato (*Lycopersicon esculentum* cv. Moneymaker): studies with the *sitiens* mutant. *Journal of Experimental Botany* 47, 89–97.

Hilhorst, H.W.M., Groot, S.P.C. and Bino, R.J. (1998) The tomato seed as a model system to study seed development and germination. *Acta Botanica Neerlandica* 47, 169–183.

Homrichhausen, T.M., Hewitt, J.R. and Nonogaki, H. (2003) Endo-β-mannanase activity is associated with the completion of embryogenesis in imbibed carrot (*Daucus carota* L.) seeds. *Seed Science Research* 13, 219–227.

Ikuma, H. and Thimann, K.V. (1963) The role of the seed-coats in germination of photosensitive lettuce seeds. *Plant and Cell Physiology* 4, 169–185.

Jacobsen, J.V., Gubler, F. and Chandler, P.M. (1995) Gibberellin action in germinated cereal grains. In: Davies, P.J. (ed.) *Plant Hormones.* Kluwer Academic Publishers, Dordrecht, The Netherlands, pp. 246–271.

Judd, W.S., Campbell, C.S., Kellog, E.A., Stevens, P.F. and Donoghue, M.J. (2002) *Plant Systematics: A Phylogenetic Approach*, 2nd edn. Sinauer Associates, Sunderland, Massachusetts.

Junttila, O. (1973) The mechanism of low temperature dormancy in mature seeds of *Syringa* species. *Plant Physiology* 29, 256–263.

Karssen, C.M. (1976) Uptake and effect of abscisic acid during induction and progress of radicle growth in seeds of *Chenopodium album. Physiologia Plantarum* 36, 259–263.

Koornneef, M., Bentsink, L. and Hilhorst, H. (2002) Seed dormancy and germination. *Current Opinion in Plant Biology* 5, 33–36.

Kucera, B., Cohn, M.A. and Leubner-Metzger, G. (2005) Plant hormone interactions during seed dormancy release and germination. *Seed Science Research* 15, 281–307.

Leubner-Metzger, G. (2002) Seed after-ripening and over-expression of class I β-1,3-glucanase confer maternal effects on tobacco testa rupture and dormancy release. *Planta* 215, 959–968.

Leubner-Metzger, G. (2003) Functions and regulation of β-1,3-glucanase during seed germination, dormancy release and after-ripening. *Seed Science Research* 13, 17–34.

Leubner-Metzger, G. and Meins, F. Jr (2000) Sense transformation reveals a novel role for class I β-1,3-glucanase in tobacco seed germination. *The Plant Journal* 23, 215–221.

Liptay, A. and Schopfer, P. (1983) Effect of water stress, seed coat restraint, and abscisic acid upon different germination capabilities of two tomato lines at low temperature. *Plant Physiology* 73, 935–938.

Liu, P.-P., Koizuka, N., Homrichhausen, T.M., Hewitt, J.R., Martin, R.C. and Nonogaki, H. (2005) Large-scale screening of *Arabidopsis* enhancer-trap lines for seed germination-associated genes. *The Plant Journal* 41, 936–944.

Mandoli, D.F. and Olmstead, R. (2000) The importance of emerging model systems in plant biology. *Journal of Plant Growth Regulation* 19, 249–252.

Manz, B., Müller, K., Kucera, B., Volke, F. and Leubner-Metzger, G. (2005) Water uptake and distribution in germinating tobacco seeds investigated *in vivo* by nuclear magnetic resonance imaging. *Plant Physiology* 138, 1538–1551.

Martin, A.C. (1946) The comparative internal morphology of seeds. *The American Midland Naturalist* 36, 513–660.

Ni, B.R. and Bradford, K.J. (1993) Germination and dormancy of abscisic acid-deficient and gibberellin-deficient mutant tomato (*Lycopersicon esculentum*) seeds: sensitivity of germination to abscisic acid, gibberellin, and water potential. *Plant Physiology* 101, 607–617.

Nonogaki, H., Gee, O.H. and Bradford, K.J. (2000) A germination-specific endo-β-mannanase gene is expressed in the micropylar endosperm cap of tomato seeds. *Plant Physiology* 123, 1235–1245.

O'Brien, T.P. and McCully, E. (1969) *Plant Structure and Development: A Pictorial and Physiological Approach.* Macmillan, Toronto, Ontario, Canada.

Pavlista, A.D. and Haber, A.H. (1970) Embryo expansion without protrusion in lettuce seeds. *Plant Physiology* 45, 636–637.

Petruzzelli, L., Müller, K., Hermann, K. and Leubner-Metzger, G. (2003) Distinct expression patterns of β-1,3-glucanases and chitinases during the germination of Solanaceous seeds. *Seed Science Research* 13, 139–153.

Schopfer, P. and Plachy, C. (1984) Control of seed germination by abscisic acid. II. Effect on embryo water uptake in *Brassica napus* L. *Plant Physiology* 76, 155–160.

Soltis, D.E. and Soltis, P.S. (2003) The role of phylogenetics in comparative genetics. *Plant Physiology* 132, 1790–1800.

Tao, K.-L. and Khan, A.A. (1979) Changes in the strength of lettuce endosperm during germination. *Plant Physiology* 63, 126–128.

Toorop, P.E., van Aelst, A.C. and Hilhorst, H.W.M. (2000) The second step of the biphasic endosperm cap weakening that mediates tomato (*Lycopersicon esculentum*) seed germination is under control of ABA. *Journal of Experimental Botany* 51, 1371–1379.

Watkins, J.T. and Cantliffe, D.J. (1983) Mechanical resistance of the seed coat and endosperm during germination of *Capsicum annuum* at low temperatures. *Plant Physiology* 72, 146–150.

Welbaum, G.E., Muthui, W.J., Wilson, J.H., Grayson, R.L. and Fell, R.D. (1995) Weakening of muskmelon perisperm envelope tissue during germination. *Journal of Experimental Botany* 46, 391–400.

Wu, C.-T., Leubner-Metzger, G., Meins, F. Jr and Bradford, K.J. (2000) Class I β-1,3-glucanase and chitinase are expressed in the micropylar endosperm of tomato seeds prior to radicle emergence. *Plant Physiology* 126, 1299–1313.

Yim, K.O. and Bradford, K.J. (1998) Callose deposition is responsible for apoplastic semipermeability of the endosperm envelope of muskmelon seeds. *Plant Physiology* 118, 83–90.

21 Dormancy Classification and Potential Dormancy-breaking Cues for Shrub Species from Fire-prone South-eastern Australia

M.K.J. Ooi

Institute for Conservation Biology, University of Wollongong, Wollongong, NSW 2522, Australia; Biodiversity Conservation Science Section, Department of Environment and Conservation (NSW), Hurstville, NSW 2220, Australia

Abstract

Ecological studies require a clear understanding of the processes of dormancy release and germination. In fire-prone environments, implying that seeds are dormant because of a lack of germination from a soil-stored seed bank can confound results of experiments aimed at identifying dormancy mechanisms of fresh seeds. In this study, dormancy classification is endorsed as a starting point for establishing mechanisms that control dormancy of species from fire-prone vegetation. Shrub species within the physical dormancy (PY), physiological dormancy (PD), morphological dormancy (MD) and morphophysiological dormancy (MPD) classes were found to occur in the south-eastern Australian region in proportions similar to that of other fire-prone regions. Seasonal temperatures, not fire cues, are the main factors that break physiologically related dormancy mechanisms and can also control the timing of germination. Seasonal emergence patterns are more likely for species with a PD component than for those with PY. This can delay seedling emergence depending on the timing of the fire event, particularly in regions without a distinct rainfall season, and potentially hinder recruitment. Species with MPD and PD may therefore be more sensitive to seasonal shifts in the fire regime, and this could have implications for community species composition and threatened species management.

Introduction

In fire-prone regions, the post-fire environment is characterized by a flush of seedling emergence. It is considered advantageous for seedlings to emerge quickly after fire as there are increased levels of nutrients, decreased levels of competition and

 Seeds: Biology, Development and Ecology (eds S. Adkins, S. Ashmore and S.C. Navie)

consequently a greater chance of survival (Whelan, 1995; Bond and van Wilgen, 1996). This is particularly important for obligate-seeding species, as population persistence after fire is dependent primarily on germination and recruitment from propagules stored in a seed bank (Keith, 1996). Seeds of species with a canopy-stored seed bank are mostly non-dormant (ND); however, those with a soil-stored seed bank often have some type of dormancy (Bell *et al.*, 1993).

For a species to take advantage of post-fire conditions, and because fire is the predominant disturbance, many ecologists assume that seed dormancy is broken by fire cues. This is true for a large number of species, namely the dormancy-breaking effect of heat shock on hard-seeded legumes (Auld and O'Connell, 1991). However, freshly dispersed seeds of many other species from fire-prone regions do not germinate during laboratory trials, even after being treated with fire cues (Dixon *et al.*, 1995; Roche *et al.*, 1997; Keeley and Fotheringham, 1998; Clarke *et al.*, 2000; van Staden *et al.*, 2000; Ooi *et al.*, 2006), and their dormancy mechanisms remain unknown. These species may still display a flush of emergence after fire *in situ* or after burial (Dixon *et al.*, 1995; Enright *et al.*, 1997; Roche *et al.*, 1997; Wills and Read, 2002; Rokich *et al.*, 2002; Ooi *et al.*, 2004). Observations of seedling emergence en masse therefore indicate that the passage of fire has provided a germination cue, but not necessarily that primary dormancy has been broken by fire.

This apparent dichotomy highlights the need for a clearer understanding of the distinction between the processes of dormancy release and germination. Implying and defining seeds as being dormant because of a lack of germination from a soil-stored seed bank is a concept that does not necessarily hinder plant ecology research (Thompson *et al.*, 2003). For example, a number of studies use this concept of dormancy and provide clear descriptions of seed bank dynamics (Enright *et al.*, 1997; Roche *et al.*, 1997). However, it can cause problems when trying to define certain research questions (Vleeshouwers *et al.*, 1995; Thompson *et al.*, 2003), and, as can be seen in this case, particularly those aimed at identifying the dormancy mechanisms of species in fire-prone environments. A recent increase in the practical ecological management applications and species-specific management has focused on the need for a greater understanding of dormancy mechanisms in fire-prone regions (Merritt and Dixon, 2003; Ooi *et al.*, 2006).

The physiological concept of dormancy identifies it as a characteristic of the seed and not of the environment (Vleeshouwers *et al.*, 1995; Fenner and Thompson, 2005). Primary dormancy may break, but germination can still be prevented until favourable environmental conditions occur. These include the right seasonal temperatures or light conditions. Seeds may also enter into a secondary dormancy (Baskin and Baskin, 1998, 2004). By taking the physiological view of dormancy, a clearer hypothesis for more complex dormancy mechanisms in fire-prone environments may be suggested. Factors required to break primary dormancy may be completely unrelated to fire; however, fire may provide subsequent germination cues and suitable environmental conditions (Bell, 1999).

Classification of dormancy type is a fundamental step for understanding the germination ecology of a species (Baskin and Baskin, 2004). Placing species into the broadest classes can help to identify likely factors required to overcome dormancy, by enabling comparison with other species within the same class. Seeds classified as ND germinate in the first few weeks under some combination of normal environmental conditions experienced in their natural habitat (Baskin and Baskin, 1998). Seeds with physical dormancy (PY) have an impermeable seed coat that needs to be broken before germination. Physiological dormancy (PD) prevents germination until a chemical change occurs within the seed (Fenner and Thompson, 2005), and this is generally brought about by stratification at some combination of seasonal temperatures (Baskin and Baskin, 1998). Morphological dormancy (MD), where the embryo is immature or not fully developed, is very rarely recorded in temperate regions, as it is usually coupled with PD (Baskin and Baskin, 1998). Freshly dispersed seeds that have a combination of PD and MD are classified as having morphophysiological dormancy (MPD).

There is a considerable gap in our knowledge about the dormancy types, other than PY, of species from fire-prone regions. In this study, the aim was to classify the dormancy of a number of shrub species from the fire-prone heaths and forests of south-eastern Australia, and subsequently identify the types of dormancy represented in these systems. Because of the lack of studies classifying dormancy in fire-prone regions, the effects of fire cues on dormancy types, other than heat shock and PY, have never been tested explicitly. Therefore, the study also aimed to examine whether the primary dormancy of species, particularly those with MPD, can be overcome by cues directly related to fire, such as smoke and heat shock.

Finally, by focusing on a more discrete geographical range, an estimation of the proportions of different dormancy types in shrub species from the east coast of Australia was made, and the role that seasonal temperatures play in controlling dormancy and germination was discussed. Unlike Mediterranean-type climates, south-eastern Australia has no distinct rainfall season (Stern *et al.*, 2000) and is climatically more similar to non-Mediterranean fire-prone regions such as Florida (Carrington and Keeley, 1999) or the south-eastern Cape regions of South Africa (Deacon *et al.*, 1992; Cowling and Lombard, 2002). Temperature is the main factor impacting on dormancy and also has a role as a germination cue (Fenner and Thompson, 2005), but this has rarely been acknowledged in the south-east Australian region. A clearer understanding of seasonal temperature effects on dormancy and germination mechanisms may therefore highlight possible effects of season on post-fire population recovery and vegetation dynamics for the many thousands of species from fire-prone habitats in this region.

Materials and Methods

Classification of dormancy types

A dormant seed is one that does not have the capacity to germinate under combinations of normal environmental factors that are otherwise favourable for its germination (Baskin and Baskin, 2004). Therefore, to assess the type of seed dormancy, it is necessary to test for germination at temperature regimes similar to those that occur naturally in the field. In addition, embryo morphology has to be investigated to identify whether or not there is a morphological component to dormancy and seed imbibition has to be assessed to identify PY. Several species are chosen, ranging from families that are considered to have more complex dormancy, including members of the *Ericaceae* (formerly *Epacridaceae*), *Rutaceae* and *Santalaceae*, to species that are likely to have no dormancy (Table 21.1).

Seed embryo morphology and imbibition

Seeds were sectioned using either a scalpel and dissecting microscope or a microtome. *Astroloma*, *Persoonia* and *Styphelia* fruit were first cracked open using a bench vice. Seed and embryo lengths were measured. To see whether freshly matured seeds could imbibe water, 20 dry seeds from each species were weighed and then placed on moist filter paper for 5 min. Excess water was then removed from the seed surface with blotting paper before being weighed again. This was used as the initial weight. Seeds were then placed on plain agar (10 g/l) and weighed every hour for at least 6 h and subsequently every 24 h for a further 4 days. To ensure passage of water through the seed coat, several seeds of each species were soaked in a solution of methylene blue for 4 days. At the end of this period, seeds were dissected to assess the region of staining.

Germination experiments

To assess whether seeds of the study species have a primary dormancy (i.e. are dormant at the time of dispersal), freshly dispersed seeds were used where possible. Either 60 or 75 seeds, depending on availability, were divided into three replicates and used for each treatment. Seeds were placed on plain agar (10 g/l) in 9 cm Petri dishes. Three

Table 21.1. Species used in this study and results from experiments assessing dormancy class. Germination percentages of each species are for all three temperature treatments.

Species	Family	Embryo type	Imbibes water	Germination (%) light	Germination (%) dark	Dormancy class
Astroloma pinifolium	*Ericaceae*	Underdeveloped linear	Yes	0	n/t	MPD
Leucopogon amplexicaulis	*Ericaceae*	Linear	Yes	0	0	PD
Leucopogon juniperinus	*Ericaceae*	Underdeveloped linear	Yes	0	0	MPD
Leucopogon parviflorus	*Ericaceae*	Underdeveloped linear	Yes	0	0	MPD
Styphelia viridis	*Ericaceae*	Underdeveloped linear	Yes	0	0	MPD
Persoonia laurina	*Proteaceae*	Non-endospermic	Yes	0	n/t	PD
Petrophile pulchella	*Proteaceae*	Non-endospermic	Yes	100	n/t	ND
Correa alba	*Rutaceae*	Linear	Yes	0	n/t	PD
Leptomeria acida	*Santalaceae*	Underdeveloped rudimentary	Yes	0	0	MPD

n/t = species not tested using dark treatment; PD = physiological dormancy; MPD = morphophysiological dormancy; ND = non-dormant.

groups of three replicate dishes were then placed at each of three seasonal temperature settings in incubators. A dark treatment was also included for species with sufficient seed numbers (Table 21.1). Dishes assigned to a dark treatment were wrapped in two layers of aluminium foil. Incubators were set at 8/16h light/dark and maximum/minimum temperature cycles. Temperatures approximating either: (i) winter (15/5°C); (ii) spring/autumn (20/10°C); or (iii) summer (25/15°C) conditions were used. Seeds were checked weekly and monitored for 6 weeks. Checking for germination in the dark treatment was conducted under green safe lights. Germination was scored on the emergence of the radicle and expressed as a proportion of total viable seed.

The effect of fire cues on morphophysiological dormancy

It is important to know whether fire cues break primary dormancy or provide germination cues once primary dormancy is overcome (Baskin and Baskin, 2003). The effect of heat shock on breaking PY is well known in this region and is excluded from this part of the study. Smoke is also known to increase levels of germination for many species from fire-prone environments (Brown, 1993; Dixon *et al.*, 1995; Bell, 1999). However, the effects of smoke or heat shock on breaking PD or MPD are not so clear. To test this, the effects of charcoal, smoke and/or heat shock on seeds of species with MPD were investigated by reviewing germination studies. MPD was chosen as the test dormancy type because a larger number of species could be identified as having MPD with a greater degree of confidence compared to species with PD, by inference from embryo morphology data. Shrub species from temperate regions that have seeds with underdeveloped embryos are very likely to have MPD (Baskin and Baskin, 1998). MPD is also overcome by similar mechanisms as PD in many regions in the world.

MPD species within Australia were identified by reviewing studies targeting dormancy classification or embryo morphology (Martin, 1946; Corner, 1976; Schatral, 1996; Schatral *et al.*, 1997; Baskin and Baskin, 1998, 2003; Tieu and Egerton-Warburton, 2000; Ooi *et al.*, 2006; M.K.J. Ooi, 2005, unpublished data). If no specific data were available, embryo morphology, and subsequently dormancy type, was inferred from closely related species. Germination studies that had used species with MPD as the subject, and included charcoal, smoke and/or heat shock treatments applied to fresh seeds, were then identified and the results collated (Bell *et al.*, 1987; Dixon *et al.*, 1995; Roche *et al.*, 1997; Campbell, 1999; Clarke *et al.*, 2000; Ooi *et al.*, 2006). Where possible, total germination values at the end of 30 days were used, as germination should have occurred by this time if dormancy were broken (Baskin and Baskin, 2004). Studies were only selected if it was possible to identify that seeds had not received pretreatments, such as long storage or chilling.

Proportion of dormancy types in the shrub layer of the fire-prone Sydney flora

A literature review approach was also used to identify dormancy types of shrub species from the east coast Australian region. Only shrub species found in fire-prone heath, woodland and forest habitats, occurring in the Central Coast Botanical

Subdivision of New South Wales (Harden, 1992), were identified for study. Species lists were obtained using the PlantNet database (http://plantnet.rbgsyd.nsw.gov.au) and other references and herbaria data (e.g. Benson and McDougall, 1995). This subdivision includes the Sydney basin. In addition to the references mentioned above, dormancy types were identified or inferred from Auld and O'Connell (1991), Pannell (1995), Campbell (1999), Kenny (2000), Morris (2000), Auld *et al.* (2000), Auld (2001), Thomas *et al.* (2003) and Bhatia *et al.* (2005), as well as a number of databases and unpublished data.

Results

Classification of dormancy types

All species imbibed water within the first few hours. Methylene blue stain had passed through the seed coats of all species after 4 days. Embryo morphologies ranged from underdeveloped rudimentary types to those that were fully developed and non-endospermic (Table 21.1). Of the five *Ericaceae* tested, only one species, *Leucopogon amplexicaulis* (Rudge) R. Br., did not have an underdeveloped embryo. Only *Petrophile pulchella* (Schrad. and J.C. Wendl.) R. Br. seeds were germinated, with total germination reaching 100% at all three temperature regimes. Rates of germination, however, varied between the three temperature regimes (data not shown). The classes of dormancy identified were MPD and PD (Table 21.1).

The effect of fire cues on morphophysiological dormancy

Studies that applied fire treatments to fresh seeds identified 62 species from around Australia that were likely to have MPD. A total of 24 genera were represented from the families *Antheriaceae*, *Apiaceae*, *Dasypogonaceae*, *Dilleniaceae*, *Ericaceae*, *Iridaceae*, *Ranunculaceae* and *Santalaceae*. Of the 62 species identified, 96% failed to germinate with or without fire cue treatments or, occasionally, produced minimal (<5%) germination, at the end of 30 days.

Proportion of dormancy types in the shrub layer of the fire-prone Sydney flora

A total of 700 shrub species from fire-prone heath and forest communities within the Central Coast Botanical Subdivision were identified. Nearly half of these species were members of the *Fabaceae*, *Proteaceae* and *Myrtaceae*. A further one-third of all shrub species came from the *Rutaceae*, *Ericaceae*, *Asteraceae* and *Lamiaceae* families, all of which are known to have many species with a physiological component to their dormancy. Dormancy type was identified, or more confidently inferred, for 608 species. Of these, 28.6% were classified as ND and 71.4% as dormant. In terms of proportion of dormant species, the largest class of dormancy represented was PY (42.9%) followed by PD (40.1%) and MPD (17.1%) (Fig. 21.1). Since little information was available on germination requirements or dormancy type for most

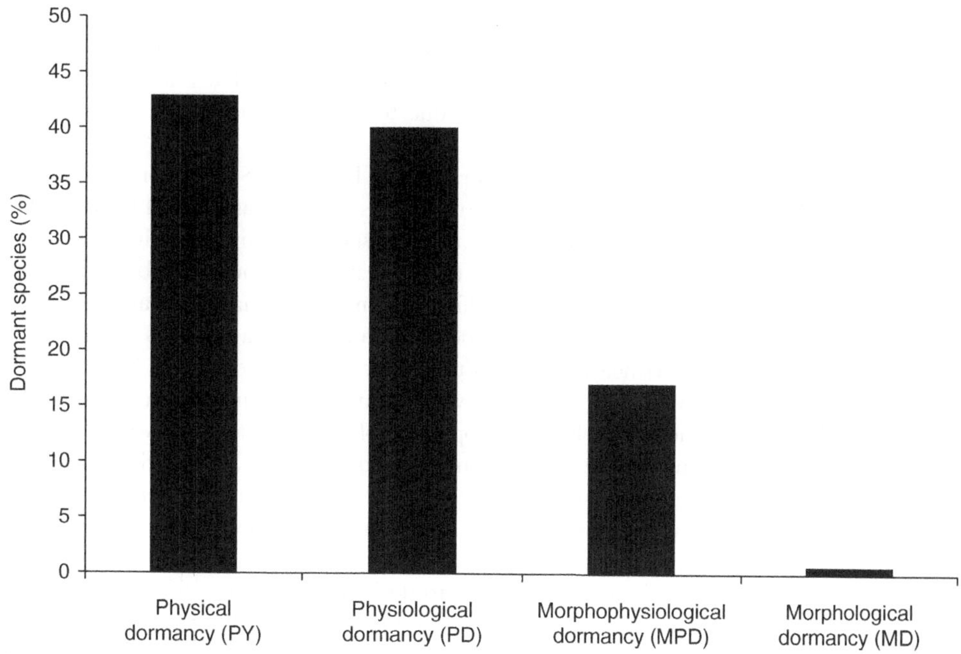

Fig. 21.1. Shrub species from the Central Coast Botanical Subdivision (New South Wales) in south-eastern Australia in each dormancy class. Values are a proportion of total species (n = 434) known or inferred to be dormant (i.e. excluding species that are non-dormant or have unknown dormancy).

of the species identified, inferences were made, sometimes at a family level. For example, data from other fire-prone regions show that the majority of species in the *Rutaceae* have PD (Bell *et al.*, 1993; Baskin and Baskin, 1998). Therefore, they were assigned to PD unless available research suggested otherwise. Most of the *Sapindaceae* and *Rhamnaceae* species were classified as PY using the same logic. Many *Asteraceae* species were left unknown due to the variability of dormancy types shown by species within this family. Although it is likely that species have either ND or PD, neither classification could confidently be assigned.

Discussion

All five classes of dormancy occur in fire-prone Australian vegetation. Germination trials conducted in this study found representative species of two dormancy types, PD and MPD. Most species within the tribe *Styphelieae* (*Ericaceae*) (Ooi *et al.*, 2006; M.K.J. Ooi, 2005, unpublished data) and *Dilleniaceae* (Schatral, 1996) have MPD, while those within the *Rutaceae*, *Lamiaceae* (Baskin and Baskin, 1998) and also *Persoonia* seem to have PD. Numerous other studies have found many species with PY, including most legumes and many *Rhamnaceae* (e.g. Auld and O'Connell, 1991; Turner *et al.*, 2005). Combinational dormancy (PY and PD) occurs in several *Hibbertia* species (Bell, 1999; Allan *et al.*, 2004) and MD is likely to occur in some mistletoes (*Loranthaceae*) (Baskin and Baskin, 1998) and several species of *Trachymene* (*Apiaceae*) (M.K.J. Ooi and B.D.E. McKenzie, 2005, unpublished data). These results are not surprising; however, they highlight the fact that a lack of previous study on dormancy classification is not due to a lack of dormancy types in fire-prone regions.

From the New South Wales Central Coast Botanical Subdivision on the east coast of Australia, the proportion of shrub species classified as PD and PY is similar to that for similar fire-prone vegetation types (described as 'matorral' in Baskin and Baskin, 1998). This vegetation type has the highest proportion of shrub species with PY in the temperate climate zone. This dormancy class may have an advantage in fire-prone vegetation as the heat shock dormancy cue translates to fast post-fire recruitment. There is nearly double the proportion of species with MPD in this region compared with other matorral vegetation (Baskin and Baskin, 1998). This could be due to the relatively high proportion of ericaceous species with this class of dormancy in Australia. *Ericaceae* from the Mediterranean and South Africa tend to have PD (Baskin and Baskin, 1998); however, many species of *Ericaceae* in south-eastern Australia, within the *Styphelieae* tribe, have underdeveloped embryos (Ooi *et al.*, 2006; M.K.J. Ooi, 2005, unpublished data).

Classification of dormancy can help to decipher the requirements for breaking dormancy. In various climatic regions around the world, MPD (and PD) is broken by stratification of seeds at seasonal temperatures (Baskin and Baskin, 1998). Species with MPD from fire-prone regions also appear to have dormancy broken by seasonal temperatures. Even though fire is the predominant disturbance, it does not provide the dormancy-breaking cue. Nearly all species identified during this study as having MPD, which were the subject of experimental treatments using fire cues such as charcoal, smoke or heat shock, did not germinate. In a recent study, Ooi *et al.* (2006) found that several *Leucopogon* (*Ericaceae*) species also did not have MPD broken by the fire cues, smoke and heat shock. Furthermore, stratification at seasonal temperatures was identified as the main dormancy-breaking factor overcoming MPD during burial. Once MPD was broken by seasonal temperatures, smoke, along with light, could enhance subsequent levels of germination. Evidence from other studies shows that burial can break primary dormancy of species with MPD and that subsequent smoke treatments can increase germination (Roche *et al.*, 1997; Keeley and Fotheringham, 1998; Tieu *et al.*, 2001). Taxa include *Leucopogon*, *Astroloma*, *Conostephium*, *Achrotriche* (all genera within the *Styphelieae* (*Ericaceae*)) and *Hibbertia* (*Dilleniaceae*) species from Australia and the Californian chaparral species *Dendromecon rigida* Benth. and *Dicentra chrysantha* (Hook. and Arn.) Walp.

Whether fire cues break primary dormancy or increase subsequent germination after breaking dormancy is an important distinction, as it clarifies the sequence of processes occurring naturally in the seed bank (Baskin and Baskin, 2003). Acknowledging this distinction can subsequently enable a more structured approach when trying to germinate species with more complex dormancy and germination mechanisms, and this is particularly important when using fresh seeds, such as for restoration or horticultural purposes. The primary dormancy of species with MPD or PD has to be broken by stratification at some combination of seasonal temperatures (Vleeshouwers *et al.*, 1995; Baskin and Baskin, 1998; Fenner and Thompson, 2005) before other factors can have an effect.

Seasonal temperatures can play a dual role, breaking dormancy and providing a germination cue (Fenner and Thompson, 2005). This can subsequently affect the timing or season of emergence. The great majority of species with a physiological component to their dormancy have seasonal temperatures acting in both roles (Baskin and Baskin, 1998). For example, species with non-deep simple MPD require a warm or cold stratification period to break dormancy, before embryo growth and germination occur at warm temperatures (Baskin and Baskin, 2004).

In a study conducted in the Sydney region, Ooi *et al.* (2004) found that emergence of several *Leucopogon* (*Ericaceae*) species occurred after fire during the late autumn or winter period, irrespective of earlier significant rainfall. These species have MPD, as well as temperature-dependent germination (Ooi *et al.*, 2006). Emergence of these species can therefore be delayed, with the magnitude of delay depending upon the season of the fire event. In the same study, the post-fire seedling emergence of many species, with either PY or ND (e.g. *Fabaceae* and species of *Banksia*, *Hakea* and *Grevillea* from the *Proteaceae*), was much faster. Many studies have found that for species with PY in south-eastern Australia, germination will occur rapidly after fire, as soon as sufficient moisture is available (Auld and O'Connell, 1991; Auld and Tozer, 1995; Ooi *et al.*, 2004).

Seasonal emergence is a common phenomenon in many Mediterranean-type habitats, where germination is considered to coincide with a distinct winter rainfall period (Whelan, 1995; Bond and van Wilgen, 1996; Bell, 1999). For example, in the Mediterranean-type climate of the Californian chaparral, the fire season is from June to October (i.e. summer to early autumn) and the majority of seedling emergence occurs in spring (Carrington and Keeley, 1999). As resprout growth is limited by drought immediately after the June to October fire season, and then by winter temperatures, delayed post-fire emergence occurs concurrently with resprouting. The seasonal delay in seedling emergence in these habitats is therefore not expected to be disadvantageous (Carrington and Keeley, 1999).

However, seasonal emergence in regions where there is no distinct rainfall season, such as south-eastern Australia and in the east of the Cape Floristic region in South Africa, could cause enough delay to disadvantage subsequent survival and growth (Jones *et al.*, 1997) of species with a seasonal requirement to their germination. Depending on the season of the fire event, seedling emergence of these species could lag behind the species that emerge more rapidly, as well as resprouters, by up to 1 year.

From a dormancy classification perspective, it is possible that in these environments species with PD or MPD (more seasonal emergers) could be more sensitive than species with PY (more rapid emergers) to any seasonal shift in the fire regime. In this study, 42.9% of the east coast shrub flora of Australia had some level of PY, while 57.2% of species had either PD or MPD. There is some evidence that fire regimes within the region are shifting from predominantly spring/summer to autumn/winter, particularly due to an increase in implemented burning (McLoughlin, 1998). Could this lead to a shift in species composition in these habitats? And does this place threatened species with PD or MPD at an even greater risk? These questions highlight the need for further research in dormancy classification, dormancy mechanisms and subsequent recruitment in fire-prone vegetation, particularly of species with PD or MPD. This includes members of *Ericaceae*, *Rutaceae*, *Apiaceae*, *Lamiaceae* and *Santalaceae*.

Acknowledgements

Tony Auld and Rob Whelan are acknowledged for their ever-helpful comments and continuing encouragement. I am grateful to staff at the Millennium Seed Bank Project, the Royal Botanic Gardens, Kew, Wakehurst Place, where part of this study was carried out. I thank Amelia Martyn and other staff at Mount Annan Botanic Garden (Botanic Gardens Trust, Sydney) for their valuable support and Emilie-Jane Ems, David Keith, Berin Mackenzie, Barbara Rice, Mark Robinson, Paul Thomas, Mark Westoby and the NSW Seedbank for access to seed collections. During this research, the author received funding from an Australian Research Council APAI Scholarship. This is contribution number 271 from the Ecology and Genetics Group at the University of Wollongong.

References

Allan, S.M., Adkins, S.W., Preston, C.A. and Bellairs, S.M. (2004) Improved germination of the Australian natives: *Hibbertia commutata*, *Hibbertia amplexicaulis* (Dilleniaceae), *Chameascilla corymbosa* (Liliaceae) and *Leucopogon nutans* (Epacridaceae). *Australian Journal of Botany* 52, 345–351.

Auld, T.D. (2001) The ecology of the Rutaceae in the Sydney region of south-eastern Australia: poorly known ecology of a neglected family. *Cunninghamia* 7, 213–239.

Auld, T.D. and O'Connell, M.A. (1991) Predicting patterns of post-fire germination in 35 eastern Australian Fabaceae. *Australian Journal of Ecology* 16, 53–70.

Auld, T.D. and Tozer, M. (1995) Patterns in emergence of *Acacia* and *Grevillea* seedlings after fire. *Proceedings of the Linnean Society of NSW* 115, 5–15.

Auld, T.D., Keith, D.A. and Bradstock, R.A. (2000) Patterns in longevity of soil seedbanks in fire-prone communities of south-eastern Australia. *Australian Journal of Botany* 48, 539–548.

Baskin, C.C. and Baskin, J.M. (1998) *Seeds: Ecology, Biogeography and Evolution of Dormancy and Germination.* Academic Press, San Diego, California.

Baskin, C.C. and Baskin, J.M. (2003) Overview and recommendations for future research priorities on native seed dormancy and germination of Australian plants. *Australian Plant Conservation* 11, 2–9.

Baskin, J.M. and Baskin, C.C. (2004) A classification system for seed dormancy. *Seed Science Research* 14, 1–16.

Bell, D.T. (1999) Turner review No. 1: the process of germination in Australian species. *Australian Journal of Botany* 47, 475–517.

Bell, D.T., Vlahos, S. and Watson, L.E. (1987) Stimulation of seed germination of understorey species of the northern jarrah forest of Western Australia. *Australian Journal of Botany* 35, 593–599.

Bell, D.T, Plummer, J.A. and Taylor, S.K. (1993) Seed germination ecology in south-western Western Australia. *The Botanical Review* 59, 24–73.

Benson, D. and McDougall, L. (1995) Ecology of Sydney plant species, Vol. 3. *Cunninghamia* 4, 217–431.

Bhatia, N.P., Nkang, A.E., Walsh, K.B., Baker, A.J.M., Ashwath, N. and Midmore, D.J. (2005) Successful seed germination of the nickel hyperaccumulator *Stackhousia tryonii*. *Annals of Botany* 96, 159–163.

Bond, W.J. and van Wilgen, B.W. (1996) *Fire and Plants.* Chapman & Hall, London.

Brown, N.A.C. (1993) Promotion of germination of fynbos seeds by plant-derived smoke. *New Phytologist* 123, 575–583.

Campbell, M. (1999) Soil seed bank dynamics of four sympatric species of sub-shrubs in grassy woodlands on the New England Tablelands, NSW. Honours thesis, University of New England, Australia.

Carrington, M.E. and Keeley, J.E. (1999) Comparison of post-fire seedling establishment between scrub communities in Mediterranean and non-Mediterranean climate ecosystems. *Journal of Ecology* 87, 1025–1036.

Clarke, P.J., Davison, E.A. and Fulloon, L. (2000) Germination and dormancy of grassy woodland and forest species: effects of smoke, heat, darkness and cold. *Australian Journal of Botany* 48, 687–700.

Corner, E.J.H. (1976) *The Seeds of Dicotyledons*. Cambridge University Press, Cambridge.

Cowling, R.M. and Lombard, A.T. (2002) Heterogeneity, speciation/extinction history and climate: explaining regional plant diversity patterns in the Cape Floristic Region. *Diversity and Distributions* 8, 163–179.

Deacon, H.J., Jury, M.R. and Ellis, F. (1992) Selective regime and time. In: Cowling, R.M. (ed.) *Ecology of Fynbos: Nutrients, Fire and Diversity*. Oxford University Press, Cape Town, South Africa, pp. 6–22.

Dixon, K.W., Roche, S. and Pate, J.S. (1995) The promotive effect of smoke derived from burnt vegetation on seed germination of western Australian plants. *Oecologia* 101, 185–192.

Enright, N.J., Goldblum, D., Ata, P. and Ashton, D.H. (1997) The independent effects of heat, smoke and ash on emergence of seedlings from the soil seed bank of a heathy *Eucalyptus* woodland in Grampians (Gariwerd) National Park, western Victoria. *Australian Journal of Ecology* 22, 81–88.

Fenner, M. and Thompson, K. (2005) *The Ecology of Seeds*. Cambridge University Press, Cambridge.

Harden, G.J. (1992) *Flora of New South Wales*, Vol. 3. University of New South Wales Press, Kensington, Australia.

Jones, R.H., Allen, B.P. and Sharitz, R.R. (1997) Why do early-emerging tree seedlings have survival advantages?: a test using *Acer rubrum* (Aceraceae). *American Journal of Botany* 84, 1714–1718.

Keeley, J.E. and Fotheringham, C.J. (1998) Smoke-induced seed germination in California chaparral. *Ecology* 79, 2320–2336.

Keith, D.A. (1996) Fire-driven extinction of plant populations: a synthesis of theory and review of evidence from Australian vegetation. *Proceedings of the Linnean Society of NSW* 116, 37–78.

Kenny, B.J. (2000) Influence of multiple fire-related germination cues on three Sydney *Grevillea* (Proteaceae) species. *Austral Ecology* 25, 664–669.

Martin, A.C. (1946) The comparative internal morphology of seeds. *American Midland Naturalist* 36, 513–660.

McLoughlin, L.C. (1998) Season of burning in the Sydney region: the historical records compared with recent prescribed burning. *Australian Journal of Ecology* 23, 393–404.

Merritt, D.J. and Dixon, K.W. (2003) Seed storage characteristics and dormancy of Australian indigenous plant species: from the seed store to the field. In: Smith, R.D., Dickie, J.B., Linington, S.H., Pritchard, H.W. and Probert, R.J. (eds) *Seed Conservation: Turning Science into Practice*. Royal Botanic Gardens Kew, UK, pp. 808–823.

Morris, E.C. (2000) Germination response of seven east Australian *Grevillea* species (Proteaceae) to smoke, heat exposure and scarification. *Australian Journal of Botany* 48, 179–189.

Ooi, M.K.J., Auld, T.D. and Whelan, R.J. (2004) Delayed post-fire seedling emergence linked to season: a case study with *Leucopogon* (Epacridaceae). *Plant Ecology* 174, 183–196.

Ooi, M.K.J., Auld, T.D. and Whelan, R.J. (2006) Dormancy and the fire-centric focus: germination of three *Leucopogon* species (Ericaceae) from south-eastern Australia. *Annals of Botany* 98(2) 421–430.

Pannell, J.R. (1995) Factors affecting seedling recruitment of *Allocasuarina distyla* and *A. nana* at burnt and unburnt sites. *Proceedings of the Linnean Society of NSW* 115, 121–133.

Roche, S., Dixon, K.W. and Pate, J.S. (1997) Seed ageing and smoke: partner cues in the amelioration of seed dormancy in selected Australian native species. *Australian Journal of Botany* 45, 783–815.

Rokich, D.P., Dixon, K.W., Sivasithamparam, K. and Meney, K.A. (2002) Smoke, mulch, and seed broadcasting effects on woodland restoration in Western Australia. *Restoration Ecology* 10, 185–194.

Schatral, A. (1996) Dormancy in seeds of *Hibbertia hyperiocides* (Dilleniaceae). *Australian Journal of Botany* 44, 213–222.

Schatral, A., Osborne, J.M. and Fox, J.E.D. (1997) Dormancy in seeds of *Hibbertia cuneiformis* and *H. huegelii* (Dilleniaceae). *Australian Journal of Botany* 45, 1045–1053.

Stern, H., de Hoedt, G. and Ernst, J. (2000) Objective classification of Australian climates. *Australian Meteorological Magazine* 49, 87–96.

Thomas, P.B., Morris, E.C. and Auld, T.D. (2003) Interactive effects of heat shock and smoke on germination of nine species forming soil seed banks within the Sydney region. *Austral Ecology* 28, 674–683.

Thompson, K., Ceriani, R.M., Bakker, J.P. and Bekker, R.M. (2003) Are seed dormancy and persistence in soil related? *Seed Science Research* 13, 97–100.

Tieu, A. and Egerton-Warburton, L.M. (2000) Contrasting seed morphology dynamics in relation to the alleviation of dormancy with soil storage. *Canadian Journal of Botany* 78, 1187–1198.

Tieu, A., Dixon, K.W., Meney, K.A. and Sivasithamparam, K. (2001) Interaction of soil burial and smoke on germination patterns in seeds of selected Australian native plants. *Seed Science Research* 11, 69–76.

Turner, S.R., Merrit, D.J., Baskin, C.C., Dixon, K.W. and Baskin, J.M. (2005) Physical dormancy in seeds of six genera of Australian Rhamnaceae. *Seed Science Research* 15, 51–58.

van Staden, J., Brown, N.A.C., Jäger, A.K. and Johnson, T.A. (2000) Smoke as a germination cue. *Plant Species Biology* 15, 167–178.

Vleeshouwers, L.M., Bouwmeester, H.J. and Karssen, C.M. (1995) Redefining seed dormancy: an attempt to integrate physiology and ecology. *Journal of Ecology* 83, 1031–1037.

Whelan, R.J. (1995) *The Ecology of Fire.* Cambridge University Press, Cambridge.

Wills, T.J. and Read, J. (2002) Effects of heat and smoke on germination of soil-stored seed in a south-eastern Australian sand heathland. *Australian Journal of Botany* 50, 197–206.

22 Quantitative Trait Loci, Epistasis and Other Interactions Associated with Dormancy in Weedy Rice (*Oryza sativa* L.)

M.E. Foley,[1] X.-Y. Gu[2] and S.F. Kianian[2]

[1]*US Department of Agriculture–Agricultural Research Service (USDA–ARS), Biosciences Research Laboratory, Fargo, ND 58105-5674, USA;* [2]*Department of Plant Sciences, North Dakota State University, Fargo, ND 58105, USA*

Abstract

A wild weedy strain (i.e. SS18-2) of *Oryza sativa* L. was selected to investigate the genetic architecture underlying seed dormancy. A framework genetic map covering the whole rice genome was constructed based on a backcross population (i.e. BC_1 [EM93-1 (non-dormant breeding line)//EM93-1/SS18-2]). Dormancy was determined by germination of seeds after 1, 11 and 21 days after ripening (DAR). Six dormancy quantitative trait loci (QTLs), designated as *qSD4*, *6*, *7-1*, *7-2*, *8* and *12*, were identified. The locus *qSD7-1* was tightly linked to the red pericarp colour gene *Rc*. In addition, the locus *qSD4* was associated with an interval imparting black hull colour. A QTL × DAR interaction was detected for *qSD12*, the locus with the highest effect at 1, 11 and 21 DAR (R^2 = 14%, 24% and 20%, respectively). Two and more (i.e. three and four) orders of epistases were detected with four and all the six QTLs, respectively. The higher-order epistases strongly suggest the presence of genetically complex networks in the regulation of variation for seed dormancy in natural populations, and make it critical to select a favourable combination of alleles at multiple loci in positional cloning of a target dormancy gene. Backcrossing techniques in combination with phenotypic and marker-assisted selection are being used to introduce the weedy rice-derived dormancy locus *qSD7-1* into the EM93-1 genetic background to develop near isogenic lines for fine-mapping and positional cloning.

Introduction

Seed dormancy is the failure of a mature seed to germinate under conditions normally favourable for germination. Dormancy distributes seed germination over time and is therefore an adaptive trait for survival of wild and weedy plants under

 Seeds: Biology, Development and Ecology (eds S. Adkins, S. Ashmore and S.C. Navie)

adverse and changing environmental conditions. It is well known that dormancy is a complex trait controlled by multiple genes and modified by environmental factors (Baskin and Baskin, 1998). Resolving the underlying molecular mechanisms for dormancy requires characterization of genes that directly affect dormancy or germinability (i.e. the propensity for immediate, intermediate or much delayed germination). The goal of our research is to elucidate mechanisms, signals and pathways that regulate dormancy in weedy grasses.

Individual genes of quantitatively inherited characteristics like dormancy are often resolved by quantitative trait loci (QTLs) mapping, which identifies genomic regions delimited by molecular markers associated with a phenotype, such as germinability, based on a segregation population. QTL analysis is a prerequisite for positional or map-based cloning to characterize the underlying genes. This analysis for dormancy or resistance to preharvest sprouting has been conducted for *Arabidopsis* (*Arabidopsis thaliana* (L.) Heynh.), barley (*Hordeum vulgare* L.), cultivated rice (*Oryza sativa* L.), sorghum (*Sorghum bicolor* (L.) Moench) and wheat (*Triticum aestivum* L.) (Gu *et al.*, 2004). The genetic constitution and available resources for most plants preclude positional cloning. However, *Arabidopsis* and rice are tractable systems for positional cloning given their genome size and a fully sequenced genome (Arabidopsis Genome Initiative, 2000; International Rice Genome Sequencing Project, 2005); cloning of dormancy QTLs is best accomplished using these experimental systems (Li and Foley, 1997; Koornneef *et al.*, 2002). Positional cloning of the major seed dormancy locus *DOG1* (delay of germination) in *Arabidopsis* is in progress (Alonso-Blanco *et al.*, 2003; Koornneef *et al.*, 2004). In contrast, map-based cloning in rice is at the stage of developing isogenic lines and fine-mapping of QTLs (Gu *et al.*, 2006).The research aimed at positional cloning of dormancy QTLs from weedy rice has yielded new insight into the genetic complexities for control of dormancy and related adaptive characteristics.

QTLs in Rice

Dormancy QTLs

Rice genotypes are greatly divergent in seed dormancy, with wild and weedy rice (*Oryza* species) usually having strongly dormant seeds in relation to cultivated varieties (Foley *et al.*, 2003; Gu *et al.*, 2005b). Dormancy QTLs have been identified on 11 of 12 chromosomes from several rice cultivars (Lin *et al.*, 1998; Dong *et al.*, 2003; Wan *et al.*, 2005), and two wild (*Oryza rufipogon* Griff.) (Cai and Morishima, 2000; Thomson *et al.*, 2003) and one weedy (*O. sativa*) rice accession (Gu *et al.*, 2004).

As a first step towards cloning dormancy genes from weedy rice, we developed a genetic map and conducted QTL analysis based on the backcross (BC_1) segregation population derived from the EM93-1//EM93-1/SS18-2 cross. SS18-2 and EM93-1 are a highly dormant weedy strain and a non-dormant breeding line, respectively (Gu *et al.*, 2003). Six dormancy QTLs, *qSD4, 6, 7-1, 7-2, 8* and *12*, were identified in the BC_1 population. Five were confirmed with the same population grown in different years (Gu *et al.*, 2005a). The proportion of phenotypic variation for dormancy represented by individual QTLs in this population ranged from 7% to 25%, depending on the QTL and the number of days after ripening (DAR) prior to germination. The major dormancy QTL *qSD12* (R^2 = 25% in the BC_1 population) appears to be unique as it is not reported for the aforementioned domesticated or wild rice accessions. The *qSD7-1* locus, which accounts for ~10% of the phenotypic variance, is tightly linked with the red pericarp gene *Rc*. The genetic effect of *qSD12* is additive, but consists of both gene additive and dominance effects for *qSD7-1* (Gu *et al.*, 2005c).

Epistasis

Epistasis is the interaction between different genes. Epistatic analysis detects whether the expression of one gene is significantly promoted or inhibited by another gene or allelic combinations of two or more genes. Epistases are involved in the regulation of genetic variation for seed dormancy with two-way interactions being detected in *Arabidopsis*, barley and wheat mapping populations (Gu *et al.*, 2004). A number of two-way, three-way and four-way epistases involving all six dormancy QTLs at different DAR in BC_1 primary segregation population were detected (Gu *et al.*, 2004). These interactions individually accounted for 1.3–2.8% of the phenotypic variance. Figure 22.1 gives a graphic example of a three-way epistasis between *qSD7-1*, *7-2* and *8* at 21 DAR. Observe both differences in magnitude and direction of response. Thus, when markers for *qSD7-2* (RM346) and *qSD8* (RM531) are heterozygous for EM93-1 alleles (M2.2 on the *x*-axis), the *qSD7-1* homozygote (M1 at RM180) and heterozygote (M2 at RM180) diverge in direction and magnitude of response when compared with other allelic combinations. The two-order or higher-order interactions

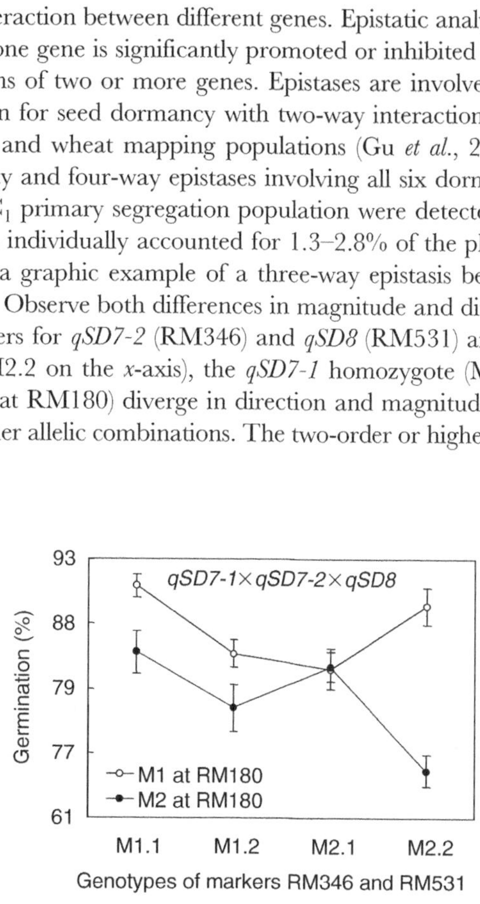

Fig. 22.1. A three-locus epistases of dormancy quantitative trait loci (QTLs) detected at 21 days after ripening (DAR) from the BC_1 population. *qSD7-1*, *qSD7-2* and *qSD8* are represented by their nearest markers RM180, RM346 and RM531, respectively. The letter M followed by a number-dot-number sequence on the horizontal axis represents the recombinant genotypes of two markers (M). The numbers 1 and 2 represent each marker locus homozygous for EM93-1 alleles and heterozygous, respectively. The lines for M1 and M2 represent the first marker locus homozygous for EM93-1 alleles and heterozygous, respectively. The solid or open circles represent the mean and the vertical bars represent the standard error for each marker combination. The vertical axis is germination percentage based on arcsine transformation data.

in the BC_1 (Gu *et al.*, 2004) and the BC_4F_2 segregation populations (Gu *et al.*, 2005c) suggest that all dormancy loci identified from SS18-2 are networked by epistases. A mutation at a particular locus could change its main effect and also the expression of one to several other dormancy loci. Epistasis was an important factor in detection of *qSD1*, a seventh SS18-2-derived dormancy QTL. In the course of backcrossing and phenotypic selection for dormancy to investigate dormancy-interrelated traits, we identified *qSD1* as an epistasis-QTL (E-QTL). E-QTLs need not have independent main effects, as they are detected by epistasis with 'normal' QTLs (e.g. *qSD7-1*), sometimes termed main (effects) QTLs (M-QTLs). The *qSD1* allele did not display a significant main effect in the BC_4F_2 population #132 (Gu *et al.*, 2005c), yet it was retained during several generations of backcrossing and phenotypic selection for dormancy; it must have benefited from its epistasis with *qSD7-1* (R^2 = 6%). Thus, the phenotypic selection for dormancy is more favourable with networked genes rather than a set of random loci. Importance of this networking will become more apparent when dormancy-interrelated QTLs are discussed.

Genotype (QTL) by environmental (E) interactions

Genotype by environmental interactions (G × E), during seed development or after ripening, affect the degree of primary seed dormancy and germinability. Thus, QTLs for seed dormancy vary in response to the environmental conditions and dormancy. QTL × E interactions have been reported for wheat, barley and sorghum (Anderson *et al.*, 1993; Oberthur *et al.*, 1995; Lijavetzky *et al.*, 2000).

After-ripening time is the major postharvest factor affecting dormancy. We detected a QTL × DAR interaction involving *qSD12* (Gu *et al.*, 2004). *qSD12* displayed a small (13%), large (24%) and moderate (19%) effect at 1, 11 and 21 DAR, respectively. In addition, we detected QTL × E interactions involving the year (i.e. seed development environment). One interaction was detected at 1 DAR involving *qSD7-2* (RM234) and two interactions at 11 and 21 DAR involving *qSD12* (RM270) (Gu *et al.*, 2005a). These two QTLs exhibit a similar QTL × E interaction pattern (Fig. 22.2) in that the phenotypic values for the heterozygous genotypes decreased and the values for the EM93-1-type homozygous genotypes increased under the 2003 experiment conditions as compared with those in 2002. In current research, we are evaluating genetic components of three dormancy QTLs in a synchronized non-dormant genetic background to identify major environmental factors during development that contribute to QTL × E interactions and dissect genetic effects for interactions. Preliminary results suggest that low temperatures tend to enhance dormancy in rice and individual genotypes respond to the environment divergently (Gu, 2006, unpublished data).

Dormancy-interrelated QTLs

Wild and weedy plants generally possess seed-associated adaptive traits beyond seed dormancy. For example, the common name for weedy rice in some parts of the world is red rice due to its red pericarp. In our initial research to investigate

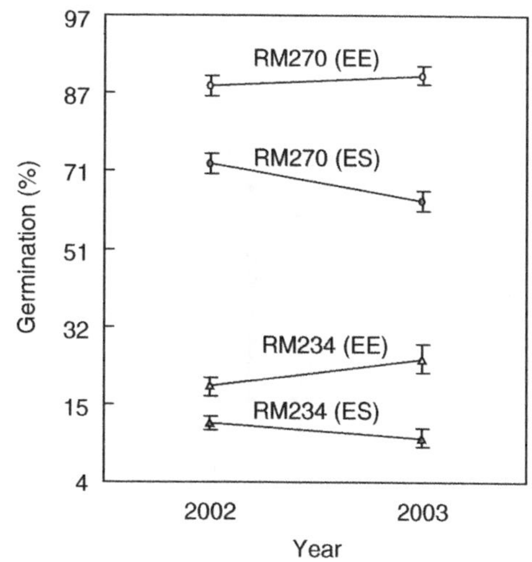

Fig. 22.2. Quantitative trait loci (QTLs) × environment (year) interactions for seed dormancy. RM234 and RM270 are the markers nearest to the dormancy loci *qSD7-2* and *qSD12*, respectively. The letters EE and ES in parentheses indicate the EM93-1-type homozygous and the heterozygous genotypes, respectively, for each marker locus. The circles (germination at 1 day) and triangles (germination at 21 days) represent the mean and the vertical bars represent the standard error for each marker genotype. The vertical axis indicates germination percentage based on arcsine transformation data.

the inheritance of dormancy, we noted associations between seed dormancy and shattering, awn length, hull colour and pericarp colour (Gu *et al.*, 2003, 2005b); the characteristics of the weedy form were associated with significantly reduced germination in BC_1 populations. The phenotypic interrelationships prompted an investigation into the genetic behaviours and QTLs for this set of adaptive traits. QTL analysis revealed five dormancy, four shattering, three awn length, two hull colour and one pericarp colour QTL in the SS18-2-derived primary segregation (BC_1) population (Gu *et al.*, 2005a). Four of the five dormancy QTLs were flanked or bracketed by 1–4 QTLs for the interrelated characteristics (Fig. 22.3). For example, *qSD7-1* is bracketed by a QTL for pigmentation on the pericarp, which itself contains the red pericarp colour locus *Rc* (Gu *et al.*, 2005a). Fine-mapping and cloning genes underlying *qSD7-1*/*Rc* may answer a long-standing question of whether pericarp colour genes have a pleiotropic effect (i.e. one locus affecting more than one phenotype) on dormancy, or whether a colour gene and a dormancy gene are tightly linked loci. The locus *qSD12* was not linked with QTLs for interrelated weedy characteristics (Fig. 22.3) and given its major and additive effect may have value for imparting resistance to preharvest sprouting. After five generations of phenotypic selection alone for low germination extremes, four dormancy loci with moderate to small effects colocated with QTLs for 1–3 of the associated traits (Gu *et al.*, 2005c). The multilocus response to the selection suggests that these dormancy genes are cumulative in effect, as well as networked by epistases, and that the network may have played a sheltering role in maintaining intact adaptive haplotypes during the evolution of weeds.

Conclusions

Seed dormancy is a genetically complex trait governed by multiple loci, various epistases and genotype by environmental interactions. These attributes dramatically

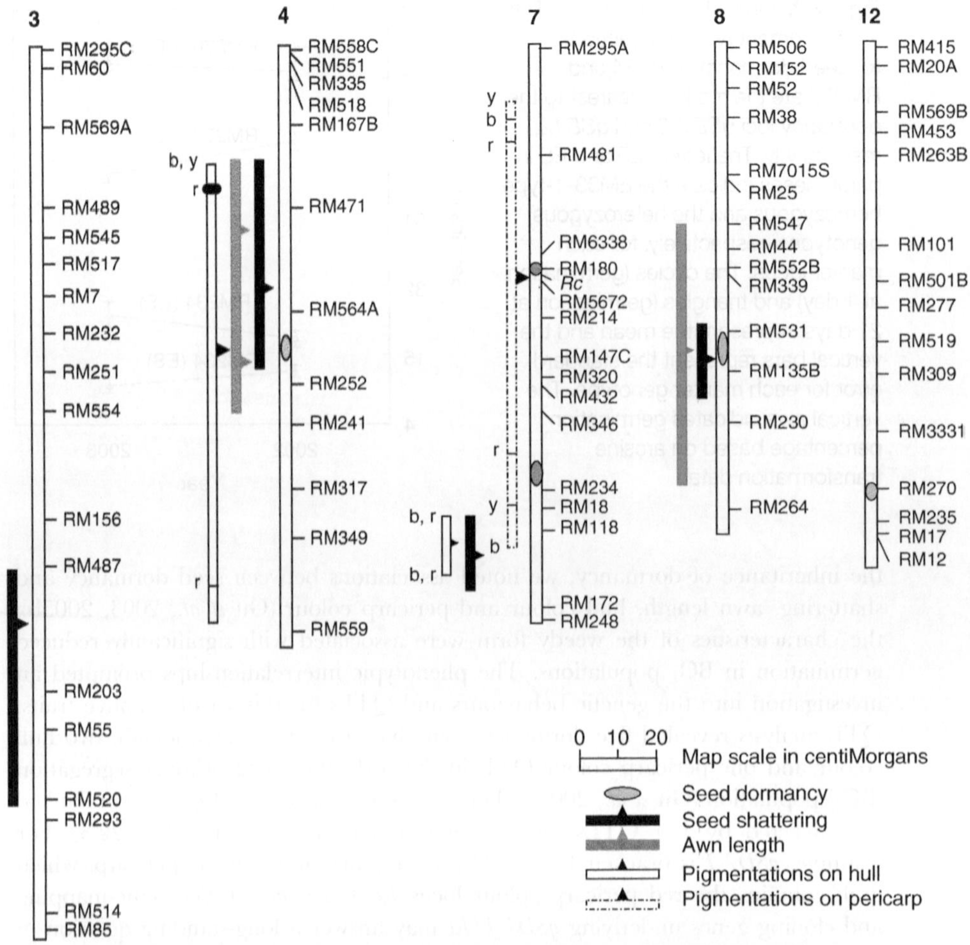

Fig. 22.3. Distribution of QTLs responsible for seed dormancy and its interrelated characteristics on rice chromosomes in the BC_1 population. Markers (RM) are labelled to the right of chromosomes. Ovals on a chromosome depict dormancy QTLs detected based on germination of seeds after ripening for 1, 11 and 21 days. The vertical boxes to the left of the chromosomes depict the 14 test statistic (TS) support limits of QTLs for seed shattering, awn length and black, red and yellow pigmentations on hull and pericarp. Arrows on the boxes indicate the TS peak positions for the QTLs. The horizontal bars on the empty boxes indicate the TS limits defined by the black (b), red (r) and yellow (y) pigmentations on the hull and the pericarp.

increase variation for dormancy and provide raw material for natural selection of this adaptive trait. The critical nature of dormancy in adaptation of weedy populations to agroecosystems is reflected in its role as an anchor for QTLs for interrelated weedy characteristics. In the near future, we will begin fine-mapping one or more SS18-2-derived dormancy QTLs as the next step towards the positional cloning and characterization of genes that control dormancy and interrelated weedy traits (e.g. *qSD7-1*) and to develop genetic resources to impart resistance to preharvest sprouting (e.g. *qSD12*).

Acknowledgement

Funding for this work was provided by USDA-National Research Initiative (0200668).

References

Alonso-Blanco, C., Bentsink, L., Hanhart, C.J., Blankestijn-de Vries, H. and Koornneef, M. (2003) Analysis of natural allelic variation at seed dormancy loci of *Arabidopsis thaliana*. *Genetics* 164, 711–729.

Anderson, J.A., Sorrells, M.E. and Tanksley, S.D. (1993) RFLP analysis of genomic regions associated with resistance to pre-harvest sprouting in wheat. *Crop Science* 33, 453–459.

Arabidopsis Genome Initiative (2000) Analysis of the genome sequence of the flowering plant *Arabidopsis thaliana*. *Nature* 408, 796–815.

Baskin, C.C. and Baskin, J.M. (1998) *Seeds: Ecology, Biogeography, and Evolution of Dormancy and Germination.* Academic Press, San Diego, California.

Cai, H.W. and Morishima, H. (2000) Genomic regions affecting seed shattering and seed dormancy in rice. *Theoretical and Applied Genetics* 100, 840–846.

Dong, Y., Tsozuki, E., Kamiunten, H., Terao, H., Lin, D., Matsuo, M. and Zheng, Y. (2003) Identification of quantitative trait loci associated with pre-harvest sprouting resistance in rice (*Oryza sativa* L.). *Field Crops Research* 81, 133–139.

Foley, M.E., Gu, X. and Nadella, D. (2003) Developing weedy rice for map-based cloning of seed dormancy genes. In: Nicolás, G., Bradford, K.J., Côme, D. and Pritchard, H.W. (eds) *The Biology of Seeds: Recent Research Advances.* CAB International, Wallingford, UK, pp. 151–159.

Gu, X.-Y., Chen, Z. and Foley, M.E. (2003) Inheritance of seed dormancy in weedy rice (*Oryza sativa* L.). *Crop Science* 43, 835–843.

Gu, X.-Y., Kianian, S.F. and Foley, M.E. (2004) Multiple loci and epistases control genetic variation for seed dormancy in weedy rice (*Oryza sativa*). *Genetics* 166, 1503–1516.

Gu, X.-Y., Kianian, S.F., Hareland, G.A., Hoffer, B.L. and Foley, M.E. (2005a) Genetic analysis of adaptive syndromes interrelated with seed dormancy in weedy rice (*Oryza sativa* L.). *Theoretical and Applied Genetics* 110, 1108–1118.

Gu, X.-Y., Kianian, S.F. and Foley, M.E. (2005b) Seed dormancy imposed by covering tissues interrelates to shattering and seed morphological characteristics in weedy rice. *Crop Science* 45, 948–955.

Gu, X.-Y., Kianian, S.F. and Foley, M.E. (2005c) Phenotypic selection for dormancy introduced a set of adaptive haplotypes from weedy into cultivated rice. *Genetics* 171, 695–704.

Gu, X.-Y., Kianian, S.F. and Foley, M.E. (2006) Isolation of three dormancy QTLs as Mendelian factors in rice. *Heredity* 96, 93–99.

International Rice Genome Sequencing Project (2005) The map-based sequence of the rice genome. *Nature* 436, 793–800.

Koornneef, M., Bentsink, L. and Hilhorst, H. (2002) Seed dormancy and germination. *Current Opinions in Plant Biology* 5, 33–36.

Koornneef, M., Bentsink, L., Clerkx, E. and Alonso-Blanco, C. (2004) The genetic analysis of seed dormancy in Arabidopsis. *Abstracts of Third International Symposium on Plant Dormancy: From Molecular Level to the Whole Plant.* Wageningen University, Wageningen, The Netherlands, p. 43.

Li, B. and Foley, M.E. (1997) Genetic and molecular control of seed dormancy. *Trends in Plant Sciences* 2, 384–389.

Lijavetzky, D., Martinez, M.C., Carrari, F. and Hopp, H.E. (2000) QTL analysis and mapping of pre-harvest sprouting resistance in sorghum. *Euphytica* 112, 125–135.

Lin, S.Y., Sasaki, T. and Yano, M. (1998) Mapping quantitative trait loci controlling seed dormancy and heading date in rice, *Oryza sativa* L., using backcross inbred lines. *Theoretical and Applied Genetics* 96, 997–1003.

Oberthur, L., Blake, T.K., Dyer, W.E. and Ullrich, S.E. (1995) Genetic analysis of seed dormancy in barley (*Hordeum vulgare* L.). *Journal of Quantitative Trait Loci* 1, 5.

Thomson, M.J., Tai, T.H., McClung, A.M., Lai, X.H., Hinga, M.E., Lobos, K.B., Xu, Y., Martinez, C.P. and McCouch, S.R. (2003) Mapping quantitative trait loci for yield, yield components and morphological traits in an advanced backcross population between *Oryza rufipogon* and the *Oryza sativa* cultivar Jefferson. *Theoretical and Applied Genetics* 107, 479–493.

Wan, J.M., Cao, Y.J., Wang, C.M. and Ikehashi, H. (2005) Quantitative trait loci associated with seed dormancy in rice. *Crop Science* 45, 712–716.

23 Differences in the *Lolium rigidum* Embryo Proteome of Seeds with a High (Light-insensitive) and Low (Light-sensitive) Level of Dormancy

K.J. Steadman[1,2] and A.H. Millar[3]

[1]*School of Plant Biology, Faculty of Natural and Agricultural Sciences, University of Western Australia, Western Australia 6009, Australia;* [2]*School of Pharmacy, The University of Queensland, Queensland 4072, Australia;* [3]*School of Biomedical, Biomolecular and Chemical Sciences, Faculty of Life and Physical Sciences, University of Western Australia, Western Australia 6009, Australia*

Abstract

A preliminary study of changes in the proteome of annual ryegrass (*Lolium rigidum* Gaud.) seeds during dormancy release is described in this chapter. Highly dormant seeds (i.e. light-insensitive, unable to germinate under light or dark conditions) were prepared by stratification in darkness at 20°C for 2 days. Seeds with a low level of dormancy (i.e. light-sensitive, able to germinate under light but not dark conditions) were prepared by stratification in darkness at 20°C for 35 days. 2D gel electrophoresis (2D-PAGE) was performed using excised embryos. Only six proteins were found to differ in abundance between light-insensitive and light-sensitive embryos. Spots 1 and 2 were too low to be analysed in light-insensitive embryos but were present in light-sensitive embryos, representing at least a 20-fold induction. Spots 3 and 4 were three- to fivefold more abundant in light-insensitive compared with light-sensitive embryos. Spots 5 and 6 were more abundant in light-sensitive embryo samples, but this difference was less than twofold. The six protein spots of interest were excised from the gels, digested with trypsin and analysed by tandem mass spectrometry (MS/MS). Spots 1 and 2 gave significant matches with glyceraldehyde 3-phosphate dehydrogenases (GAPDH). Spot 4 could not be matched either against the NCBInr Viridiplantae (Green Plants) or a translation of the *Lolium* gene index from The Institute for Genomic Research (TIGR). *De novo* sequencing of two peptides was performed, but matching of these sequences to *Lolium* or total plant protein data-sets with the basic local alignment search tool (BLAST) has been unsuccessful. Data from spots 3, 5 and 6 were not sufficient to allow pattern matching identification, or successful *de novo* sequencing analysis, and remain unidentified.

Introduction

The isolation of processes involved in dormancy release from those involved in germination is required for the investigation of the biochemical and molecular control of dormancy. Hilhorst (1997) promotes the use of *Sisymbrium officinale* (L.) Scop. as secondary dormancy can be manipulated for hydrated seeds in darkness because seeds do not begin to germinate unless they receive a flash of light in combination with nitrate (Derkx *et al.*, 1993). Similarly, *Arabidopsis* (*Arabidopsis thaliana* (L.) Heynh.) requires a flash of light and nitrate to stimulate germination, so dormancy can be manipulated in darkness (Derkx and Karssen, 1993).

Annual ryegrass (*Lolium rigidum* Gaud.) seeds present an opportunity to separate the release of primary dormancy from germination in a monocot species. Winter annual grasses commonly lose dormancy during dry after-ripening, and non-dormant seeds subsequently germinate when water becomes available (Steadman *et al.*, 2003). However, the separation of tissues of dry seeds for tissue-specific biochemical work is difficult, if not impossible. *L. rigidum* seeds also experience dormancy release during stratification (hydrated storage) in warm, dark conditions (Steadman, 2004; Steadman *et al.*, 2004). Dormant seeds do not germinate under dark or light conditions. Seeds that have been dark-stratified for around 4 weeks do not germinate in darkness, but they will germinate following exposure to more than 5 h of light. Thus, seeds that are in an imbibed but non-germinating state can be prepared at a variety of stages of dormancy release, from a high level of dormancy (light-insensitive) to a low level of dormancy (light-sensitive), enabling the excision and separation of embryonic and endospermic tissues.

Several genes and proteins have been isolated that are differentially expressed in dormant and non-dormant seeds; however, none are yet direct candidates for involvement in the maintenance or termination of dormancy (Li and Foley, 1997; Koornneef *et al.*, 2002). Many proteins such as heat shock proteins (HSPs), late embryogenesis abundant (LEA) proteins and peroxiredoxin antioxidants appear to be stress-related, and many of these are also abscisic acid (ABA)-responsive. Proteomics is a potentially powerful means to investigate the biochemical and molecular control of seed dormancy. Proteomic analysis has been used to identify proteins involved with different stages of development and germination of non-dormant *Arabidopsis* seeds (Gallardo *et al.*, 2001, 2002, 2003). The current work represents preliminary steps towards the investigation of changes in the proteome of *L. rigidum* seeds during dormancy release. The comparison of dormant with light-sensitive seeds, which are fully imbibed but not germinating, may enable the identification of only those proteins that are involved in the dormancy mechanism.

Materials and Methods

Plant material

Mature annual ryegrass spikes (13.6 ± 0.3 g H_2O/100 g FW) were collected from wheat (*Triticum aestivum* L.) crop in long-term cropped fields from the Department of Agriculture research station at Wongan Hills, Western Australia (30°50'S,

116°43'E) on 6 November 2000. Spikes were threshed and florets (hereafter called seeds) separated from chaff by sieving. Seeds were equilibrated over 100% glycerol for 2 weeks and sealed inside aluminium foil bags. The seeds (7.8 ± 0.1 g H_2O/100 g FW) were stored at −20°C until use to maintain primary dormancy.

Seed stratification

Seeds were removed from the freezer and allowed to warm to room temperature for 2 h. They were then stratified at full hydration (seed water content >60 g H_2O/100 g FW) by placing on 6 mm deep solidified agar–water (1% w/v) in 9 cm diameter Petri dishes (~50 per dish). Dishes were individually wrapped in aluminium foil to provide darkness and placed at 20°C ± 1°C for either 2 days or 35 days to produce seeds in a light-insensitive (i.e. dormant) and light-sensitive state, respectively (Steadman, 2004). Nine of the dishes of each type were used for a germination test and 24 dishes of each type were used for proteomic analysis of the embryo.

Embryo excision

Embryos were excised from seeds using a dissection microscope and placed immediately into an Eppendorf tube surrounded by ice. Plates remained wrapped in foil and were opened to remove each ungerminated seed for excision. Excision of the embryo took a maximum of 30 s per seed. Since these seeds require more than 5 h of similar light intensity to stimulate germination (Steadman *et al.*, 2004), this was not expected to start germination-associated biochemistry. Once 100 embryos were in a tube, it was capped and stored at −80°C until use.

Seed germination and viability tests

Seed dormancy status was measured by conducting a germination test at 12 h alternating 25°C and 15°C, the optimal temperature for *L. rigidum* germination (Steadman, 2004). Three germination tests were performed by using three replicate dishes for each of the 2- and 35-day stratification treatments:

1. Germination test in light: aluminium foil was removed, germinated seeds counted and removed, and dishes placed inside clear plastic bags at 25/15°C; Osram 40 W fluorescent white light tubes provided 45 μmol/m^2/s at shelf level during the 12 h at 25°C.
2. Germination test in darkness: aluminium foil was not removed, and dishes were placed directly into the incubator set at 12 h alternating 25/15°C.
3. Germination test in darkness following light exposure similar to that of seeds undergoing embryo excision: aluminium foil was removed and dishes were exposed to the microscope lighting conditions for 1 min before being rewrapped with foil and placed at 25/15°C. During the light exposure, germinated seeds were counted and removed.

Germination was assessed after 14 days; the criterion for germination was visible radicle protrusion. Inviability of soft, ungerminated seeds was confirmed by cutting through the seed. Samples of firm, ungerminated seeds were assessed for viability with tetrazolium chloride solution (Steadman *et al.*, 2003).

Protein extraction and 2D gel electrophoresis

Embryo material (200 or 600 embryos) was thoroughly suspended in 550 µl of an isoelectric focusing (IEF) denaturation solution (1% w/v 3-[*N*,*N*-dimethyl (3-myristoylaminopropyl)-ammonio]propanesulfonate (ASB14), 7 M urea, 2 M thiourea, 40 mM unbuffered Tris, 0.05% v/v tributyl phosphine (TBP), 0.5% v/v ampholytes (Pharmacia Biotech Pharmalytes, 3–10)). Roche Complete Mini protease inhibitor cocktail was also included (one tablet per 10 ml solution).

After clarification by centrifugation, 500 µl aliquots were loaded into Pharmacia 17 cm immobilized pH gradient (IPG) drystrips, p*I* 3–10 (linear) by the whole strip hydration method. IEF was carried out on a Pharmacia Multiphor II electrophoresis unit following a stepped voltage programme to a final voltage of 3.5 kV for 18 h, with a total of 66 kVh. Prior to the sodium dodecyl sulphate polyacrylamide gel electrophoresis (SDS-PAGE) dimension, each IEF strip was soaked for 20 min in 16 ml SDS re-equilibration buffer (380 mM Tris-HCl pH 8.8, 2.7% acrylamide, 6 M urea, 2% SDS, 20% v/v glycerol, 0.05% v/v tributylphosphine). Polyacrylamide gels (20 × 20 cm, 12% polyacrylamide with acrylamide: piperazine diacrylamide (PDA) cross-linker ratio = 40:1) were made in a BioRad gel pouring apparatus with a gel buffer containing 380 mM Tris-HCl pH 8.8. IEF strips were secured to the top of the polyacrylamide gels with 0.5% w/v agarose in gel-running buffer. Electrophoresis was carried out using a standard Tris–glycine buffer system for 5 h at 25 mA per gel (constant current) in a Bio-Rad Multi-Protean II apparatus. Three replicate gels were run for samples of light-insensitive and light-sensitive embryos, each gel containing protein from ~200 embryos. One preparative gel was run for each of the light-insensitive and light-sensitive embryos, each containing protein from ~600 embryos. Gels were stained with colloidal Coomassie Brilliant Blue G250 according to Sweetlove *et al.* (2002). Images were digitized using a PerkinElmer ProXpress scanner, and ProFinder software version 3.1 (PerkinElmer). Images were processed using ProGenesis v 2005 (Non-Linear Dynamics, Manchester, UK).

Mass spectrometry

Protein samples were trypsin-digested and extracted according to standard techniques (see Sweetlove *et al.*, 2001). Peptides were analysed by electrospray time-of-flight mass spectrometry (LC/TOF/MS) using a Q-Star Pulsar *i* (Applied Biosystems). Each extracted peptide sample was loaded in turn with the Agilent 1100 series capillary liquid chromatography (LC) system onto a 0.5 × 50 mm 5 µm C18 reverse-phase column in 5% acetonitrile and 0.1% formic acid. Peptides were eluted from the C18 reverse-phase column by an acetonitrile gradient (5–80%) under constant formic acid concentrations of 0.1%. The collision-induced disassociation (CID) data from each sample were exported from ANALYST QS using a purpose-built script obtained from Matrix Sciences. Data were used for searching the Mascot search engine (Matrix Sciences) for protein identifications. Search parameters at Mascot employed a peptide tolerance of ±2 Da and tandem mass spectrometry (MS/MS) tolerance of ±1.2 Da, no variable modifications, allowed up to one miss cleavage for trypsin digestion and the instrument type was set to ESI-QUAD-TOF. Searches were performed against the NCBInr 20050211 database with a taxonomy restriction to Viridiplantae (Green Plants) or against a translation of the *Lolium* gene index from The Institute for Genomic Research (TIGR). Individual CID spectra from some gel spots were also chosen for *de novo* sequence determination using BIOANALYST (Applied Biosystems). Ile(I)/Leu(L) are indistinguishable by mass spectrometry (MS) and as such are interchangeable within the *de novo* sequence presented. Oxidized-Met can be produced under sample handling conditions and this modified amino acid has a near identical mass to Phe (F); hence, assignment of F may represent the presence of M in the original sequence. Lys (K) and Glu (Q) have near-identical masses; hence, an internal Q may represent an internal K cleavage site that has been missed.

Results

Seed dormancy status

Seeds that were dark-stratified for 2 days were dormant, with <2% germination in light or darkness (Fig. 23.1a). In contrast, seeds that were dark-stratified for 35 days subsequently germinated when exposed to a 12 h light regime. No seeds germinated in darkness, either during the dark-stratification process or subsequently during the germination stage of the test. Germination following exposure to a short period of light, similar to that experienced by seeds used for embryo excision, was very low (Fig. 23.1b). All seeds remaining at the end of the germination test were viable according to tetrazolium staining.

Protein profiles

IEF/SDS-PAGE gel patterns revealed a typical profile of protein spots with a majority of the distribution between 10 and 75 kDa in mass and had p*I* values of 4–8 (Fig. 23.2). A series of smaller, abundant, basic proteins were found in both samples, which represented a more unusual feature of the ryegrass embryo proteome. Gels from light-insensitive and light-sensitive embryo tissues were very similar in general appearance, as both contained ~350 proteins of similar location and abundance between replicated samples. Analysis with ProGenesis involved background subtraction and normalization between replicate gels on total spot volume. In this process, a series of six relatively faint protein spots were found to differ significantly in abundance between light-insensitive and light-sensitive samples (Fig. 23.2). Quantitation of these spot volumes showed that spots 1 and 2 were too low to be analysed in light-insensitive embryos, but were present in light-sensitive embryos (Table 23.1). With regard to the background levels, this represents at least a 20-fold induction of these

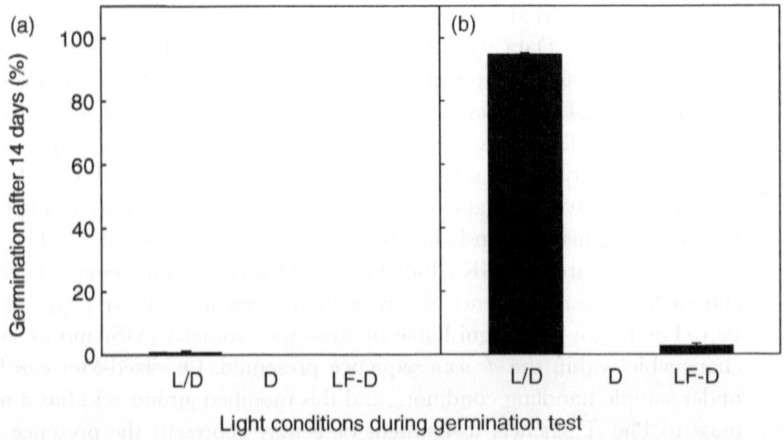

Fig. 23.1. Germination of *Lolium rigidum* seeds at 12 h alternating 25/15°C with a 12 h light regime (L/D), darkness (D) or exposure to a light flash and then darkness (LF-D). The proportion of seeds germinated was assessed after 14 days under these conditions. Before starting the germination test, seeds were subjected to stratification in darkness for either (a) 2 days or (b) 35 days, during which no seeds germinated. Bars indicate ± SE of the mean for three replicates of 50 seeds each.

proteins. Spots 3 and 4 were three- to fivefold more abundant in light-insensitive embryos when compared with light-sensitive ones. Spots 5 and 6 were more abundant in light-sensitive embryo samples, but the difference was less than twofold.

Protein identification

The six protein spots of interest were excised from gels, in-gel digested with trypsin and analysed by MS/MS. Three ions from spot 1 were doubly charged peptides that matched a *Lolium multiflorum* Lam. sequence (gi|46507768) with a combined molecular weight search (MOWSE) score of 68 (Table 23.2). The 95% confidence level for this match has a MOWSE score of 37. Two ions were also found to match this same sequence in spot 2, again with the MOWSE score representing significant matches. The peptide of mass 1160.6 kDa with a matched sequence of AGIALNDNFVK is highly conserved in the sequence of glyceraldehyde 3-phosphate dehydrogenases (GAPDH) across many plant species. Using the basic local alignment search tool (BLAST) for the gi|46507768 sequence showed that this represented a plant-like GAPDH protein across the length of the protein sequence. Spot 4 could not be matched either against the NCBInr Viridiplantae (Green Plants) or against a translation of the *Lolium* gene index from TIGR. Instead, *de novo* sequencing of two peptides was performed (Table 23.2). BLAST matching of these sequences to *Lolium* or total plant protein data-sets has been unsuccessful to date, suggesting that sequence variation is too high to reliably identify these proteins. Data from spots 3, 5 and 6 were not sufficient to allow pattern-matching identification or successful *de novo* sequencing analysis, and hence they remain unidentified.

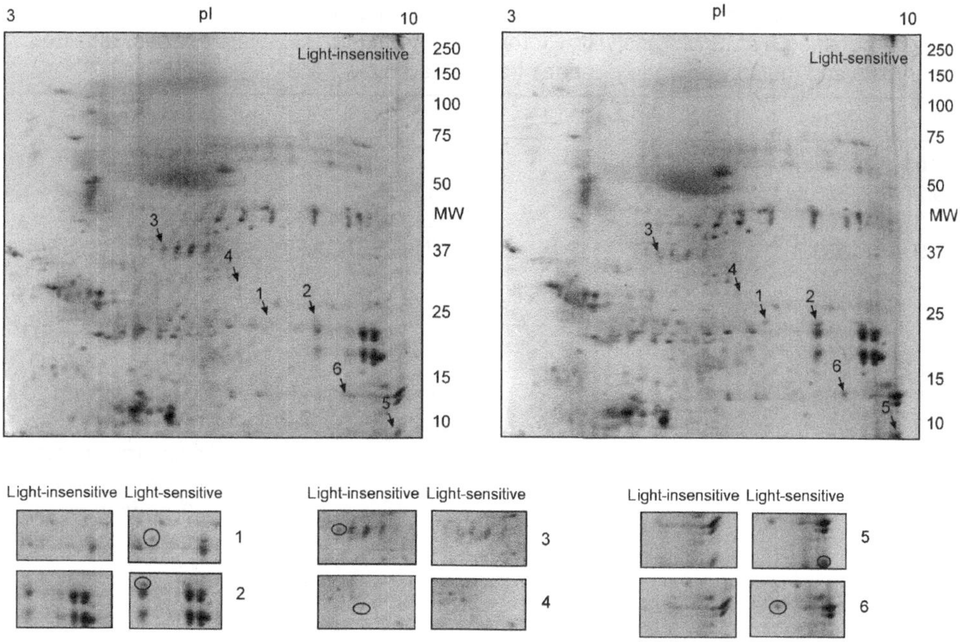

Fig. 23.2. IEF/SDS-PAGE gels showing the protein profile for light-insensitive (seeds stratified for 2 days in darkness) and light-sensitive (seeds stratified for 35 days in darkness) *Lolium rigidum* embryos. Six spots showed differential abundance: spots 1, 2, 5 and 6 were more abundant in light-sensitive embryos, and spots 3 and 4 were more abundant in light-insensitive embryos.

Table 23.1. Quantification of spots changing in abundance between protein profiles from light-insensitive and light-sensitive embryos.

Spot number	MW (kDa)	p*I*	Light-insensitive			Light-sensitive			Fold change
			Rep 1	Rep 2	Rep 3	Rep 1	Rep 2	Rep 3	
1	25	6.8	ND	ND	ND	0.519	0.269	0.179	>20
2	25	7.8	ND	ND	ND	0.136	0.086	0.089	>20
3	40	5.4	0.290	0.257	0.250	0.085	0.073	0.062	0.3
4	33	6.5	0.050	0.061	0.055	0.012	0.010	ND	0.2
5	10	9.5	0.989	1.102	ND	1.890	2.131	1.432	1.7
6	14	8.5	0.194	0.252	ND	0.399	0.272	0.331	1.5

Spot intensity is in arbitrary units, three independent replicates (Rep) of ~200 embryos are shown, along with the apparent molecular weight (MW) and the apparent isoelectric point (p*I*) of the proteins. Fold changes are based on averages of the ratio of paired replicates intensities. ND = not detected.

Discussion

The two dark-stratification treatments produced seeds with contrasting levels of dormancy. Those stratified for 2 days had a high level of dormancy because they

Table 23.2. Identification of peptides derived from light-insensitive and light-sensitive embryos obtained by MS/MS analysis. Protein spots from Fig. 23.2 were excised, digested with trypsin and multiply charged peptides fragmented by MS/MS.

M/z	Mass	Delta	MOWSE	Peptide	Match
Spot 1					
435.2	868.5	−0.03	9	VLPELNGK	gi|46507768|gb|
581.3	1160.6	−0.05	47	AGIALNDNFVK	Mass: 21,613
717.9	1433.7	−0.05	12	AASFNIIPSSTGAAK	*Lolium multiflorum* Glyceraldehyde 3-phosphate dehydrogenase
Spot 2					
581.3	1160.6	−0.04	31	AGIALNDNFVK	gi|46507768|gb|
717.9	1433.7	−0.04	11	AASFNIIPSSTGAAK	Mass: 21,613 *Lolium multiflorum* Glyceraldehyde 3-phosphate dehydrogenase
Spot 4					
429.2	856.4	–	–	(Q/K)(L/I)WASPR	
495.8	989.5	–	–	ATDS(L/I)(L/I)TAAK	

The mass/charge ratio (M/z), the predicted mass of the peptides (mass), the mass difference to matched sequence (Delta), the molecular weight search (MOWSE) score and the peptide sequence matched are shown for spots 1 and 2. The mass/charge ratio (M/z), the predicted mass of the peptides (mass) and the *de novo* sequence deduced are shown for spot 4. Leucine and isoleucine residues have identical mass and are shown as alternatives. Lysine (K) and glutamine (Q) residues only differ in mass by 0.05 Da, given that a tryptic peptide would not be expected to contain an N-terminal K, Q is preferred but it is designated Q/K.

were unresponsive to germination stimulation by light. In contrast, those stratified for 35 days did respond when light was provided. These cannot be termed 'non-dormant' because they were unable to germinate without additional germination stimulants (Vleeshouwers *et al.*, 1995). In *L. rigidum* seeds, a non-dormant state is not reached by dark stratification, but appears to be possible only by extended periods of dry after-ripening (Steadman *et al.*, 2003).

Both the light-insensitive and light-sensitive embryos can be expected not to begin germination-related biochemistry. Seeds that were exposed to a light flash, similar to that experienced during embryo excision, did not subsequently germinate in darkness, and embryos that were excised were quickly placed into subzero temperatures to stop any biochemical changes. Furthermore, proteomic analysis established that very few proteins exhibited significant differential abundance between imbibed light-insensitive (i.e. dormant) and light-sensitive seeds. Large numbers of proteins change in abundance during the early stages of germination (Gallardo *et al.*, 2001), so germination-related events can be presumed not to have begun in *L. rigidum*. Thus, with *L. rigidum* we can easily excise the embryo from the rest of the seed at a variety of dormancy levels without germination-related biochemistry complicating the picture.

Proteomic analysis indicated that GAPDH was undetectable in light-insensitive seed embryos but present in light-sensitive seeds. Previously, an increase in glycolysis was the evidence of the start of germination. For example, a rise in fructose 2,6-bisphosphate is one of the earliest metabolic events in the germination of *Avena sativa* L. seeds (Larondelle *et al.*, 1987). Similarly, GAPDH and other metabolic enzymes increase upon thidiazuron-induced dormancy release in apple (*Malus pumila* Mill.) buds, correlating with the initial stages of resumption of growth (Wang *et al.*, 1991). In *Arabidopsis* seeds, GAPDH is one of the proteins upregulated in seeds imbibed for 1 day, prior to initiation of radicle growth (Gallardo *et al.*, 2002). In this case, the seeds appear to be preparing for germination by increasing the capacity for energy production even before they switch into a germinative mode. GAPDH is commonly observed to be stress-responsive (Velasco *et al.*, 1994; Laxalt *et al.*, 1996), raising the possibility that the 35-day period of dark stratification was a stressful event that elicited the increase in abundance over that of the 2-day stratified seeds, rather than the change in dormancy status per se.

This initial study has shown that there are some differences in the protein profile between the embryos of light-insensitive and light-sensitive seeds. This work needs to be extended in a number of ways:

1. Currently, only the more abundant portion of the proteins present in the seeds has been visualized, and there may be many changes occurring in the less abundant proteins, so protein loading on the gels should be increased so that they can be detected.
2. Membrane-bound proteins have been hypothesized, which may be involved in the dormancy mechanism (Hilhorst, 1998; Hallett and Bewley, 2002; Steadman, 2004), so separation of membrane-bound and cytosolic proteins is required.
3. The endosperm in the region of the radicle tip has an important role in dormancy in a number of species (Leubner-Metzger, 2003), so particular emphasis on these tissues is necessary.
4. Light (red or white) can inhibit dormancy release during warm stratification, while darkness and far-red light are permissive for dormancy release (Steadman, 2004), presenting an alternative control treatment that would allow light-insensitive and light-sensitive seeds to be imbibed for equal periods of time, avoiding the potential for differences in the imbibition period to influence protein changes.

Acknowledgements

This research project was funded by a grant from the University of Western Australia to K.J. Steadman. A.H. Millar is supported by an ARC QEII Fellowship and funded by the ARC Discovery Programme.

References

Derkx, M.P.M. and Karssen, C.M. (1993) Effects of light and temperature on seed dormancy and gibberellin-stimulated germination in *Arabidopsis thaliana*: studies with gibberellin-deficient and insensitive mutants. *Physiologia Plantarum* 89, 360–368.

Derkx, M.P.M., Smidt, W.J., van der Plas, L.H.W. and Karssen, C.M. (1993) Changes in dormancy of *Sisymbrium officinale* seeds do not depend on changes in respiratory activity. *Physiologia Plantarum* 89, 707–718.

Gallardo, K., Job, C., Groot, S.P.C., Puype, M., Demol, H., Vandekerckhove, J. and Job, D. (2001) Proteomic analysis of *Arabidopsis* seed germination and priming. *Plant Physiology* 126, 835–848.

Gallardo, K., Job, C., Groot, S.P.C., Puype, M., Demol, H., Vandekerckhove, J. and Job, D. (2002) Proteomics of *Arabidopsis* seed germination: a comparative study of wild-type and gibberellin-deficient seeds. *Plant Physiology* 129, 823–827.

Gallardo, K., Le Signor, C., Vandekerckhove, J., Thompson, R.D. and Burstin, J. (2003) Proteomics of *Medicago truncatula* seed development establishes the time frame of diverse metabolic processes related to reserve accumulation. *Plant Physiology* 133, 664–682.

Hallett, B.P. and Bewley, J.D. (2002) Membranes and seed dormancy: beyond the anaesthetic hypothesis. *Seed Science Research* 12, 69–82.

Hilhorst, H.W.M. (1997) Seed dormancy. *Seed Science Research* 7, 221–223.

Hilhorst, H.W.M. (1998) The regulation of secondary dormancy: the membrane hypothesis revisited. *Seed Science Research* 8, 77–90.

Koornneef, M., Bentsink, L. and Hilhorst, H. (2002) Seed dormancy and germination. *Current Opinion in Plant Biology* 5, 33–36.

Larondelle, Y., Corbineau, F., Dethier, M., Come, D. and Hers, H.G. (1987) Fructose 2,6-bisphosphate in germinating oat seeds: a biochemical study of seed dormancy. *European Journal of Biochemistry* 166, 605–610.

Laxalt, A.M., Cassia, R.O., Sanllorenti, P.M., Madrid, E.A., Andreu, A.B., Daleo, G.R., Conde, R.D. and Lamattina, L. (1996) Accumulation of cytosolic glyceraldehyde-3-phosphate dehydrogenase RNA under biological stress conditions and elicitor treatments in potato. *Plant Molecular Biology* 30, 961–972.

Leubner-Metzger, G. (2003) Functions and regulation of β-1,3-glucanases during seed germination, dormancy release and after-ripening. *Seed Science Research* 13, 17–34.

Li, B. and Foley, M.E. (1997) Genetic and molecular control of seed dormancy. *Trends in Plant Science* 2, 384–389.

Steadman, K.J. (2004) Dormancy release during hydrated storage in *Lolium rigidum* seeds is dependent on temperature, light quality and hydration status. *Journal of Experimental Botany* 55, 929–937.

Steadman, K.J., Crawford, A.D. and Gallagher, R.S. (2003) Dormancy release in *Lolium rigidum* seeds is a function of thermal after-ripening time and water content. *Functional Plant Biology* 30, 345–352.

Steadman, K.J., Bignell, G.P. and Michael, P.J. (2004) Stimulating dormancy release and emergence of annual ryegrass (*Lolium rigidum*) seeds using short-term hydrated storage in darkness. *Australian Journal of Agricultural Research* 55, 787–795.

Sweetlove, L.J., Mowday, B., Hebestreit, H.F., Leaver, C.J. and Millar, A.H. (2001) Nucleoside diphosphate kinase III is localised to the inter-membrane space in plant mitochondria. *FEBS Letters* 508, 272–276.

Sweetlove, L.J., Heazlewood, J.L., Herald, V., Holtzapffel, R., Day, D.A., Leaver, C.J. and Millar, A.H. (2002) The impact of oxidative stress on Arabidopsis mitochondria. *The Plant Journal* 32, 891–904.

Velasco, R., Salamini, F. and Bartels, D. (1994) Dehydration and ABA increase mRNA levels and enzyme activity of cytosolic GAPDH in the resurrection plant *Craterostigma plantagineum*. *Plant Molecular Biology* 26, 541–546.

Vleeshouwers, L.M., Bouwmeester, H.J. and Karssen, C.M. (1995) Redefining seed dormancy: an attempt to integrate physiology and ecology. *Journal of Ecology* 83, 1031–1037.

Wang, S.Y., Jiao, H.J. and Faust, M. (1991) Changes in metabolic enzyme activities during thidiazuron-induced lateral budbreak of apple. *HortScience* 26, 171–173.

24 Transcriptomic and Proteomic Profiling of *FsPP2C1*-overexpressing *Arabidopsis* Plants

M.P. González-García, D. Rodríguez, C. Nicolás, G. Nicolás and O. Lorenzo

Departamento de Fisiología Vegetal, Centro Hispano-Luso de Investigaciones Agrarias, Facultad de Biología, Universidad de Salamanca, Plaza de los Doctores de la Reina s/n, 37007 Salamanca, Spain

Abstract

FsPP2C1 is a functional protein phosphatase type-2C (PP2C) isolated from beech (*Fagus sylvatica* L.) seeds. Previously, this gene was shown to be upregulated by abscisic acid (ABA) treatment and specifically expressed in ABA-treated seeds. It was demonstrated that constitutive FsPP2C1 overexpression in *Arabidopsis* (*Arabidopsis thaliana* (L.) Heynh.) confers ABA insensitivity to seeds and a reduced degree of seed dormancy. Transgenic *35S:FsPP2C1* plants resistant to different stresses (i.e. paclobutrazol, NaCl and mannitol) showed ABA-resistant early root growth and diminished induction of *RAB18* and *KIN2* suggesting that FsPP2C1 is a negative regulator of ABA signalling. To fully characterize this insensitive phenotype and the contribution of FsPP2C1 to promoting the transition from seed dormancy to germination, transcriptomic and proteomic tools have been used. A complete transcriptomic analysis in *Arabidopsis* highlights the insensitive phenotype of the FsPP2C1-overexpressing plants, the cross-talk of ABA with other signalling pathways and the conservation of the ABA transduction pathway between beech and *Arabidopsis*. Similarly, proteomic analysis in ABA-treated seeds exhibits significant changes in the protein pattern of transgenic *35S:FsPP2C1* compared with wild-type (WT) seeds. These results are discussed to confirm and open new insights into the role of FsPP2C1 and the possible interactions between ABA and other plant hormones.

Introduction

Seed germination begins with the uptake of water and concludes with the emergence of the embryo through the surrounding tissues. The failure of an intact,

 Seeds: Biology, Development and Ecology (eds S. Adkins, S. Ashmore and S.C. Navie)

viable seed to complete germination under favourable conditions is defined as seed dormancy. Both key survival developmental states are tightly regulated and reversible protein phosphorylation is a fundamental mechanism suggested to control these processes in response to hormonal and environmental cues (Trewavas, 1988; Walker-Simmons, 1998; Campalans *et al.*, 1999).

The phosphorylation status of a protein is determined by the balance between the activities of protein kinases and protein phosphatases. In plants, protein phosphatases type-2C (PP2Cs) represent an important class of proteins highlighted by the large number of *PP2C* genes found in the *Arabidopsis* (*Arabidopsis thaliana* (L.) Heynh.) genome (Kerk *et al.*, 2002). Some members of the PP2C family have been identified as key elements in abscisic acid (ABA) signal transduction and play a crucial role as regulators of seed dormancy (Leung *et al.*, 1994; Meyer *et al.*, 1994; González-García *et al.*, 2003).

Coordinated interactions between several positive and negative regulators are required for proper control of ABA responses operating in a cell. Genetic studies of *ABA-insensitive 1* (*abi1*) and *abi2* mutants, transient expression and analysis of transgenic *A. thaliana PP2CA* (*AtPP2CA*) antisense plants (Sheen, 1998; Gosti *et al.*, 1999; Merlot *et al.*, 2001; Tahtiharju and Palva, 2001), as well as gain-of-function phenotypes with *homology to ABI1* (*HAB1*) and *FsPP2C1* (González-García *et al.*, 2003; Sáez *et al.*, 2004) characterize PP2Cs as negative regulators of ABA signalling. In addition to the PP2Cs, other proteins have been identified as negative regulators of the pathway, including the *A. thaliana* homeodomain protein (AtHB6) (Himmelbach *et al.*, 2002) and the farnesyl transferase β-subunit-enhanced response to ABA1 (ERA1) (Cutler *et al.*, 1996; Pei *et al.*, 1998). Therefore, a complex regulatory mechanism seems to have evolved to attenuate ABA signalling and avoid undesirable effects due to sustained activation of the ABA pathway.

Further investigation of *FsPP2C1*-overexpressing plant lines entails the study of transcriptional changes using *Arabidopsis* whole genome microarrays to discover the extent to which this protein is involved in ABA signalling pathway and to identify new genes regulated by FsPP2C1 alone or in concert with known signalling processes. This genomics-based approach will be complemented at the protein level by a proteomics-based analysis of Columbia-0 (Col-0); *35S:FsPP2C1* vs wild-type (WT) germinating seeds that will distinguish the insensitive phenotypes found in transgenic plants.

Materials and Methods

Biological materials and growth conditions

A. thaliana Col-0 is the genetic background for all WT and transgenic *FsPP2C1*-expressing plants used in this work, except for the *abi1-1* mutants (and their corresponding WT control) that are of *Landsberg erecta* (Ler) background. Plants were grown *in vitro* (in MS media; Murashige and Skoog, 1962) and in soil as previously described (González-García *et al.*, 2003).

Microarray analysis

Slide treatment

Superamine Telechem slides containing more than 26,000 spots corresponding to the *A. thaliana* oligo set from Qiagen-Operon were obtained from David Galbraith (University of Arizona, Tucson, Arizona, USA). More information about printing and the oligo set can be found on the following web site (http://ag.arizona.edu/microarray/). Printed slides were rehydrated over a 40°C water bath for 5 s and dried on a 65°C heating block for 5 s. This hydration step was repeated twice. Oligonucleotides were fixed by UV radiation of 120 mJ. The slides were washed in 1% sodium dodecyl sulphate (SDS) for 5 min and in absolute ethanol for 30 s. Finally, the slides were dried by centrifugation at 141 *g* for 3 min.

Prehybridization was performed at 42°C for 30–45 min in 6X saline sodium citrate (SSC), 0.5% SDS and 1% bovine serum albumin (BSA), and then the slides were rinsed five times with distilled water. Cyanine 5 (Cy5) and Cy3 amplified RNA (aRNA)-fragmented probes (200 pmol of each label) were mixed with 20 μg of PolyA (Sigma) and 20 μg of yeast tRNA (Sigma) in a final volume of 90 μl of hybridization buffer (50% formamide, 6 × SSC, 0.5% SDS, 5X Denhardt's solution). The probes were denatured at 95°C for 5 min and applied to the slides using a LifterSlip (Erie Scientific). Slides were then incubated at 37°C for 16 h in hybridization chambers (Array-It). After incubation, slides were washed twice with 0.5 × SSC, 0.1% SDS for 5 min each, twice with 0.5 × SSC for 5 min and finally in 0.05 × SSC for 2 s. Slides were dried by centrifugation at 563 *g* for 1 min.

Images from Cy3 and Cy5 channels were equilibrated and captured with a ScanArray Lite (Perkin Elmer) and spots quantified using ScanArray Express software (Perkin Elmer). To enhance accuracy of scan measurements at extreme intensity values, two different scans were obtained at two laser light conditions (low intensity for non-saturated spots and high intensity for <1% saturated spots). Thus, the few saturated spots at high intensity were reanalysed using the low-intensity scan.

Four independent biological 'dye-swapped' replicates of each microarray hybridization were performed. For example, replicates one and three of *35S:FsPP2C1* were labelled with Cy5, whereas replicates two and four were labelled with Cy3. Data analysis of replicates was done using SOLAR software (ALMA Bioinformatica). A lowess correction was applied to each replicate and then the logarithm of ratios merged with the corresponding standard deviation and z-score.

RNA amplification and labelling

RNA was extracted (using RNAwiz, Ambion) from 12-day-old WT seedlings and *35S:FsPP2C1* plants grown on MS liquid media and treated with ABA (50 μM). RNA was purified with RNeasy™ columns (Qiagen). Total RNA (3 μg of each) was amplified using Amino Allyl MessageAmp™ aRNA kit (Ambion) following the manufacturer's instructions and ~50–80 μg of aRNA were obtained. One aliquot from each sample containing 10 μg of aRNA was labelled with either Cy3 or Cy5 mono *N*-hydroxysuccinimide (NHS) Ester (Cy™Dye Post-labelling Reactive Dye Pack, Amersham) and purified following the manufacturer's instructions for the aRNA kit. Cy3 and Cy5 labels were detected in 1 µl of the probe using a Nanodrop spectrophotometer (Nanodrop Technologies, Inc.). For each hybridization, 200 pmol per sample of Cy3 and Cy5 probes were mixed, speed-vac dried and resuspended in 9 µl of RNase-free water. Labelled aRNA was fragmented by adding 1 µl of 10X fragmentation buffer (Ambion) and incubating at 70°C for 15 min. The reaction was stopped with 1 µl of Stop solution (Ambion). Integrity and medium size of total RNA, aRNA and fragmented aRNA was evaluated using a Bioanalyzer 2100 (Agilent). The average size of aRNAs was ~1500 nucleotides and that of fragmented aRNAs ~100 nucleotides. The probe was diluted to a final volume of 90 µl in hybridization solution.

Proteomic analysis

Protein extraction

Mature seeds were imbibed for 3 days in MS, treated with 50 µM ABA for 3 h and ground to a fine powder in liquid nitrogen with a mortar and pestle. Powder (200 mg) was resuspended in a 10% (w/v) trichloroacetic acid (TCA)/acetone solution containing 20 mM dithiothreitol (DTT), and proteins were precipitated at −20°C for 1 h. After centrifugation at 35,000 *g* for 15 min, the supernatant was discarded and the pellet rinsed twice with −20°C cold acetone containing 20 mM DTT. The final pellet was vacuum dried and solubilized in lysis buffer containing 7 M (w/v) urea, 2 M (w/v) thiourea, 4% (w/v) 3-[(3-cholamidopropyl)-dimethylammonio]-1-propanesulfonate (CHAPS), 50 mM DTT and 1% (v/v) biolytes.

Protein concentration was determined using the RC-DC Protein Assay (Bio-Rad) with ovoalbumin as standard and samples were stored at −20°C until used.

2D electrophoresis

Protein samples (100–300 µg each) were first separated by isoelectrofocusing using gel strips forming an immobilized non-linear pH gradient from 3 to 11 (Immobiline DryStrip, pH 3–11 NL, 13 cm; Amersham Biosciences).

Strips were rehydrated in the IPGphor system (Amersham Pharmacia Biotech Biosciences) for 13 h at 20°C with the thiourea/urea lysis buffer containing 4% (w/v) CHAPS, 50 mM DTT and the protein extracts. Isoelectrofocusing was performed at 20°C in a Multiphor II system (Amersham Biosciences) as follows: 500 V 1 h, 1000 V 1 h and 8000 V 5 h. The gel strips were equilibrated twice for 15 min in 5 ml of equilibration solution containing 6 M urea, 20% (v/v) glycerol, 2% (w/v) SDS, 375 mM Tris(hydroxymethyl)aminomethane-hydrochloride (Tris-HCl). DTT (50 mM) or Tris(2-carboxyethyl)phosphine (TCEP; 5 mM) was added to the first equilibration solution, and iodoacetamide (2.5% [w/v]) was added to the second one. Equilibrated gel strips were placed on top of vertical gels for second dimension to separate proteins according to size. Sodium dodecyl sulphate polyacrylamide gel electrophoresis (SDS-PAGE) was performed on 10% polyacrylamide gels (Amersham Pharmacia Biotech system) and run at 85 V constant current at 25 mA for 30 min until the dye reached the bottom of the gel. Two-dimensional (2D) gels were made in triplicate and from two independent protein extractions for each condition analysed.

Protein staining and image analysis of 2D gels

Gels were stained with Sypro Ruby Protein Gel Stain (molecular probes) and scanned with a scanner (FLA-3000, Fujifilm). Image analysis was carried out with ImageReader and ImageGauge V.5 software (Fujifilm), according to the manufacturer's instructions. Spot detection and background subtraction on detected spots were performed using the mode of average on boundary function, 2D gels were aligned, matched and the quantitative determination of the spot volumes carried out with MELANIE 5 software. For each analysis, statistical data showed a high level of reproducibility between normalized spot volumes of gels produced in triplicate from the two independent protein extractions.

Results and Discussion

Background

We have previously reported the isolation of FsPP2C1 as a functional PP2C from *Fagus sylvatica* seeds (Lorenzo *et al.*, 2001). FsPP2C1 displays a high phylogenetic relationship with a cluster of *Arabidopsis* protein phosphatases 2C (AtPP2Cs) physiologically characterized and proposed to participate in ABA signalling, in particular ABI1, ABI2 (Merlot *et al.*, 2001), HAB1, HAB2 (Sáez *et al.*, 2004) and PP2CA (Sheen, 1998; Tahtiharju and Palva, 2001). *FsPP2C1* expression is upregulated upon ABA treatment specifically in seeds and correlates with the induction of seed dormancy, while it is abolished by treatments that break dormancy (Lorenzo *et al.*, 2001). FsPP2C1 appears to be the first reported member of the PP2C family that specifically accumulates in ABA-treated seeds, but not in ABA-treated vegetative tissues of *F. sylvatica* seedlings, suggesting that FsPP2C1 plays a specific role in ABA-induced seed dormancy.

After the isolation and characterization of FsPP2C1 from *F. sylvatica* nuts, it was necessary to gain genetic evidence to firmly establish whether this PP2C was a positive or negative regulator of seed dormancy and germination. It has also been demonstrated that constitutive *FsPP2C1* overexpression in *Arabidopsis* confers ABA insensitivity in seeds and a reduced degree of seed dormancy (González-García *et al.*, 2003). Transgenic *35S:FsPP2C1* plants were resistant to different stresses (i.e. paclobutrazol, NaCl and mannitol), showed ABA-resistant early root growth and diminished induction of ABA-regulated genes (i.e. *RAB18* and *KIN2*), suggesting that FsPP2C1 is a negative regulator of ABA signalling.

The functional evidence that FsPP2C1 is a negative regulator of ABA signalling (González-García *et al.*, 2003), together with the ABA inducibility of *FsPP2C1* expression specifically in seeds (Lorenzo *et al.*, 2001), suggests an important role of FsPP2C1 in the modulation of the ABA signalling in *F. sylvatica* seeds through a negative feedback loop, and the probable involvement of this protein in the regulation of the transition from seed dormancy to germination during the early weeks of stratification.

These results were corroborated using *HAB1*, one of the orthologous genes of *FsPP2C1* in *Arabidopsis*, where gain-of-function and loss-of-function conferred ABA insensitivity and ABA hypersensitivity in seed germination assays, respectively, revealing its role as a negative regulator of ABA signalling (Sáez *et al.*, 2004).

Gene expression profiling of *35S:FsPP2C1* in response to ABA

In order to further investigate the involvement of FsPP2C1 in the expression control of ABA-regulated genes, and to further understand why *35S:FsPP2C1* transgenic plants are highly insensitive to ABA, a complete transcriptome analysis of Col-0;*35S:FsPP2C1* plants was performed using the *Arabidopsis* slides containing more than 26,000 spots corresponding to the *A. thaliana* oligo set from Qiagen-Operon that was obtained from David Galbraith (University of Arizona, Tucson, Arizona, USA).

Four independent biological 'dye-swapped' replicates were analysed. The results obtained from the whole-genome transcriptional profiling uncovered the existence of two groups of genes that are differentially regulated by FsPP2C1 (Fig. 24.1). The first group includes genes involved in defence responses against pathogens (data not shown), and is activated by FsPP2C1. The second group, characterized by genes involved in ABA biosynthesis and ABA-mediated responses (data not shown), is repressed by FsPP2C1, highlighting that ABA-induced transcription was affected in *35S:FsPP2C1* plants and confirmed the results obtained by González-Garcia *et al.* (2003).

The defence-related genes whose expression is upregulated by FsPP2C1 are also regulated by the ethylene/jasmonic acid (ET/JA) pathway, which is fully consistent with recent results, demonstrating that the ABA and ET/JA signalling pathways antagonize each other and modulate defence- and stress-responsive gene expression elicited by biotic and abiotic stress (Anderson *et al.*, 2004). These results also suggest a role of FsPP2C1 in the regulation of ET/JA-dependent defence gene expression, and may help to understand this cross-talk.

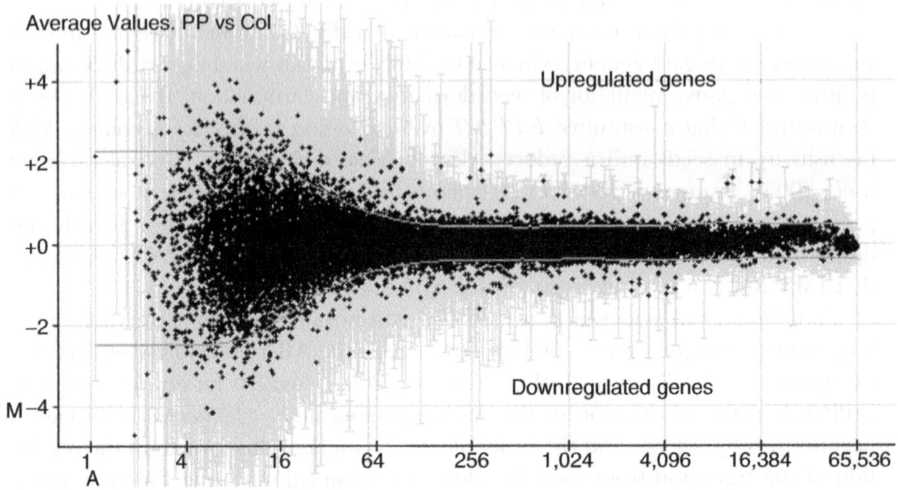

Fig. 24.1. Transcriptomic analysis of ABA-treated wild type vs *35S:FsPP2C1 Arabidopsis* seedlings. MA plot of averaged values: M values (vertical axis) are the log-differential expression ratios; A values (horizontal axis) are the log intensity of the spot. Vertical bars represent the standard deviation of M. The two grey lines are plots of the |z-score| = 2 line (see Materials and Methods section).

The reliability of the microarray data is being corroborated by Northern blot analysis of the genes expressed in the different groups in response to ABA in WT and *35S:FsPP2C1* plants. In general, a very good correlation between chip and Northern blot data has been observed (data not shown).

In summary, the transcriptional profiling analysis indicates that FsPP2C1 plays a key role in the regulation of the expression of genes induced by ABA and defence-related genes.

Proteomic profiling of FsPP2C1-overexpressing seeds in ABA

For a further understanding of this insensitive phenotype and to extend this study to the protein level, a proteome analysis of *35S:FsPP2C1* vs WT seeds was achieved through the analysis of total proteins from mature, imbibed *Arabidopsis* seeds using high-resolution 2D gel electrophoresis. This proteomic analysis revealed significant differences in the protein pattern of ABA-treated seeds from transgenic *35S:FsPP2C1* compared with the WT seeds (Fig. 24.2). The identification of the proteins, whose levels differ after overexpression of FsPP2C1, is currently under study and may give relevant information about the role of this protein in ABA-regulated seed dormancy or germination.

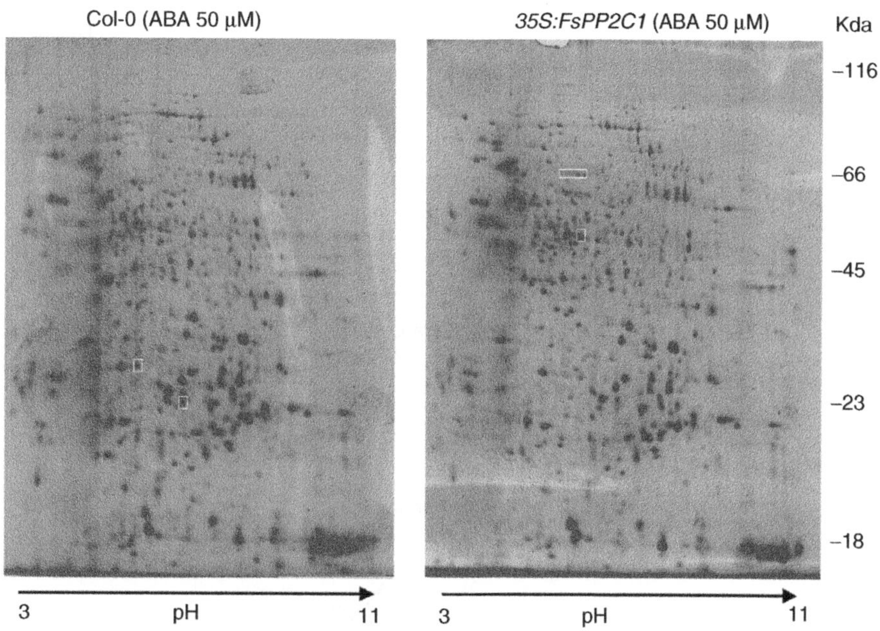

Fig. 24.2. Proteomic analysis of ABA-treated wild type (Col-0) vs *35S:FsPP2C1* seeds. Total seed proteins from Col-0 and *35S:FsPP2C1* treated with 50 μM ABA for 3 h were extracted and separated in two-dimensional (2D) gels. Isoelectric focusing (IEF) was performed on 7 cm immobilized pH gradient (IPG) strips of pH 3–11 using 100–300 μg of proteins. Examples of selected proteins are boxed.

Proposed role of FsPP2C1 in ABA signalling

Figure 24.3 presents a hypothetical model, which shows the role of FsPP2C1 in the ABA signal transduction pathway. Different types of abiotic stresses, such as salt, drought or the exogenous application of ABA, induce the synthesis and subsequent

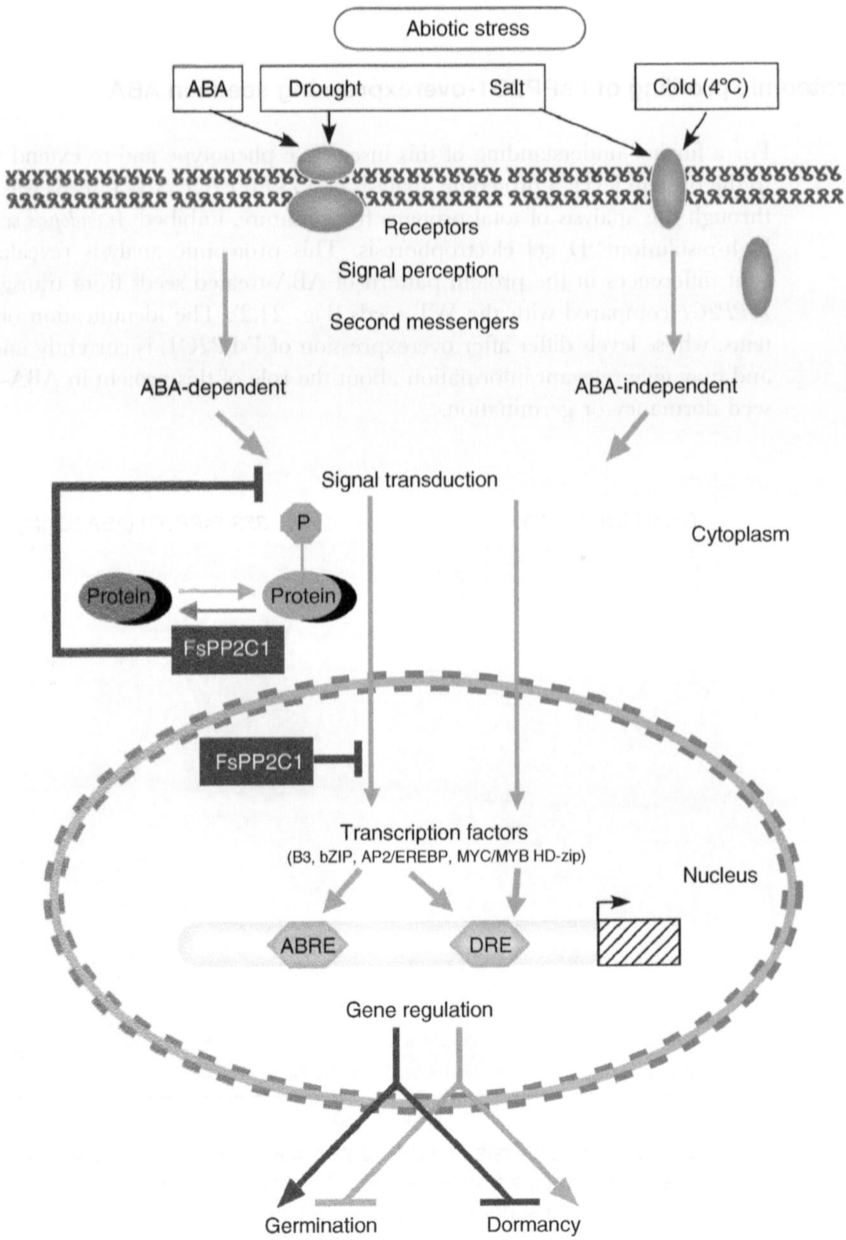

Fig. 24.3. Schematic representation of the function of FsPP2C1 in the ABA signalling pathway, including its putative role in seed dormancy and germination.

activation of the ABA pathway. An increase in ABA will induce the expression of FsPP2C1, which would be responsible for the negative regulation of the ABA signal transduction pathway and for the repression of ABA-response genes.

Conclusions

Our findings provide the following conclusions: (i) FsPP2C1 functions as a negative regulator of ABA signalling in seeds and early seedling growth; (ii) FsPP2C1 overexpression in transgenic plants leads to a reduced seed dormancy compared with untransformed plants, and a promotion of germination under inhibitory concentrations of ABA or high-osmoticum media. This ABA-insensitive phenotype, together with the ABA-induced expression of FsPP2C1 transcripts specifically in seeds, suggests that this PP2C is involved in promoting the transition from seed dormancy to germination during the early weeks of stratification through modulation of ABA signalling; (iii) FPP2C1 regulates ABA responses and cross-talk with biotic stress; and (iv) FsPP2C1 overexpression in *Arabidopsis* produces significant changes in the protein profile during seed imbibition.

Acknowledgements

We thank Dr Roberto Solano and Dr Gloria García for critical discussion and excellent technical assistance in the transcriptomic analysis. We also thank Gloria González for her help in the proteomic experiments. This work was financed by grant BFI2003-01755 to D. Rodriguez and grant BIO2005-08473 to O. Lorenzo from the Spanish Ministerio de Educación y Ciencia. O. Lorenzo is supported by a 'Ramón y Cajal' research contract.

References

Anderson, J.P., Badruzsaufari, E., Schenk, P.M., Manners, J.M., Desmond, O.J., Ehlert, C., Maclean, D.J., Ebert, P.R. and Kazan, K. (2004) Antagonistic interaction between abscisic acid and jasmonate-ethylene signalling pathways modulates defence gene expression and disease resistance in *Arabidopsis*. *The Plant Cell* 16, 3460–3479.

Campalans, A., Messeguer, R., Goday, A. and Pagés, M. (1999) Plant responses to drought, from ABA signal transduction events to the action of the induced proteins. *Plant Physiology and Biochemistry* 37, 327–340.

Cutler, S., Ghassemian, M., Bonetta, D., Cooney, S. and McCourt, P. (1996) A protein farnesyl transferase involved in abscisic acid signal transduction in *Arabidopsis*. *Science* 273, 1239–1241.

González-Garcia, M.P., Rodriguez, D., Nicolas, C., Rodriguez, P.L., Nicolas, G. and Lorenzo, O. (2003) Negative regulation of abscisic acid signaling by the *Fagus sylvatica* FsPP2C1 plays a role in seed dormancy regulation and promotion of seed germination. *Plant Physiology* 133, 135–144.

Gosti, F., Beaudoin, N., Serizet, C., Webb, A.A.R., Vartanian, N. and Giraudat, J. (1999) ABI1 protein phosphatase 2C is a negative regulator of abscisic acid signalling. *The Plant Cell* 11, 1897–1909.

Himmelbach, A., Hoffmann, T., Leube, M., Hohener, B. and Grill, E. (2002) Homeodomain protein ATHB6 is a target of the protein phosphatase ABI1 and regulates hormone responses in *Arabidopsis*. *The EMBO Journal* 21, 3029–3038.

Kerk, D., Bulgrien, J., Smith, D.W., Barsam, B., Veretnik, S. and Gribskov, M. (2002) The complement of protein phosphatase catalytic subunits encoded in the genome of *Arabidopsis. Plant Physiology* 129, 908–925.

Leung, J., Bouvierd-Durand, M., Morris, P.C., Guernier, D., Chefdor, F. and Giraudat, J. (1994) *Arabidopsis* ABA response gene *ABI1*: features of a calcium-modulated protein phosphatase. *Science* 264, 1448–1452.

Lorenzo, O., Rodríguez, D., Nicolás, G., Rodríguez, P.L. and Nicolás, C. (2001) A new protein phosphatase 2C (FsPP2C1) induced by ABA is specifically expressed in dormant seeds of *Fagus sylvatica* L. *Plant Physiology* 125, 1949–1956.

Merlot, S., Gosti, F., Guerrier, D., Vavasseur, A. and Giraudat, J. (2001) The ABI1 and ABI2 protein phosphatases 2C act in a negative feedback regulatory loop of the abscisic acid signaling pathway. *The Plant Journal* 25, 295–303.

Meyer, K., Leube, M.P. and Grill, E. (1994) A protein phosphatase 2C involved in ABA signal transduction in *Arabidopsis thaliana. Science* 264, 1452–1455.

Murashige, T. and Skoog, F. (1962) A revised medium for rapid growth and bioassays with tobacco. *Physiologia Plantarum* 15, 473–497.

Pei, Z.M., Ghassemian, M., Kwak, C.M., McCourt, P. and Schroeder, J.I. (1998) Role of farnesyltransferase in ABA regulation of guard cell anion channels and plant water loss. *Science* 282, 287–290.

Sáez, A., Apostolova, N., González-Guzmán, M., González-García, M.P., Nicolás, C., Lorenzo, O. and Rodríguez, P.L. (2004) Gain of function and loss of function phenotypes of the protein phosphatase 2C HAB1 reveal its role as a negative regulator of abscisic acid signaling. *The Plant Journal* 37, 354–369.

Sheen, J. (1998) Mutational analysis of protein phosphatase 2C involved in abscisic acid signal transduction in higher plants. *Proceedings of the National Academy of Sciences of the United States of America* 95, 975–981.

Tahtiharju, S. and Palva, T. (2001) Antisense inhibition of protein phosphatase 2C accelerates cold acclimation in *Arabidopsis thaliana. The Plant Journal* 6, 461–470.

Trewavas, A.J. (1988) Timing and memory processes in seed embryo dormancy: a conceptual paradigm for plant development questions. *Bioessays* 6, 87–92.

Walker-Simmons, M.K. (1998) Protein kinases in seeds. *Seed Science Research* 8, 193–200.

25 Constitutive Expression of a *Fagus* ABA-induced PP2C (FsPP2C2) in *Arabidopsis* Suggests Interactions Between ABA and Gibberellins in Seed Dormancy

D. REYES, C. NICOLÁS, M.P. GONZÁLEZ-GARCÍA, O. LORENZO, G. NICOLÁS AND D. RODRÍGUEZ

Departamento de Fisiología Vegetal, Facultad de Biología, Universidad de Salamanca-Centro Hispano Luso de Investigaciones Agrarias, Plaza de los Doctores de la Reina s/n, 37007 Salamanca, Spain

Abstract

FsPP2C2 was previously characterized in *Fagus sylvatica* L. seeds as a functional protein phosphatase type-2C (PP2C). The expression of the *FsPP2C2* gene is strongly upregulated by abscisic acid (ABA) in seeds and vegetative tissues. Since transgenic work is not feasible in *Fagus*, we have used an overexpression approach in *Arabidopsis* (*Arabidopsis thaliana* L., Heynh.) to provide genetic evidence on FsPP2C2 function in seed dormancy and other ABA responses. In contrast with other PP2Cs, we have found that constitutive expression of FsPP2C2 in *Arabidopsis* under the cauliflower mosaic virus (CaMV) 35S promoter led to enhanced sensitivity to ABA and osmotic stress. It also resulted in a deeper degree of seed dormancy in the wild type (WT), as well as plants with dwarf phenotype, and these effects were reversed by gibberellins (GAs). These results suggest an interaction between ABA signalling and GA biosynthesis through this PP2C from *Fagus*.

Introduction

The plant hormone abscisic acid (ABA) plays an important regulatory role in several aspects of plant growth and development, including embryogenesis, seed maturation and dormancy, as well as the ability to sense and respond to environmental stresses by controlling stomatal closure and expression of specific genes (Zeevaart and Creelman, 1988; Leung and Giraudat, 1998). Genetic analyses performed mainly on *Arabidopsis* (*Arabidopsis thaliana* L., Heynh.), because of its excellent suitability for genetic and molecular studies (Koornneef *et al.*, 1984), have demonstrated the crucial role of ABA in seed dormancy and the requirement of gibberellins (GAs) for germination (Finkelstein *et al.*, 2002; Koornneef *et al.*, 2002), as well as the antagonistic effects of ABA and GAs in the regulation of seed germination (Koornneef and Karssen, 1994; Steber *et al.*, 1998). Moreover, we have previously shown that *Fagus sylvatica* L. seeds represent a suitable model to study the biochemical and molecular mechanisms involved in seed dormancy of woody plants, since they exhibit a particular deep dormancy, which is maintained by ABA and overcome by stratification at 4°C or gibberellic acid (GA_3) treatment (Nicolás *et al.*, 1996, 1997).

Substantial progress has been made in the characterization of the ABA signal transduction pathway (Leung and Giraudat, 1998). Genes that respond to exogenous ABA have been identified (Skriver and Mundy, 1990) and secondary messengers such as Ca^{2+}, cADPR and inositol triphosphates have been implicated in ABA-mediated responses (Wu *et al.*, 1997; Pandey *et al.*, 2000; Viswanathan and Zhu, 2002). ABA signalling also appears to involve certain RNA-binding proteins, such as hyponastic leaves 1 (HYL1), ABA-hypersensitive 1 (ABH1) and supersensitive to ABA and drought 1 (SAD1) (Lu and Fedoroff, 2000; Hugouvieux *et al.*, 2001; Xiong *et al.*, 2001), and a complex network of positive and negative regulators, including kinases, phosphatases and transcriptional regulators (reviewed by Finkelstein *et al.*, 2002; Abe *et al.*, 2003).

In addition, evidence indicates that ABA regulates different processes, including seed dormancy and germination, by phosphorylation or dephosphorylation events through the activation and/or expression of specific protein kinases (PKs) and phosphatases (PPs) (Trewavas, 1988; Walker-Simmons, 1998; Finkelstein *et al.*, 2002). Several ABA- and stress-responsive PKs have been described in seeds and other plant tissues, including mitogen-activated protein kinases (MAPKs) (Hirt, 2000), calcium-dependent protein kinases (CDPKs) (Urao *et al.*, 1994), ABA-activated protein kinases (AAPKs) (Li *et al.*, 2000) and OPEN STOMATA1 (OST1) (Mustilli *et al.*, 2002), which are involved in stomatal closure; ABA-induced protein kinase 1 (PKABA1), which suppresses GA-inducible gene expression in barley (*Hordeum vulgare* L.) aleurone layers (Holappa and Walker-Simmons, 1995; Gómez-Cadenas *et al.*, 1999, 2001); and *Fagus sylvatica* protein kinase 1 (FsPK1), a dual serine/threonine (ser/thr) and tyrosine (tyr) PK that is upregulated by ABA and calcium in *F. sylvatica* seeds (Lorenzo *et al.*, 2003). There is also genetic evidence of the involvement of PPs in ABA-regulated processes, including dormancy and germination. In particular, protein phosphatases type-2C (PP2Cs), including four *Arabidopsis* ser/thr PP2Cs (i.e. ABI1, ABI2, AtPP2CA and HAB1) and one PP2C from *F. sylvatica* (i.e. FsPP2C1), have been shown to act as negative regulators of ABA signalling (Gosti *et al.*, 1999; Merlot *et al.*, 2001; Tahtiharju and Palva, 2001; González-García *et al.*, 2003; Sáez *et al.*, 2004).

The isolation and characterization of FsPP2C2 has previously been reported as a functional PP2C from *F. sylvatica* nuts, and it has also been shown to be upregulated by ABA in seeds and other plant tissues (Lorenzo *et al.*, 2002). Since genetic transformation is still not possible in *Fagus*, the current work aims to study the possible role of this PP from *F. sylvatica* in ABA regulation of seed dormancy and germination in *Arabidopsis*. This will be achieved by overexpression of *FsPP2C2* and functional analysis of the corresponding protein in this heterologous system.

Materials and Methods

Plant material

The *Arabidopsis* wild-type (WT; Columbia ecotype) and transgenic plants were grown in a greenhouse under long-day conditions (i.e. in a 16h light/8h dark cycle) in pots containing a 1:3 vermiculite/soil mixture. Sterilized seeds were plated on MS medium (Murashige and Skoog, 1962) containing 1% (w/v) sucrose and solidified with 1% (w/v) agar for *in vitro* culture.

Vector construction and plant transformation

To analyse the function of FsPP2C2 in the regulation of dormancy and germination by ABA, and since genetic transformation of *Fagus* is not possible at the moment, the genetic tools available in *Arabidopsis* were used. Thus, to overexpress the *FsPP2C2* gene in this heterologous system, a construction containing the cDNA clone and including the *NPTII* gene to confer resistance to kanamycin was used to transform *Arabidopsis* plants using the *Agrobacterium* system.

To construct the transgenic plants, the coding region of the *FsPP2C2* cDNA was cloned into the pBIN121 vector, which contains the modified 35S promoter of cauliflower mosaic virus (CaMV). 5'-GGATCCATGTTTTCGGA-'3 and 5'-GAGCTCTTAGGTGCTGCCAAC-'3 were used as primers to amplify the *FsPP2C2* cDNA fragment containing the *Bam*HI and *Sac*I cloning sites and subcloned into the corresponding sites of the pBIN121 vector. The pBIN121–FsPP2C2 construction was introduced into *Agrobacterium tumefaciens* C58C1 (pGV2260; Deblaere *et al.*, 1985) by heat shock. *Arabidopsis* plants (Col-0 ecotype) were transformed by the floral dip method (Clough and Bent, 1998). Transgenic seeds were selected on kanamycin medium (50μg/ml). T_2 plants that produced 100% kanamycin-resistant plants in the T_3 generation were considered homozygous and used for further studies.

Germination assays

To test the sensitivity to ABA and high osmotic environment, *35S:FsPP2C2* and WT seeds were plated on an MS medium containing different concentrations of ABA (i.e. 0.1, 0.3 and 0.5μM), 100mM mannitol or 50mM NaCl.

To determine the degree of seed dormancy in our transgenic lines, compared with the WT (Columbia ecotype), seeds were harvested at the same time and cold-stratified for 0, 24 or 120h at 4°C. This was done to release the endogenous dormancy present in *Arabidopsis* seeds, and the germination percentage was calculated 5 days after sowing.

For the other hormonal treatments, 100μM GA_3, 10–100μM brassinosteroids (BR) or 10μM 1-aminocyclopropane-1-carboxylic acid (ACC) were used as indicated.

In order to score seed germination, the percentage of seeds that had germinated and produced fully expanded green cotyledons was determined after 5–10 days. At least three replicates of ~100 seeds were included in each germination assay.

Results and Discussion

Generation of transgenic *Arabidopsis* plants overexpressing FsPP2C2

In the different transgenic lines obtained, the insertion of the gene in the *Arabidopsis* genome was confirmed using Southern blots and the levels of expression were also checked using Northern blots (data not shown).

Constitutive expression of FsPP2C2 in *Arabidopsis* confers ABA and osmotic stress hypersensitivity in seeds

WT dormant seeds stratified for 24 or 120h lost dormancy and their germination percentage increased with the period of stratification used for pretreatment. However, the germination percentages of the transgenic seeds remained very low, which showed that the FsPP2C2 transgenic seeds exhibited a deeper degree of dormancy than the WT seeds (Fig. 25.1a).

The sensitivity of both the WT and the transgenic seeds to ABA was determined on MS media supplemented with low concentrations of ABA. After 8 days, most of the WT stratified seeds were able to germinate even in the highest ABA concentration (0.5 μM) used. In contrast, only 50%, 20% and 5% of the FsPP2C2 transgenic seeds germinated in the presence of 0.1, 0.3 and 0.5 μM ABA, respectively (Fig. 25.1b). These results indicate that the overexpression of FsPP2C2 in *Arabidopsis* plants confers hypersensitivity to ABA in the seeds, which is in agreement with the increased seed dormancy observed (Fig. 25.1a).

To further investigate this effect, since ABA has an important role in stress responses, the effects of osmotic stress imposed by NaCl and mannitol on seed germination were also tested. Germination of *35S:FsPP2C2* transgenic seeds was significantly lower than in the WT (Fig. 25.1c). The inhibitory effect of osmotic stress on germination is mediated by ABA; therefore, these results are consistent with those described earlier.

These results indicate that the transgenic seeds are hypersensitive to ABA and suggest that FsPP2C2 acts as a positive regulator of ABA signalling, in contrast to the function described for other plant PP2Cs (Gosti *et al.*, 1999; Merlot *et al.*, 2001; Tahtiharju and Palva, 2001; González-García *et al.*, 2003; Sáez *et al.*, 2004).

Effect of hormonal treatments on germination and growth of *35S:FsPP2C2* transgenic plants

In order to check the involvement of other hormonal factors in the phenotype observed in the transgenic seeds overexpressing FsPP2C2, the effects of gibberellin (GA_3), ethylene (using ACC as a precursor of ethylene biosynthesis) and BR on seed germination were analysed in both WT and transgenic seeds.

All the treatments slightly increased germination percentage in the WT, but only the addition of GA_3 clearly increased the germination percentages of the transgenic seeds up to values similar to those observed in WT seeds, both in the

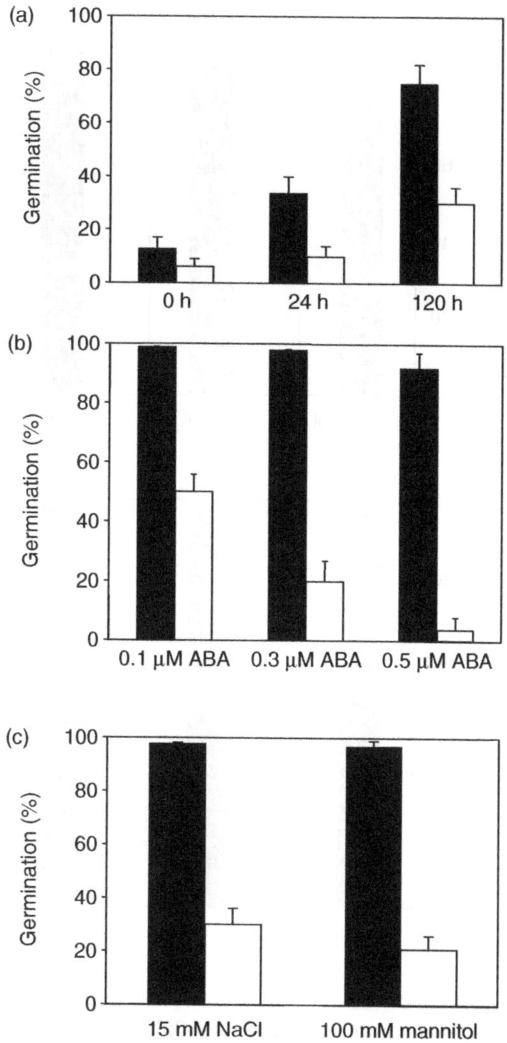

Fig. 25.1. Mean germination percentages of wild-type (Col-0; black bars) and *FsPP2C2*-overexpressing seeds (white bars). Only the seeds developing green cotyledons were considered as germinated seeds. (a) Dormancy assay: germination percentages of dormant seeds at 5 days after 0, 24 and 120 h of stratification at 4°C. (b) ABA inhibition assay: germination percentages of seeds after 8 days in 0.1, 0.3 and 0.5 μM ABA. (c) Stress response assay: germination percentages of seeds after 5 days in 50 mM NaCl and 100 mM mannitol. Error bars represent mean ± SD of three independent experiments with ~100 seeds per plate.

absence (Fig. 25.2a) and presence (Fig. 25.2b) of ABA, indicating that GA_3 is able to counteract the effect of the expression of this ABA-induced PP2C on seed germination. These results suggest a role for GAs in the effects observed after the overexpression of FsPP2C2 and a possible interaction between ABA and GAs.

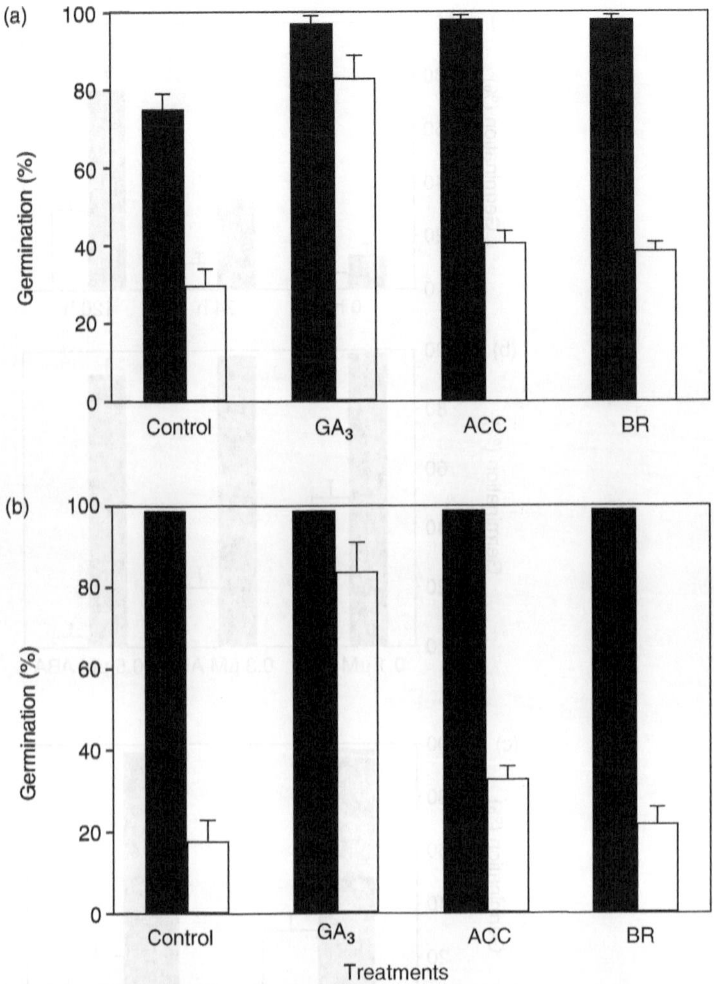

Fig. 25.2. Effects of 100 μM GA_3, 10 μM ACC and 10 μM brassinosteroids (BR) on the mean germination of wild-type (Col-0; black bars) and *FsPP2C2*-overexpressing plants (white bars). (a) Seeds sown in MS for 5 days. (b) Seeds sown in MS supplemented with 0.3 μM ABA for 10 days. Error bars represent mean ± SD of three independent experiments with ~100 seeds per plate.

In addition, transgenic plants overexpressing FsPP2C2 show a dwarf phenotype, which is characteristic of GA-deficient and GA-insensitive mutants (Ross *et al.*, 1997). The stem length of adult FsPP2C2 transgenic plants was ~60% smaller than that of the WT (Fig. 25.3), and this dwarfism was also reverted by the addition of GA_3 to soil-grown transgenic seedlings, while ACC and BR had no effect on the stem growth (Fig. 25.3). Since the reversion of the dwarf phenotype by GAs also applies for GA-deficient mutants, but not for GA-insensitive mutants, these results indicate that the ectopic expression of FsPP2C2 in *Arabidopsis* plants may reduce the levels of GAs by affecting GA biosynthesis.

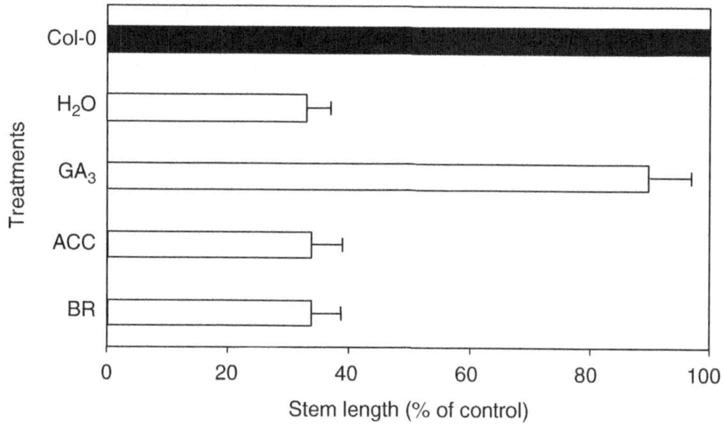

Fig. 25.3. Effects of 100 μM GA_3, 100 μM ACC and 10 μM brassinosteroids (BR) on the stem growth of adult *35S:FsPP2C2* transgenic plants (white bars) calculated as a percentage of the wild-type (Col-0) growth (black bar). Error bars represent mean ± SD of three independent experiments with ~100 seeds per plate.

Conclusions

All these data suggest that FsPP2C2 acts as a positive regulator of the ABA signalling pathway in *Arabidopsis*, unlike other PP2Cs described so far, and that this regulation may be inhibiting some steps in the GA biosynthesis pathway.

Acknowledgements

This work was supported by grant BFI2003-01755 from the Ministerio de Ciencia y Tecnología (Spain) and grant SA046A05 from Junta de Castilla y León. O. Lorenzo is supported by a 'Ramón y Cajal' research contract.

References

Abe, H., Urao, T., Seki, M., Shinozaki, K. and Yamaguchi-Shinozaki, K. (2003) Arabidopsis AtMYC2 (bHLH) and AtMYB2 (MYB) function as transcriptional activators in abscisic acid signalling. *The Plant Cell* 15, 63–78.

Clough, S.J. and Bent, A.F. (1998) Floral dip: a simplified method for *Agrobacterium*-mediated transformation of *Arabidopsis thaliana*. *The Plant Journal* 16, 735–743.

Deblaere, R., Bytebier, B., De Greve, H., Deboeck, F., Schell, J., Van Montagu, M. and Leemans, J. (1985) Efficient octopine Ti plasmid-derived vectors for *Agrobacterium*-mediated gene transfer to plants. *Nucleic Acids Research* 13, 4777–4788.

Finkelstein, R.R., Gampala, S.S. and Rock, C.D. (2002) Abscisic acid signalling in seeds and seedlings. *The Plant Cell* 14 (Suppl.), S15–S45.

Gómez-Cadenas, A., Verhey, S.D., Holappa, L.D., Shen, Q., Ho, T.H.D. and Walker-Simmons, M.K. (1999) An abscisic acid induced protein kinase, PKABA1, mediates abscisic acid-suppressed gene expression in barley aleurone. *Proceedings of the National Academy of Sciences of the United States of America* 96, 1767–1772.

Gómez-Cadenas, A., Zentella, R., Walker-Simmons, M.K. and Ho, T.H.D. (2001) Gibberellin/abscisic acid antagonism in barley aleurone cells: site of action of the protein kinase PKABA1 in relation to gibberellin signalling molecules. *The Plant Cell* 13, 667–769.

González-García, M.P., Rodríguez, D., Nicolás, C., Rodríguez, P.L., Nicolás, G. and Lorenzo, O. (2003) Negative regulation of abscisic acid signalling by the *Fagus sylvatica* FsPP2C1 plays a role in seed dormancy regulation and promotion of seed germination. *Plant Physiology* 133, 135–144.

Gosti, F., Beaudoin, N., Serizet, C., Webb, A.A., Vartanian, N. and Giraudat, J. (1999) ABI1 protein phosphatase 2C is a negative regulator of abscisic acid signalling. *The Plant Cell* 11, 1897–1910.

Hirt, H. (2000) MAP kinases in plant signal transduction. In: Hirt, H. (ed.) *MAP Kinases in Plant Signal Transduction.* Springer, Berlin, Germany, pp. 1–10.

Holappa, L.D. and Walker-Simmons, M.K. (1995) The wheat abscisic acid-responsive protein kinase mRNA, PKABA1, is upregulated by dehydration, cold temperature, and osmotic stress. *Plant Physiology* 108, 1203–1210.

Hugouvieux, V., Kwak, J.M. and Schroeder, J.I. (2001) An mRNA cap binding protein, ABH1, modulates early abscisic acid signal transduction in *Arabidopsis. Cell* 106, 477–487.

Koornneef, M. and Karssen, C.M. (1994) Seed dormancy and germination. In: Meyerowitz, E.M. and Somerville, C.R. (eds) *Arabidopsis.* Cold Spring Harbour Laboratory Press, Cold Spring Harbour, New York, pp. 313–334.

Koornneef, M., Reuling, G. and Karssen, C.M. (1984) The isolation and characterization of abscisic acid-insensitive mutants of *Arabidopsis thaliana. Physiologia Plantarum* 61, 377–383.

Koornneef, M., Bentsink, L. and Karssen, C.M. (2002) Seed dormancy and germination. *Current Opinion in Plant Biology* 5, 33–36.

Leung, J. and Giraudat, J. (1998) Abscisic acid signal transduction. *Annual Review of Plant Physiology and Plant Molecular Biology* 19, 199–222.

Li, J., Wang, X.Q., Watson, M.B. and Assmann, S.M. (2000) Regulation of abscisic acid-induced stomatal closure and anion channels by guard cells AAPK kinase. *Science* 287, 300–303.

Lorenzo, O., Nicolás, C., Nicolás, G. and Rodríguez, D. (2002) Molecular cloning of a functional protein phosphatase 2C (FsPP2C2) with unusual features and synergistically upregulated by ABA and calcium in dormant seeds of *Fagus sylvatica. Physiologia Plantarum* 114, 482–490.

Lorenzo, O., Nicolás, C., Nicolás, G. and Rodríguez, D. (2003) Characterization of a dual protein kinase (FsPK1) up-regulated by abscisic acid and calcium and specifically expressed in dormant seeds of *Fagus sylvatica* L. *Seed Science Research* 13, 261–271.

Lu, C. and Fedoroff, N. (2000) A mutation in the *Arabidopsis* HYL gene encoding a dsRNA binding protein affects responses to abscisic acid, auxin and cytokinin. *The Plant Cell* 12, 2351–2366.

Merlot, S., Gosti, F., Guerrier, D., Vavasseur, A. and Giraudat, J. (2001) The ABI1 and ABI2 protein phosphatases 2C act in a negative feedback regulatory loop of the abscisic acid signalling pathway. *The Plant Journal* 25, 295–303.

Murashige, T. and Skoog, F. (1962) A revised medium for rapid growth and bioassays with tobacco. *Physiologia Plantarum* 15, 473–497.

Mustilli, A.C., Merlot, S., Vavasseur, A., Fenzi, F. and Giraudat, J. (2002) *Arabidopsis* OST1 protein kinase mediates the regulation of stomatal aperture by abscisic acid and acts upstream of reactive oxygen species production. *The Plant Cell* 14, 3089–3099.

Nicolás, C., Nicolás, G. and Rodríguez, D. (1996) Antagonistic effects of abscisic acid and gibberellic acid on the breaking of dormancy of *Fagus sylvatica* seeds. *Physiologia Plantarum* 96, 244–250.

Nicolás, C., Rodríguez, D., Poulsen, F., Eriksen, E.N. and Nicolás, G. (1997) The expression of an abscisic acid-responsive glycine rich protein coincides with the level of seed dormancy in *Fagus sylvatica. Plant and Cell Physiology* 38, 1303–1310.

Pandey, S., Tiwary, S.B., Upadhyaya, K.C. and Sopory, S.K. (2000) Calcium signalling: linking environmental signals to cellular functions. *Critical Reviews in Plant Science* 19, 291–318.

Ross, J.J., Murfet, I.C. and Reid, J.B. (1997) Gibberellin mutants. *Physiologia Plantarum* 100, 550–560.

Sáez, A., Apostolova, N., González-Guzmán, M., González-García, M.P., Nicolás, C., Lorenzo, O. and Rodríguez, P.L. (2004) Gain of function and loss of function phenotypes of the protein phosphatase 2C HAB1 reveal its role as a negative regulator of abscisic acid signalling. *The Plant Journal* 37, 354–369.

Skriver, K. and Mundy, J. (1990) Gene expression in response to abscisic acid and osmotic stress. *The Plant Cell* 2, 503–512.

Steber, C.M., Cooney, S.E. and McCourt, P. (1998) Isolation of the GA-response mutant *sly1* as a suppressor of ABI1-1 in *Arabidopsis thaliana. Genetics* 149, 509–521.

Tahtiharju, S. and Palva, T. (2001) Antisense inhibition of protein phosphatase 2C accelerates cold acclimation in *Arabidopsis thaliana. The Plant Journal* 26, 461–470.

Trewavas, A.J. (1988) Timing and memory processes in seed embryo dormancy: a conceptual paradigm for plant development questions. *Bioessays* 6, 87–92.

Urao, T., Katagiri, T., Mizoguchi, T., Yamaguchi-Shinozaki, K., Hayashida, N. and Shinozaki, K. (1994) Two genes that encode Ca^{2+}-dependent protein kinases are induced by drought and high-salt stresses in *Arabidopsis thaliana. Molecular and General Genetics* 244, 331–340.

Viswanathan, C. and Zhu, J.K. (2002) Molecular genetic analysis of cold-regulated gene transcription. *Philosophical Transactions of the Royal Society of London. Series B, Biological Sciences* 357, 877–886.

Walker-Simmons, M.K. (1998) Protein kinases in seeds. *Seed Science Research* 8, 193–200.

Wu, Y., Kuzma, J., Marechal, E., Lee, H.C., Foster, R. and Chua, N.H. (1997) Abscisic acid signalling through cyclic ADP-ribose in plants. *Science* 278, 2126–2130.

Xiong, L., Gong, Z., Rock, C.D., Subramanian, S., Guo, Y., Xu, W., Galbraith, D. and Zhu, J.K. (2001) Modulation of abscisic acid signal transduction and biosynthesis by an Sm-like protein in Arabidopsis. *Developmental Cell* 1, 771–781.

Zeevaart, J.A.D. and Creelman, R.A. (1988) Metabolism and physiology of abscisic acid. *Annual Review of Plant Physiology and Plant Molecular Biology* 39, 439–473.

26 Nucleotide Triphosphate Synthesis and Energy Metabolism in Primary Dormant and Thermodormant Oat Seeds

F. Corbineau, D. Vinel, J. Leymarie and D. Côme

Physiologie Végétale Appliquée, Université Pierre et Marie Curie-Paris 6, 75252 Paris, Cedex 05, France

Abstract

The objective of the present study was to investigate whether the energy metabolism of the embryo, evaluated by adenylic triphosphate nucleotide (ATP) and non-adenylic triphosphate nucleotide (NTP) levels and energy charge (EC), is involved in primary dormancy and the induction of thermodormancy in oat (*Avena sativa* L.) seeds. ATP content and EC were similar in embryos of primary dormant and non-dormant seeds incubated at 30°C, a temperature at which primary dormant seeds do not germinate and enter a secondary dormancy (i.e. thermodormancy). The induction of thermodormancy required an embryo EC of at least 0.6, and was seen as an active phenomenon. NTP content sharply increased during the first 3 h of imbibition of non-dormant seeds at 30°C, and then decreased up to the time of radicle protrusion. Such a transient peak in NTP content was not observed in the embryos of primary dormant seeds, in which NTP content increased during the first 15 h of imbibition and then remained constant. An exogenous supply of adenosine resulted in a similar increase in ATP content in the embryos of both types of seeds, but it induced an increase in NTP content only in the embryos of non-dormant seeds. This chapter discusses NTP metabolism as an early biochemical marker of oat seed dormancy.

Introduction

Freshly harvested seeds of numerous species are considered to be in a state of primary dormancy because they do not germinate, or germinate poorly, in environmental conditions (temperature, oxygen availability and water potential of the medium, etc.) that are normally favourable for germination. In oat (*Avena sativa* L.), as in other cereals such as barley (*Hordeum vulgare* L.) and wheat (*Triticum aestivum* L.), this primary dormancy corresponds to the inability of the seeds to germinate above

15–20°C, and disappears progressively during dry storage (Corbineau *et al.*, 1986). Incubation of dormant seeds at 30°C, a temperature at which they do not germinate, induces loss of ability to germinate subsequently at 20°C, and is considered to be thermodormancy (Corbineau *et al.*, 1993).

Germination of dormant cereal seeds in the presence of respiratory inhibitors has led to the theory that the pentose phosphate pathway might play a role in the metabolic regulation of germination and dormancy (Roberts and Smith, 1977; Côme and Corbineau, 1990). On the other hand, the transient burst of fructose 2,6-bisphosphate, which has been observed during the first hours of imbibition of non-dormant seeds of oat (Larondelle *et al.*, 1987) and red rice (*Oryza sativa* L.) (Cohn and Footitt, 1993), and in dormant apple (*Malus pumila* Mill.) embryos treated with cyanide (Bogatek, 1995), has suggested that glycolysis might also be involved. Recently, Gallais *et al.* (1998) reported a quick rise in the catabolic redox charge (CRC), expressed as NADH/(NADH + NAD^+), in the embryo of non-dormant oat seeds during imbibition, which was probably due to the activation of glycolysis. These authors have concluded that the lack of increase in CRC, and the subsequent weak synthesis of pyridine nucleotides, might be one of the limiting factors for germination of dormant seeds. Although energy metabolism has been cited as being essential in the regulation of germination (Raymond *et al.*, 1985), dormancy does not result from an inability of the embryo to synthesize adenylic triphosphate nucleotides (ATPs) (Côme and Corbineau, 1990). However, the possible role played by energy metabolism in the induction of thermodormancy has not been investigated.

The involvement of purine metabolism in dormancy is also poorly documented. However, the capacity of synthesis of non-adenylic triphosphate nucleotides (NTPs) in the presence of adenosine, a precursor of this metabolism, seems to be a good marker of dormancy in the buds of Jerusalem artichoke (*Helianthus tuberosus* L.) tubers (Gendraud, 1977; Le Floc'h and Lafleuriel, 1981), in the buds of ash (*Fraxinus excelsior* L.) (Lavarenne *et al.*, 1982), oak (*Quercus* spp.) (Barnola *et al.*, 1986) and peach (*Prunus persica* (L.) Batsch.) (Balandier *et al.*, 1993) trees, and in the cotyledons of apple embryos (Thomas *et al.*, 1985). NTPs synthesized by phosphate transfer from ATPs to corresponding nucleosides are involved in numerous biochemical changes (e.g. glucose transport, sugar interconversion and RNA and DNA synthesis), which are essential for growth (Buchanan *et al.*, 2000). ATPs and purine derivatives might also function as signalling molecules (Demidchik *et al.*, 2003).

The aim of the present work was to investigate whether primary dormancy and thermodormancy in oat seeds are related to the energy metabolism of the embryo, evaluated by ATP and NTP content and energy charge (EC), and to the ability of the embryo to synthesize triphosphate nucleotides (ATPs and NTPs) in the presence of a purine precursor.

Materials and Methods

Plant material

Naked *A. sativa* L. 'Moyencourt' seeds (i.e. seeds without glumellae) were used for this study. In order to maintain the initial primary dormancy, freshly harvested seeds were stored at −20°C until experiments started. Dormancy was broken by storing the seeds dry for at least 6 months at 20°C in open air (Corbineau *et al.*, 1986). Thermodormancy was induced by incubating primary dormant seeds at 30°C for 42 h (Corbineau *et al.*, 1993).

Germination assays

Germination assays were performed with 100 naked seeds placed in Petri dishes 10 cm in diameter (i.e. four replicates of 25 seeds per dish) on a layer of cotton wool wetted with deionized water or various solutions of polyethylene glycol-8000 (PEG-8000) or NaN_3. The water potential of various solutions was adjusted by the addition of PEG-8000 and calculated according to Michel and Kaufmann (1973). A seed was considered to have germinated as soon as the radicle protruded from the grain envelopes (i.e. pericarp and seed coat).

Adenosine phosphate and non-adenylic triphosphate nucleotide determinations

ATP, adenylic diphosphate nucleotide (ADP), adenylic monophosphate nucleotide (AMP) and NTP were extracted from embryos isolated from primary dormant, non-dormant and thermodormant seeds incubated for various durations on water or PEG-8000 solutions at 20°C or 30°C, according to the method described by Lecat *et al.* (1992). They were also extracted after a subsequent 3 h incubation of the seeds in the presence of 10 mM adenosine. Adenylic nucleotide and NTP levels were determined using a bioluminescence method (Gendraud, 1977; Lecat *et al.*, 1992) with a pico-ATP biophotometer (Jobin et Yvon, France). The results obtained are expressed as picomol ATP, ADP, AMP or NTP per milligram dry weight (DW), and are the means of values obtained with five extracts from two embryos each ± standard deviation (SD). The ATP/ADP and NTP/ATP ratios and the EC, expressed as (ATP + 0.5 ADP)/(ATP + ADP + AMP), were also calculated.

Results and Discussion

Germination of dormant, non-dormant and thermodormant seeds

The effects of temperature on the germination of dormant and after-ripened (i.e. non-dormant) naked oat seeds are shown in Fig. 26.1a. As for other *Poaceae* originating from temperate climates (Simpson, 1990), dormancy of oat seeds was hardly expressed at relatively low temperatures, because 90–100% of the seeds germinated within 7 days at 10–20°C, but it increased with temperatures above 20°C. Approximately 40% of the seed population germinated at 25°C, but almost no germination occurred at 30°C and 35°C. Breaking of dormancy by dry storage for 6 months resulted in a widening of the temperature range within which germination

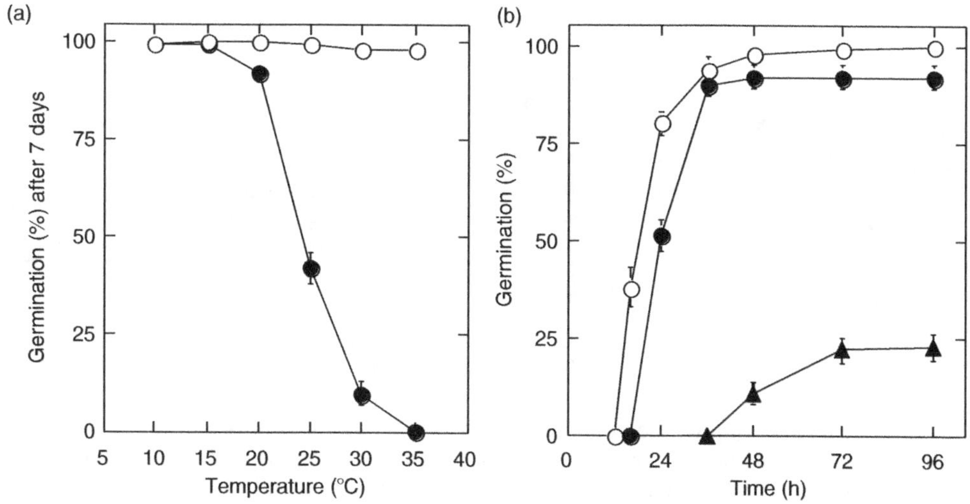

Fig. 26.1. Germination of naked oat seeds. (a) Effects of temperature on the mean germination percentages obtained after 7 days with primary dormant (●) and non-dormant seeds (○). (b) Time course of germination at 20°C of primary dormant (●), non-dormant (○) and thermodormant (▲) seeds. Data presented are the means of four measurements of 25 seeds ± SD.

occurred, and 98–100% of these non-dormant seeds germinated within 7 days at temperatures ranging from 10°C to 35°C.

Incubation of primary dormant seeds for 42 h at 30°C, a temperature at which they did not germinate, resulted in a loss of their ability to subsequently germinate at 20°C (Fig. 26.1b). This phenomenon, called thermodormancy (Corbineau *et al.*, 1993), required a moisture content of the embryo higher than 40% DW, and it increased as the embryo moisture content increased up to 80% DW (Corbineau and Côme, 2003).

Both primary dormancy and thermodormancy result from an inhibitory action of the grain-covering structures and, probably, of the endosperm, since the isolated embryos germinate in a wide range of temperatures (Corbineau *et al.*, 1986, 1993). However, the embryo itself appears to play a role in dormancy, as embryos isolated from dormant seeds are more sensitive to exogenous abscisic acid (ABA) and the water potential of the medium than those isolated from non-dormant seeds (Corbineau and Côme, 2000, 2003).

Energy metabolism and primary dormancy

In both primary dormant and non-dormant dry seeds, ATP, ADP and AMP levels of the embryo represented 7–9%, 18–21% and 70–75% of the adenylate pool, respectively (data not shown), and EC was very low (0.16–0.19) (Fig. 26.2b). The ATP content of the embryo of both types of seeds increased rapidly during the first hours of incubation at 30°C (Fig. 26.2a), thus resulting in a large increase in EC that reached a maximum value (0.75–0.80) after 12 h, the time at which non-dormant seeds germinated (Fig. 26.2b). This rapid activation of energy metabolism, which

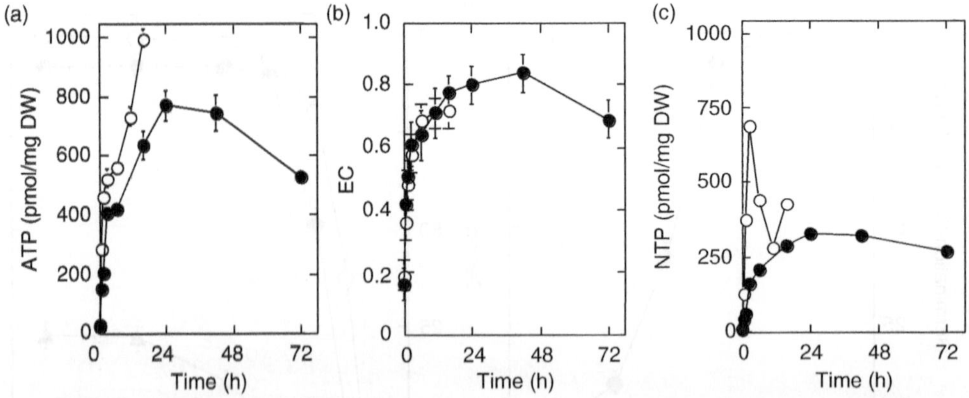

Fig. 26.2. Changes in (a) ATP content, (b) EC and (c) NTP content of the embryo during incubation at 30°C of primary dormant (•) and non-dormant (○) seeds. Data presented are the means of measurements obtained with five extracts from two embryos each ± SD. SD is less than the size of the symbols where no bars are shown. Non-dormant seeds germinated after ~12 h of incubation.

was probably linked to an oxygen-dependent cyanide-sensitive respiratory pathway, as shown by Raymond *et al.* (1982) in other seeds, demonstrates that oxidative phosphorylation occurred immediately after seed imbibition. In primary dormant seeds, the embryo EC was highest (~0.80) from 16 to 42 h of incubation, and then decreased, reaching 0.69 after 72 h (Fig. 26.2b). ATP content was slightly higher in embryos from non-dormant seeds than in those from dormant ones (Fig. 26.1a), but the ATP/ADP ratio (data not shown) and EC increased similarly as long as the non-dormant seeds had not yet germinated (Fig. 26.2b). Therefore, dormancy is not due to the inability of the embryo to synthesize ATP from ADP and AMP.

In contrast, change in embryo NTP content, during incubation of seeds at 30°C, depended on the dormancy state (Fig. 26.2c). In non-dormant seeds, NTP increased rapidly during the first 3 h of imbibition, and then decreased during a period that roughly ended with radicle protrusion. In dormant seeds, the NTP level of the embryo progressively increased during the first 12 h of imbibition, but to a lesser extent than in the embryos of non-dormant seeds, and then remained constant. The NTP/ATP ratio was high (0.8–1.4) during the realization of germination *stricto sensu* in non-dormant seeds, whereas it remained close to ~0.3–0.5 in dormant seeds (data not shown). An exogenous supply of 10 mM adenosine to primary dormant and non-dormant seeds always resulted in an increase in embryo ATP and NTP levels when compared with incubation in water (Table 26.1). However, NTP content markedly increased in embryos of non-dormant seeds placed at 30°C, and in embryos of both types of seeds placed at 20°C (i.e. when seeds were capable of germinating). The ΔNTP/ΔATP ratio was similar (0.6–0.7) for primary dormant and non-dormant seeds placed at 20°C, but it was much higher in the embryos of non-dormant seeds (0.8) than in those of dormant ones (0.24) incubated at 30°C. As in the vegetative buds of trees and Jerusalem artichoke tubers (Gendraud, 1977; Lavarenne *et al.*, 1982; Barnola *et al.*, 1986; Balandier *et al.*, 1993), the absence of dormancy in oat seeds was associated with a better ability to synthesize NTP.

Table 26.1. Increase in nucleotide triphosphate levels (ΔATP and ΔNTP) in the embryo during a 3h incubation of seeds at 20°C and 30°C on adenosine compared with incubation in water.

Temperature of incubation (°C)	Period of incubation (h)	Dormancy stage	ΔATP (pmol per mg DW)	ΔNTP (pmol per mg DW)	ΔNTP/ΔATP
20	0–3	D_1	963	696	0.72
	0–3	ND	740	505	0.68
	0–3	D_2	848	528	0.62
	24–27	D_2	593	290	0.49
30	0–3	D_1	1551	371	0.24
	0–3	ND	1764	1411	0.80

Primary dormant (D_1), non-dormant (ND) and thermodormant (D_2) seeds were placed in the presence of 10mM adenosine at 20°C or 30°C either directly (0–3h) or after imbibition in water for 24h (24–27h). The data presented are the means of measurements obtained with five extracts from two embryos each.

Energy metabolism and induction of thermodormancy

Thermodormancy was reduced when primary dormant seeds were incubated at 30°C in the presence of NaN_3, and it was not induced at all when the concentration of this respiratory inhibitor was sufficiently high (1 mM) (Table 26.2). A low water potential of the incubation medium was also shown to reduce the induction of thermodormancy (Table 26.2). Increased NaN_3 and PEG solutions resulted in a marked decrease in the EC of the embryo, suggesting that induction of thermodormancy required respiratory function. This thermodormancy effect of a pretreatment at 30°C required an embryo EC of at least 0.5, and increased as the embryo EC rose to ~0.8 (Table 26.2). A sublinear relationship exists between embryo EC higher than 0.5 and the degree of seed thermodormancy evaluated by the germination percentages obtained after 3 days at 20°C (data not shown). However, the effect of NaN_3 might also be related to ABA synthesis, as it was associated with a decrease in embryo ABA content (Corbineau and Côme, 2003).

Thermodormant seeds were able to synthesize NTP from adenine at the beginning of their transfer to 20°C in the presence of this precursor, but lost this ability thereafter (Table 26.1).

Conclusions

As in other *Poaceae* (Simpson, 1990), primary dormant naked oat seeds germinated easily at temperatures up to 20°C, but were unable to do so at higher temperatures. Incubation of these seeds at 30°C, a temperature at which almost no germination occurred, resulted in a reinforcement of dormancy, considered to be a thermodormancy. Modulation of embryo energy metabolism during incubation of seeds at 30°C demonstrated that the induction of thermodormancy is an active phenomenon requiring an EC of the embryo higher than ~0.5.

Table 26.2. Effects of a 42 h pretreatment of primary dormant seeds at 30°C in water (control), NaN_3 solutions at various concentrations and PEG-8000 solutions at various water potentials on subsequent germination percentages obtained after 3 days incubation at 20°C and on the embryo EC at the moment of transfer of seeds from 30°C to 20°C.

Incubation at 30°C for 42 h	Mean germination (%) after 3 days at 20°C	Energy charge (EC)
Water (control)	27 ± 5	0.82 ± 0.06
NaN_3 (mM)		
0.1	42 ± 6	0.77 ± 0.05
0.5	88 ± 2	0.58 ± 0.04
1.0	99 ± 1	0.54 ± 0.03
PEG (MPa)		
–1.0	54 ± 5	0.74 ± 0.05
–2.0	78 ± 5	0.65 ± 0.05
–2.5	94 ± 3	0.61 ± 0.04

Data presented are the means ± SD of four replicates of 25 seeds each (germination) or measurements obtained with five extracts from two embryos (EC).

Primary seed dormancy did not result from a lack of ATP synthesis by the embryo, but was marked by an inability of the latter to increase the NTP pool, even in the presence of a purine precursor. The very early peak of NTP content, which was observed in the embryo of non-dormant seeds before radicle protrusion, might therefore be regarded as a possible biochemical marker for the absence of dormancy. However, the respective involvement of various NTPs (i.e. CTP, GTP and UTP) and the significance of their metabolism in this process have to be further investigated. In addition to the role of NTP and ATP in various syntheses, their possible activity as signals must be taken into account.

References

Balandier, P., Gendraud, M., Rageau, R., Bonhomme, M., Richard, J.P. and Parisot, E. (1993) Bud break delay on single node cutting and bud capacity for nucleotide accumulation as parameters for endo and paradormancy in peach trees in tropical climate. *Scientia Horticulturae* 55, 249–261.

Barnola, P., Crochet, A., Payan, E., Gendraud, M. and Lavarenne, S. (1986) Modifications du métabolisme énergétique et de la perméabilité dans le bourgeon apical et l'axe sous-jacent au cours de l'arrêt de croissance momentanée de jeunes plantes de chêne. *Physiologie Végétale* 24, 307–314.

Bogatek, R. (1995) The possible role of fructose 2,6-bisphosphate in the cyanide-mediated removal of embryonic dormancy in apple. *Physiologia Plantarum* 94, 460–464.

Buchanan, B.B., Gruissem, W. and Jones, R.L. (2000) *Biochemistry and Molecular Biology of Plants.* American Society of Plant Physiologists, Rockville, Maryland.

Cohn, M.A. and Footitt, S. (1993) Initial signal transduction steps during the dormancy breaking process. In: Côme, D. and Corbineau, F. (eds) *Fourth International Workshop on Seeds: Basic and Applied Aspects of Seed Biology.* ASFIS, Paris, France, pp. 599–605.

Côme, D. and Corbineau, F. (1990) Some aspects of metabolic regulation of seed germination and dormancy. In: Taylorson, R.B. (ed.) *Recent Advances in the Development and Germination of Seeds.* Plenum Press, New York, pp. 165–179.

Corbineau, F. and Côme, D. (2000) Dormancy of cereal seeds as related to embryo sensitivity to ABA and water potential. In: Viémont, J.D. and Crabbé, J. (eds) *Dormancy in Plants: From Whole Plant Behaviour to Cellular Control.* CAB International, Wallingford, UK, pp. 183–194.

Corbineau, F. and Côme, D. (2003) Involvement of energy metabolism and ABA in primary and secondary dormancies in oat (*Avena sativa* L.) seeds: a physiological approach. In: Nicolás, G., Bradford, K.J., Côme, D. and Pritchard, H.W. (eds) *The Biology of Seeds: Recent Research Advances.* CAB International, Wallingford, UK, pp. 113–120.

Corbineau, F., Lecat, S. and Côme, D. (1986) Dormancy of three cultivars of oat seeds (*Avena sativa* L.). *Seed Science and Technology* 14, 725–735.

Corbineau, F., Black, M. and Côme, D. (1993) Induction of thermodormancy in *Avena sativa* seeds. *Seed Science Research* 3, 111–117.

Demidchik, V., Nichols, C., Oliynyk, M., Dark, A., Glover, B.J. and Davies, J.M. (2003) Is ATP a signaling agent in plants? *Plant Physiology* 133, 456–461.

Gallais, S., Pou de Crescenzo, M.A. and Laval-Martin, D.L. (1998) Pyridine nucleotides and redox charges during germination of non-dormant and dormant caryopses of *Avena sativa* L. *Journal of Plant Physiology* 153, 664–669.

Gendraud, M. (1977) Etude de quelques aspects du métabolisme des nucléotides de pousses de Topinambour en relation avec leurs potentialités morphogénétiques. *Physiologie Végétale* 15, 121–132.

Larondelle, Y., Corbineau, F., Dethier, M., Côme, D. and Hers, H.G. (1987) Fructose 2,6-bisphosphate in germinating oat seeds: a biochemical study of seed dormancy. *European Journal of Biochemistry* 166, 605–610.

Lavarenne, S., Champciaux, M., Barnola, P. and Gendraud, M. (1982) Métabolisme des nucléotides et dormance des bourgeons chez le frêne. *Physiologie Végétale* 20, 371–376.

Lecat, S., Corbineau, F. and Côme, D. (1992) Effects of gibberellic acid on the germination of dormant oat (*Avena sativa* L.) seeds as related to temperature, oxygen, and energy metabolism. *Seed Science and Technology* 20, 421–433.

Le Floc'h, F. and Lafleuriel, J. (1981) The purine nucleosidases of Jerusalem Artichoke shoots. *Phytochemistry* 20, 2127–2129.

Michel, B.E. and Kaufmann, M.R. (1973) The osmotic potential of polyethylene glycol 6000. *Plant Physiology* 51, 914–916.

Raymond, P., Hourmant A., Leblanc, J.M., Al-Ani, A. and Pradet, A. (1982) Mécanismes régénérateurs d'ATP au cours des premières phases de la germination. *Bulletin de la Société Botanique de France* 129, 91–97.

Raymond, P., Al-Ani, A. and Pardet, A. (1985) ATP production by respiration and fermentation, and energy charge during aerobiosis and anaerobiosis in twelve fatty and starchy germinating seeds. *Plant Physiology* 79, 879–884.

Roberts, E.H. and Smith, R.D. (1977) Dormancy and the pentose phosphate pathway. In: Khan, A.A. (ed.) *The Physiology and Biochemistry of Seed Dormancy and Germination.* Elsevier/North-Holland Biochemical Press, Amsterdam, The Netherlands, pp. 385–411.

Simpson, G.M. (1990) *Seed Dormancy in Grasses.* Cambridge University Press, Cambridge.

Thomas, F., Thévenot, C. and Gendraud, M. (1985) Métabolisme des nucléosides triphosphates et dormance des cotylédons des embryons de pommier. *Comptes Rendus de l'Académie des Sciences Paris* 300, Série III 9, 409–412.

27 Dormancy and Germination in *Eucalyptus globulus* Seeds

T.S. Nair, S. Wilson and C. Spurr

Tasmanian Institute of Agricultural Research, School of Agricultural Science, University of Tasmania, Hobart, Tasmania 7001, Australia

Abstract

Under laboratory test conditions commercial *Eucalyptus globulus* Labill. seed lots generally have a high level of viability and rapid, uniform germination. However, commercial nurseries occasionally report seed lots in which germination is spread over a period of up to 10 weeks. The present study examined apparent dormancy induction in seeds that were initially imbibed for periods ranging from 0 to 48 h in darkness followed by air drying and open storage for 1 week at 25°C ± 1°C. Germination tests were carried out in a 12 h light/12 h dark (12/12) photoperiod or in complete darkness. Longer periods of imbibition prior to drying reduced the percentage of seeds that germinated and slowed the rate of germination in seeds germinated under a 12/12 photoperiod. Seeds with the same pretreatment germinated normally in continuous darkness. These data suggest a previously unreported inducible, light-regulated dormancy in *E. globulus* seeds.

Introduction

Eucalyptus globulus Labill. is widely used in plantations in southern Australia. Generally seedlings are raised in Lannen® type trays and grown under nursery conditions for ~6 months before planting out in winter. Under most circumstances *E. globulus* seeds used in commercial nurseries have high viability (95%) and usually complete germination in 5–6 days. Occasionally, germination is erratic. Anecdotal evidence suggests that this problem occurs following a brief period of desiccation or high temperatures between sowing and germination.

Germination characteristics of *Eucalyptus* have not been widely studied but Clifford (1953) noted that *E. globulus* might germinate better in the dark. Wilson *et al.* (2005) found that *E. ovata* Labill. seed, hydrated and subsequently dehydrated during pelleting, exhibited a type of secondary dormancy broken by germination in complete darkness. A hydration–dehydration treatment has been used to promote germination in the seeds of several species, including tomato (*Lycopersicon esculentum* Mill.), oats (*Avena sativa* L.) (Berrie and Drennan, 1971), wheat (*Triticum aestivum* L.) (Hanson, 1973) and *Aster kantoensis* Kitam. (Kagaya *et al.*, 2005). The main purposes of this study were to investigate the effects of hydration–dehydration treatments under laboratory conditions, and light conditions during germination, on the induction and expression of secondary dormancy in *E. globulus* seeds.

Materials and Methods

Plant material and standard germination trials

Four commercial, size-graded seed lots used in experiments were from *E. globulus* orchards located near Mount Gambier in South Australia. The orchard consists of trees originating from the geographical range of *E. globulus* Labill. species and populations intergrading with other *E. globulus* subspecies. The majority (~90%) of trees present in seed orchards are from Western Otways, Strezelecki ranges, Furneaux and south-eastern Tasmanian races. For further details on the racial classification, refer to Dutkowski and Potts (1999). Seed lot 'h' (1000–1200 μm) and 'Li' (1588–1936 μm) were from orchard mixtures, whereas seed lots 'KI' (1693–1954 μm) and 'Ki' (1954–2183 μm) were from maternal parents recorded as King Island within the seed orchard.

All experiments were randomized complete block designs, with blocks arranged according to position in the germination cabinet. There were four replicates of 50 seeds in each trial. An analysis of variance (ANOVA) was conducted on the results using the general linear models package of SPSS (version 12.0.1, 2003) and means were compared using Fisher's least significant difference (LSD) test (Steel and Torrie, 1981). In all cases, individual seed lots were subjected to separate ANOVAs. Error bars shown on graphs are standard errors (SEs) of the mean. Laboratory experiments were carried out according to the International Seed Testing Association (ISTA, 1999) guidelines for the species. Germination was carried out at 25°C ± 1°C in a Lindner and May model LMRIL-1-SD germination cabinet with 12/12 photoperiod where lighting was provided by two GE fluorescent tubes (F30W/33). Seeds were germinated on Petri dishes with two layers of Type 2 Advantec 80 mm filter paper moistened with 7 ml of deionized water. Additional deionized water was added as required during the germination test to avoid drying. Seeds were considered to be germinated when the radicle was longer than 4 mm. At each count, germinated seeds were removed from the Petri dishes after scoring. A squash test was performed at the end of the germination period to determine the viability of the ungerminated seeds (Yates *et al.*, 1996).

Presoaking

Seeds from the 'h' and 'Li' seed lots were surface-sterilized with 10 ml of 12% commercial bleach for 10 min before being rinsed thoroughly in distilled water

4–5 times and blotted dry with a paper towel. Four replicates of 50 seeds were then soaked in separate vials of distilled water for 0 (control), 6, 12, 24 and 48 h at room temperature of 18–20°C in natural light and dark conditions (~16/8 h light/dark period). The distilled water was changed every 6 h to prevent hypoxia. Seeds were then placed on filter paper in Petri dishes and transferred immediately to the germination cabinet with a 12/12 photoperiod. Germination was scored on days 5, 8 and 10.

Presoaking and drying back

Seeds from the 'h' and 'Li' seed lots were soaked as above for 0, 18, 24 and 30 h in complete darkness at 25°C, dried back and then stored at room temperature (18–20°C) under the same natural light conditions for 7 days before germination in 12/12 photoperiod. Germination was scored on days 5 and 11.

Responses to light during germination

Seeds from the 'KI' and 'h' seed lots were pre-imbibed for 0 (control), 7, 14, 24, 30 and 35 h in the dark on wet filter paper in Petri dishes at 25°C ± 1°C, surface-dried with a paper towel before being dried back and stored. Germination was carried out in complete darkness or in a 12/12 photoperiod. Covering Petri dishes with two layers of aluminium foil created complete darkness and the dishes were unwrapped under a green safe light for germination counts. Dishes were left unwrapped for the 12/12 treatments. Germination counts were conducted on days 6, 10, 13 and 19. The experiment was analysed as a factorial design with six times of imbibition by two light treatments (i.e. 12/12 photoperiod and complete darkness).

Glasshouse trial

Seeds from the 'KI' seed lot were imbibed for 30 h in the dark at 25°C ± 1°C, dried with a paper towel and stored at room temperature (18–20°C in 16/8 h light/dark period) for 7 days. After storage, seeds were germinated in six Lannen® trays filled with Richgro® seed-raising mixture. Each treatment contained 60 seeds, with 5 seeds per cell. After sowing, trays were irrigated before half of each tray was covered with an opaque reflective insulation sheet. The remaining half of each tray was covered with a transparent polythene sheet. Sensors attached to 'Tiny Talk' temperature loggers (Gemini Data Logger, Chichester, West Sussex, UK) were inserted just below the surface of the potting mix in both treatments to record temperatures near the germinating seeds. The trial was a factorial design with two light and two imbibition (i.e. 0 and 30 h) treatments with four replicates. Germinated seeds were scored between days 13 and 20 in natural light conditions.

Results and Discussion

Presoaking

Soaking seeds in laboratory conditions for >24 h significantly reduced germination on day 5 ($P < 0.05$) when compared with the control in seed lot 'h' (Fig. 27.1), but in seed lot 'Li' soaking did not adversely affect germination (data not shown). Battaglia (1993) reported a similar result in *E. delegatensis* seeds, where a longer period of imbibition decreased germination capacity, but significant interprovenance differences in germination capacity were observed.

Presoaking and drying back

Dark soaking for 18 h or more significantly reduced germination on day 5 and 11 at $P < 0.05$ (Fig. 27.2). In addition, this pretreatment contributed to a greater spread in germination through the test period (data not shown). There was a consistent reduction in initial germination with increased soak time up to 25 h, which suggests that the response is related to changes in the seed during the very early stages of germination.

Responses to light during germination

Studies carried out by Bell *et al.* (1995, 1999) with *E. marginata* and by Wilson *et al.* (2005) with *E. ovata* demonstrated that darkness may increase germination in some *Eucalyptus* species. In both seed lots used in the present trial, there was a significant ($P < 0.05$) interaction between germination conditions and imbibition time, with a marked decrease in day 6 germination when time of imbibition increased for seed germinated in alternating light and dark conditions (Fig. 27.3a and c). In contrast,

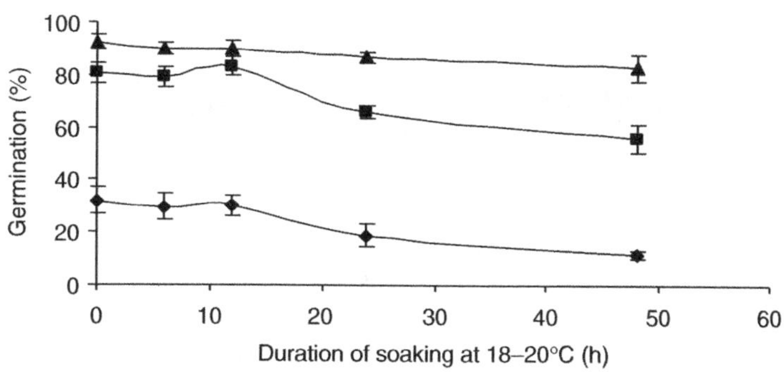

Fig. 27.1. Mean germination percentage on days 5 (◆), 8 (■) and 10 (▲) for seed lot 'h', germinated after soaking in laboratory conditions at 18–20°C for 0–48 h. Data points are results of four replicates of 50 seeds ± standard error (SE) of the mean.

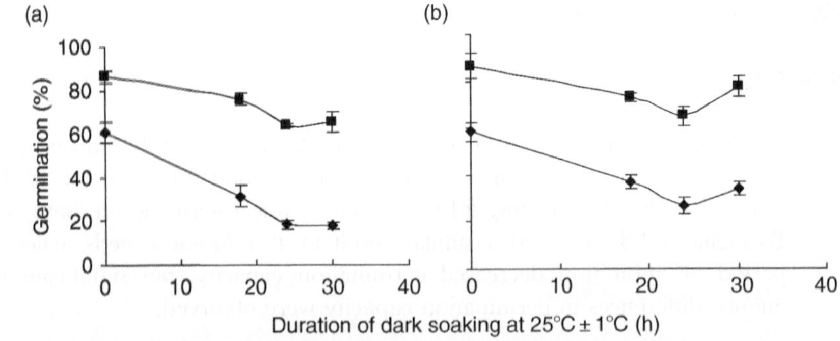

Fig. 27.2. Mean germination percentage on days 5 (◆) and 11 (■) for seed lot (a) 'h' and (b) 'Li' germinated after different periods of dark soaking at 25°C ± 1°C, followed by 7 days drying back at room temperature (18–20°C) in natural light conditions. Data points are results of four replicates of 50 seeds ± standard error (SE) of the mean.

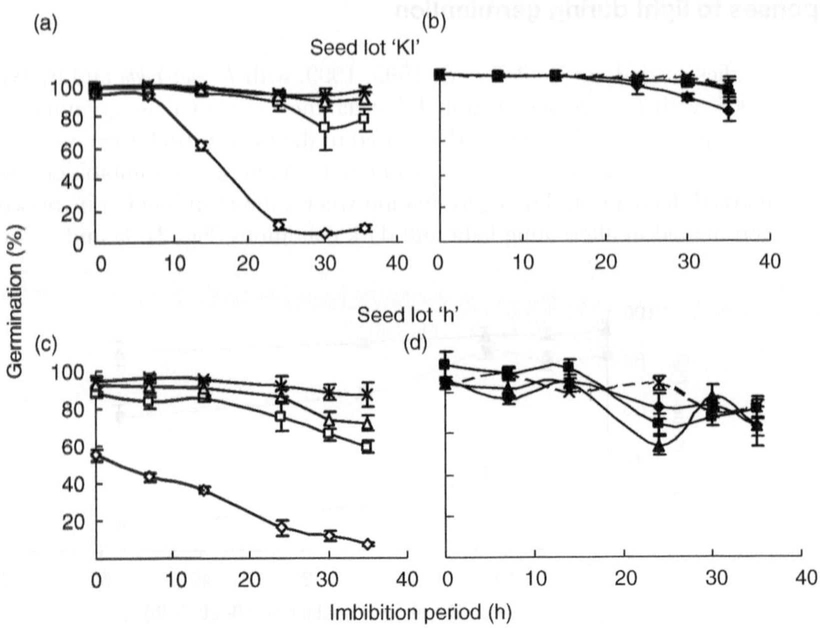

Fig. 27.3. Mean germination (%) in alternating light and dark (open symbols) and darkness (solid symbols) on days 6 (◇, ◆), 10 (□, ■), 13 (△, ▲) and 19 (×, –×–) for seeds imbibed for different time periods, followed by drying back for 7 days at 18–20°C in natural light conditions. Data points are results of four replicates of 50 seeds ± standard error (SE) of the mean.

seed germinated in the dark showed no significant response to pretreatment (Fig. 27.3b and d). There were no significant effects of either pretreatment or germination conditions on days 11–19 of germination.

The results clearly indicate delayed germination or possibly a secondary dormancy induced by pretreatment, and an effect of continuous darkness during germination in overcoming this imposed inhibition of germination.

Glasshouse trial

Under transparent plastic covering, temperature ranged from daytime maxima of 47–50°C to night-time minima of 9–10°C. In trays under the opaque cover, temperatures ranged from maxima of 38–40°C to minima of 9–10°C. There was a marked difference in germination between pretreated and non-treated seed, with pretreatment causing a significant reduction ($P < 0.05$) in germination (Fig. 27.4). Further, germination was significantly higher ($P < 0.05$) in darkness compared with natural light, but there was no significant interaction ($P > 0.05$) between pretreatment and light during germination.

The experiments described show that pretreatment, by allowing the seed to imbibe water in darkness for around 20–30 h, followed by drying back, leads to reduced germination or possibly secondary dormancy in this species. From the results obtained, especially the significant reduction in germination with soaking without drying back in the first experiment, it is difficult to separate soaking (or perhaps soaking conditions) from the drying back treatment. The response to light during germination indicates that the induced dormancy is broken by germination in darkness. Dark-promoted germination of dormant seeds is relatively uncommon, but the induction of dark-responsive dormancy reported here suggests some re-examination of the possible influences of environmental conditions, particularly during early germination, in seed testing and dormancy studies. From a practical

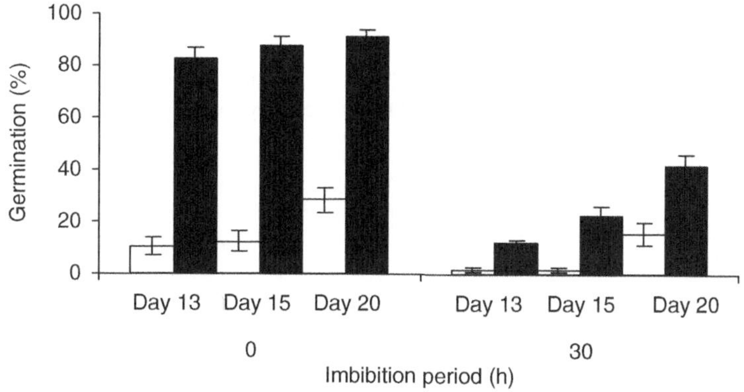

Fig. 27.4. Glasshouse germination on days 13, 15 and 20 for seeds imbibed in darkness for 0 or 30 h at 25°C ± 1°C and dried back and stored for 7 days at room temperature (18–20°C) and germinated in alternating natural light and dark (□) or continuous dark (■) conditions. Bars are standard errors (SE) of the mean (*n* = 6).

point of view, appropriate management of light during the early stages of germination appears to be critical in commercial nurseries. The implications of such a dormancy response system in natural ecosystems remain to be explored.

Acknowledgements

The authors wish to acknowledge the support of Seed Energy P/L, who supplied the seeds for all the experiments.

References

Battaglia, M. (1993) Seed germination physiology of *Eucalyptus delegatensis* RT Baker in Tasmania. *Australian Journal of Botany* 41, 119–136.

Bell, D.T., Rokich, D.P., McChesney, C.J. and Plummer, J.A. (1995) Effects of temperature, light and gibberellic acid on the germination of 43 species native to Western Australia. *Journal of Vegetative Science* 6, 797–806.

Bell, D.T., King, L.A. and Plummer, J.A. (1999) Ecophysiological effects of light quantity and nitrate on seed germination in species from Western Australia. *Australian Journal of Ecology* 24, 2–10.

Berrie, A.M.M. and Drennan, S.H. (1971) The effect of hydration–dehydration on seed germination. *New Phytologist* 70, 135–142.

Clifford, H.T. (1953) A note on germination of *Eucalyptus* seed. *Australian Forestry* 17, 17–20.

Dutkowski, G.W. and Potts, B.M. (1999) Geographic patterns of genetic variation in *Eucalyptus globulus* ssp. *globulus* and a revised racial classification. *Australian Journal of Botany* 47, 237–263.

Hanson, A.D. (1973) The effects of imbibition drying treatments on wheat seeds. *New Phytologist* 72, 1063–1073.

ISTA (1999) International rules for seed testing. *Seed Science and Technology* 27(Suppl.).

Kagaya, M., Tani, T. and Kachi, N. (2005) Effect of hydration and dehydration cycles on seed germination of *Aster kantoensis* (Compositae). *Canadian Journal of Botany* 83, 329–334.

Steel, R.G.D. and Torrie, J.H. (1981) *Principles and Procedures of Statistics: A Biometrical Approach*, 2nd edn. McGraw-Hill, New York.

Wilson, S., Cover, S., Nair, T.S. and Sprod, D. (2005) Clay/humus pellets reduce germination in *Eucalyptus ovata*. In: *Eighth International Workshop on Seeds: Germinating New Ideas.* Workshop Abstracts, Brisbane, Australia, p. 87.

Yates, C.J., Hobbs, R.J. and Bell, W.R. (1996) Factors limiting the recruitment of *Eucalyptus salmonophloia* in remnant woodlands. III. Conditions necessary for seed germination. *Australian Journal of Botany* 44, 283–296.

28 The Effect of Hydropriming on Germination Barriers in Triploid Watermelon Seeds

X.Y. Zheng, X.Q. Li and Y. Xu

The Vegetable Research Centre, Academy of Agricultural Sciences of Beijing, China

Abstract

In order to overcome the germination barriers present in triploid watermelon (*Citrullus lanatus* (Thunb.) Matsum. & Nakai), seeds from six cultivars were subjected to priming by soaking them in 0.1% H_2O_2 or water for 2 h, followed by 24, 34 or 48 h incubation at 100% relative humidity (RH). Germination, viability and seedling growth rates were significantly increased (38–56%) by the priming treatments. The germination percentage of the six cultivars was increased by varying degrees (14–66%), depending on the cultivar. The H_2O_2 priming treatment, when combined with either 34 or 48 h incubation, gave the best results. The beneficial effects of the priming treatment were maintained after the primed seeds were re-dried back to their original moisture content and stored. A soft X-ray investigation of seed and embryo structure indicated that the germination barriers present in the triploid watermelon seeds were related to seed coat, and that the priming treatments overcame these barriers.

Introduction

Triploid watermelons (*Citrullus lanatus* (Thunb.) Matsum. & Nakai) have become popular in recent times because they are seedless, have a high sugar content, are disease-resistant and have a long shelf life. Although triploid watermelons have great market potential, production remains low. Seed production is difficult and costly, and germination is poor due to a relatively thick seed coat (a result of their tetraploid parent). The germination percentages observed for most cultivars are only 20–60% of their potential germination capacity. In addition, the seedling growth rate of triploid watermelons is relatively slow.

Seed coat nicking, splitting or removal have all been reported to be successful in enhancing the germination performance of triploid watermelon seeds (Duval and NeSmith, 2000; Xi, 2000). However, the possibility of embryo injury and the time and labour involved with these treatments are considerable disadvantages.

 Seeds: Biology, Development and Ecology (eds S. Adkins, S. Ashmore and S.C. Navie)

Therefore, research is underway to develop a practical and easy-to-use technique to improve the germination capacity of triploid watermelon seeds.

Seed priming treatments, including osmoconditioning and hydropriming, have been reported to effectively enhance seed germination, vigour and field emergence of various species of field crops and vegetables (Brocklehurst and Dearman, 1983; Thornton and Powell, 1992; Tylkowska and van den Bulk, 2001; Mehra *et al.*, 2003). In addition, H_2O_2 treatments (in which seeds are imbibed in low concentrations of H_2O_2) have been shown to enhance the germination performance of some vegetable species with seed germination barriers (Zheng *et al.*, 1986; Duval and NeSmith, 2000). In cucurbits, priming has been shown to increase the germination rate or germination percentage of muskmelon (*Cucumis melo* L.), diploid watermelon and bitter gourd seeds (Demir and van de Venter, 1999; Nascimento and West, 2000; Demir and Oztokat, 2003). However, few reports have been made on such treatments being applied to triploid watermelon seeds.

Most varieties of triploid watermelon have germination barriers present, but the degree to which these barriers affect germination varies. It is currently thought that these germination barriers are caused by the thick seed coat and/or embryo agenesis (i.e. the improper development of the embryo).

In the present study, the effects of hydropriming and H_2O_2 priming treatments, and subsequent re-drying conditions, on the germination performance of triploid watermelon seeds were investigated. In order to identify the barriers to germination, seed and embryo structure were also investigated using soft X-ray analysis.

Materials and Methods

Plant material and seed germination

The seeds of six triploid watermelon cultivars were obtained from the Beijing Vegetable Research Centre (i.e. Jingxin3 and Jingxin4), Tianjin Vegetable Research Institute (i.e. Jinmi1 and Jinmi3) and Zhengzhou Fruit Tree Research Institute (i.e. Zhengkang3 and Zhengkang5). Seed germination conditions were essentially the same as those described in the 'International Rules for Seed Testing' (ISTA, 2003). Seeds were placed on three filter papers inside plastic germination boxes. The filter papers were moistened with 10ml of water and incubated at 25°C in the dark. Germination was deemed to have occurred when radicle length had reached 2mm.

Seed treatments

Each of the following treatments involved three replicates with 100 seeds each.

Hydropriming and H_2O_2 priming treatments

Seeds were placed in water or a 0.1% H_2O_2 solution for 2h and then surface-sterilized in 1% sodium hypochlorite solution for 5min, rinsed with water for 10min, and then surface-dried with filter paper. After hydration, the seeds were incubated in saturated humidity for 24, 34 or 48h at 20°C.

Hydration treatments
Seeds were soaked in water for 3h or a 0.1% H_2O_2 solution for 6h.

Cracking treatment
Seeds were soaked in water for 2h and then the seed coat was artificially cracked.

Seed re-drying

After incubation, the primed seeds were re-dried by exposing them to one of three separate drying regimes. These regimes involved placing the seeds in one or more sets of controlled conditions for a period of 48h, each with a constant temperature (i.e. 23°C) and relative humidity (RH). Seeds that underwent 'slow drying' were exposed to 70% RH, followed by 40%. 'Moderate drying' exposed the seeds to 70% RH, followed by 40%, and then 20%. In 'fast drying', seeds were exposed to 10–15% RH.

Soft X-ray embryo structure measurements

A SOFTEX ISTV-25-1 soft X-ray instrument was used to take images of embryo structure in 100 seeds of both the Jingxin3 and Jingxin4 cultivars. Seeds were classified into eight types according to the proportion of the seeds that the embryo covered when viewed under the soft X-ray (i.e. 1, 0.9, 0.8, 0.7, 0.5–0.6, 0.4, 0.1–0.3 and 0). The percentage of seeds in each seed type was then determined for each cultivar. Each seed type was subdivided into two groups based on whether the seeds were intact or cracked. The germination of each group (i.e. intact or cracked) within each seed type was tested and the percentage of germination calculated.

Data processing

The seed vigour index (VI) was calculated according to the following formula:

$$VI = GI \times S_x$$

where GI represents the germination index ($GI = \Sigma G_t/D_t$, where G_t is the number of germinants at day t and D_t is the number of days that have elapsed) and S_x represents the average dry weight of seedlings after germination for x days.

SAS (version 6.03) software was used to analyse the data in the present experiment.

Results

Effects of different priming methods on the germination capacity and vigour of triploid watermelon seeds

Under standard conditions, 37% of Jingxin3 seeds were germinated. However, germination of this cultivar could be enhanced to 97% if the seed coat was cracked.

This indicates that the potential germination capacity of the seed lot was at least 97%. The aim of the priming treatments is to increase the actual germination to the point where it reaches or approaches the potential germination capacity of the seed lot. The results of the present study showed that the germination of Jingxin3 seeds could be enhanced to different degrees (between 20% and 60%) by various H_2O_2 priming treatments (i.e. 0.1% H_2O_2 for 6h or 0.1% H_2O_2 for 2h, followed by maintenance at 100% RH for 24, 34 or 48h) and hydropriming treatments (i.e. water for 2h followed by maintenance at 100% RH for 24, 34 or 48h) (Table 28.1). The H_2O_2 priming treatments involving 34 and 48h incubation at 100% RH resulted in the highest germination percentages (i.e. 93% and 92% germination, respectively). This was 55–56% higher than the control, and approached the potential germination capacity of the seed lot. When comparing the H_2O_2 priming treatments with the hydropriming treatments, the germination percentages in the H_2O_2 treatments were slightly higher. Germination was also enhanced by 20% when seeds were imbibed in 0.1% H_2O_2 for 6h (Table 28.1). Thus, the results indicate that a 0.1% H_2O_2 treatment may promote the germination capacity of triploid watermelon seeds to a certain degree, and this combined with high-humidity incubation for 34 or 48h is the best way to overcome germination barriers and improve subsequent germination.

When comparing the results of the three time courses of high-humidity incubation (i.e. 24, 34 and 48h), using germination percentage and the VI, the seeds incubated for 34h had significantly higher values than those incubated for 24h (Table 28.1). Although the seeds incubated for 48h had very high germination percentages regardless of the priming treatment used, the VI in the H_2O_2 treatment was lower than in the water treatment. This was mainly due to 20% lower germination after the first day. This reduction in germination resulted because the

Table 28.1. The effect of several priming treatments on seed germination, seedling dry weight (dw) and vigour index (VI).

Treatment	Germination (%)	Comparison with control (%)	Mean dry weight (g)	Vigour Index
Control	37 g	–	0.0613	2.88 h
H_2O_2 – 24h	89 b	+52	0.0627	8.71 d
H_2O_2 – 34h	93 ab	+56	0.0631	10.90 b
H_2O_2 – 48h	92 ab	+55	0.0628	9.03 c
W – 24h	75 e	+38	0.0618	6.67 f
W – 34h	89 b	+52	0.0627	8.71 d
W – 48h	87 bc	+50	0.0626	10.76 b
H_2O_2 – 6h	57 f	+20	0.0614	4.17 g
CR	97 a	+60	0.0630	11.85 a

All treatments involved three replicates of 100 seeds, and values in the same column followed by different letters are significantly different ($P < 0.01$). H_2O_2 – 24h, H_2O_2 – 34h and H_2O_2 – 48h: seeds were soaked in 0.1% H_2O_2 for 2h and incubated at 100% RH for 24, 34 or 48h; W – 24h, W – 34h and W – 48h: seeds were soaked in water for 2h and incubated at 100% RH for 24, 34 or 48h; CR: seed coats were cracked.

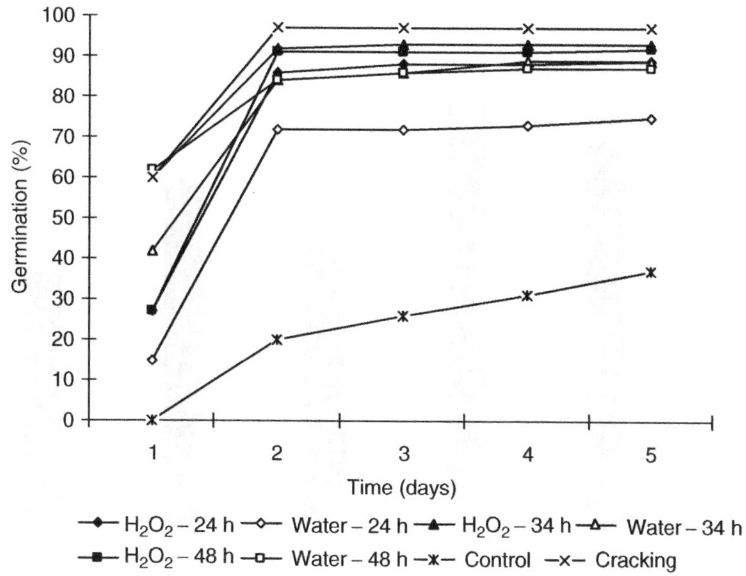

Fig. 28.1. Germination rate of triploid watermelon seeds (cv. Jingxin3) after different priming treatments.

seeds may have started to sprout during this treatment, and hence they were very sensitive to dehydration.

The germination percentage and VI results were multicompared with priming treatment using an analysis of variance (ANOVA) procedure. In all incubation durations, the two priming treatments (i.e. hydropriming and H_2O_2 priming) were significantly different ($P \leq 0.01$). Regardless of the priming treatment, the 34 and 48 h incubation durations were not significantly different from each other, but both were significantly different from the 24 h incubation duration (Fig. 28.1).

The effect of priming treatment on the germination of different triploid watermelon cultivars

The germination capacity of the six different triploid watermelon cultivars was dissimilar due to the different sources of the parent lines. These cultivars, with variable germination capacity, were primed with H_2O_2 for 48 h and re-dried after priming. The results of germination tests indicate that the germination capacity of all six varieties was enhanced with priming and the degree of enhancement varied from 14% to 66%. The varieties in which the original germination capacity was lower, and had the greatest discrepancy with their potential germination capacity, showed the greatest increase in germination (i.e. better priming efficiency). For example, the germination percentage of Jingxin3 was increased to 68%, from 13% to 81%. Conversely, the varieties in which the original germination capacity was highest, and closer to their potential germination capacity, showed lower increases

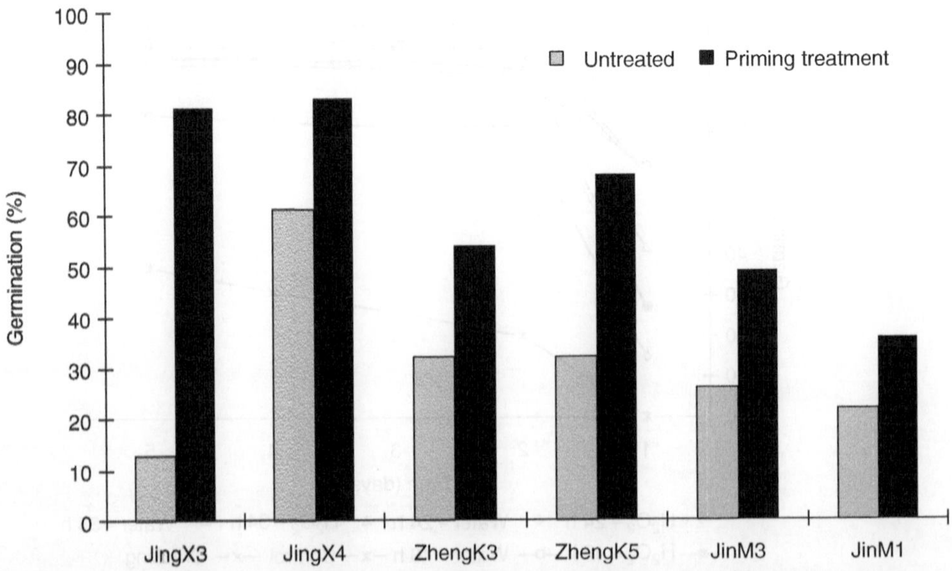

Fig. 28.2. Effect of priming treatment on the germination of seeds of six triploid watermelon cultivars.

in germination (i.e. lower priming efficiency). For example, the germination percentage of Jingxin4 increased from 61% to 83%, an increase of only 22% (Fig. 28.2).

Maintenance of the beneficial effects of priming relative to the rate of seed dehydration

Dehydration conditions are very important for maintaining the beneficial effects in the primed seeds, especially for 48 h incubated seeds. Table 28.2 shows the degree of maintenance of the beneficial effect of priming after three different sets of seed re-drying conditions. The beneficial effects of priming for seeds incubated for 24 h were well maintained after dehydration in all three re-drying conditions. Seeds incubated for 48 h generally maintained the beneficial effects of priming after

Table 28.2. Effect of seed re-drying rate on maintenance of the benefits of priming.

	Germination (%)		
Priming treatment	Slow re-drying	Moderate re-drying	Fast re-drying
H_2O_2 – 48 h	92	89	59
W – 48 h	87	86	64
H_2O_2 – 24 h	89	89	80
W – 24 h	73	73	73
Control	37	–	–

H_2O_2 – 24 h and H_2O_2 – 48 h: seeds were soaked in 0.1% H_2O_2 for 2 h and incubated at 100% RH for 24 or 48 h; W – 24 h and W – 48 h: seeds were soaked in water for 2 h and incubated at 100% RH for 24 or 48 h.

exposure to 'slow re-drying' and 'moderate re-drying'; however, the germination percentage of these seeds dropped significantly after 'fast re-drying'. This indicates that seeds exposed to longer incubation periods, and with strong priming benefits, are very sensitive to re-drying conditions. Embryos incubated for 48 h may have started to sprout and the radicle may have been injured by the fast drying rate, leading to a reduced germination percentage. However, seeds that were exposed to shorter incubation durations, and had a weaker priming benefit, could tolerate a relatively fast re-drying rate and maintain the benefits of priming.

Therefore, in order to maintain the benefits of priming in triploid watermelon seeds, longer seed re-drying conditions are required. The re-drying experiment indicates that the benefits of priming can be maintained if primed seeds are re-dried in slow or middle rate treatments, particularly where 40% RH is maintained in a sealed container at 10°C.

Mechanism analysis of germination barriers and hydropriming

The relationship between the degree of embryo integrity and germination capacity

Triploid watermelon seeds could be divided into seven embryo size classes when observed under a soft X-ray machine (Fig. 28.3, Table 28.3). The results of germination tests (with cracked and non-cracked seeds) showed that the germination percentage of non-cracked seeds was generally much lower than that of cracked seeds. This was the case for both cultivars studied and in all embryo size classes where germination occurred. The maximum difference in germination capacity between cracked and non-cracked seeds was 59%, and this was observed in Jingxin3 seeds in the 0.5–0.6 embryo size class.

Embryo size did not markedly affect seed germination capacity when it was greater than 0.5–0.6. Seeds that had poorly developed embryos did not germinate, and no germination was recorded below the 0.5–0.6 embryo size class. However, these size classes accounted for <10% of the seed lots for both cultivars. These results demonstrate that mechanical resistance, due to the thick seed coat, is the main barrier to germination in the two seedless watermelon cultivars studied. However, seeds with higher levels of embryo integrity showed a greater ability to overcome the mechanical resistance of the seed coat.

The effect of embryo structure after hydropriming treatment

The effects of priming and re-drying on the embryo structure of watermelon seeds were investigated using a soft X-ray apparatus. The embryos of seeds that

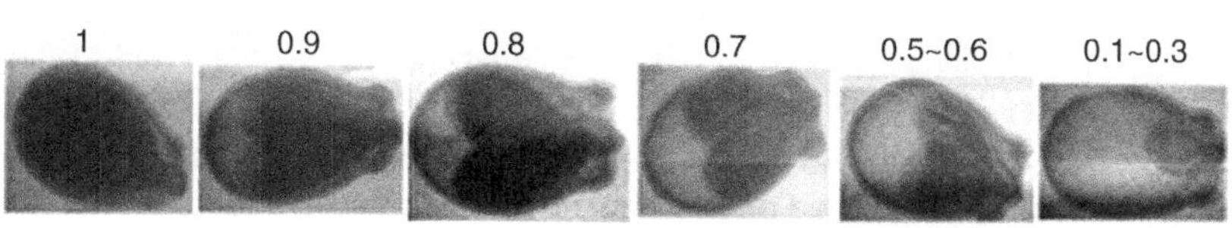

Fig. 28.3. Embryo configuration and classification of Jingxin3 triploid watermelon seeds as observed under a soft X-ray.

Table 28.3. Effect of cracking the seed coat and embryo integrity on the germination capacity of seeds from two triploid watermelon cultivars.

Embryo size class	Jingxin3 Proportion of sample (%)	Jingxin3 Germination (%) Intact	Cracked	Difference	Jingxin4 Proportion of sample (%)	Jingxin4 Germination (%) Intact	Cracked	Difference
1	22	54	92	+38	21	83	100	+17
0.9	17	46	89	+43	17	71	100	+29
0.8	24	80	100	+20	29	72	97	+25
0.7	23	50	88	+38	24	63	83	+20
0.5–0.6	11	33	92	+59	4	58	83	+25
0.4	0	0	0	0	2	0	0	0
0.1–0.3	1	0	0	0	2	0	0	0
0	2	0	0	0	1	0	0	0

Seeds were placed into different embryo size classes based on the proportion of the seed that the embryo covered (i.e. 1: full intact embryo; 0: no embryo).

were primed and then incubated for 24 or 34 h displayed no distinct changes. However, the radicle tips of seeds primed and incubated for 48 h started to expand and emerge from the germination aperture (Fig. 28.4b). Seeds that were primed and incubated for 24 or 34 h were not affected, but those that were primed and incubated for 48 h were very sensitive to the re-drying rate. When an excessive re-drying rate is used, cracks may form in the radicle tip or cotyledons (Fig. 28.4c), and result in the formation of abnormal seedlings.

Discussion

Priming treatments (i.e. hydropriming and H_2O_2 priming) were shown to effectively enhance the vigour of triploid watermelon seeds by enabling their embryos to overcome the mechanical resistance imposed by the thick seed coat, thereby greatly enhancing their germination capacity. The germination percentage of seeds could

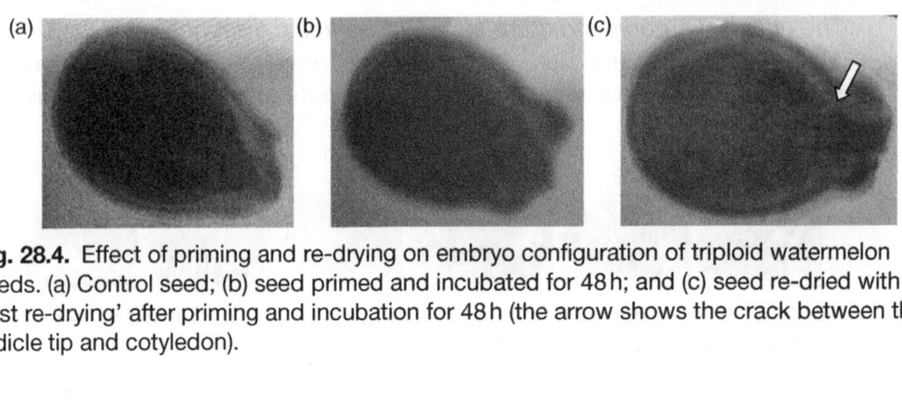

Fig. 28.4. Effect of priming and re-drying on embryo configuration of triploid watermelon seeds. (a) Control seed; (b) seed primed and incubated for 48 h; and (c) seed re-dried with 'fast re-drying' after priming and incubation for 48 h (the arrow shows the crack between the radicle tip and cotyledon).

be increased by more than 60% in some cases, and often approached the potential germination capacity of the cultivar. The H_2O_2 priming treatment combined with either 34 or 48 h incubation at saturated RH gave the best results in the present experiment. The advantage of these treatments is that they do not require the considerable time and effort involved in cracking the seed coat, thereby reducing the handling of triploid watermelon seeds. It is envisaged that these priming procedures could be used for batch processing of large numbers of triploid watermelon seeds, and re-dried seeds could be preserved for use in large-scale production.

All seven priming treatments used in the present research were shown to significantly enhance the germination percentage of triploid watermelon seeds. However, the water content in hydroprimed seeds was higher than in seeds primed by osmotic conditioning. The VI of seeds that were H_2O_2-primed and incubated for 34 h was significantly higher than that of seeds that were H_2O_2-primed and incubated for 48 h. One possible reason for this is that the seed moisture content was at saturation during the priming treatment, thereby increasing the metabolic activity in the seeds, and leading to the initiation of germination if the seeds are then exposed to a prolonged incubation period. In the current experiments, the radicle tips of seeds incubated for 48 h were very sensitive to subsequent seed re-drying. Although the germination percentage of these seeds was not reduced, their VI was lower and a few of the seedlings exhibited an abnormal appearance. Therefore, in this study, an incubation time of 34 h gave the best priming result in order to avoid injury to the seeds during re-drying.

The beneficial effects of the priming treatments are due to the initiation of metabolic activity and induction of DNA cell cycle-related events during priming. Bino *et al.* (1992) observed that the cell cycle progresses to the G_2 phase prior to germination in imbibing tomato (*Lycopersicon esculentum* Mill.) seeds. This accumulation of cells in the G_2 phase takes place because mitosis and cell division do not finish before protrusion of the root tip through the seed coat (Argerich and Bradford, 1989). Moreover, by preconditioning seeds in an osmotic solution (i.e. priming) followed by re-drying, tomato seeds can be stably arrested in the G_2 phase (Bino *et al.*, 1992). Therefore, priming may improve both the rate and the uniformity of seed germination upon subsequent imbibition of water (Heydecker and Coolbear, 1977). This may be due to prior activation of pregerminative processes, including cell cycle activation. In the present study, priming may have induced both DNA replication and metabolic activity, thereby increasing the capacity of the embryo to overcome the mechanical resistance imposed by the seed coat, and accelerating the germination rate and vigour upon subsequent imbibition of water.

It is important to study the mechanisms that underlie barriers to germination in order to find the specific reasons for their occurrence. Bradford (1990) developed a model in which radicle emergence will not occur if ψ_b is higher than the actual seed ψ. However, the lower (more negative) the ψ_b for radicle emergence relative to the ambient ψ, the more rapid will be the germination (Bradford, 1990, 1996). The soft X-ray detection showed that the mechanical resistance imposed by the thick seed coat was the main barrier to the germination of triploid watermelon seeds. Therefore, if the radicle tip of a triploid watermelon seed is going to overcome this barrier, it needs to either accumulate more energy in order to break through the micropylar tissues or reduce the mechanical resistance of the tissues (i.e. reduce ψ_b) by increasing the activity of cell wall-hydrolysing enzymes (such as endo-β-mannanase). In the present study, untreated triploid watermelon seeds could not reach their potential germination percentage, even if they were germinated for 14 days. The priming treatments seem to have lowered the mechanical resistance of the seed coat and decreased the puncture force required by the radicle tips. Therefore, the germination rate and percentage of triploid watermelon seeds were greatly enhanced by these priming treatments.

References

Argerich, C.A. and Bradford, K.J. (1989) The effects of priming and aging on seed vigour in tomato. *Journal of Experimental Botany* 40, 599–607.

Bino, R.J., De Vries, J.N., Kraak, H.L and Van Pijlen, J.G. (1992) Flow cytometric determination of nuclear replication stages in tomato seeds during priming and germination. *Annals of Botany* 69, 231–236.

Bradford, K.J. (1990) A water relations analysis of seed germination rates. *Plant Physiology* 94, 840–849.

Bradford, K.J. (1996) Population-based models describing seed dormancy behaviour: implications for experimental design and interpretation. In: Lang, G. (ed.) *Plant Dormancy: Physiology, Biochemistry, and Molecular Biology*. CAB International, Wallingford, UK, pp. 313–339.

Brocklehurst, P.A. and Dearman, J. (1983) Interactions between seed priming treatments and nine seed lots of carrot, celery and onion. I. Laboratory germination. *Annals of Applied Biology* 102, 577–584.

Demir, I. and Oztokat, C. (2003) Effect of salt priming on germination and seedling growth at low temperatures in watermelon seeds during development. *Seed Science and Technology* 31, 765–770.

Demir, I. and van de Venter, H.A. (1999) The effect of priming treatments on the performance of watermelon (*Citrullus lanatus* (Thunb.) Matsum. & Nakai) seeds under temperature and osmotic stress. *Seed Science and Technology* 27, 871–875.

Duval, J.R. and NeSmith, D.S. (2000) Treatment with hydrogen peroxide and seed coat removal or clipping improve germination of 'Genesis' triploid watermelon. *HortScience* 35, 85–86.

Heydecker, W. and Coolbear, P. (1977) Seed treatments for improved performance: survey and attempted prognosis. *Seed Science and Technology* 5, 353–425.

ISTA (2003) *International Rules for Seed Testing.* International Seed Testing Association, Zurich, Switzerland.

Mehra, V., Tripathi, J. and Powell, A.A. (2003) Aerated hydration treatment improves the response of *Brassica juncea* and *Brassica campestris* seeds to stress during germination. *Seed Science and Technology* 31, 57–70.

Nascimento, W. and West, S.H. (2000) Drying during muskmelon (*Cucumis melo* L.) seed priming and its effects on seed germination and deterioration. *Seed Science and Technology* 28, 211–215.

Thornton, J.M. and Powell, A.A. (1992) Short term aerated hydration for the improvement of seed quality in *Brassica oleracea* L. *Seed Science Research* 2, 41–49.

Tylkowska, K. and van den Bulk, R.W. (2001) Effects of osmo- and hydropriming on carrot (*Daucus carota* L.) seed germination and health. *Seed Science and Technology* 29, 365–375.

Xi, Z.X. (2000) Further exploration for enhancing budding ways of triploid watermelon seeds. *Journal of Changde Teachers University (Natural Science Edition)* 12, 78–83.

Zheng, X.Y., Kong, X.H. and Zheng, G.H. (1986) Preliminary studies on osmotic preconditioning with some vegetable seeds. *Scientia Agricultura Sinica* 2, 36–41.

29 Molecular Mechanisms of Protein Degradation in Germinating Seeds

J. Wang, Y. Miao, Y.C. Tse and L. Jiang

Department of Biology, The Chinese University of Hong Kong, Shatin, New Territories, Hong Kong, China

Abstract

Plants accumulate and store proteins in protein storage vacuoles (PSVs) during seed development and maturation. Upon seed germination, these storage proteins are mobilized to provide nutrients for seedling growth. There are at least three possible mechanisms of protein mobilization in germinating seeds: (i) proteases stored in PSVs during seed development and maturation remain inactive until seed germination commences when stored proteases are activated to degrade the storage proteins in PSVs; (ii) proteases and storage proteins stored in physically separated sub-compartments during seed maturation mix together upon seed germination, leading to storage protein digestion; and (iii) *de novo*-synthesized proteases in response to germination are transported into PSVs through a secretory pathway to participate in protein mobilization. This study focuses on how *de novo*-synthesized proteases are delivered to PSVs for protein degradation. In particular, we were interested in determining the functional roles of vacuolar sorting receptor (VSR) proteins and identifying their ligands in germinating seeds.

Introduction

Plant cells contain lytic vacuoles (LVs) and protein storage vacuoles (PSVs) (Okita and Rogers, 1996). Proteins reach vacuoles because they contain vacuolar sorting determinants (VSDs) that are recognized by vacuolar sorting receptor (VSR) proteins and are mediated by transport vesicles. During seed development and maturation, seeds accumulate large amounts of starch, proteins or lipids, and proteins are stored in a specialized compartment termed a protein body or PSV. Upon seed rehydration and germination, these stored reserves are degraded and used for seedling growth (Bewley and Black, 1994). We are interested in understanding the molecular mechanisms by which proteins are degraded by hydrolytic enzymes within PSVs during seed germination.

 Seeds: Biology, Development and Ecology (eds S. Adkins, S. Ashmore and S.C. Navie)

Seed PSVs and Protein Sorting

Seeds that are rich in proteins usually contain numerous PSVs (Fig. 29.1a). These PSVs are surrounded by a single membrane and can be defined by the presence of abundant levels of tonoplast intrinsic protein (TIP) in their tonoplasts (Fig. 29.1b). When viewed under a transmission electron microscope (TEM), PSVs in most seeds, including tobacco (*Nicotiana tabacum* L.), contain three morphologically and functionally distinct sub-compartments (Fig. 29.1c and d): crystalloid, matrix and globoid (Jiang *et al.*, 2000, 2001, 2002). The matrix and crystalloid contain storage proteins, while the globoid contains salts of phytic acid. However, recent studies have indicated that the globoid, which is surrounded by a single unique membrane, might represent a lytic compartment within the PSV because it contains marker proteins for the LV pathway (Jiang *et al.*, 2001).

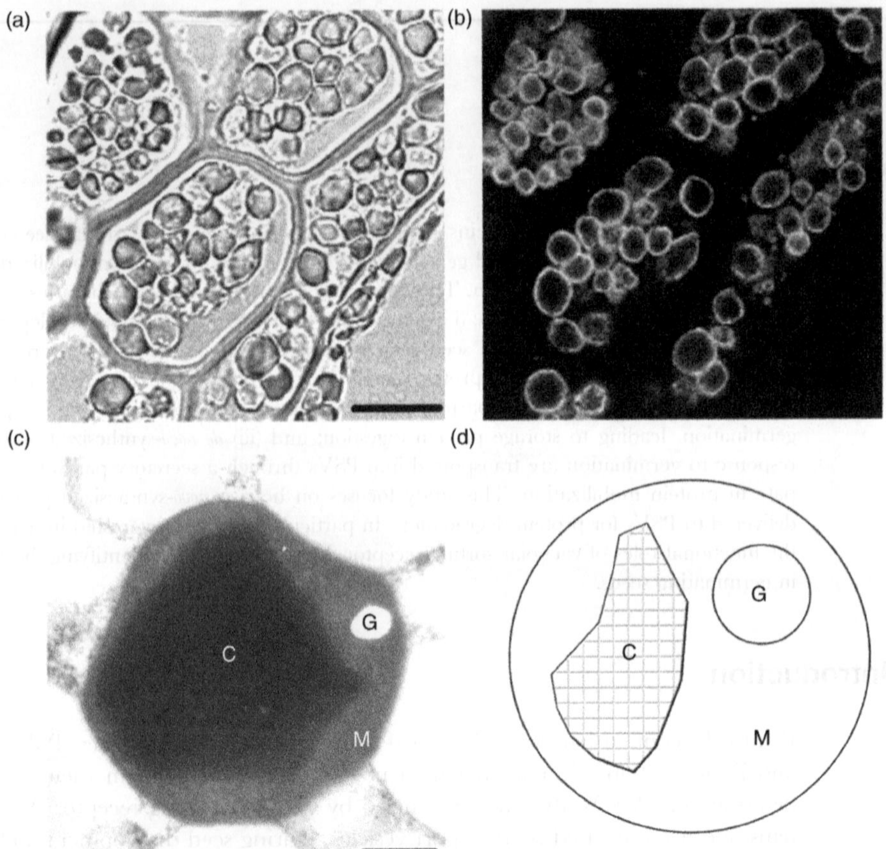

Fig. 29.1. The structure of protein storage vacuoles (PSVs) in mature tobacco seeds. Paraffin-embedded sections were prepared from mature seeds. The top panels demonstrate PSVs in cells visualized by differential interference contrast (a) and by staining with anti-α-TIP visualized with epifluorescence (b). Panel c shows a TEM image of a single PSV that contains three distinct sub-compartments, which is also represented diagrammatically (d). M = matrix; C = crystalloid; G = globoid. Scale bar = 50 μm in a and b and 2 μm in c.

Multiple mechanisms are responsible for transporting storage proteins to PSVs. In rice (*Oryza sativa* L.), storage proteins reach protein bodies either through the Golgi apparatus or directly from the endoplasmic reticulum (ER) (Okita and Rogers, 1996). Aggregation of storage proteins has also been suggested to play a role in transporting the storage protein 7S vicillin to PSVs in pea (*Pisum sativum* L.) (Robinson *et al.*, 1998). Recently, a member of the *Arabidopsis* (*Arabidopsis thaliana* (L.) Heynh.) VSR group of proteins and the *Arabidopsis* receptor-like protein RMR (Jiang *et al.*, 2000), have been shown to function as receptors in the transport of the storage proteins globulin and phaseolin, respectively, to PSVs in *Arabidopsis* (Shimada *et al.*, 2003; Park *et al.*, 2005).

Mechanisms of Storage Protein Degradation in Germinating Seeds

Germination occurs upon rehydration of seeds in suitable environments and temperatures. It is defined by the emergence of the radicle as it breaks through the seed coat, and this marks the onset of seedling growth (Bewley and Black, 1994). During seed germination, and post-germinative seedling growth, storage proteins are degraded by hydrolytic enzymes to generate amino acids to be used by the seedlings. However, little is known about the molecular mechanisms by which the storage proteins in PSVs are mobilized during seed germination.

Figure 29.2 shows three possible mechanisms of protein mobilization in germinating seeds. The first model suggests that both proteases and storage proteins are transported to the same PSV for storage during seed development, but protease activity is inhibited by one or more factors. Upon seed germination the inhibition factors are removed, and this leads to the breakdown of the storage proteins by the activated proteases (Fig. 29.2a). For example, globulin degradation in germinating buckwheat (*Fagopyrum esculentum* Moench) is initiated by a stored metalloendopeptidase, which is inactivated by association with its inhibitor (Elpidina *et al.*, 1991). Its dissociation from the inhibitor during germination may activate the enzyme. In addition, changes in the concentration of free metal cations and/or the pH may contribute to a further control of protease activation during germination, because both endopeptidases and carboxypeptidases are found to be present in the PSVs of dry quiescent seeds (Muntz *et al.*, 2001).

Studies using transgenic tobacco seeds demonstrated that the globoid might represent a lytic compartment within the PSV. The globoid in transgenic mature tobacco seeds contains a unique membrane that is defined by the presence of two marker proteins for the LV: γ-TIP and BP-80 reporter (Jiang *et al.*, 2001). Therefore, during seed development and maturation, proteases could be transported to the globoid and kept physically separated from the storage protein-containing matrix and the crystalloid. Upon seed germination, the broken globoid would release its proteases into the matrix and the crystalloid would serve as a ready source of digestive enzymes to initiate the degradative processes during early germination (Jiang *et al.*, 2001) (Fig. 29.2b).

De novo-synthesized proteases are also known to play a role in mediating protein mobilization in germinating seeds (Fig. 29.2c). In germinating mungbean (*Vigna mungo* (L.) Hepper) seeds, the newly synthesized cysteine protease SH-EP was transported directly from the ER to PSVs through a KDEL-vesicle (KV) compartment for

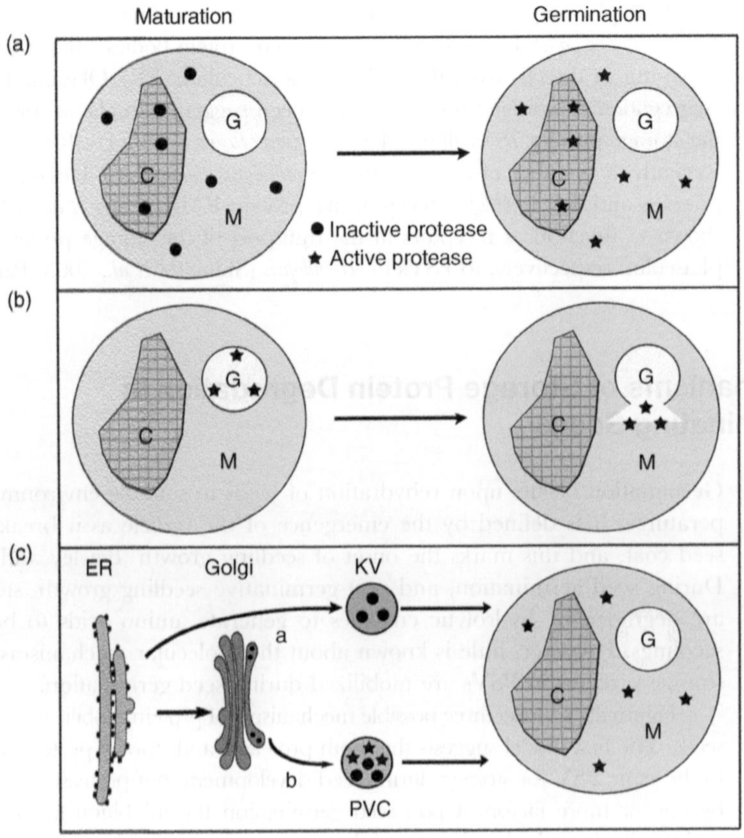

Fig. 29.2. Working models of storage protein degradation during seed germination. There are at least three possible mechanisms responsible for mediating protein degradation within protein storage vacuoles (PSVs) in germinating seeds. (a) Both proteases and storage proteins are transported to the same PSV during seed development and maturation. However, the proteases remain inactive until germination occurs, when rehydration and germination activate the stored proteases to digest the storage proteins within the PSV. (b) Proteases and storage proteins are transported to distinct PSV sub-compartments during seed development, and thus they are kept physically separated. Upon seed germination, the protease-containing globoids are broken and release the proteases into the PSV, which then function to digest the proteins. (c) In response to germination, newly synthesized proteases are transported either (a) directly from the ER to the PSV through a KV compartment or (b) from the ER to the Golgi apparatus and then to the PSV through a prevacuolar compartment (PVC).

protein degradation (Toyooka *et al.*, 2000). Alternatively, VSR proteins may transport newly synthesized proteases through the Golgi apparatus and a prevacuolar compartment (PVC) to the PSV for protein degradation (Fig. 29.2c). This notion is supported by a recent study demonstrating that VSR proteins are associated with seed germination in *Arabidopsis*, because seeds with antisense VSR knockout are not capable of germinating upon rehydration (Laval *et al.*, 2003).

How do Vacuolar Sorting Receptor Proteins Transport Proteases to the Lytic Vacuoles?

De novo-synthesized proteases have to be transported to vacuoles for the digestion of storage proteins upon seed germination. Proteins targeted to the LV are synthesized in the ER, then transported through the Golgi apparatus where they are recognized by VSR proteins, such as the pea BP-80, and then packed into clathrin-coated vesicles (CCVs) for their subsequent delivery to the LV through the lytic PVC (Fig. 29.3a) (Jiang and Rogers, 2003). In tobacco cells, the lytic PVC is a multivesicular body (MVB) enriched with VSR proteins (Li *et al.*, 2002; Tse *et al.*, 2004). The molecular components and their interaction within a CCV are shown in Fig. 29.3b, where the N-terminal (NT) of the receptor interacts with the cargo protein, while its C-terminal (CT) interacts specifically with adaptor.

Current and Future Research

At present, we are interested in determining the functional roles of VSR proteins and identifying their ligands in germinating seeds. Here, we hypothesize that VSR proteins transport newly synthesized proteases through the Golgi apparatus and

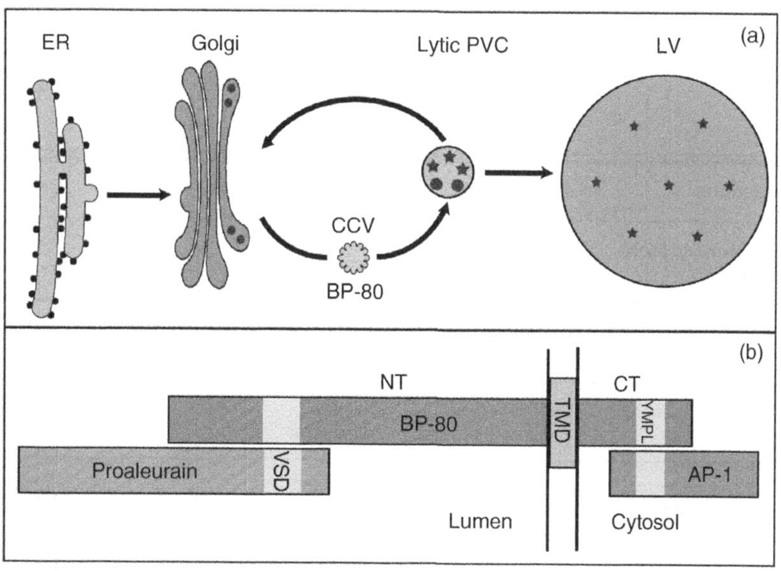

Fig. 29.3. Receptor-mediated transport of proteases to lytic vacuoles (LVs) in plant cells. (a) Proteases are recognized by BP-80, a vacuolar sorting receptor, in the late Golgi apparatus and packed into clathrin-coated vesicles (CCVs) before they are delivered into the lytic vacuole through a lytic prevacuolar compartment (PVC). Mis-sorted proteins can be recycled back to the Golgi apparatus through the PVC. (b) The N-terminal (NT) of the integral membrane receptor is proposed to interact with a cargo protein through a vacuolar sorting determinant (VSD), while its C-terminal (CT) interacts with an adaptor protein (AP) through a conserved tyrosine YMPL motif.

lytic PVCs to PSVs for protein degradation, where the PSV transits into an LV in germinating seeds (Fig. 29.4).

To test this hypothesis, several experiments will be carried out to address the following questions: Are VSR proteins present in both mature and germinating seeds? Where are the VSR proteins localized in mature and germinating seeds? What are the ligands of VSR proteins? How are these ligands transported to their final destinations? Are VSR proteins and their ligands *de novo*-synthesized in germinating seeds? Figure 29.4 summarizes the planned experiments and their predicted results, as well as the conclusions that may be made from them.

As a first step to test this hypothesis a morphological study of PSVs has been undertaken, in which germinating mungbean seeds were collected for fixation and subsequent TEM study over time. The appearance of the PSV changed dramatically from day 0 to 5 (Fig. 29.5). In mature dry seeds (day 0), PSVs have dense

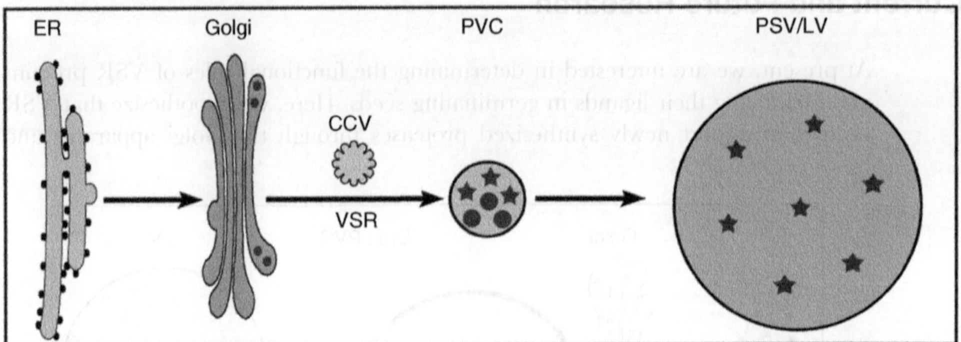

Questions	Approaches	Expected results	Conclusions
1. Do germinating seeds contain VSRs?	Western blot	Presence of VSRs	Germinating seeds contain VSRs
2. Where are VSRs localized?	ImmunoEM and confocal	PVCs	VSRs are localized in MVBs
3. Which proteins interact with VSRs?	Crosslinking and Immunoprecipitation	Identify proteases	Proteases interact with VSRs
4. Which are ligands of VSRs localized?	ImmunoEM and confocal	PVCs and PSVs	Proteases co-localize with VSRs in MVBs
5. Are ligands and VSRs newly synthesized?	Western blot and 2D gel	Absence in mature seeds but presence in germinating seeds	*De novo* synthesis of ligands and VSRs

Fig. 29.4. Hypothesis, experiments and expected results in the current research. The top panel shows the hypothesis that newly synthesized proteases are transported by vacuolar sorting receptors (VSRs) to a protein storage vacuole (PSV), for protein degradation in germinating seeds, through a prevacuolar compartment (PVC). These proteases can be activated upon reaching a PVC or PSV. The table shows the questions being asked to address the hypothesis, and their expected results and conclusions from this research.

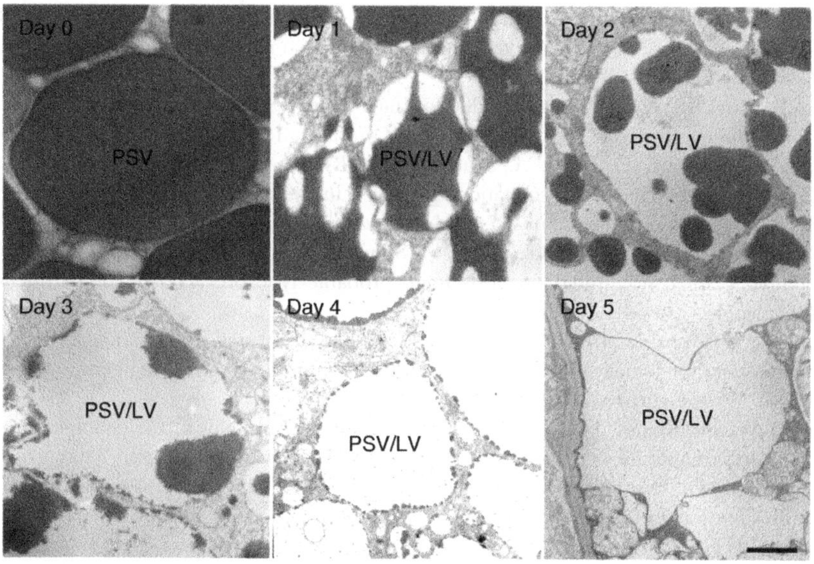

Fig. 29.5. Morphological changes of protein storage vacuoles (PSVs) in germinating mungbean seeds. The mungbean seeds were germinated in water followed by collection and preparation of the samples, at the indicated number of days after imbibition, for morphological study by TEM. Typical examples of the observed changes in PSVs are shown. The digestive PSV is also defined as a lytic vacuole (LV). Scale bar = 2 μm.

contents that represent storage proteins. However, these dense contents gradually become reduced upon germination, from day 1 to 5, and the PSV develops a translucent appearance. These results suggest that the contents of the PSV are being transferred to an LV during seed germination.

Acknowledgements

This work is supported by the Research Grants Council of Hong Kong (CUHK4156/01M, CUHK4260/02M, CUHK4307/03M and CUHK4580/05M) and a Hong Kong University Grants Committee (UGC)-Area of Excellence grant to L. Jiang.

References

Bewley, J.D. and Black, M. (1994) *Seeds: Physiology of Development and Germination*, 2nd edn. Plenum Press, New York.

Elpidina, E.N., Voskoboynikova, N.E., Belozersky, M.A. and Dunaevsky, Y.E. (1991) Localization of a metalloproteinase and its inhibitor in the protein bodies of buckwheat seeds. *Planta* 185, 46–52.

Jiang, L. and Rogers, J.C. (2003) Sorting of lytic enzymes in the plant Golgi apparatus. *Annual Plant Review* 9, 114–140.

Jiang, L., Phillips, T.E., Rogers, S.W. and Rogers, J.C. (2000) Biogenesis of the protein storage vacuole crystalloid. *The Journal of Cell Biology* 150, 755–770.

Jiang, L., Phillips, T.E., Hamm, C.A., Drozdowicz, Y.M., Rea, P.A., Maeshima, M., Rogers, S.W. and Rogers, J.C. (2001) The protein storage vacuole: a unique compound organelle. *The Journal of Cell Biology* 155, 991–1002.

Jiang, L., Erickson, A. and Rogers, J. (2002) Multivesicular bodies: a mechanism to package lytic and storage functions in one organelle? *Trends in Cell Biology* 12, 362–367.

Laval, V., Masclaux, F., Serin, A., Carriere, M., Roldan, C., Devic, M., Pont-Lezica, R.F. and Galaud, J.P. (2003) Seed germination is blocked in *Arabidopsis* putative vacuolar sorting receptor (atbp80) antisense transformants. *Journal of Experimental Botany* 54, 213–221.

Li, Y.B., Rogers, S.W., Tse, Y.C., Lo, S.W., Sun, S.S., Jauh, G.Y. and Jiang, L. (2002) BP-80 and homologs are concentrated on post-Golgi, probable lytic prevacuolar compartments. *Plant Cell Physiology* 43, 726–742.

Muntz, K., Belozersky, M.A., Dunaevsky, Y.E., Schlereth, A. and Tiedemann, J. (2001) Stored proteinases and the initiation of storage protein mobilization in seeds during germination and seedling growth. *Journal of Experimental Botany* 52, 1741–1752.

Okita, T.W. and Rogers, J.C. (1996) Compartmentation of proteins in the endomembrane system of plant cells. *Annual Review Plant Physiology and Plant Molecular Biology* 47, 327–350.

Park, M., Lee, D., Lee, G.J. and Hwang, I. (2005) AtRMR1 functions as a cargo receptor for protein trafficking to the protein storage vacuole. *The Journal of Cell Biology* 170, 757–767.

Robinson, D.G., Baumer, M., Hinz, G. and Hohl, I. (1998) Vesicle transfer of storage proteins to the vacuole: the role of the Golgi apparatus and multivesicular bodies. *Journal of Plant Physiology* 152, 659–667.

Shimada, T., Fuji, K., Tamura, K., Kondo, M., Nishimura, M. and Hara-Nishimura, I. (2003) Vacuolar sorting receptor for seed storage proteins in *Arabidopsis thaliana*. *Proceedings of the National Academy of Sciences of the United States of America* 100, 16095–16100.

Toyooka, K., Okamoto, T. and Minamikawa, T. (2000) Mass transport of proform of a KDEL-tailed cysteine proteinase (SH-EP) to protein storage vacuoles by endoplasmic reticulum-derived vesicle is involved in protein mobilization in germinating seeds. *The Journal of Cell Biology* 148, 453–464.

Tse, Y.C., Mo, B., Hillmer, S., Zhao, M., Lo, S.W., Robinson, D.G. and Jiang, L. (2004) Identification of multivesicular bodies as prevacuolar compartments in *Nicotiana tabacum* BY-2 cells. *The Plant Cell* 16, 672–693.

30 A Role for Reactive Oxygen Species in Endosperm Weakening

K. Müller, B. Hess and G. Leubner-Metzger

Institute for Biology II, Botany/Plant Physiology, Albert-Ludwigs-University Freiburg, Schänzlestr. 1, D-79104 Freiburg i. Br., Germany

Abstract

In many seeds, including the established model species *Arabidopsis* (*Arabidopsis thaliana* (L.) Heynh.) and the emerging *Brassicaceae* model system garden cress (*Lepidium sativum* L.), the process of germination features testa rupture and endosperm rupture as two separate events. Endosperm rupture requires cell wall weakening in the endosperm layer. Puncture force measurements are a useful tool for quantifying this endosperm weakening. We have established such measurements for *L. sativum* and have shown that endosperm weakening takes place prior to endosperm rupture. Various mechanisms have been proposed to promote endosperm weakening. A novel mechanism proposed by us involves the cleavage of cell wall polymers in the endosperm by reactive oxygen species, or more specifically, by apoplastic hydroxyl radicals (•OH) formed when superoxide (O_2^{-}•) and hydrogen peroxide (H_2O_2) undergo a Fenton reaction in the presence of peroxidases.

Introduction

Seed germination and endosperm weakening

The process of seed germination ends when the radicle has protruded through all the seed-covering layers. In many seeds, including economically important species, the endosperm is a germination-limiting tissue barrier. For an endospermic seed to complete germination, the growth potential of the radicle must be high enough to overcome the tissue resistance of the endosperm (Bewley, 1997a). Factors that influence germination, e.g. plant hormones, can thus influence the resistance of the endosperm tissue by promoting or inhibiting endosperm weakening.

In larger seeds, changes in the solidity of the endosperm tissue can be directly quantified by puncture force measurements. This method has proved to be useful

 Seeds: Biology, Development and Ecology (eds S. Adkins, S. Ashmore and S.C. Navie)

in model species like lettuce (*Lactuca sativa* L.) (Abeles, 1986; Sung, 1998) and tomato (*Lycopersicon esculentum* Mill.) (Groot and Karssen, 1987; Chen and Bradford, 2000; Toorop *et al.*, 2000), and has recently been used to characterize changes in the seed coat of coffee (*Coffea arabica* L.) (Da Silva *et al.*, 2004, 2005). In all these species, a decline in the endosperm puncture force precedes endosperm rupture.

Endosperm weakening requires cell wall weakening. This process involves cleavage of cell wall polymers or loosening of bonds between the polymers. Several molecular mechanisms have been proposed for endosperm weakening. Most prominent among them is the digestion of cell wall polysaccharides by hydrolytic enzymes. The main focus so far has been on β-1,4-mannanase (Bewley, 1997b) and β-1,3-glucanase (Leubner-Metzger, 2003). Chen and Bradford (2000) further proposed that expansins also play a role in the endosperm weakening process. It can be safely assumed that endosperm weakening is not caused by one single mechanism, but by interaction of a variety of mechanisms (see Leubner-Metzger *et al.*, Chapter 20, this volume).

Action of reactive oxygen species on cell walls

In addition to cell wall modifying proteins, the action of reactive oxygen species (ROS), such as H_2O_2, O_2^- and ·OH, on cell wall polymers has also been proposed to cause cell wall loosening during fruit softening (Brummell *et al.*, 2004) and elongation growth (Schopfer, 2001). The amount of hemicelluloses, mostly xyloglucan, in cell walls is strongly reduced during auxin-induced elongation growth (Bret-Harte and Talbott, 1993). Schopfer (2001) has postulated that this reduction in hemicelluloses and the corresponding cell wall loosening are caused by ·OH. These radicals are highly reactive and therefore extremely short-lived. As their range of action is limited by diffusion, they must be produced directly in the cell wall in order to cleave cell wall polymers.

The following mechanism has been proposed for the formation of ·OH (Fig. 30.1): NADH oxidases located in the plasma membrane catalyse the formation of apoplastic $O_2^-\cdot$ anions. $O_2^-\cdot$ is dismutated by the antioxidant enzyme superoxide dismutase (SOD), leading to the formation of H_2O_2 and molecular oxygen. Thus $O_2^-\cdot$ and H_2O_2 are both present in the apoplast. In the presence of bivalent cations (e.g. Fe^{2+} or Cu^{2+}), ·OH can be formed from H_2O_2 in the so-called Fenton reaction, and regeneration of these catalytic bivalent cations can be achieved by the oxidation of $O_2^-\cdot$ (Vianello and Macri, 1991). The generation of ·OH in the cell wall by a Fenton-type reaction can also take place in the presence of peroxidases, which are abundant in the plant cell wall (Chen and Schopfer, 1999), or in the presence of ascorbate and bivalent cations (Fry, 1998). ·OH are able to cleave hemicelluloses and have been shown to cause *in vitro* cleavage of cell wall polysaccharides (Fry, 1998; Schweikert *et al.*, 2002). ROS have been shown to play a major role in the elongation growth of maize (*Zea mays* L.) roots, maize coleoptiles and sunflower (*Helianthus annuus* L.) hypocotyls (Liskay *et al.*, 2004). We propose that this mechanism is also involved in endosperm cell wall loosening during endosperm weakening and radicle elongation.

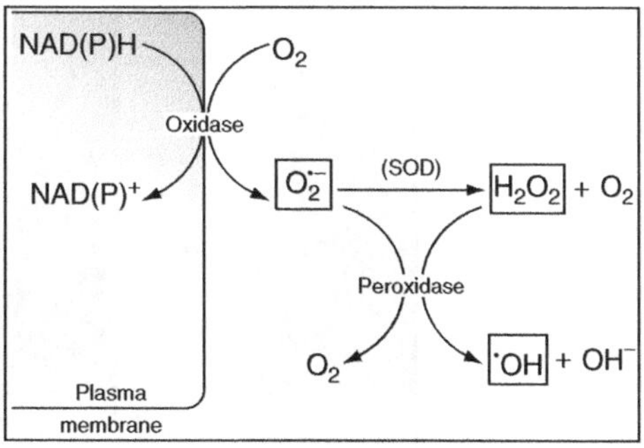

Fig. 30.1. Qualitative model for the generation of apoplastic hydroxyl radicals. NADH oxidases located in the plasma membrane catalyse the formation of apoplastic $O_2^{-}\cdot$ anions. $O_2^{-}\cdot$ is dismutated to H_2O_2 and molecular oxygen by superoxide dismutase (SOD). A Fenton-type reaction can take place in the presence of peroxidases, leading to the formation of ·OH. (From Schopfer *et al*., 2001.)

Materials and Methods

Seeds and germination assays

Lepidium sativum L. 'Gartenkresse einfache' seeds (Juliwa, Heidelberg, Germany) were incubated in Petri dishes containing 6ml 1/10 Murashige-Skoog medium without hormones or vitamins (Duchefa, Haarlem, The Netherlands) and two layers of filter paper. The Petri dishes were sealed with Parafilm, placed in a Sanyo Versatile Environmental Test Chamber MLR-350 (Sanyo, Loughborough, UK) and incubated at 18°C in continuous white light (8.35 μmol/m^2.s). Germination was scored under a binocular microscope. Testa rupture was defined as a stage in which the whitish endosperm was visible through the cracked reddish-brown testa, and endosperm rupture as a stage in which the endosperm cap enveloping the radicle was no longer intact. For each data point, at least three replicates of 50 seeds were used in at least two independent experiments. If indicated, 10 μM (+/−*cis*−*trans*) abscisic acid (ABA; Sigma, Taufkirchen, Germany) and 10 mM H_2O_2 were added to the germination medium.

Puncture force measurements

Puncture force was measured using a custom-made machine (Fig. 30.2a and b). *L. sativum* seeds were cut in half, the embryo and remnants of testa stuck to the endosperm cap were carefully removed and the empty but intact endosperm cap placed in a seed-shaped mould. A metal probe (0.3 mm diameter) was slowly driven into the endosperm cap (2 mm/min) and the force it took to rupture the tissue was measured and registered as a peak on an attached recorder (Fig. 30.2c). These peaks were measured and the corresponding force was calculated. A calibration was performed using defined volumes of water. The method produces a certain background caused by friction between the metal probe and the endosperm. This background can be calculated by using the small peaks that result when endosperm tissues from already ruptured seeds are punctured. This background was subtracted from the individual values. At least 40 seeds from at least two independent experiments were used for each data point.

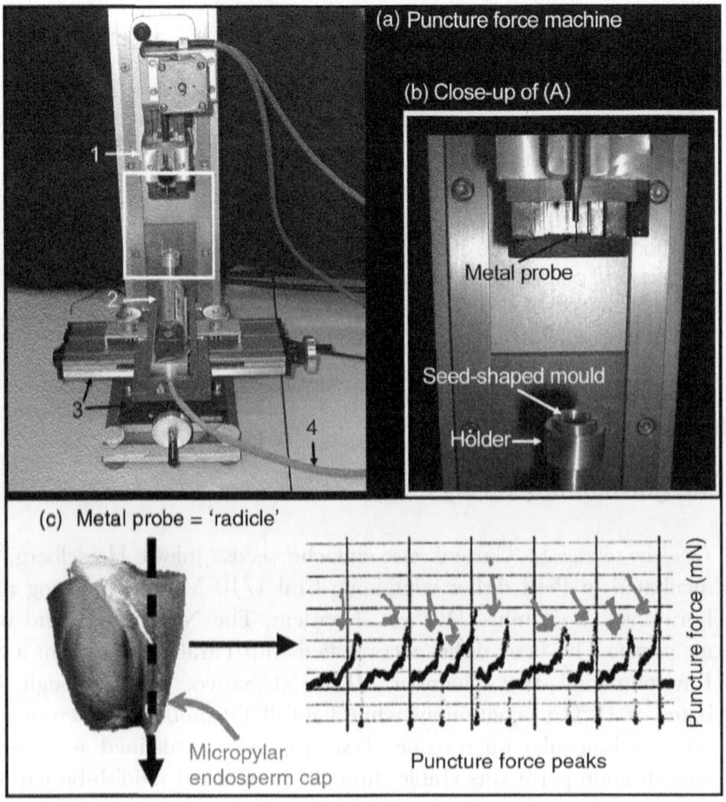

Fig. 30.2. Puncture force measurements. (a) and (b) The custom-made machine that was used to measure puncture force. Its main elements are labelled in the pictures (1 = slide for lowering the metal probe, 2 = strain gauge attached to beam, 3 = centering screws, 4 = connection to recorder). (c) A dissected *Lepidium* seed and a graph showing puncture force peaks. For the measurements, *Lepidium* seeds were cut in half and the embryo and remnants of testa stuck to the endosperm cap were removed, leaving the empty but intact endosperm cap into which the metal probe could be lowered.

Results and Discussion

Puncture force measurements – a useful method requiring a suitable seed system

The choice of the seed system for puncture force experiments strongly influences the experimental options offered by the method. It is essential that the seeds are large enough to be fixed properly and also leave enough space for the metal probe to be lowered on to the tissue with no or very little friction. This shows that the tiny seeds of the model plant *Arabidopsis thaliana* (L.) Heynh. are not suitable for puncture force measurements (see Leubner-Metzger *et al.*, Chapter 20, this volume).

In an optimal seed system for puncture force experiments, it should be possible to puncture the seed coat layers separately in order to assign the weakening to a

Table 30.1. Changes in puncture force during the germination of *Lepidium* seeds. Seeds were incubated at 18°C in continuous white light.

Treatment	Time (h)	Testa rupture (%)	Endosperm rupture (%)	Puncture force (mN)
(a) Control	8	86 ± 4	2 ± 1	38 ± 3[a]
(b) Control	18	99 ± 1	59 ± 1	19 ± 2[a]
(c) 10 μM ABA	18	99 ± 1	0	38 ± 2[a]
(d) 10 μM ABA	48	99 ± 1	0	45 ± 3[b]
(e) 10 Mm ABA + 10 mM H_2O_2	48	99 ± 1	41 ± 1	30 ± 2[b]

(a) and (b) control – puncture force declines over time before endosperm rupture takes place. (c) and (d) 10 μM ABA – ABA delays endosperm rupture and the decline in puncture force. (e) 10 μM ABA plus 10 mM H_2O_2 – the addition of H_2O_2 partially reverses the ABA-induced delay in endosperm rupture and weakening. Mean values ± standard error (SE) are presented.

[a]First seed batch.

[b]Second seed batch.

specific seed coat layer. When more than one covering layer or even the whole seed including the embryo is punctured, as has been done with lettuce (Abeles, 1986), it is hard to tell which layer of the seed is responsible for the observed changes in the puncture force.

Another relevant technical issue is the sensitivity of the measuring system in relation to the actual values measured. While most published results have been produced using commercially available texture-analysing machines, our custom-made machine proved sensitive enough to cope with the approximately three cell layers of the endosperm of *Lepidium* seeds. Our results range from ~40 mN to 20 mN (Table 30.1), while puncture forces measured for harder seed coats, like those for coffee or tomato, range from ~1400 mN to 600 mN (Da Silva *et al.*, 2004; Da Silva *et al.*, 2005) or from ~600 mN to 300 mN (Toorop *et al.*, 2000), respectively. In addition to differences in the sensitivity of the machines used, the size and shape of the metal probe has a significant impact on the absolute values measured. It may therefore be impossible to compare absolute puncture force values from different publications.

Another factor that might influence the choice of a suitable model system for endosperm weakening is the availability of molecular data and applicability of molecular and biochemical methods. Puncture force experiments alone, while delivering valuable results and giving clues to the mechanisms responsible for the observed processes, cannot be used to understand how these processes work. For most seeds on which puncture force measurements have been performed, especially for lettuce and tomato, molecular data and a large range of methods are available.

Lepidium sativum endosperm weakening occurs prior to endosperm rupture

While hardly any molecular data exists for *Lepidium*, it is closely related to *Arabidopsis*, and this offers the possibility of using the extensive molecular data available for the latter. With *Lepidium* seeds, a complete study of endosperm weakening might be possible on different experimental levels, from puncture force measurements to transcriptome analysis using *Arabidopsis* microarray chips. While this has to be proven in our future experiments, it is already clear on the physiological level that *Lepidium* has distinct advantages over other established model systems (see Leubner-Metzger *et al.* Chapter 20, this volume): (i) *Lepidium* exhibits a two-step germination process with separate testa and endosperm rupture, which make it possible, at least after testa rupture, to manipulate the endosperm directly (Fig. 30.3); and (ii) ABA inhibits endosperm rupture, but does not inhibit testa rupture.

We measured the puncture force required at different stages during *Lepidium* seed germination (Table 30.1): (i) control after 8 h when most seeds have undergone testa rupture, but almost no endosperm rupture; (ii) control after 18 h when the majority of seeds have already progressed to endosperm rupture and the remaining seeds will do so within the next 2–4 h; (iii) ABA after 18 h; (iv) ABA after 48 h (the addition of 10 μM ABA delays endosperm rupture, thus none of the seeds has undergone endosperm rupture at 48 h); and (v) after 48 h in 10 μM ABA plus 10 mM H_2O_2, which partially reverts this delay in endosperm rupture.

Endosperm weakening is evident in *Lepidium* seeds prior to endosperm rupture (Table 30.1). ABA delays the onset of both endosperm weakening and endosperm rupture. Table 30.1 shows that the ABA-mediated delay of both endosperm weakening and endosperm rupture can be partially reverted by the addition of H_2O_2. Although 10 mM H_2O_2 did not cause visible oxidative stress responses, concentrations more than 50 mM H_2O_2 provoked visible oxidative stress symptoms in *Lepidium* seeds (i.e. a negative effect on germination and a bluish colour in the emerging radicles due to the presence of stress-induced anthocyanins).

There are no publications investigating the effect of ROS on endosperm weakening, although ROS have been postulated as a molecular mechanism for endosperm weakening (Leubner-Metzger, 2003; Bailly, 2004). Our results suggest that the H_2O_2-induced reversion of the ABA-mediated delay in endosperm rupture is, at least in part, caused by the promotion of endosperm weakening. This evidence supports our hypothesis that ROS are a novel molecular mechanism for endosperm weakening during seed germination.

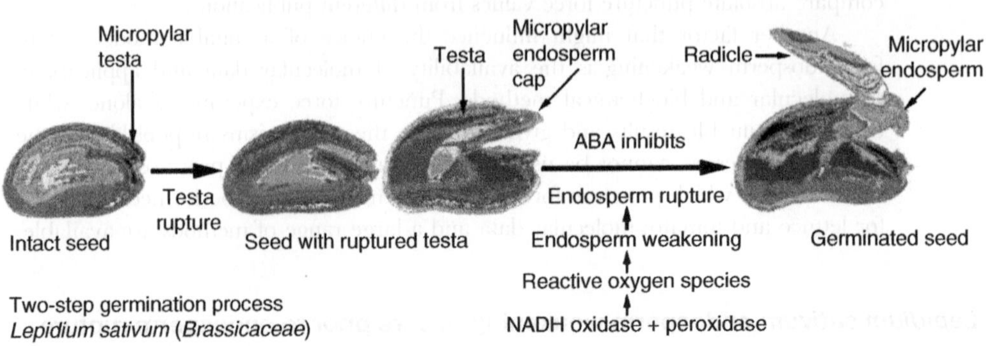

Fig. 30.3. *Lepidium* seed germination is a two-step process (i.e. testa rupture precedes endosperm rupture). ABA delays endosperm rupture, but not testa rupture. It is proposed that reactive oxygen species (ROS) produced in the cell wall play a role in endosperm weakening, which takes place before endosperm rupture.

Reactive oxygen species in seed germination

It is known that H_2O_2 promotes germination in various species. H_2O_2 can be used in high concentrations to promote germination of seeds with hard seed coats by scarification, it also has a germination-promoting effect in lower concentrations. Germination studies on *Zinnia elegans* Jacq. seed showed a dose-dependent promotion of germination by H_2O_2 (Ogawa and Iwabuchi, 2001). Inhibition of catalase in lettuce seeds led to higher concentrations of H_2O_2 in the seeds and to faster germination (Hendricks and Taylorson, 1975).

Our interpretation of this promotion is based on the model presented in Fig. 30.1: H_2O_2 reacts withE $O_2^{-}\cdot$ in the presence of cell wall peroxidases, leading to the formation of ·OH (Chen and Schopfer, 1999). These ·OH then act on the cell wall by causing polysaccharide cleavage resulting in endosperm weakening. This model and our hypothesis for endosperm weakening could also explain why tobacco (*Nicotiana tabacum* L.) seeds overexpressing a cell wall peroxidase germinate faster in the presence of osmotica than the corresponding wild type (Amaya *et al.*, 1999). Peroxidase activity increases in the micropylar endosperm of tomato seeds prior to endosperm rupture (i.e. during endosperm weakening) (Morohashi, 2002).

In addition to its effects on seeds, H_2O_2 promotes elongation growth requiring cell wall loosening in other plant parts. Overexpression of horseradish-peroxidase under the control of the CaMV-35S-promotor leads to faster elongation growth of zucchini (*Cucurbita pepo* L.) hypocotyls (Dunand *et al.*, 2003). In the roots of onion (*Allium cepa* L.), the highest peroxidase activity is found in elongating tissues (Cordoba-Pedregosa *et al.*, 2003). These tissues also show the highest concentrations of H_2O_2. However, H_2O_2 can also have an adverse effect (i.e. growth reduction). For example, addition of ABA to the growth medium leads to a higher activity of peroxidases, a higher concentration of H_2O_2 and reduced growth in the roots of rice (*Oryza sativa* L.) seedlings (Lin and Kao, 2001), and ROS production is essential for lignification and cross-linking of cell wall polymers in vascular tissue (Ogawa *et al.*, 1997).

We propose that cleavage of cell wall polymers by ·OH not only takes place in the endosperm, but also plays a role in radicle elongation. Here, cell walls have to be loosened in order to allow cell elongation, caused by water uptake, which takes place when the water potential in the embryo is lower than that of the surrounding medium.

When working with ROS, it is important to differentiate between ROS signalling effects (Laloi *et al.*, 2004) and direct action, such as the cleavage of cell wall polymers. The only way to prove the latter would be to experimentally show both the presence of ROS and the products of reactions caused by the ROS.

Acknowledgements

We gratefully acknowledge travel support for attending the Eighth International Seed Workshop in Brisbane, Australia, May 2005, by the Wissenschaftliche Gesellschaft Freiburg (WGF) and International Seed Science Society (ISSS) to K. Müller, and by the Deutsche Forschungsgemeinschaft (DFG) to G. Leubner-Metzger. Our research is supported by grants from the DFG and the WGF.

References

Abeles, F.B. (1986) Role of ethylene in *Lactuca sativa* cv. Grand rapids seed germination *Plant Physiology* 81, 780–787.

Amaya, I., Botella, M., de la Calle, M., Medina, M., Heredia, A., Bressan, R., Hasegawa, P., Quesada, M. and Valpuesta, V. (1999) Improved germination under osmotic stress of tobacco plants overexpressing a cell wall peroxidase. *FEBS Letters* 457, 80–84.

Bailly, C. (2004) Active oxygen species and antioxidants in seed biology. *Seed Science Research* 14, 93–107.

Bewley, J.D. (1997a) Seed germination and dormancy. *The Plant Cell* 9, 1055–1066.

Bewley, J.D. (1997b) Breaking down the walls – a role for endo-β-mannanase in release from seed dormancy? *Trends in Plant Science* 2, 464–469.

Bret-Harte, M. and Talbott, L. (1993) Changes in composition of the outer epidermal cell wall of pea stems during auxin-induced growth. *Planta* 190, 369–378.

Brummell, D., Dal Cin, V., Crisosto, C. and Labavitch, J. (2004) Cell wall metabolism during maturation, ripening and senescence of peach fruit. *Journal of Experimental Botany* 55, 2029–2039.

Chen, F. and Bradford, K.J. (2000) Expression of an expansin is associated with endosperm weakening during tomato seed germination. *Plant Physiology* 124, 1265–1274.

Chen, S. and Schopfer, P. (1999) Hydroxyl-radical production in physiological reactions: a novel function of peroxidase. *European Journal of Biochemistry* 260, 726–735.

Cordoba-Pedregosa, M., Cordoba, F., Villalba, J. and Gonzalez-Reyes, J. (2003) Differential distribution of ascorbic acid, peroxidase activity, and hydrogen peroxide along the root axis in *Allium cepa* L. and its possible relationship with cell growth and differentiation. *Protoplasma* 221, 57–65.

Da Silva, E.A.A., Toorop, P.E., van Aelst, A.C. and Hilhorst, H.W.M. (2004) Abscisic acid controls embryo growth potential and endosperm cap weakening during coffee (*Coffea arabica*) seed germination. *Planta* 220, 251–261.

Da Silva, E.A.A., Toorop, P., Nijsse, J., Bewley, J. and Hilhorst, H.W.M. (2005) Exogenous gibberellins inhibit coffee (*Coffea arabica* cv. rubi) seed germination and cause cell death in the embryo. *Journal of Experimental Botany* 56, 1029–1038.

Dunand, C., De Meyer, M., Crévecoeur, M. and Penel, C. (2003) Expression of a peroxidase gene in zucchini in relation with hypocotyl growth. *Plant Physiology and Biochemistry* 41, 805–811.

Fry, S. (1998) Oxidative scission of plant cell wall polysaccharides by ascorbate-induced hydroxyl radicals. *Biochemical Journal* 332, 507–515.

Groot, S.P.C. and Karssen, C.M. (1987) Gibberellins regulate seed germination in tomato by endosperm weakening: a study with gibberellin-deficient mutants. *Planta* 171, 525–531.

Hendricks, S. and Taylorson, R. (1975) Breaking of seed dormancy by catalase inhibition. *Proceedings of the National Academy of Sciences of the United States of America* 72, 306–309.

Laloi, C., Apel, K. and Danon, A. (2004) Reactive oxygen signalling: the latest news. *Current Opinion in Plant Biology* 7, 323–328.

Leubner-Metzger, G. (2003) Functions and regulation of β-1,3-glucanase during seed germination, dormancy release and after-ripening. *Seed Science Research* 13, 17–34.

Lin, C.C. and Kao, C.H. (2001) Abscisic acid induced changes in cell wall peroxidase activity and hydrogen peroxide level in roots of rice seedlings. *Plant Science* 160, 323–329.

Liskay, A., van der Zalm, E. and Schopfer, P. (2004) Production of reactive oxygen intermediates by maize roots and their role in wall loosening and elongation growth. *Plant Physiology* 136, 3114–3123.

Morohashi, Y. (2002) Peroxidase activity develops in the micropylar endosperm of tomato seeds prior to radicle protrusion. *Journal of Experimental Botany* 53, 1643–1650.

Ogawa, K. and Iwabuchi, M. (2001) A mechanism for promoting the germination of *Zinnia elegans* seeds by hydrogen peroxide. *Plant Cell Physiology* 42, 286–291.

Ogawa, K., Kanematsu, S. and Asada, K. (1997) Generation of superoxide anion and localization of CuZn-superoxide dismutase in the vascular tissue of spinach hypocotyls: their association with lignification. *Plant Cell Physiology* 38, 1118–1126.

Schopfer, P. (2001) Hydroxyl radical-induced cell-wall loosening *in vitro* and *in vivo*: implications for the control of elongation growth. *The Plant Journal* 28, 679–688.

Schopfer, P., Plachy, C. and Frahry, G. (2001) Release of reactive oxygen intermediates (superoxide radicals, hydrogen peroxide and hydroxyl radicals) and peroxidase in germinating radish seeds controlled by light, gibberellin, and abscisic acid. *Plant Physiology* 125, 1591–1602.

Schweikert, C., Liszkay, A. and Schopfer, P. (2002) Polysaccharide degradation by Fenton reaction- or peroxidase generated hydroxyl radicals in isolated plant cell walls. *Phytochemistry* 61, 31–35.

Sung, Y. (1998) Using a puncture test to identify the role of lettuce seed coverings on thermotolerant lettuce seed germination. *Journal of the American Society for Horticultural Science* 123, 1102–1106.

Toorop, P.E., van Aelst, A.C. and Hilhorst, H.W.M. (2000) The second step of the biphasic endosperm cap weakening that mediates tomato (*Lycopersicon esculentum*) seed germination is under control of ABA. *Journal of Experimental Botany* 51, 1371–1379.

Vianello, A. and Macri, F. (1991) Generation of superoxide anion and hydrogen peroxide at the surface of plant cells. *Journal of Bioenergetics and Biomembranes* 23, 409–423.

31 Characterization of a Dioxygenase Gene with a Potential Role in Steps Leading to Germination of the Root Parasite *Orobanche aegyptiaca*

D.M. JOEL,[1] H. BAR,[1] A.M. MAYER,[2] V. VERDOUCQ,[3] G. WELBAUM[3] AND J. WESTWOOD[4]

[1]*Department of Weed Research, Agricultural Organization, Newe-Ya'ar Research Center, Ramat Yishai 30095, Israel;* [2]*Department of Botany, The Hebrew University of Jerusalem, Jerusalem 91904, Israel;* [3]*Department of Horticulture, Virginia Tech, Blacksburg, VA 24061, USA;* [4]*Department of Plant Pathology, Physiology and Weed Science, Virginia Tech, Blacksburg, VA 24061, USA*

Abstract

Seeds of parasitic plants in the genus *Orobanche* require a chemical signal in order to germinate. The seeds are, however, receptive only after several days of conditioning in the presence of water, when metabolic activities take place. Using differential display polymerase chain reaction (PCR) analysis, 62 gene fragments were differentially expressed during conditioning, of which 53 were sequenced and 32 resembled known genes. The complementary DNA (cDNA) of a putative dioxygenase, expressed during conditioning, encodes a predicted protein of 365 amino acids. Immunoelectron microscopy located the putative dioxygenase in embryo dictyosomes.

Introduction

Seed germination is a fundamental event in the life of all flowering plants. Many seeds have the ability to detect environmental stimuli such as light, water, temperature and CO_2 levels to ensure that germination occurs only at the most appropriate times. The most specialized and intriguing germination mechanisms belong to some parasitic angiosperms that do not germinate until they detect the presence of a host plant.

Orobanche species are obligate root parasites, devoid of leaves and chlorophyll (dePamphilis and Palmer, 1990). Lacking functional roots of their own, *Orobanche* development is totally dependent on its host plant (Kuijt, 1969; Musselman, 1980). Most angiospermous root parasites produce exceptionally high numbers of seeds, which remain viable and dormant in the soil for many years. The seeds of *Orobanche* are extremely small, between 200 and 400 μm. The embryo is rudimentary, lacking cotyledons and a root cap, and surrounded by endosperm. The radicle develops during germination and can grow 1–4 mm towards a host root. Growth towards the host is believed to be by chemotropism (Parker and Riches, 1993). The parasite radicle penetrates the host root and establishes a vascular connection by means of a haustorium (Joel and Losner-Goshen, 1994; Dörr and Kollmann, 1995).

Seed germination in the *Orobanchaceae* is characterized by two phases: the first is conditioning (sometimes referred to as preconditioning), during which the seeds take up water and show active metabolism, but are unable to respond to germination stimulants; the second is germination itself, which may begin after the termination of conditioning and depends on detection of specific chemicals (Joel *et al.*, 1991). The sharp transition between these two phases appears to be unique to *Orobanche* and similar to root parasites like *Striga* species (Joel *et al.*, 1995).

The optimum environment for conditioning varies among species and even seed populations within species. 'Conditioned' seeds are quiescent, but metabolically active. Their sensitivity to root exudates that stimulate germination increases with conditioning time (Hsiao *et al.*, 1981; Chae *et al.*, 2004; Song *et al.*, 2005).

The rate of seed respiration, which reflects general metabolic activity, dramatically changes during conditioning, with the highest rates occurring in the first few days. A respiration peak was documented 3 days after the onset of conditioning in *Orobanche aegyptiaca* Pers. (Bar-Nun and Mayer, 1993). Respiration rises quite sharply during the first 4 days of conditioning and then decreases markedly. The peak of oxygen uptake was significantly higher when gibberellin was applied to the seeds (Bar-Nun and Mayer, 1993), and there is evidence to prove that gibberellin synthesis takes place during conditioning (Joel *et al.*, 1991; Zehhar *et al.*, 2002), and that exogenous gibberellin promotes seed germination (Nash and Wilhelm, 1960; Hsiao *et al.*, 1988; Chae *et al.*, 2004). Part of the oxygen uptake during conditioning of *O. aegyptiaca* seeds is mediated through the cyanide-resistant pathway, often called the alternative-oxidase (AOX) pathway (Bar-Nun *et al.*, 2003).

To complement the research on seed behaviour during conditioning, a molecular genetic approach was used in order to identify genes expressed during the conditioning of *Orobanche* seeds. The ultimate objective was to characterize this critical period, correlating events in gene expression, protein synthesis and physiological processes to describe temporal as well as spatial processes leading to the state of readiness to perceive the germination signal.

Materials and Methods

Plant material

Seeds of *O. aegyptiaca* were collected by hand from tomato (*Lycopersicon esculentum* Mill.) fields in Israel (Lots #3288 and #3290).

Differential display polymerase chain reaction

Total RNA was extracted from seeds of *O. aegyptiaca* 1, 2, 3, 4, 5, 6 or 8 days after the start of conditioning (imbibition) using phenol-SDS (Sambrook *et al.*, 1989), and poly (A^+) RNA was isolated from total RNA (Oligotex mRNA Spin-Columns, QIAGEN, Valencia, California, USA). The resultant RNA was treated with DNase and used in differential display PCR analysis. Differential display was performed using the restriction fragment differential display protocol of Display Systems Biotech. Briefly, RNA was reverse-transcribed and second-strand synthesis was performed. The double-stranded cDNA was digested with *Taq*I and the adaptors were ligated. Primers specific to these adaptors were used in the PCR. Of them, one primer was universal and acted as the downstream primer. The other primer had a three-base 3' extension that anchored it to a specific cDNA fragment. There were 64 such primers and each was used in combination with the downstream universal primer. PCR was performed at 55°C annealing temperature for 25 cycles using α-^{33}P-ATP as a label. PCR products were run on a 5% polyacrylamide urea gel. After drying, the gels were exposed to Kodak biomax film for 27 h at −70°C. Differentially displayed bands were identified, cut from the gel and placed in 20 µl of PCR buffer (500 mM KCl; 100 mM Tris-HCl, pH 8.3). Cut bands were reamplified by the PCR at 58–60°C for 45 cycles for further analysis. Bands with interesting expression patterns during preconditioning were sequenced and compared to known genes to identify homologues. One such gene encoded a putative dioxygenase that was further characterized.

A full-length version of the dioxygenase cDNA sequence was obtained. RNA was extracted from *O. aegyptiaca* seeds that were conditioned for 3 days. The RNA was treated with DNase and converted to cDNA using a 3' rapid amplification of cDNA ends (RACE) kit (Gibco-BRL). PCR with a dioxygenase-specific primer and abridged universal amplification primer (AUAP) (55°C 1 min, 72°C 1 min, 96°C 1 min; 40 cycles) yielded a band that was cloned and sequenced to determine the 3' end. The 5' end of the gene was identified using 5' RACE. RNA was ligated to an oligonucleotide provided by the kit manufacturer (Invitrogen) and reverse-transcribed. The PCR product (conditions as for 3' RACE) was cloned and sequenced.

Cloning of the full-length dioxygenase from Orobanche

RNA BLOTTING. Total RNA was extracted from *O. aegyptiaca* seeds (as described above) 1, 2, 3, 4, 5, 6 or 8 days after the start of conditioning, and 8 days after conditioning, with the germination stimulator GR-24 added on day 6 (Salzman *et al.*, 1999). For each treatment, 20 µg total RNA was separated by electrophoresis on a 0.8% (w/v) agarose formaldehyde gel and transferred to an uncharged nylon membrane as described by Sambrook *et al.* (1989). The membranes were hybridized in Church buffer (Church and Gilbert, 1984) at 60°C in the presence of digoxygenin-labelled probe for *O. aegyptiaca* dioxygenase. The RNA was finally washed at 23°C in 1 mM ethylenediaminetetraacetic acid (EDTA) and 40 mM $NaHPO_4$. Detection was with CDP-Star™ (Boehringer Mannhein, Indianapolis, Indiana, USA) according to manufacturer's instructions.

DNA HYBRIDIZATION. DNA hybridization analysis was performed on total genomic DNA isolated from *O. aegyptiaca* flowers. The DNA was digested overnight at 37°C with *Sac*I, separated on a 0.8% agarose gel, and transferred to a Hybron-N^+ charged nylon membrane (Amersham Bioscience, Piscataway, New Jersey, USA) according to Sambrook and Russell (2001) with some modifications. The gel was soaked in 250 ml alkaline transfer buffer (0.4 M NaOH and 1 M NaCl) with gentle agitation on a rotary shaker for 15 min at room temperature, followed by an additional 20 min in fresh alkaline transfer buffer. DNA was transferred from the gel to the membrane by capillary action overnight. Following transfer, the membrane was soaked in neutralization buffer (0.5 M Tris-HCl, pH 7.2 and 1 M NaCl) for 15 min at room temperature, and the DNA was immobilized by irradiating the membrane at 254 nm in a Spectrolinker XL-1000 UV crosslinker (Spectronics Corp., Westbury, New York, USA) at 1200 × 100 $\mu J/cm^2$. The probe was labelled with digoxygenin as described above.

The membrane was prehybridized at 57°C with 15 ml of hybridization buffer and 200 µl salmon testes DNA (Sigma). The hybridization buffer contained 5X Denhardt's reagent (Denhardt, 1966; Sambrook and Russell, 2001) (i.e. 0.01% (w/v) ficoll, 0.01% (w/v) polyvinylpyrrolidone and 0.01% (w/v) acetylated bovine serum albumin (BSA)) with 5X saline-sodium citrate (SSC) buffer and 0.5% sodium dodecyl sulphate (SDS). The membranes were prehybridized for 1 h prior to addition of the probe. After blocking, the prehybridization buffer was removed and a fresh buffer was added. The probe was added directly to the buffer and replaced in the hybridization oven (National Lab Net, Edison, New Jersey, USA) at 57°C overnight. The membrane was washed twice with 1 × SSC and 0.1% SDS for 20 min, and then twice with 0.1 × SSC and 0.1% SDS for 15 min at 57°C.

PROTEIN IMMUNODETECTION. The full-length dioxygenase gene was inserted as an *Nde*I *Xho*I fragment into the pET-28a(+) vector (Novagen) and transformed into BL21 $codon^+$ bacteria. Protein expression was induced by the addition of isopropyl-β-D-thiogalactopyranoside (IPTG) (0.2 mM, 4 h). The protein associated with the pellet of extracted cells was extracted with the urea buffer for insoluble protein as directed by a histidine (HIS) tag purification kit (Novagen). Purified protein (3 mg/ml) was used to produce antibodies in chicken (Cocalico Biologicals Inc., Reamstown, Pennsylvania, USA).

Total protein was extracted by homogenization with 4 ml of extraction buffer (i.e. 0.1 M K-phosphate buffer, pH 7.0, 5 mM dithiothreitol and protease inhibitor cocktail Complete™ Mini (Roche Diagnostics, Mannheim, Germany)). The extract was clarified by centrifugation at 10,000 rpm for 15 min. The supernatant was filtered through Miracloth (EMD Bioscience Inc., Darmstadt, Germany) and concentrated using a Centricon YM-10 column (Millipore, Bedford, Massachusetts, USA). Protein concentration was determined by the Bradford assay (Bradford, 1976) using BSA as a standard.

The protein immuno-blotting protocol was as described by Sambrook *et al.* (1989) with some modifications. Samples were boiled in SDS buffer (62.5 mM Tris-HCl (pH 6.8), 2% SDS, 10% glycerol, 5% β-mercaptoethanol (v/v), 0.025% bromophenol blue) for 5 min, and resolved (30 μg of total protein per lane) by 12% SDS-polyacrylaminde gel electrophoresis. The benchmark prestained protein ladder (Invitrogen) was used as molecular standards for protein size determination. Separated proteins were transferred to a 0.2 μm nitrocellulose membrane (Trans-Blot transfer medium, Bio-Rad) overnight (25 V, 4°C) in transfer buffer (25 mM Tris, 192 mM glycine and 20% (v/v) methanol (pH 8.3)) using a Bio-Rad miniblot apparatus. The membrane was blocked with 3% BSA in phosphate buffered saline Tween-20 (PBST) (i.e. 80 mM Na_2HPO4, 100 mM NaCl, 0.3% Tween, pH 7.5) for 1 h at room temperature and then incubated in primary antibody (chicken anti-dioxygenase antibody, diluted 1:100) in 3% BSA in PBST for 1 h, and washed three times for 15 min each in PBST. The membrane was then incubated in secondary antibody, goat anti-chicken IgG alkaline phosphatase conjugated (diluted 1:3000 in PBST, Promega, Madison, Wisconsin, USA). CDP-star™ was used as a chemiluminescent substrate for alkaline phosphatase. The membrane was incubated twice for 5 min in detection buffer (0.1 M Tris-HCl, 0.1 M NaCl, pH 9.5) and then incubated for 5 min in 4 ml CDP-Star™ solution (1:100 dilution in detection buffer) containing 200 μl Nitroblock Enhancer II (Tropix, Bedford, Massachusetts, USA).

ELECTRON MICROSCOPY. For electron microscopy, plant material was fixed in 2% paraformaldehyde and 2.5% glutaraldehyde in 0.1 M sodium cacodylate buffer pH 7.2 for 2 h, and embedded in London resin (LR) white. The material was sectioned with an LKB ultramicrotome (LKB, Bromma, Sweden), and stained with uranyl acetate and lead citrate (Hall and Hawes, 1991), before examination under an electron microscope (Jeol 100CX at 80 KV). Immunostaining followed the methods used by Losner-Goshen *et al.* (1998).

Results and Discussion

Gene expression

Differential display PCR analysis revealed 92 DNA fragments that were differentially expressed over the course of seed conditioning. A total of 62 bands were reamplified and 53 fragments were sequenced (Table 31.1). From these candidates, for further investigation, we selected a gene fragment showing higher expression after 3 days of conditioning, and whose sequence showed similarity to several known dioxygenase genes.

Because differential display PCR only produced a gene fragment, 5'-RACE and 3'-RACE were used to obtain additional sequence information. We recovered a 1400 bp section of cDNA containing this gene, which encodes a predicted protein of 365 amino acids. Alignment of the predicted protein sequence with others in the National Center for Biotechnology Information (NCBI) database indicated a strong homology to a putative *Arabidopsis thaliana* (L.) Heynh. 1-aminocyclopropane-1-carboxylate oxidase (ACCase)-like protein (50% identity; 67% similar amino acids)

Table 31.1. Details of bands sequenced from differential display PCR.

Band	Sequence homology	Expression pattern[a]				Size (bp)
		2	3	4	6	
1	Unknown		–		+	226
2	Unknown		+		++	210
3	Unknown		+		++	258
4	Unknown		–		++	89
5	Unknown		–		++	184
6	Putative dioxygenase (ACC oxidase)		++		+	570
7	Ribosomal protein		+		++	400
8	*A.t* protein with Leu-rich repeat of unknown function		+		++	309
9	*A.t* 2-oxoglutarate dh E2 subunit		+		++	384
10	Unknown		++		+	245
11	Unknown	+	+	+	+	117
12	Unknown		++		+	230
13	Unknown	+	+	+	+	74
14	Unknown		+		+	402
15	Unknown	–	+	+	++	215
16	Unknown	+	+	+	+	173
17	Similar to *A.t* chromosome II	+	++		+++	236
18	Similar to nucleoporin-like protein (*A.t*)		+		++	291
19	Similar to putative phosphate/PEP translocator protein (*A.t*)		+		++	198
20	Similar to putative tRNA synthetase or oligopeptide transporter (*A.t*)	+	+	+	+	287
21	Unknown	+	+	+	+	222
22	Unknown	++	++	+	+	416
23	Unknown					305
24	Few similarities with prolyl endopeptidase (*A.t*) and nodulin-like protein	++	+++	++	+	153
25	Similar to glycine rich cell wall protein precursor (*A.t*)	–	–	++	++	250
26	Similar to phosphate/phosphoenolpyruvate translocator protein (pb)	–	–	–	++	198
27	Similar to *A.t* chromosome IV		+		–	354
28	Similar to *A.t* chromosome II and human retroelement transcriptase	++	+++	–	–	86
29	Similar to chromosome III (*A.t*) and Gbox binding protein	+	+	+	+	293
30	Putative aminopeptidase	++	+++	+	++	322
31	(Same as 29)	+	+	+	+	293
32	Ribosomic RNA 28S	++	++	+	+	299
33	Unknown	+	+	–	+++	433
34	Similar to receptorkinase-like protein		+		++	243
35	Unknown		+++		–	236
36	Similar to *A.t* chromosome IV		+	+++	+++	398
37	Similar to unknown protein of *Saccharomyces cerevisiae*	+	+	+	++	385

Continued

Table 31.1. *Continued*

Band	Sequence homology	Expression pattern[a]				Size (bp)
		2	3	4	6	
38	Similar to CHP rich zinc protein, to human genome, to *A.t* chromosome I and glutathione peroxydase mRNA	+	++	++	++	176
39	Similar to human genome		+		+	191
40	Similar to *A.t* chromosome IV		+		+	278
41	Similar to *A.t* BAC		+		++	280
42	Similar to *A.t* chromosome I YAC	+	++	++	–	372
43	Similar to *A.t* chromosome I and zinc finger helicase	–	–	–	+	229
44	Similar to *A.t* chromosome IV, to *Caenorhabitis* and *Nicotiana tabacum*	+	+	++	++	353
45	Similar to *A.t* chromosome I and to human tyrosine kinase	+	+	+	++	184
46	Similar to *Mus musculus* adult male thymus cDNA, to *Homo sapiens*, to *A.t* chromosome V	–	+	–	++	215
47	Unknown	+	++	++	–	219
48	Similar to *A.t* chromosome I and human DNA	++	+	+	+	336
49	Unknown	+	++	–	+++	289
50	(Same as 45)	+	+	+	++	314
51	Similar to P-glycoprotein-like protein pgp3, mdr multi-drug resist. protein (*A.t*) and *A.t* chromosomes III and V	–	–	–	++	428
52	Similar to *Caulobacter crescentus* section 247	–	–	–	+	435
53	Similar to *A.t*chromosome V and to human sequence	+	++	–	+++	462

Expression code: no expression –, some expression +, high expression ++, maximum expression +++.
A.t = *Arabidopsis thaliana*.
[a]Days of conditioning.

and to an oxidoreductase (50% identity; 65% similar amino acids). Given the role of ACCase in the last step in the synthesis of the plant hormone ethylene, this gene may prepare the seed for germination during conditioning. Another possibility is that it belongs to the 2-oxoacid-dependent group of dioxygenases, which are known to be involved in the biosynthetic pathway of gibberellic acid (GA) in various plant systems (Prescott and John, 1996). Since GA biosynthesis was demonstrated in the seed during conditioning (Joel *et al.*, 1991; Zehhar *et al.*, 2002), we cannot exclude the possibility that *Orobanche* dioxygenase may be involved in GA synthesis.

An important task in characterizing the putative *Orobanche* dioxygenase was to independently confirm its expression pattern during conditioning. We used two approaches to do this, the first being real-time PCR to quantify mRNA based on the kinetics of PCR amplification. The advantage of this technique is that it uses extremely low amounts of RNA, which was ideal considering the limited yield from *Orobanche* seeds. However, the technique may also produce inconsistent results because the PCR reactions can be sensitive to very small variations in the

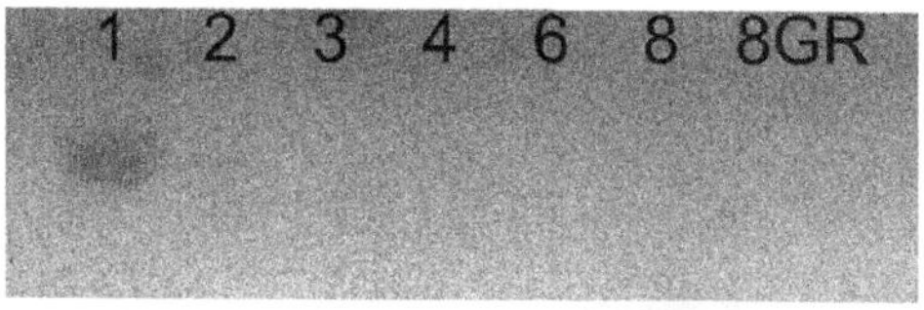

Fig. 31.1. RNA blot hybridization analysis to show expression of *Orobanche aegyptiaca* dioxygenase during conditioning. Lanes represent 1, 2, 3, 4, 6 and 8 days of conditioning. The lane labelled 8GR had the germination stimulant GR-24 added to seeds on day 6.

system. An actin gene from *O. aegyptiaca* was cloned based on homology to actin from the related parasite, *Striga* species, using PCR to serve as an internal control (i.e. a second gene that does not change between time points) to normalize starting RNA concentrations. Comparison of the dioxygenase to actin and the unknown *O. aegyptiaca* genes indicated greater expression of the dioxygenase gene on day 3 as opposed to day 6 of conditioning (results not shown).

RNA blot hybridization analysis was also performed to verify the results from real-time PCR. This traditional method requires larger amounts of RNA, which was an additional challenge considering the small size of *Orobanche* seeds. The dioxygenase gene was originally selected because it showed higher expression at day 3 of conditioning than at day 6 by differential display PCR. RNA hybridization analysis suggests that the greatest expression of the putative dioxygenase gene was actually on the first day of conditioning (Fig. 31.1). The signal was barely detectable by this method on day 2 and not detected thereafter. While the different techniques yielded different expression patterns, both indicated that the dioxygenase is expressed during *Orobanche* seed conditioning.

To investigate dioxygenase protein levels in *O. aegyptiaca* seeds during conditioning, monoclonal antibodies were generated to the dioxygenase fusion protein and used to probe protein extracts from conditioned seeds. Western blotting revealed that dioxygenase increased with seed hydration and could be detected on day 1 of conditioning but not in dry seeds (data not shown). Protein expression remained consistent throughout conditioning with the highest concentrations on day 6. However, three distinct bands of different molecular weight were visible during conditioning so the monoclonal antibody may have lacked specificity (data not shown).

Because very little is known about the *O. aegyptiaca* genome, we also wanted to know how many copies of the dioxygenase gene were present. To determine the number of copies of the dioxygenase gene in *O. aegyptiaca*, total genomic DNA was probed with a fragment of the dioxygenase gene. Results indicated a single copy of the dioxygenase gene (Fig. 31.2), but at least four homologous genes also cross-hybridized with the probe.

Immuno-localization of dioxygenase in the seed

The immunocytochemical localization of the putative dioxygenase protein using the monoclonal antibody described above showed immunogold labelling in cisternae and vesicles of dictyosomes in the embryo (Fig. 31.3). Similar localization of

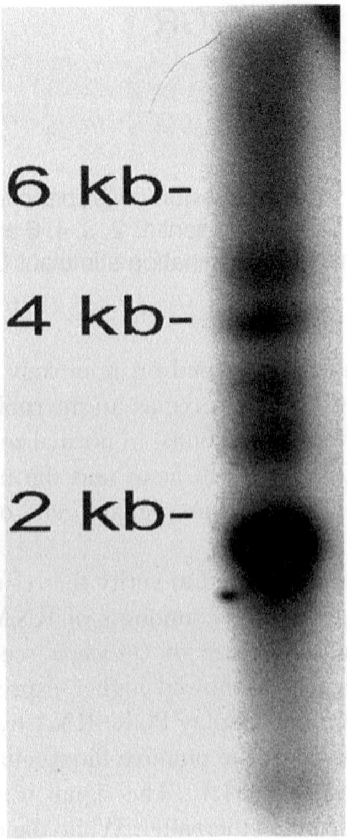

Fig. 31.2. DNA hybridization of total *Orobanche aegyptiaca* genomic DNA digested with *Sac*I. The probe was made from dioxygenase RNA.

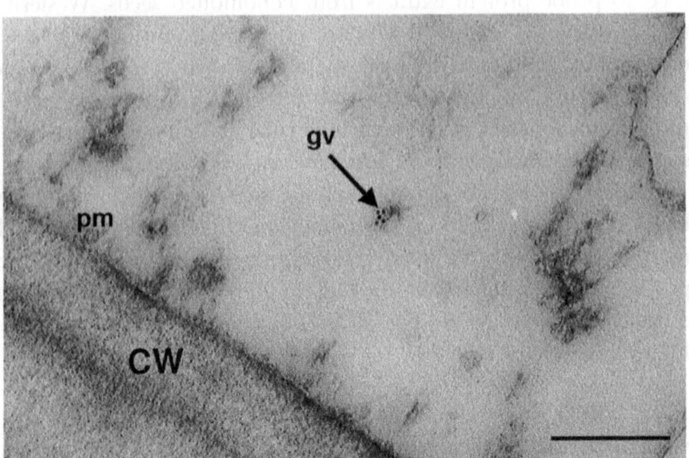

Fig. 31.3. Immuno-localization of the dioxygenase protein in embryo cell of *Orobanche aegyptiaca* during seed conditioning. gv = Golgi vesicle, CW = cell wall, pm = plasma membrane. Scale bar = 5 μm.

dioxygenases in dictyosomes and Golgi vesicles was previously described by Wojtaszek *et al.* (1999) for *Phaseolus vulgaris* L. and in some other organisms. Using immunoelectron microscopy and confocal laser scanning microscopy Neve *et al.* (1999, 2003) showed, for example, that the intracellular localization of the *Tetrahymena* homolog of 4-hydroxyphenylpyruvate dioxygenase is associated with membranes of the Golgi apparatus and its transport vesicles.

Limited resources have not allowed us to characterize additional genes. Differential display PCR is a laborious technique not well suited for characterizing all the genes expressed during *Orobanche* seed conditioning. However, Table 31.1 gives some idea of the possible genes that may be involved. Other genes in addition to the dioxygenase described above might eventually be characterized to give a complete representation of how seed conditioning occurs and enable the development of strategies to control this costly parasitic weed.

Acknowledgements

We thank Star Gao for her technical assistance. This work was supported by US-Israel Binational Science Foundation (BSF) grant no. 1999372. Additional support was provided by NRICGP 2001-35320-10900 and USDA Hatch project no. 135657 to J. Westwood.

References

Bar-Nun, N. and Mayer, A.M. (1993) Preconditioning and germination of *Orobanche* seeds: respiration and protein synthesis. *Phytochemistry* 34, 39–45.

Bar-Nun, N., Plakhine, D., Joel, D.M. and Mayer, A.M. (2003) Changes in the activity of the alternative oxidase in *Orobanche* seeds during conditioning and their possible physiological function. *Phytochemistry* 64, 235–241.

Bradford, M. (1976) A rapid and sensitive method for the quantification of microgramme quantities of protein utilizing the principle of protein-dye binding. *Analytical Biochemistry* 72, 248–254.

Chae, S.H., Yoneyama, K., Takeuchi, Y. and Joel, D.M. (2004) Fluridone and norflurazon, carotenoid-biosynthesis inhibitors, promote seed conditioning and germination of the holoparasite *Orobanche minor*. *Physiologia Plantarum* 120, 328–337.

Church, G.M. and Gilbert, W. (1984) Genomic sequencing. *Proceedings of the National Academy of Sciences of the United States of America* 81, 1991–1995.

Denhardt, D.T. (1966) A membrane-filter technique for the detection of complementary DNA. *Biochemistry and Biophysics Research Communications* 23, 641–646.

dePamphilis, C.W. and Palmer, J.D. (1990) Loss of photosynthetic and chlororespiratory genes from the plastid genome of a parasitic flowering plants. *Nature* 348, 337–339.

Dörr, I. and Kollmann, R. (1995) Symplasmic sieve element continuity between *Orobanche* and its host. *Botanica Acta* 108, 47–56.

Hall, J.L. and Hawes, C. (1991) *Electron Microscopy of Plant Cells*. Academic Press, London.

Hsiao, A.I., Worsham, A.D. and Moreland, D.E. (1981) Effects of sodium hypochlorite and certain plant growth regulators on germination of witchweed (*Striga asiatica*) seeds. *Weed Science* 29, 98–100.

Hsiao, A.I., Worsham, A.D. and Moreland, D.E. (1988) Effects of chemicals often regarded as germination stimulants on seed conditioning and germination of witchweed (*Striga asiatica*). *Annals of Botany* 62, 17–24.

Joel, D.M. and Losner-Goshen, D. (1994) The attachment organ of the parasitic angiosperms *Orobanche cumana* and *O. aegyptiaca* and its development. *Canadian Journal of Botany* 72, 564–574.

Joel, D.M., Back, A., Kleifeld, Y. and Gepstein, S. (1991) Seed conditioning and its role in *Orobanche* seed germination: inhibition by paclobutrazol. In: Wegmann, K. and Musselman, L.J. (eds) *Progress in* Orobanche *Research: Proceedings of the International Workshop on* Orobanche *Research*. Eberhard-Karls-Universitat, Tubingen, Germany, pp. 147–156.

Joel, D.M., Steffens, J.C. and Matthews, D.E. (1995) Germination of weedy root parasites. In: Kigel, J. and Galili, G. (eds) *Seed Development and Germination*. Marcel Dekker, New York, pp. 567–598.

Kuijt, J. (1969) *The Biology of Parasitic Flowering Plants*. University of California Press, Berkeley, California.

Losner-Goshen, D., Portnoy, V., Mayer, A.M. and Joel, D.M. (1998) Pectolytic activity by the haustorium of the parasitic plant *Orobanche* L. (Orobanchaceae) in host roots. *Annals of Botany* 81, 319–326.

Musselman, L.J. (1980). The biology of *Striga*, *Orobanche*, and other root-parasitic weeds. *Annual Review of Phytopathology* 18, 463–489.

Nash, S.M. and Wilhelm, S. (1960) Stimulation of broomrape seed germination. *Phytopathology* 50, 772–774.

Neve, S., Aarenstrup, L., Tornehave, D., Rahbek-Nielsen, H., Corydon, T.J., Roepstorff, P. and Kristiansen, K. (2003) Tissue distribution intracellular localization and proteolytic processing of rat 4-hydroxyphenylpyruvate dioxygenase. *Cell Biology International* 27, 611–624.

Neve, S., Tornehave, D., Corydon, T.J. and Kristiansen, K. (1999) The *Tetrahymena* homolog of bacterial and mammalian 4-hydroxyphenylpyruvate dioxygenases localizes to membranes of the endoplasmic reticulum. *Cell Biology International* 23, 719–728.

Parker, C. and Riches, C.R. (1993) *Parasitic Weeds of the World, Biology and Control*. CAB International, Wallingford, UK.

Prescott, A.G. and John, P. (1996) Dioxygenases: molecular structure and role in plant metabolism. *Annual Review of Plant Physiology and Plant Molecular Biology* 47, 245–271.

Salzman, R.A., Fujita, T., Zhu-Salzman, K., Hasegawa, P.M. and Bressan, R.A. (1999) An improved RNA isolation method for plant tissues containing high levels of phenolic compounds or carbohydrates. *Plant Molecular Biology Reporter* 17, 11–17.

Sambrook, J., Fritsch, E.F. and Maniatis, T. (1989) *Molecular Cloning: A Laboratory Manual*, 2nd edn. Cold Springs Harbor Laboratory Press, Cold Springs Harbor, New York.

Sambrook, J. and Russell, D.W. (2001) *Molecular Cloning: A Laboratory Manual*, 3rd edn. Cold Springs Harbor Laboratory Press, Cold Springs Harbor, New York.

Song, W.J., Zhou, W.J., Jin, Z.L., Cao, D.D., Joel, D.M., Takeuchi, Y. and Yoneyama, K. (2005) Germination response of *Orobanche* seeds subjected to conditioning temperature, water potential and growth regulator treatments. *Weed Research* 45, 467–476.

Wojtaszek, P., Smith, C.G. and Bolwell, G.P. (1999) Ultrastructural localisation and further biochemical characterisation of prolyl 4-hydroxylase from *Phaseolus vulgaris*: comparative analysis. *International Journal of Biochemistry and Cell Biology* 31, 463–477.

Zehhar, N., Ingouff, M., Bouya, D. and Fer, A. (2002) Possible involvement of gibberellins and ethylene in *Orobanche ramosa* germination. *Weed Research* 42, 464–469.

32 Computer Imaging to Assess Seed Germination Performance

M.B. McDonald

Department of Horticulture and Crop Science, The Ohio State University, Columbus, OH 43210, USA

Abstract

Seed vigour testing provides valuable information for assessing seed lot quality. However, vigour testing has not experienced widespread use because it is labour-intensive, high in cost and the test results often vary from laboratory to laboratory. An automated seed vigour assessment system is presented for lettuce (*Lactuca sativa* L.), soybean (*Glycine max* (L.) Merr.) and maize (*Zea mays* L.) that is objective, economical and easy to perform. The system interfaces a flatbed scanner, which captures digital images of germinating seedlings, with a computer. The images are processed by a computer to generate numerical values (i.e. a vigour index) that collectively represent the quality of a seed lot based on various statistics acquired from morphological features of the imaged seedlings. These statistics indicate the speed and uniformity of seedling development. They include the sample mean of hypocotyl and radicle lengths, as well as the sample standard deviation of the hypocotyl length, radicle length, total length (hypocotyl length plus radicle length) and radicle-to-hypocotyl length ratio. The system was tested on lettuce, soybean and maize seedlings grown for 3 days in the dark. The results indicated that the imaging system accurately quantified those parameters to yield reproducible, objective vigour assessments.

Introduction

Seed vigour is a quantitative and qualitative value that describes the quality of a seed lot and can be based on sampled observations of seedling growth. Traditionally, seed analysts determine seed vigour by visual inspection of the speed and uniformity of seedling growth or manual measurements of certain seed/seedling features. Various specifications for seed vigour testing exist, including those listed in the Association of Official Seed Analysts' (AOSA) *Vigor Testing Handbook* (AOSA, 1983) and the International Seed Testing Association (ISTA) *Seed Vigor Testing Handbook* (Hampton and TeKrony, 1995).

The objectives of this study were to: (i) develop an imaging platform that could be rapidly adapted in a routine seed testing laboratory at little cost to enhance the standardization of vigour testing; and (ii) establish an imaging platform that could capture multiple images of lettuce (*Lactuca sativa* L.), soybean (*Glycine max* (L.) Merr.) and maize (*Zea mays* L.) seedlings enabling simultaneous measurements of hypocotyl, radicle and overall seedling length that could be further processed by more advanced computer methods to minimize human intervention, as well as increase accuracy and reproducibility in seedling measurements for vigour assessment.

Materials and Methods

For seed vigour determinations of lettuce, soybean and maize the following procedures were used.

Germination

Germination procedures for lettuce (Sako *et al.*, 2001), soybean (Hoffmaster *et al.*, 2003) and maize (Hoffmaster *et al.*, 2005) were as described earlier.

Image acquisition

An Epson GT-15000 flatbed scanner with a scanning area of 43 × 29 cm was used to acquire all seedling images. The flatbed scanner offered several advantages over other imaging devices, such as a video camera or a digital camera (McDonald *et al.*, 2001).

Software processing of the images

After seedling images were acquired, they were processed by a software developed by Sako (2000) for lettuce, Hoffmaster (2002) for soybean and Hoffmaster *et al.* (2005) for maize. On the basis of the separation point and skeleton for lettuce, the lengths of the hypocotyl and the radicle could be computed for each seedling. In the case of soybean, root hairs were not present to mark the separation point, so the entire seedling length was computed. For maize, the entire growth of the seedling was determined (i.e. primary roots, seminal roots and coleoptile).

Seedling length determinations

Lettuce

After measurements of the hypocotyl and the radicle were made for all seedlings, the results were combined to obtain a vigour index. The index for lettuce was defined as follows:

$$\text{Vigour} = w_G \times \text{growth} + w_U \times \text{uniformity}$$

where the w's represented associated weights with the parameters being multiplied;

$$\text{Growth} = \min(w_h \times \bar{l}_h + w_r \times \bar{l}_r, 1000)$$

where the maximum value is 1000 and $\bar{l}_h$ and $\bar{l}_r$ are the sample means of the hypocotyl length and the radicle length and

$$\text{uniformity} = \max\,(1000 - (w_h \times s_h + w_r \times s_r + w_{\text{total}} \times s_{\text{total}} + w_{r/h} \times s_{r/h}) - w_{\text{d}} \times \text{numdead}, 0)$$

where numdead refers to the number of dead seeds, the minimum value is 0, and s_h, s_r, s_{total} and $s_{r/h}$ are the sample standard deviations of the hypocotyl length, radicle length, total length and the ratio of the hypocotyl and radicle lengths.

Soybean

Before processing, soybean seedlings were classified into alphabetical types based on the characteristics of their skeletons. The six types recognized by the program were the I, T, Y, P, A and M seedling types and are also examples of differing seedling shapes (Hoffmaster, 2002; Hoffmaster *et al.*, 2003).

Maize

Maize seedlings possess a differing root structure that is composed of multiple roots (i.e. seminal and primary roots) connected to each seed when compared to soybean seedlings. The multiple roots present connection problems to the seed because of shadows and create an overlap problem in determining which root belongs to which seed. To address these issues, a different approach was used for maize seedlings than for lettuce and soybean seedlings.

As a first step, maize seeds were identified based on their yellowish-red colour. To accomplish this, a synthesized image of maize seeds was used as a training image. In SUV colour space, the range of s-, u- and v-value colours that maize seeds fall into was established. The red, green, blue (RGB) maize image was then transformed into SUV colour space.

Roots were separated by their white colour against the brown background of the germination paper. When disconnected, roots were encountered and assigned to the seed that was nearest to it.

Results and Discussion

Lettuce

The vigour assessment system was used to numerically evaluate four lettuce seed lots, with different levels of seed vigour, 3 days after the initiation of germination (Fig. 32.1). The software marked hypocotyls in red and radicles in green. For most seedlings, these features were detected correctly. For the purpose of this study, lettuce was selected as a model species because its seeds typically produce seedlings that are straight with only one primary root. Moreover, lettuce has seedling structures that are well defined: hypocotyl (shoot) and radicle (root). Furthermore, it has been reported that both embryo elongation and germination rates are good

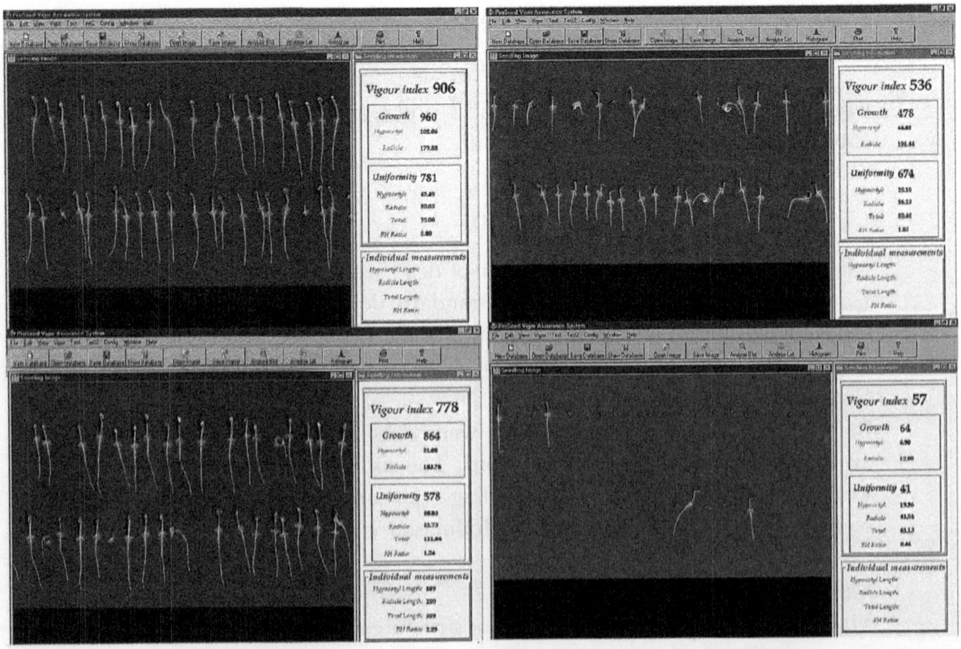

Fig. 32.1. Vigour test results for four different lettuce seed lots 3 days after the initiation of germination. Note the differences in the overall vigour index values and their relationship with seed lot performance.

predictors of lettuce seed vigour (Smith *et al.*, 1973; Wurr and Fellows, 1985; Hacisalihoglu *et al.*, 1999). Both Penaloza *et al.* (2005) and Contreras and Barros (2005) utilized the seed vigor imaging system (SVIS) and showed that the results correlated better than any other vigour test with lettuce seedling emergence in the greenhouse. Although this system was developed specifically for lettuce, it serves as a foundation for building automated vigour assessment systems for other species with similar seedling structures.

Soybean

By using the program on a sample image of 50 soybean seedlings 3 days after the initiation of germination, the following pictorial results were obtained (Fig. 32.2). A green line overlaid on each normal seedling reflected the length measured for each seedling. Abnormal seedlings were marked in red. The speed and uniformity of growth values and the vigour index were shown in the output window below the seedling image (Fig. 32.2). Correlation coefficients have been run on standard germination, early growth rate, accelerated ageing, cold test, field emergence, and SVIS 3-day vigour, SVIS 3-day uniformity, SVIS 3-day growth and accelerated ageing SVIS vigour, growth and uniformity for 169 soybean seed lots. All vigour tests except accelerated ageing vigour,

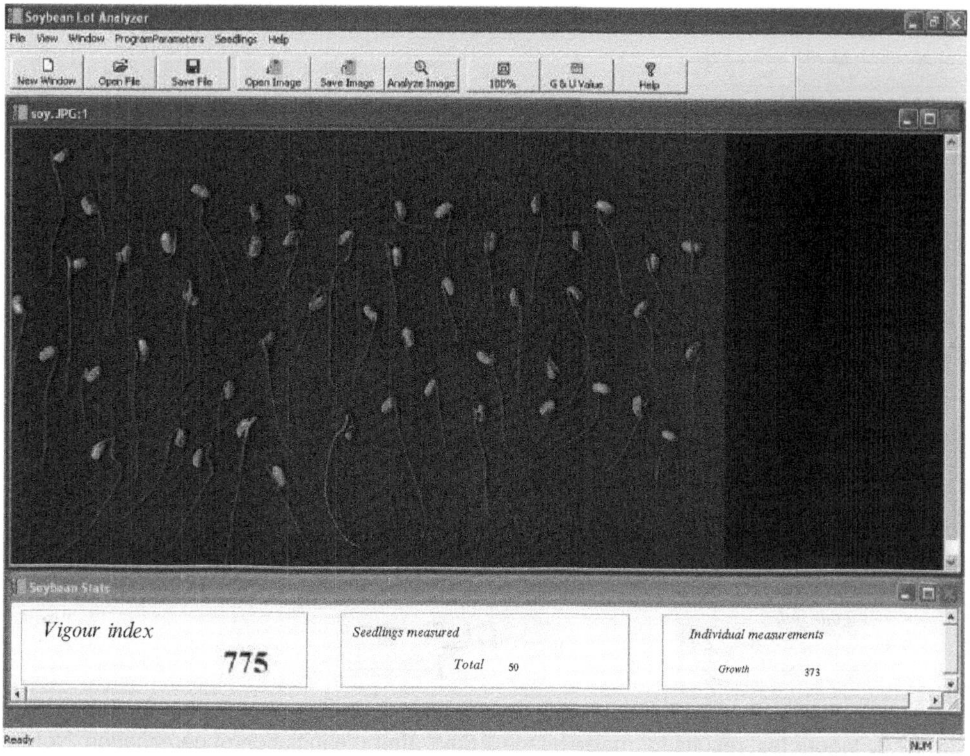

Fig. 32.2. Vigour test results for soybean seeds 3 days after the initiation of germination. Seedling lengths are measured but cotyledons are not measured. Individual seedling measurements can also be acquired.

ageing uniformity, ageing growth and standard germination were not significantly different in predicting field performance (Hoffmaster *et al.*, 2005).

Through the development of this soybean system, a fast, objective and reproducible method to perform a seed vigour test on soybean seedlings was achieved. For every input example presented to the software for processing, a seed vigour index was derived in less than 20 s. This is a significant advancement compared to manually measuring the seedlings, which may take up to 30 min. Another advantage of this soybean imaging system is the objective and reproducible results.

Even though the developed system was designed specifically for soybean seedlings, it shows promise for other seedlings with similar structure. The principles of the soybean system were also applied to 3-day-old cotton (*Gossypium hirsutum* L.) and impatiens (*Impatiens walleriana* Hook. f.) seedlings to produce vigour indices and/or length measurements with success (Hoffmaster, 2002).

Maize

Maize seedlings were imaged at 3 days after the initiation of germination followed by an SVIS analysis (Fig. 32.3). The results demonstrated that the nearest neighbour

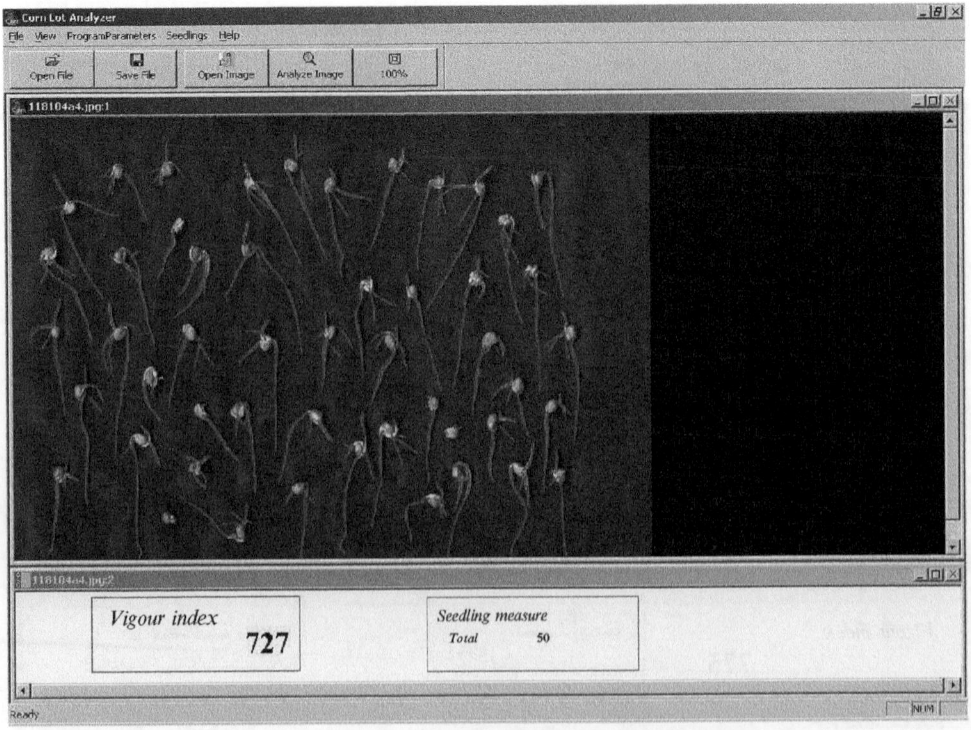

Fig. 32.3. Vigour test results for maize seeds 3 days after the initiation of germination. Note that the primary and seminal roots are measured but the seed structure is not determined. Individual seedling lengths can also be assessed.

connection and overlapping algorithms successfully attached roots and separated overlapping seedlings.

Maize seedlings pose a greater challenge for software development of seedling imaging when compared to soybean seedlings. They possess both seminal and primary roots that often grow horizontally, separated by a seed producing a coleoptile (Fig. 32.3). This creates three dilemmas. First, the size of the seed and the number of roots create shadows that make connection of the roots to the correct seed difficult. Second, the horizontal nature, particularly of the seminal roots, causes an overlapping syndrome that makes it difficult to ascertain which root belongs to which seed. Third, the coleoptile must be measured as a part of the growing seedling, but it is separated in distance from the roots by a substantial seed. These three issues were solved by developing two principal mathematical solutions. The first involved connecting the seedlings using nearest neighbour analysis that was applied both to the roots and the coleoptile. Overlapping of seedling parts is undesired in image analysis. One approach to solving this problem was to simply germinate the seeds and acquire images of the seedlings before they grew to a stage where overlap occurs. The second approach has been to develop an algorithm using angular features of the roots that minimized overlap when it occurred. This area of software development is continuing.

Conclusions

In conclusion, the research presented here advances the use of computer software to simplify seed technology tasks. The work of Sako *et al.* (2001) for lettuce seedlings was extended to support more complicated seedlings such as soybean (Hoffmaster *et al.*, 2003) and maize (Hoffmaster *et al.*, 2005). The soybean and maize systems analysed images of 3-day-old seedlings and derived length measurements for each seedling that were combined into a 0–1000 index representing the overall vigour of the seed lot. The vigour index was designed to reflect the speed and uniformity of growth measures that indicated the seedlings' potential for rapid and uniform emergence upon planting in the field. Through extensions of the principles of the lettuce, soybean and maize systems, other seed vigour assessment systems for different crops can be developed.

Acknowledgements

Funding for this research was provided by the Ohio Agricultural Research and Development Center Research Enhancement Competitive Grants Program and the American Seed Research Foundation. Salaries and additional research support were provided in part by state and federal funds appropriated to the Ohio Agricultural Research and Development Center, The Ohio State University.

References

AOSA (1983) *Seed Vigor Testing Handbook*, Contribution No. 32. Association of Official Seed Analysts, Lincoln, Nebraska.

Contreras, S. and Barros, M. (2005) Pruebas de vigor en semillas de lechuga y su correlacion con emergencia. *Ciencia e Investigacion Agraria* 32, 3–11.

Hacisalihoglu, G., Taylor, A.G., Paine, D.H., Hilderbrand, M.B. and Khan, A.A. (1999) Embryo elongation and germination rates as sensitive indicators of lettuce seed quality: priming and ageing studies. *Horticultural Science* 34, 1240–1243.

Hampton, J.G. and TeKrony, D.M. (1995) *Handbook of Vigour Test Methods*. International Seed Testing Association, Zurich, Switzerland.

Hoffmaster, A.L. (2002) An automated system for seed vigor testing three-day-old soybean seedlings. MSc thesis, The Ohio State University, Ohio.

Hoffmaster, A.L., Fujimura, K., McDonald, M.B. and Bennett, M.A. (2003) An automated system for vigor testing three-day-old soybean seedlings. *Seed Science and Technology* 31, 701–713.

Hoffmaster, A.F., Xu, L., Fujimura, K., McDonald, M.B., Bennett, M.A. and Evans, A.F. (2005) The Ohio State University Seed Vigor Imaging System (SVIS) for soybean and corn seedlings. *Seed Technology* 27, 7–24.

McDonald, M.B., Evans, A.F. and Bennett, M.A. (2001) Using scanners to improve seed/seedling evaluations. *Seed Science and Technology* 29, 683–689.

Penaloza, P., Ramierz-Rosales, G., McDonald, M.B. and Bennett, M.A. (2005) Lettuce (*Lactuca sativa* L.) seed quality evaluation using seed physical attributes, saturated salt accelerated ageing and the Seed Vigor Imaging System. *Electronic Journal of Biotechnology*[online], 15 December 2005, Vol. 8, No. 3. Available at: http://www.ejbiotechnology.info/content/vol8/issue3/full/11/index.html

Sako, Y. (2000) Systems for seed vigor assessment and seed classification. MSc thesis, The Ohio State University, Ohio.

Sako, Y., McDonald, M.B., Fujimura, K., Evans, A.F. and Bennett, M.A. (2001) A system for automated seed vigour assessment. *Seed Science and Technology* 29, 625–636.

Smith, O.E., Welch, N.C. and McCoy, O.D. (1973) Studies on lettuce seed quality: II. Relationship of seed vigor to emergence, seedling weight and yield. *Journal of the American Society for Horticultural Science* 98, 552–556.

Wurr, D.C.E. and Fellows, J.R. (1985) A determination of the seed vigour and field performance of crisp lettuce seedstocks. *Seed Science and Technology* 13, 11–17.

33 Development of a Sequential Digital Imaging System for Evaluating Seed Germination

R.L. Geneve, M. Dutt and A.B. Downie

Department of Horticulture, University of Kentucky, Lexington, KY 40546, USA

Abstract

A sequential imaging system using a flatbed scanner interfaced to a personal computer (PC) was developed as a research tool to study seed germination. The utility of the system was demonstrated in various germination-related studies including: (i) imbibition of physically dormant seeds; (ii) germination rate calculations; (iii) mutant assessment; and (iv) changes in seedling growth rates following dormancy release. Results from these studies revealed changes that were not previously observed or were too tedious to measure using conventional methods. This simple sequential imagery system offers an alternative research tool to study time-sensitive processes related to seed germination.

Introduction

High quality seed lots display rapid, uniform germination under laboratory and commercial conditions. Seed lot quality is typically described by germination percentage, speed and the spread between early and late germinating seeds. Additional quality information can be gained by measuring early seedling growth. The precision of these measurements is limited by the seed analyst's ability to repeatedly monitor the progression of germination and seedling growth. Computer-assisted measurements of digital images can help alleviate potential errors inherent in hand analysis and increase the number of samples an analyst can measure.

Two alternatives have been developed for capturing digital images for analysis of seed germination – charged coupled device (CCD) cameras (Howarth and Stanwood, 1993; Dell'Aquila *et al.*, 2000) and flatbed scanners (Geneve and Kester, 2001; Sako *et al.*, 2001). Both generally capture images with the quality required for computer-assisted analysis. However, flatbed scanners have the advantage of providing consistent lighting and the ability to capture usable images from very small seeds (Geneve and Kester, 2001).

Development of a non-destructive system for capturing sequential digital images for analysis could provide additional precision and insight concerning the aspects of seed germination. It would also provide a way to observe germination and seedling growth of individual seeds making it possible to select seeds displaying a particular behaviour or to monitor seedling growth following germination. Systems developed to take sequential images include time sequence photography (Tomas *et al.*, 1992), machine vision (Howarth and Stanwood, 1993) and computerized automated seed analysis using a hand potentiometric calliper (Keys *et al.*, 1984). We have developed a simple sequential imaging system that uses a flatbed scanner interfaced with a personal computer (PC), which captures images on an hourly basis. The objective of the current contribution is to demonstrate the range of germination-related applications, on which sequential imagery can be applied to using a flatbed scanner system. Applications include an example for imbibition, time to radicle protrusion, analysis of a mutant germination phenotype and seedling growth rate.

Materials and Methods

Sequential imaging system

The flatbed scanner system for capturing images during seed germination and seedling growth has been described in detail earlier (Geneve and Kester, 2001; Oakley *et al.*, 2004). Seeds were sown into 6 cm diameter Petri dishes containing one piece of sterile transparent cellulose acetate film wetted with 1 ml of sterile water or a thin layer of 1% (w/v) agarose. Petri dishes were sealed with Parafilm or Nexcare gentle paper tape (3, St. Paul, Minnesota, USA) and placed on a HP Scanjet 5370C flatbed scanner with a transparency adapter (Hewlett Packard, Palo Alto, California, USA) inside a growth chamber. The flatbed scanner was interfaced with a PC using Windows 98SE (Microsoft, Seattle, Washington, USA) operating system. A Visual Basic macro in SigmaScan Pro (SPSS, Chicago, Illinois, USA) captured images (300 or 600 dpi, colour TIFF file format) every hour for up to 5 days.

Imbibition patterns in honey locust seeds

Honey locust (*Gleditsia triacanthos* L.) seeds were acid scarified for 2 h in concentrated H_2SO_4 or hot water (100°C) for 1 min. The pattern of water uptake was determined as the portion of the seed swelling upon initial imbibition. Changes in seed size were measured using SigmaScan Pro by creating an overlay and using its area function. There were six seeds per Petri dish and 60 seeds were evaluated per treatment. Time to 50% germination was compared by single degree of freedom *F*-test.

Germination rate

Time to radicle protrusion was determined in petunia (*Petunia* × *hybrida* hort. ex E. Vilm.) seeds. Twenty-five seeds were used in five replicate Petri dishes containing cellulose film and were placed on the flatbed scanner in a growth chamber set at 25°C. Seeds were considered to have germinated when the radicle was visible on the digital image. Actual time to radicle protrusion was compared with germination rate calculated as the time for 50% of the germinating seed population to reach radicle protrusion (i.e. T_{50}). Germination rates (T_{50}) were calculated using sigmoidal equations and compared against the actual time to radicle protrusion using increasing evaluation intervals between 1 and 16 h beginning after seeds had imbibed for 24 h.

Characterization of cold temperature mutants

Four replicates of ~50 seeds from several overexpressing mutants of *Arabidopsis* (*Arabidopsis thaliana* (L.) Heynh.), that complete germination faster than the Columbia wild type (Col), were surface sterilized along with wild-type control seeds, and sown on 1% (w/v) agarose solid media in a Petri dish. Each of the four 60 × 15 mm Petri dishes was bifurcated by a line and, in each of the four plates, one replicate of 50 wild-type seeds was sown on one side of the line and 50 mutant seeds were sown on the other side. The plates were sealed and the four dishes were arranged side by side on the flatbed scanner inside a germinator maintaining a constant temperature at 10°C ± 1°C and constant light (135 µmol/m^2/s). Images were captured every hour starting on the third day of the experiment (Salaita *et al.*, 2005). Careful evaluation of the exact timing of radicle protrusion between the mutant putatively exhibiting faster completion of germination and the wild type was conducted from the chronology of images.

Seedling growth rate

Redbud (*Cercis canadensis* L.) seeds were treated with concentrated H_2SO_4 for 30 min and stratified at 4°C for 4 weeks then fully imbibed (i.e. for 48 h). Non-stratified seeds were acid scarified and fully imbibed, but not chilled. Embryos were surgically removed from non-stratified and stratified seeds and evaluated for radicle elongation for 115 and 70 h, respectively. The growth chamber was set at 25°C with a 16 h photoperiod supplied by cool-white fluorescent lamps (providing ~80 µmol/m^2/s). Images were captured every 5 h and the data were compared on the basis of analysis after 5 or 24 h intervals. Time taken by the radicle to reach a length of 10 mm was compared by a single degree of freedom *F*-test.

Results and Discussion

Imbibition patterns in honey locust seeds

Scarified honey locust seeds showed a typical triphasic increase in seed size that was indicative of water uptake during imbibition (Figs 33.1 and 33.2). Acid-scarified seeds had a faster imbibition rate compared with hot water-treated seeds. Acid-treated

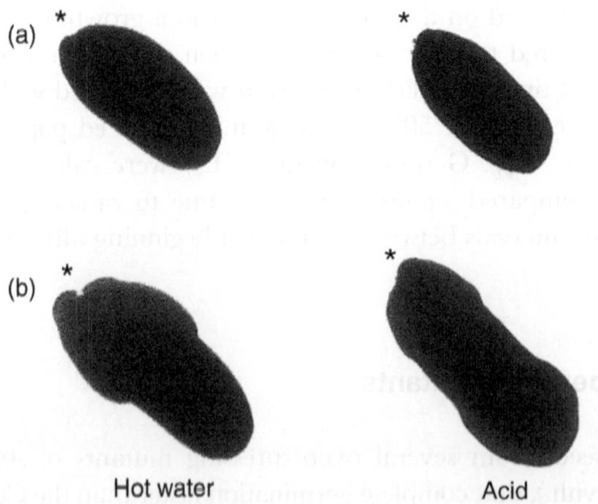

Fig. 33.1. Increase in seed size in honey locust seeds treated with acid for 2 h or hot water for 1 min. (a) is initial seed size; (b) is seed shape after 50% imbibition; and * indicates the micropylar end of the seed.

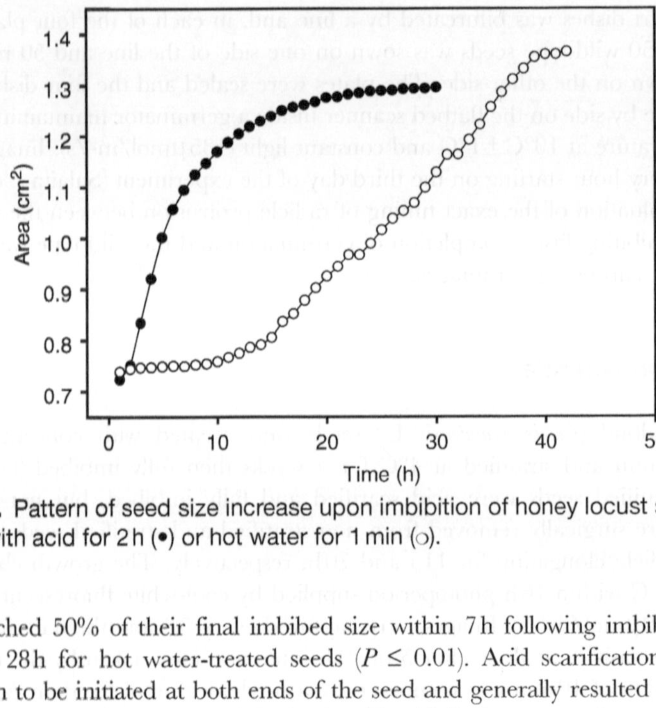

Fig. 33.2. Pattern of seed size increase upon imbibition of honey locust seeds treated with acid for 2 h (•) or hot water for 1 min (○).

seeds reached 50% of their final imbibed size within 7 h following imbibition compared to 28 h for hot water-treated seeds ($P \leq 0.01$). Acid scarification permitted imbibition to be initiated at both ends of the seed and generally resulted in a typical 'dumbbell' shape midway through hydration (Fig. 33.1). In contrast, hot water-treated seeds consistently showed imbibition initiating only at the micropylar end of the seed (Fig. 33.1).

Physical dormancy restricts the ability of seeds to imbibe when exposed to water and is typical of members of the *Fabaceae* (Baskin and Baskin, 1998). Honey locust seeds have a typical palisade epidermal layer with thick-walled macrosclerieds cells that are responsible for restricting imbibition. Baskin *et al.* (2000) suggested that in legume seeds, the lens is the first place on the seed coat for water entry when hard seeds become permeable under natural conditions.

It is becoming apparent that high temperature is the most probable cause for breaking physical dormancy (Morrison *et al.*, 1998). Sequential images of hot water-treated seeds support the lens as the natural entry point for imbibition in honey locust.

In acid-scarified legume seeds, Liu *et al.* (1981) showed a general reduction in the materials covering macrosclerieds throughout the seed. Therefore, rather than a single entry point for water, it would be anticipated that acid-treated seeds would show uniform water uptake over the entire seed surface. However, when water entry was followed on an hourly basis, acid-treated honey locust seeds showed asymmetric water uptake across the seed with more water initially entering at the polar ends (Fig. 33.1). This suggests that the cells in the polar regions of the seed were more susceptible to acid scarification than cells in the middle of the seed or that they are more adept at allowing water entry. It should also be considered that the extremely rapid increase in seed size suggests that water could be imbibed across the entire surface of the seed in contact with water, but at a lower rate for the non-polar regions of the seed.

Sequential imagery was adequate for determining the general region of the seed where imbibition initiated and provided a compelling case that honey locust seeds exposed to high temperature initiated imbibition at the micropylar end of the seed. However, the specific region of cells (i.e. lens) responsible for initial water entry could not be resolved from these images. Another limitation for strictly observational evaluation of water uptake was that it was not possible to determine if water was directly entering particular cells or if the size increase was due to lateral movement of water from cell to cell. It is anticipated that sequential imagery would provide good insight into specific water movement into seeds if water soluble dyes were used (Wilson and Geneve, 2004).

Germination rate

Germination usually approximates a normal distribution and can be described by total germination percentage, the median (i.e. time to reach 50% germination or T_{50}), and uniformity expressed as the standard deviation or the difference between the upper and lower quartile (Hara, 1999). The Association of Official Seed Analysts (AOSA) considers germination rate to be an indicator of seed vigour (AOSA, 1983). Seed lots with similar total germination percentages often vary in their rate of germination and growth. However, it is difficult to determine the actual time of radicle protrusion accurately because the time interval between evaluations is limited by the researcher's availability and can be as long as 24 h.

It is common to use non-linear equations to estimate germination rate and uniformity. Using images captured every hour, the actual time to radicle protrusion was determined for a petunia seed lot (Table 33.1). Actual germination rate and uniformity was compared to estimates derived from sigmoidal equations using data sampled with increasing intervals. T_{50} was accurately estimated until the sampling interval reached 16 h. In contrast, mathematical estimates of uniformity (i.e. T_{75} minus T_{25}) were less accurate and were only statistically similar to actual values when sampled after 1 or 10 h (Table 33.1).

Table 33.1. Comparison between actual and calculated germination rates depending on the frequency of observations in a petunia seed lot.

Hours between observations	Visual T_{50}	Sigmoidal T_{50}	Visual T_{75-25}	Sigmoidal T_{75-25}
1	33.0 ± 1.2 b	33.2 ± 1.5 b	10.8 ± 2.2 b	9.3 ± 1.6 ab
2	33.0 ± 1.2 b	33.1 ± 1.6 b	10.5 ± 1.9 b	9.1 ± 1.6 ab*
4	34.0 ± 2.3 b	33.1 ± 2.0 b	10.0 ± 2.3 b	9.1 ± 1.7 ab*
6	35.0 ± 3.5 b	33.7 ± 2.1 b	10.5 ± 3.0 b	9.0 ± 1.6 ab*
8	36.0 ± 0.0 b	33.0 ± 1.6 b	10.0 ± 4.0 b	9.7 ± 1.2 ab
10	40.0 ± 0.0 a	33.5 ± 1.3 b	7.5 ± 5.0 c	10.6 ± 1.2 a
12	38.0 ± 6.9 b	33.2 ± 2.3 b	15.0 ± 2.0 a	8.9 ± 2.6 b*
14	41.0 ± 8.1 a	32.8 ± 0.8 b	10.0 ± 0.0 b	6.9 ± 1.8 c*
16	36.0 ± 0.0 b	34.8 ± 0.6 a*	12.0 ± 8.0 b	4.6 ± 3.4 d*

Observations began after 24 h and, on average, germination was complete after 65 h. Means ± one standard deviation followed by the same letter within a column were not different by Tukey's test ($\alpha = 0.05$).

*Difference between calculated and actual germination according to a single degree of freedom *F*-test ($P \leq 0.05$).

Analysis of cold temperature mutants

After the initial screening, 36 mutant lines putatively resulting in faster than usual germination at 10°C were identified. However, after retesting these lines, only five were found to be demonstrably faster than the wild type. Despite using a suboptimal germination temperature (10°C), which is competent to alleviate *Arabidopsis* seed dormancy (Salaita *et al.*, 2005) and capable of accentuating differences in commencement of radicle protrusion, the longer the seeds had been after-ripened (AR) prior to the analysis, the more restricted became the period in which a measurable difference in germination percentage was observed (compare Fig. 33.3 (6 months AR) and Salaita *et al.* (2005) (~1 month AR)). Statistically significant differences in percentage germination were observed at fewer time points at the optimal (25°C) germination temperature (Salaita *et al.*, 2005). However, while percentage germination has been depicted every 12 h in Fig. 33.3, data was collected every hour. The ability to examine the progression of radicle protrusion every hour facilitated the discrimination between *bona fide ctg* mutants and false positives, particularly at 25°C.

Seedling growth rate

Redbud seeds have intermediate physiological dormancy. Embryos displaying this type of dormancy show an increased growth potential following chilling stratification (Hartmann *et al.*, 2002). Geneve (1991) showed that isolated redbud embryos from chilled seeds grew faster than non-chilled ones. However, these measurements were performed by hand and made every 24 h. In contrast, using the computer-aided imaging system, radicle length could be measured every hour and a precise growth rate calculated with little researcher investment in time. As predicted, radicles of

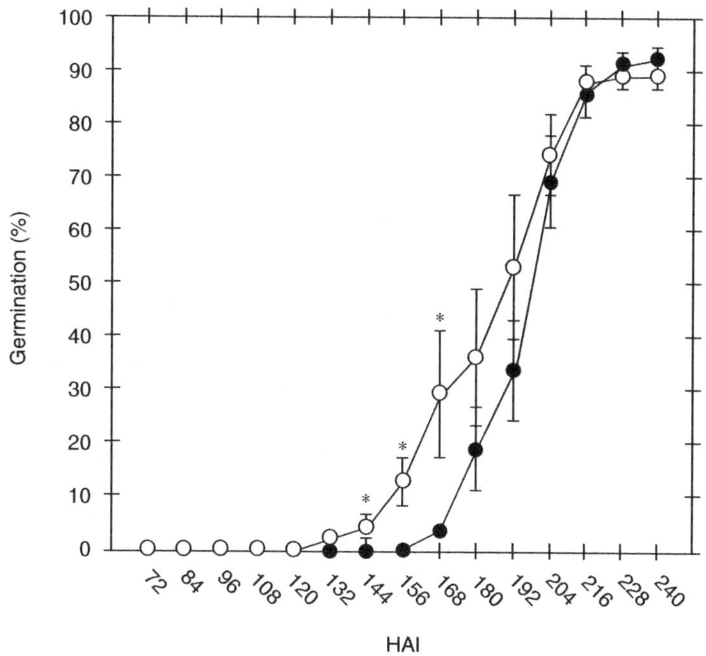

Fig. 33.3. Germination time course of *ctg144-D* (○) and Columbia wild-type (•) *Arabidopsis thaliana* seeds at 10°C after 6 months dry after-ripening (AR) at 25°C. Each symbol represents the mean of four replicates of 50 seeds and the bars represent the standard error. An asterisk above a symbol depicts statistically significant differences in percentage germination determined by a single degree of freedom *F*-test ($P \leq 0.05$). HAI = hours after imbibition.

non-chilled redbud embryos took 90h to reach 10mm in length, while embryos chilled for 4 weeks reached a radicle length of 10mm in only 45h (Fig. 33.4; $P \leq 0.01$). If the data was evaluated on a 24h basis, as has been done earlier (Geneve, 1991), a clear quadratic increase in seedling size was seen (Fig. 33.4a). However, using a shorter interval image analysis, redbud embryos showed three distinct phases of growth (Fig. 33.4b). Following removal from the testa, there was a lag period prior to initiation of a slow increase in radicle size. This was followed by a rapid linear increase in size. In non-chilled embryos, growth did not begin until 50h after removal from the testa. The subsequent slow growth phase required an additional 30h before embryos entered the rapid linear phase of growth. In contrast, embryos from chilled seeds required only 15h to initiate growth and began the rapid linear phase only 25h later. These results show that the major difference between embryos isolated from non-chilled and chilled seeds was the time to initiate radicle growth. After growth was initiated the growth patterns were comparable.

Conclusions

Analysis of sequential digital images offers an alternative research tool to study time-sensitive processes related to seed germination. Analysis of these images revealed

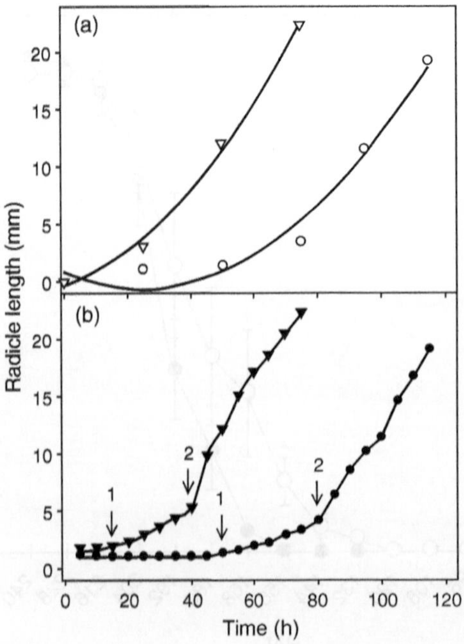

Fig. 33.4. Radicle length in isolated redbud embryos from seeds that have received 4 weeks of chilling stratification (Δ) or no chilling (○). (a) Length measured every 24 h. Regression lines were fitted to a second order polynomial; $r^2 = 0.98$ and 0.99 for chilled and non-chilled embryos, respectively. (b) Length measured every 5 h. Arrows indicate the start of the first and second growth phases.

changes during seed germination that were not previously observed or were too tedious to measure using conventional methods. This technique would allow the researcher to more precisely identify key stages of development during seed germination for future physiological or biochemical analyses and facilitate identification of unique phenotypes during mutant screens.

References

AOSA (1983) *Seed Vigor Testing Handbook, Contribution No. 32.* Association of Official Seed Analysts, Lincoln, Nebraska.

Baskin, C.C. and Baskin, J.M. (1998) *Seeds: Ecology, Biogeography, and Evolution of Dormancy and Germination.* Academic Press, New York.

Baskin, J.M., Baskin, C.C. and Li., X. (2000) Taxonomy, anatomy and evolution of physical dormancy in seeds. *Plant Species Biology* 15, 139–152.

Dell'Aquila, A., van Eck, J.W. and van der Heijden, G.W.A.M. (2000) The application of image analysis in monitoring the imbibition process of white cabbage (*Brassica oleracea* L.) seeds. *Seed Science Research* 10, 163–169.

Geneve, R.L. (1991) Seed dormancy in eastern redbud (*Cercis canadensis* L.). *Journal of the American Society for Horticultural Science* 116, 85–88.

Geneve, R.L. and Kester, S.T. (2001) Evaluation of seedling size following germination using computer-aided analysis of digital images from a flat-bed scanner. *HortScience* 36, 1117–1120.

Hara, Y. (1999) Calculation of population parameters using Richards function and application of indices of growth and seed vigor to rice plants. *Plant Production Science* 2, 129–135.

Hartmann, H.T., Kester, D.E., Davies, F.T., Jr and Geneve R.L. (2002) *Hartmann and Kester's Plant Propagation: Principles and Practices*, 7th edn. Prentice-Hall, Englewood Cliffs, New Jersey.

Howarth, M.S. and Stanwood, P.C. (1993) Measurement of seedling growth by machine vision. *Transactions of the American Society of Agricultural Engineers* 36, 959–963.

Keys, R.D., Margapuram, R.G. and Reusche, G.A. (1984) Automated seedling length measurement for germination/vigor estimation using a CASAS (computerized automated seed analysis system). *Journal of Seed Technology* 9, 40–53.

Liu, N.Y., Khatamian, H. and Fretz, T.A. (1981) Seed coat structure of three woody legume species after chemical and physical treatments to increase seed germination. *Journal of the American Society for Horticultural Science* 106, 691–694.

Morrison, D.A., McClay, K. Porter, C. and Rish, S. (1998) The role of the lens in controlling heat-induced breakdown of testa-imposed dormancy in native Australian legumes. *Annals of Botany* 82, 35–40.

Oakley, K., Kester, S.T. and Geneve, R.L. (2004) Computer-aided digital image analysis of seedling size and growth rate for assessing seed vigour in impatiens. *Seed Science and Technology* 32, 907–915.

Sako, Y., McDonald, M.B., Fujimura, K., Evans, A.F. and Bennett, M.A. (2001) A system for automated seed vigour assessment. *Seed Science and Technology* 29, 625–636.

Salaita, L., Kar, R.K., Majee, M. and Downie, A.B. (2005) Identification and characterization of activation tagged Arabidopsis mutants exhibiting rapid seed germination. *Journal of Experimental Botany* 56, 2059–2069.

Tomas, T.N., Taylor, A.G. and Ellerbrock, L.A. (1992) Time-sequence photography to record germination events. *HortScience* 27, 372.

Wilson, T.T. and Geneve, R.L. (2004) The impact of film coating on initial water uptake and imbibitional chilling injury in high and low vigor *sh2* sweet corn seeds. *Seed Science and Technology* 32, 271–281.

34 Seed Quality and Germination

L. Rajjou,[1] Y. Lovigny,[1] C. Job,[1] M. Belghazi,[2]
S. Groot[3] and D. Job[1]

[1]*Centre National de la Recherche Scientifique (CNRS)-Bayer CropScience Joint Laboratory, (UMR CNRS 2847), Bayer CropScience, 14/20 rue Pierre Baizet, BP 9163, F-69263 Lyon, Cedex 09, France;* [2]*Institut National de la Recherche Agronomique, Unité Mixte de Recherche N° 6175, Service de Spectrométrie de Masse Pour la Protéomique, Nouzilly, France;* [3]*Plant Research International, Wageningen University and Research Centre, PO Box 16, 6700 AA Wageningen, The Netherlands*

Abstract

Seed storage is often accompanied by a progressive loss of germination vigour and viability. In the present study, we have used *Arabidopsis thaliana* (L.) Heynh. seeds as a model, and carried out differential proteomics to investigate seed vigour. In our system, based on a controlled deterioration treatment (CDT), we compared seed lots treated for different time periods up to 7 days. Germination tests showed a progressive decrease of seed vigour depending on the duration of CDT. Proteomic analyses revealed that loss in seed vigour can be accounted for by protein changes in the dry seed and by an inability of the low vigour seeds to display a normal proteome during germination. Furthermore, the CDT strongly increased the extent of protein oxidation (i.e. carbonylation), which will in turn induce a loss of functional properties of proteins and enzymes and/or enhance their susceptibility towards proteolysis. These results highlighted essential mechanisms for germinative quality such as translational capacity and mobilization of seed storage reserves.

Introduction

Before ageing irreparably leads to the death of the seed, the deterioration accumulated during storage is likely to affect its potential ability to germinate. This deterioration process occurs even under ideal storage conditions. The lifespan of seeds is determined by their genetic and physiological storage potential and by any deteriorating events that occur prior to or during storage, as well as by interaction with environmental factors. A recent investigation reported large differences in the response to storage of seeds descended from different plant species (Walters *et al.*, 2005). We are interested in determining the molecular basis of seed storability in the dry state. The model plant *Arabidopsis* (*Arabidopsis thaliana* (L.) Heynh.) is a very good reference species for this purpose, because it allows a molecular dissection of storage response. Indeed, disclosure of the *Arabidopsis* genome sequence (Arabidopsis Genome Initiative, 2000; Somerville and Koornneef, 2002) increased markedly our knowledge and understanding of the great complexity in regulation of plant growth and development. Genetic and global approaches such as transcriptomic profiling (Ogawa *et al.*, 2003; Clerkx *et al.*, 2004; Nakabayashi *et al.*, 2005) have proved useful for the characterization of potential biomarkers of seed quality and germinative capacity. However, the functional components of a biological system are proteins and not genes or mRNAs. Thanks to the availability of genomic sequence information, the progress achieved in sensitive and rapid separation of proteins as well as their high-throughput identification by electrophoresis and mass spectrometry, proteomic approaches have opened up new perspectives to analyse the complex functions of model plants and crop species (Canovas *et al.*, 2004). In this way, we have used proteomics to unravel the requirements in terms of RNA and protein synthesis for *Arabidopsis* seed germination (Gallardo *et al.*, 2001, 2002a,b; Rajjou *et al.*, 2004). In particular, these studies have revealed that proteins and mRNAs stored in dry mature seeds are sufficient for germination *sensu stricto* (Rajjou *et al.*, 2004).

In the present study we have used such proteomic tools and a seed deterioration treatment, known as CDT, which is presumed to mimic natural ageing (Clerkx *et al.*, 2004). CDT is widely used as a vigour assay for numerous seed species and has been described for *Arabidopsis* seeds (Tesnier *et al.*, 2002). We compared eight *Arabidopsis* seed lots treated for different time periods up to 7 days. A comparison of the dry seed proteome for each sample was carried out to reveal changes in the accumulation of specific proteins during the treatment. The proteome of 1-day-imbibed seeds was also characterized for all seed samples to analyse the behaviour of the treated seeds during the early steps of the germination process. Since the CDT is presumed to entail an oxidative stress, which can lead to the formation of oxidatively modified proteins, we also analysed the oxidized proteome in the treated seeds.

Materials and Methods

Plant material and germination experiments

Non-dormant seeds of *Arabidopsis*, accession Landsberg *erecta* (*Ler*), were used in all experiments. Germination assays were carried out at 25°C, with a 16 h light/8 h dark daily regime, as described in Rajjou *et al.* (2004).

Controlled deterioration test (CDT)

The CDT was performed according to Tesnier *et al.* (2002). Seeds were briefly equilibrated at 85% relative humidity (RH) (20°C) and day 0 controls were immediately dried back at 32% RH. Treatment involved storing the seeds (at 85% RH) for various time periods (i.e. 0 h, 4 h, 16 h, 1 day, 2 days, 3 days, 5 days and 7 days) at 40°C. Seeds were then dried back at 32% RH (20°C) and stored at 4°C.

Preparation of total protein extracts and two-dimensional electrophoresis

Total protein extracts were prepared from dry mature seeds and seeds at different stages of germination as described earlier (Rajjou *et al.*, 2004; Job *et al.*, 2005). Proteins, with an equivalent to an extract of 100 seeds, corresponding to about 200 μg protein for all samples, were analysed by two-dimensional polyacrylamide gel electrophoresis (2D-PAGE) as described earlier (Görg *et al.*, 1987; Rajjou *et al.*, 2004). Two-dimensional gels were stained with silver nitrate according to either Blum *et al.* (1987), for densitometric analyses, or Shevchenko *et al.* (1996), for the mass spectrometry analyses. Image analysis was carried out with the ImageMaster 2D Elite version 4.01 software (Amersham Biosciences).

Detection of oxidized proteins and western blotting

Detection of oxidized proteins by carbonylation was performed by derivatization of protein extracts with 2–4 dinitrophenylhydrazine (DNPH) and immunological detection of the DNP adducts with monoclonal anti-DNP antibody (OxyBlot™ oxidized protein detection kit, Chemicon, France) as described earlier (Job *et al.*, 2005; Nyström, 2005).

De novo protein synthesis

Labelled-proteins were synthesized *in vivo* by imbibing seeds in water for 1 day in the presence of [^{35}S]methionine (1.85 MBq; ICN Biomedicals, S.A.R.L.). Protein synthesis was measured by trichloroacetic acid (TCA) precipitation of aliquots of reaction mixtures spotted on Whatmann GF/C filters; after ten washing steps in cold 5% TCA and 0.04 M sodium pyrophosphate, and two washing steps in absolute ethanol, filters were dried and counted for radioactivity in a liquid scintillation counter.

Protein identification by mass spectrometry

Spots of interest were excised from 2D SDS-PAGE gels with sterile tips and put in 1.5 ml sterile tubes. Each spot was rinsed, reduced with 10 mM dithiothreitol (DTT), alkylated with 55 mM iodoacetamide and incubated overnight at 37°C with 12.5 ng/μl trypsin (sequencing grade; Roche Diagnostics) in 25 mM NH_4HCO_3. Analysis of tryptic peptides by tandem mass spectrometry (MS/MS) was performed on a nanoelectrospray ionization quadrupole time-of-flight hybrid mass spectrometer (Q-TOF Ultima Global; Waters Micromass, Manchester, UK) coupled with a nano-HPLC (Cap-LC; Waters), as described in Job *et al.* (2005). The peptide masses and sequences obtained were either matched automatically to proteins in a non-redundant database (NCBI) using the Mascot MS/MS Ions Search algorithm (http://www.matrixscience.com) or blasted manually against the current databases.

Results and Discussion

Germination parameters of *Arabidopsis* seed samples

The germination parameters of the presently studied seed samples are listed in Table 34.1. There was a dramatic decline in germinability of the seeds submitted to the CDT. Not only was the seed vigour affected, but there was also a strong reduction in seed viability as indicated by a marked decrease of the maximum germination percentage (G_{max}).

Proteome variation of dry mature seeds

To explore the molecular mechanisms associated with the loss of seed vigour during accelerated ageing, a differential proteomic approach was used. To achieve this, total soluble proteins extracted from all seed lots were separated by 2D-PAGE. Following silver nitrate staining, protein patterns were analysed by image analysis. A typical gel is presented in Fig. 34.1. We first investigated the effect of the CDT on the dry seed proteome (i.e. from the proteome of seeds collected after conducting the treatment). Only a few protein spots showed reproducible variations in their accumulation level (Fig. 34.1).

Our differential proteomic approach reveals that 12 protein spots were more abundant in the aged seeds and six protein spots were less abundant in the deteriorated seeds. These results highlight that protein modifications can occur during artificial seed ageing.

Characterization of oxidized proteins in the dry mature seeds

Ageing in all organisms, notably in plants, is associated with oxidative stress (Bailly, 2004) that entails oxidation, by carbonylation, of specific proteins (Berlett and Stadtman, 1997; Toda, 2000). Since highly sensitive methods have been described

Table 34.1. Germination parameters of *Arabidopsis* seed lots subjected to different periods of controlled deterioration treatment (CDT).

Seed lot	$t_{1\%}$ (h)	$t_{10\%}$ (h)	$t_{25\%}$ (h)	$t_{50\%}$ (h)	$t_{75\%}$ (h)	$t_{90\%}$ (h)	G_{max} (%)
0 h	39.6	40.9	41.9	43.2	45.1	49.0	100
4 h	41.1	42.5	43.6	45.1	47.8	63.1	98
16 h	41.0	42.5	43.5	45.0	47.5	57.3	98
1 day	43.0	44.5	45.7	47.4	50.4	64.0	96
2 days	45.2	46.8	48.0	49.7	52.6	–	85
3 days	47.9	49.8	51.1	53.2	57.8	–	79
5 days	59.8	62.5	64.9	–	–	–	28
7 days	102.7	–	–	–	–	–	2

$t_{1\%}$, $t_{10\%}$, $t_{25\%}$, $t_{50\%}$, $t_{75\%}$ and $t_{90\%}$ are the times to reach 1%, 10%, 25%, 50%, 75% and 90% germination, respectively; G_{max} = maximum germination after 1 week.

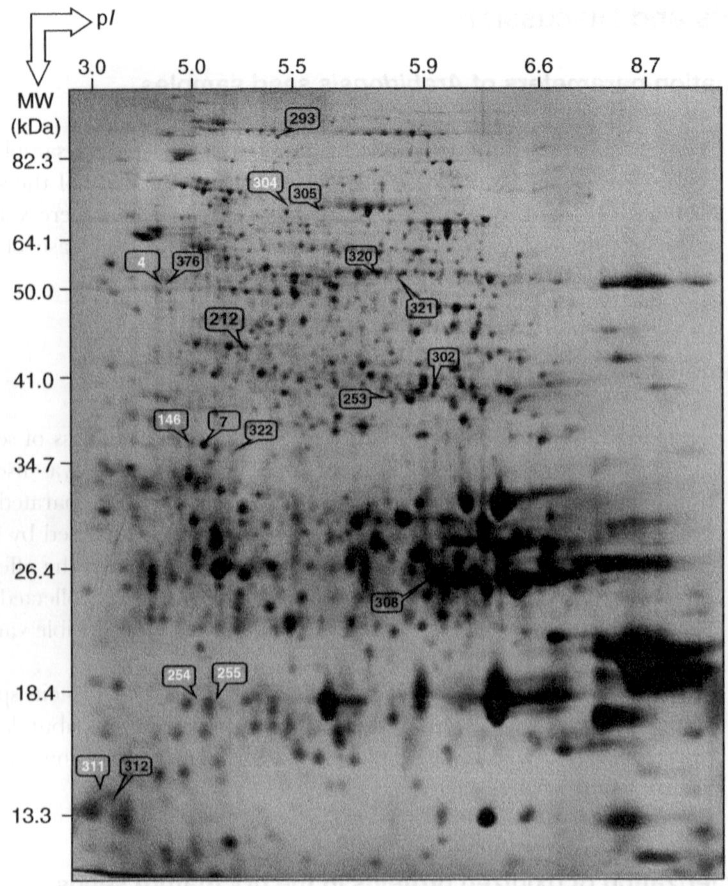

Fig. 34.1. Detection of age-related protein alterations by 2D-PAGE. This figure shows a 2D gel of total soluble proteins from dry mature seeds after 7 days of CDT. Proteins were first separated by electrophoresis according to charge. Isoelectrofocusing (IEF) was carried out with protein samples with the equivalent to an extract of ~100 seeds, corresponding to ~200 µg protein for all samples. Proteins were then separated according to size by SDS-PAGE using 10% polyacrylamide gels. Proteins were visualized by silver nitrate staining. Black and white numbers indicate proteins with increased or decreased levels following the CDT, respectively.

for the detection of carbonylated proteins (Levine *et al.*, 1990), we characterized the influence of the CDT on the oxidized proteome of *Arabidopsis* seeds (Fig. 34.2).

Carbonylated proteins were identified by matching the 2,4-dinitrophenylhydrazone (DNP)-derivatized protein spots to master gel maps of *Arabidopsis* seed proteins (Gallardo *et al.*, 2001, 2002a; Rajjou *et al.*, 2004; http://seed.proteome.free.fr). The results revealed that protein carbonylation strongly increased in deteriorated seeds, indicative of the progressive accumulation of reactive oxygen species (ROS) during the CDT. The example presented in Fig. 34.2 shows that several polypeptides corresponding to the α- and β-subunits of the 12S cruciferins (legumin type seed storage proteins) are heavily carbonylated in the aged seeds. It must be stressed that in non-deteriorated seeds, carbonylation of

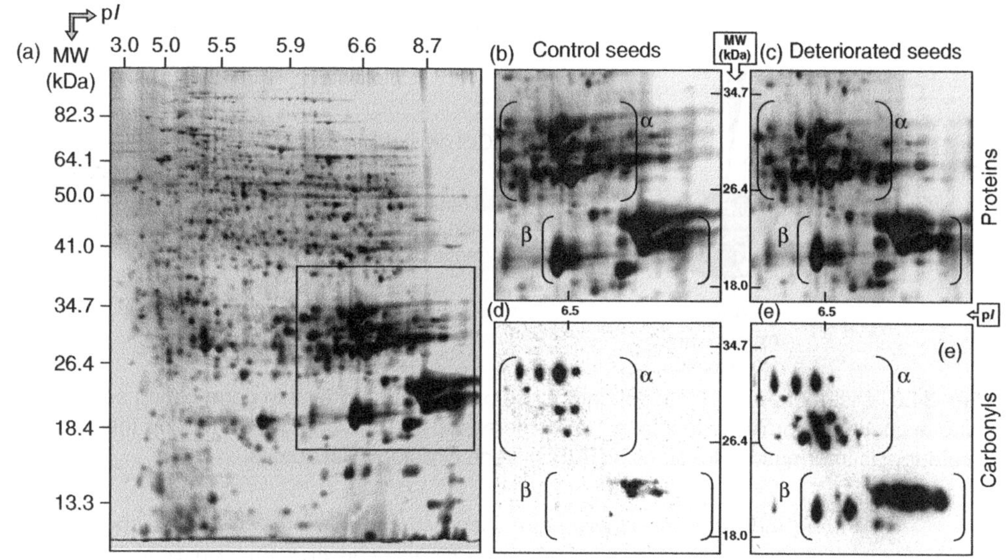

Fig. 34.2. Increased protein carbonyl levels in β-subunits of 12S cruciferins after a CDT. Protein extracts were prepared from the dry mature seeds (control seeds) and seeds submitted to a CDT (deteriorated seeds) and analysed by 2D-PAGE. The portion of the 2D gels shown correspond to the window in panel (a). Protein silver stains (panels (b) and (c)) and anti-DNP immunoassays (panels (d) and (e)) are shown.

12S-cruciferin β-subunits was much lower than for the α-subunits (compare Figs. 34.2b and 34.2d). In marked contrast, the CDT entailed a strong increase in the extent of carbonylation of the 12S-cruciferin β-subunits, up to a level similar of that of the α-subunits (compare Figs. 34.2c and 34.2e). In addition to the 12S-cruciferin subunits, several other proteins ought to be oxidized in the aged *Arabidopsis* seeds (data not shown). In conclusion, the present data strongly support the finding that loss in seed vigour afforded by the CDT arises from overproduction of ROS associated with oxidative protein damage.

Proteome variation of germinating seeds

The proteome of the artificially aged seed lots was also analysed after 1 day imbibition in water. This stage corresponds to the germination *sensu stricto* of *Arabidopsis* (*Ler*) non-deteriorated seeds (none of the seeds showed radicle protrusion at that time; see Table 34.1). This analysis revealed the proteome evolution of differentially aged seeds during imbibition. Among 54 protein spots presenting reproducible variations in their accumulation level, 34 were less abundant in germinating deteriorated seeds and 20 were more abundant. Functional categorization of these genes is presented in Fig. 34.3. It appears that several protein functions were affected by the CDT. One of the specific features observed from this clustering analysis is a larger abundance of storage proteins in germinating deteriorated seeds,

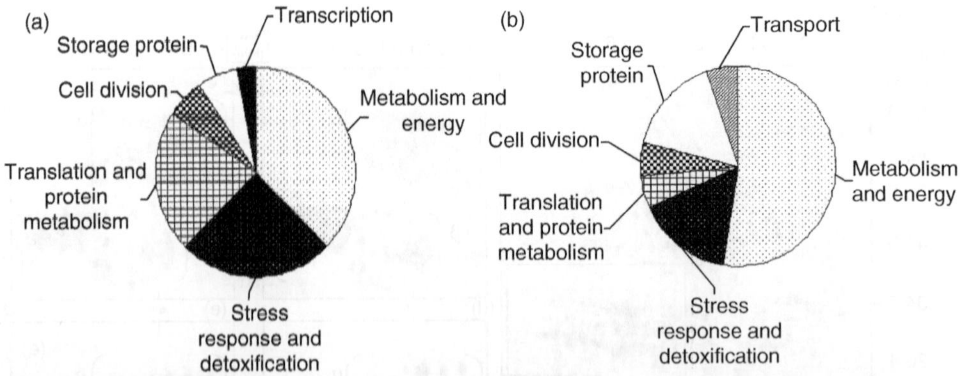

Fig. 34.3. Distribution of functional categories of the seed proteins in germinating seeds whose accumulation level decreased (a) or increased (b) following the CDT using the Arabidopsis Information Resource's (TAIR's) Gene Ontology Resources.

a finding indicative of a correlation between the loss of seed vigour and storage mobilization ability. Another interesting feature concerns the apparent correlation between protein metabolism and translation, and the reduction of seed germinability induced by the CDT (Fig. 34.3).

To get a direct insight into protein synthesis during germination, *sensu stricto* proteins that were newly synthesized *in vivo* following seed imbibition in water for 1 day were labelled in the presence of radioactive [^{35}S]methionine. The control seed lot, which had a maximum germination of 100%, supported a very active [^{35}S]methionine incorporation, testifying to a high translational activity during germination *sensu stricto*. As shown in Fig. 34.4, the extent of [^{35}S]methionine incorporation declined dramatically in the deteriorated seed lots.

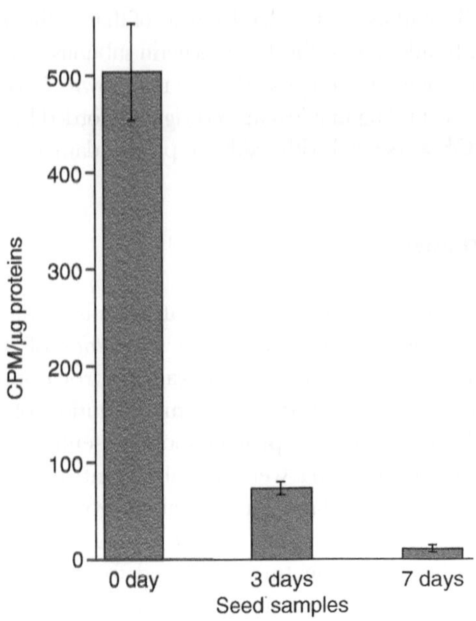

Fig. 34.4. Influence of the CDT on *de novo* protein synthesis. Seeds were incubated for 1 day in the presence of [^{35}S]methionine. Protein synthesis was measured by trichloroacetic acid (TCA) precipitation of aliquots of reaction mixtures spotted on Whatmann GF/C filters; after ten washing steps in cold 5% TCA and 0.04 M sodium pyrophosphate and two washing steps in absolute ethanol, filters were dried and counted for radioactivity (CPM) in a liquid scintillation counter.

For example, seed lots that were deteriorated for 3 days presented an eightfold decrease in [^{35}S]methionine incorporation compared with control seeds, although under these conditions the aged seeds still kept a high vigour with a maximum germination of about 80%. This result demonstrated that translation capacity can be an excellent criterion for the estimation of seed vigour, a finding that is in good agreement with the previous work on seed ageing in soybean (*Glycine max* (L.) Merr.) (Pillay, 1977). The seed lot that was deteriorated for 7 days had a maximum germination of ~2% and showed almost no translational activity. However, very low residual translation was detectable, suggesting that these seeds were not dead. The consequences of the observed reduction of protein synthesis can be diverse. This may affect the systems necessary for the maintenance, repair and normal resumption of metabolism, the efficiency of detoxification, the efficiency of the signalling pathways, and/or the production and secretion of several metabolites as well as plant hormones like gibberellins.

Similar events occur during accelerated and natural ageing

Uncertainty prevails as to why the CDT mimics natural ageing. This is a major concern of seed companies because, for practical reasons, they rely on the CDT and germination assays to predict seed storability (Delouche and Baskin, 1973). It is therefore of importance to compare the biochemical behaviour of seeds submitted to the CDT and of seeds that have been naturally aged. For that purpose, three naturally aged *Arabidopsis* seed lots were examined. Two of them, aged 7 and 8 years, presented a maximum germination of about 45% and 23%, respectively. A third one, aged 11 years, did not germinate at all after 14 days of imbibition. A proteome analysis revealed common features between the artificially and naturally aged seeds. Furthermore, the extent of protein carbonylation strongly increased during natural ageing, as also occurs in artificial ageing. Finally, protein translation capacity was strongly repressed in naturally aged seeds, a specific feature also observed with the CDT (Fig. 34.4). Our data thereby provides the first molecular indication supporting the usefulness of the CDT for prediction of seed storability.

Conclusions

Proteomics provided an innovative and powerful tool for investigating the molecular mechanisms of seed vigour and seed viability during ageing. Changes in the regulation of protein synthesis, post-translational modifications and protein turnover are crucial determinants of age-related decline in the maintenance, repair and survival of the seed.

References

Arabidopsis Genome Initiative (2000) Analysis of the genome sequence of the flowering plant *Arabidopsis thaliana*. *Nature* 408, 796–815.

Bailly, C. (2004) Active oxygen species and antioxidants in seed biology. *Seed Science Research* 14, 93–107.

Berlett, B.S. and Stadtman, E.R. (1997) Protein oxidation in aging, disease and oxidative stress. *The Journal of Biological Chemistry* 272, 20313–20316.

Blum, H., Beier, H. and Gross, H.J. (1987) Improved silver staining of plant proteins, RNA and DNA in polyacrylamide gels. *Electrophoresis* 8, 93–99.

Canovas, F.M., Dumas-Gaudot, E., Recorbet, G., Jorrin, J., Mock, H.P. and Rossignol, M. (2004) Plant proteome analysis. *Proteomics* 4, 285–298.

Clerkx, E.J., El-Lithy, M.E., Vierling, E., Ruys, G.J., Blankestijn-De Vries, H., Groot, S.P., Vreugdenhil, D. and Koornneef, M. (2004) Analysis of natural allelic variation of *Arabidopsis* seed germination and seed longevity traits between the accessions Landsberg *erecta* and Shakdara, using a new recombinant inbred line population. *Plant Physiology* 135, 432–443.

Delouche, J.C. and Baskin, C.C. (1973) Accelerated aging techniques for predicting the relative storability of seed lots. *Seed Science and Technology* 1, 427–452.

Gallardo, K., Job, C., Groot, S.P.C., Puype, M., Demol, H., Vandekerckhove, J. and Job, D. (2001) Proteomic analysis of *Arabidopsis* seed germination and priming. *Plant Physiology* 126, 835–848.

Gallardo, K., Job, C., Groot, S.P.C., Puype, M., Demol, H., Vandekerckhove, J. and Job, D. (2002a) Proteomics analysis of *Arabidopsis* seed germination: a comparative study of wild-type and GA-deficient seeds. *Plant Physiology* 129, 823–837.

Gallardo, K., Job, C., Groot, S.P.C., Puype, M., Demol, H., Vandekerckhove, J. and Job, D. (2002b) Importance of methionine biosynthesis for *Arabidopsis* seed germination and seedling growth. *Physiologia Plantarum* 116, 238–247.

Görg, A., Postel, W., Weser, J., Günther, S., Strahler, J.R., Hanash, S.M. and Somerlot, L. (1987) Elimination of point streaking on silver stained two-dimensional gels by addition of iodoacetamide to the equilibration buffer. *Electrophoresis* 8, 122–124.

Job, C., Rajjou, L., Lovigny, Y., Belghazi, M. and Job, D. (2005) Patterns of protein oxidation in *Arabidopsis* seeds and during germination. *Plant Physiology* 138, 790–802.

Levine, L.R., Garland, D., Oliver, C.N., Amici, A., Climent, I., Lenz, A.-G., Ahn, B.-W., Shaltiel, S. and Stadtman, E.R. (1990) Determination of carbonyl content in oxidatively modified proteins. *Methods in Enzymology* 186, 464–478.

Nakabayashi, K., Okamoto, M., Koshiba, T., Kamiya, Y. and Nambara, E. (2005) Genome-wide profiling of stored mRNA in *Arabidopsis thaliana* seed germination: epigenetic and genetic regulation of transcription in seed. *The Plant Journal* 41, 697–709.

Nyström, T. (2005) Role of oxidative carbonylation in protein quality control and senescence. *The EMBO Journal* 24, 1311–1317.

Ogawa, M., Hanada, A., Yamauchi, Y., Kuwahara, A., Kamiya, Y. and Yamaguchi, S. (2003) Gibberellin biosynthesis and response during *Arabidopsis* seed germination. *The Plant Cell* 15, 1591–1604.

Pillay, D.T. (1977) Protein synthesis in aging soybean cotyledons: loss in translational capacity. *Biochemical and Biophysical Research Communications* 79, 796–804.

Rajjou, L., Gallardo, K., Debeaujon, I., Vandekerckhove, J., Job, C. and Job, D. (2004) The effect of α-amanitin on the *Arabidopsis* seed proteome highlights the distinct roles of stored and neosynthesized mRNAs during germination. *Plant Physiology* 134, 1598–1613.

Shevchenko, A., Wilm, M., Vorm, O. and Mann, M. (1996) Mass spectrometric sequencing of proteins from silver-stained polyacrylamide gels. *Analytical Chemistry* 68, 850–858.

Somerville, C. and Koornneef, M. (2002) A fortunate choice: the history of *Arabidopsis* as a model plant. *Nature Reviews Genetics* 3, 883–889.

Tesnier, K., Strookman-Donkers, H.M., van Pijlen, J.G., van der Geest, A.H.M., Bino, R.J. and Groot, S.P.C. (2002) A controlled deterioration test of *Arabidopsis thaliana* reveals genetic variation in seed quality. *Seed Science and Technology* 30, 149–165.

Toda, T. (2000) Current status and perspectives of proteomics in aging research. *Experimental Gerontology* 35, 803–810.

Walters, C., Wheeleer, L.M. and Grotenhuis, J.M. (2005) Longevity of seeds stored in a genebank: species characteristics. *Seed Sciences Research* 15, 1–20.

35 Effects of Bean Seed Production Conditions on Germination and Hypocotyl Elongation Responses to Temperature and Water Potential

P. Moreau-Valancogne,[1] F. Coste,[1] C. Dürr[2] and Y. Crozat[1]

[1]*Laboratoire d'Ecophysiologie Végétale et Agronomie, Ecole Supérieure d'Agriculture, 55 rue Rabelais, BP 30748, 49007 Angers, Cedex 01, France;* [2]*INRA Agronomie LRM, rue Fernand Christ, 02007 Laon Cedex, France*

Abstract

In order to test whether production conditions are responsible for variations in emergence behaviour between seed lots, germination and hypocotyl elongation responses to temperature and water potential were measured in the laboratory for bean (*Phaseolus vulgaris* L.) seed lots produced under different conditions. Germination and hypocotyl elongation responses were studied on eight seed lots in temperatures ranging from 10°C to 40°C, and the response of germination to water potentials ranging from 0 to −1 MPa was also observed. Germination rates differed most between seed lots at low (10°C) and supra-optimal (40°C) temperatures. Differences in germination rate at low temperatures could be explained by differences in germination base temperature (Tb). Hypocotyl elongation was almost nil at 10°C and 40°C, although germination was still observed. Base temperature was lower and varied more between seed lots for germination (3.5–8.9°C) than for hypocotyl elongation (7.8–9.1°C). Base water potential differed slightly between seed lots (−1.9 to −2.5 MPa) and was correlated with germination Tb. These differences in germination and hypocotyl elongation responses between seed lots are likely to lead to field emergence differences, especially at low temperatures.

Introduction

It is well known that emergence percentage, rate and homogeneity vary according to seed lot. In order to avoid emergence problems, seed testing organizations, such as the International Seed Testing Association (ISTA), have attempted to establish vigour tests to characterize emergence ability of seed lots. Many studies, based on the results of vigour tests, have shown that production conditions can affect certain seed characteristics and are responsible for considerable variation between seed lots (Zanakis *et al.*, 1994; Dornbos, 1995). However, results of these tests are not well correlated with field emergence data (e.g. Kulik and Yaklich, 1982; Egli and TeKrony, 1995). These tests cannot be used to assess highly variable seed responses to the large range of conditions that can be encountered in the field.

Emergence modelling is used for gathering knowledge and predicting seed lot emergence in different sowing environments. Emergence models (e.g. Bouaziz and Bruckler, 1989a,b,c; Finch-Savage *et al.*, 1998; Dürr *et al.*, 2001) analyse two steps of emergence separately (i.e. germination and seedling growth before emergence), for which responses to environmental factors can differ. In these models, temperature and water potential are among the main driving variables of germination and early growth. Each of the two phases is thus characterized by Tb and a thermal time needed for its achievement, which enable prediction of germination and emergence in the field environment under fluctuating soil temperatures. Germination is also characterized by a base water potential (Ψb), above which thermal time is cumulated and below which no germination occurs.

A complementary approach to the use of seed vigour tests could be the characterization of seed lot responses using the input variables of emergence models. Base temperatures and base water potential are generally characterized for each species. Seed response to temperature and water potential are not often characterized for different seed lots of a given species produced under different environments. For common bean (*Phaseolus vulgaris* L.) in particular, desiccation conditions seem to greatly affect the results of vigour tests (Sanhewe and Ellis, 1996; Coste *et al.*, 2002). In this study, we tested the effect of various production conditions, and in particular desiccation conditions, on bean seed responses to temperature and water potential during germination and seedling elongation.

Materials and Methods

Seed production conditions

Production conditions of the seed lots are summarized in Table 35.1. Seed lots were produced from common bean crops, *P. vulgaris* 'Booster', sown on three occasions in 2003 and two occasions in 2004 in Angers (in the west of France). To obtain seed lots that were as homogeneous as possible, the pods were selected according to their colour, which is an indicator of the end of seed filling stage (Coste *et al.*, 2005), using a spectrophotometer (Minolta CM-503i). Subsequently, pods were either left to dry in the field, or harvested and dried under controlled conditions.

Table 35.1. Production characteristics of the seed lots: year of production and sowing date, desiccation rate and temperature.

Seed lot	Year	Sowing date	Desiccation rate (mg/g/h)	Desiccation temperature (°C)
Desiccation in the field				
F1	2003	23 May	6.4	22
F2	2003	2 June	5.7	22
F3	2003	11 June	7.5	23
F4	2004	4 June	5.0	18
F5	2004	18 June	2.2	16
Desiccation in chambers				
C1	2003	23 May	6.5	27
C2	2003	2 June	11.7	27
C3	2003	2 June	16.1	27

Five seed lots of about 10,000 seeds were desiccated in the field, one from each sowing date (i.e. F1, F2 and F3 in 2003; F4 and F5 in 2004). Three seed lots of about 12,000 seeds produced in 2003 from the first (C1) and second sowing dates (C2 and C3), were desiccated in small closed chambers. These chambers were placed in rooms with controlled temperature at 27°C. Inside the chambers, a range of relative humidities (RHs) were obtained by putting saturated salt solutions (NaCl or $CaCl_2$) or silica gel at the bottom of each chamber.

Temperature and RH were recorded near the pods throughout desiccation. Desiccation was precisely recorded by frequent measurements of seed water content on ten pod samples. Desiccation rates, expressed in mg/g/h (on a fresh weight basis), were calculated from the slope of the linear regression between water content and time interval between the end of the seed filling stage and end of desiccation (0.54 to ~0.15 g/g seed water content). All the seed lots were harvested by hand to avoid mechanical damage, a harvest feature that would not be due to the production conditions. Seed lots were hand-shelled and stored for 4 months at 5°C and 50% RH in paper bags. At the beginning of the measurements, the seed lots had reached a stable water content of ~0.11–0.13 g/g.

Measurements of seed lot characteristics

Germination

Germination was recorded at six different temperatures (i.e. 10°C, 15°C, 20°C, 25°C, 30°C and 40°C). Four replicates of 25 seeds were sown in plastic boxes (55 × 120 × 180 mm) on pleated blotting paper imbibed with 50 ml distilled water (one seed for two pleats). The boxes were then incubated in darkness in temperature regulated chambers. Germination was also recorded at four water potentials at 20°C: 0 MPa (demineralized water); and −0.2, −0.5 and −1.0 MPa (created by using various polyethylene glycol (PEG) 8000 MW solutions) (Michel and Kaufmann, 1973). Three replicates of 25 seeds were used in this experiment. For each replicate, seeds were sown in the same boxes as described above, on a layer of cotton wool (15 g) imbibed with 50 ml of PEG solution. The seeds were observed every 1 to 3 h, according to the temperature or water potential, and were considered to have germinated when their radicle protruded through the seed coat. Germinated seeds were progressively removed from the boxes.

Hypocotyl elongation

The six seed lots produced in 2003 (i.e. F1 to F3 and C1 to C3) were studied at six different temperatures (i.e. 10°C, 15°C, 20°C, 25°C, 30°C and 40°C), and the two seed lots produced in 2004 (i.e. F4 and F5) were studied at four temperatures (i.e. 15°C, 20°C, 25°C and 30°C). For each seed lot, three replicates of ten seedlings were grown in darkness in plastic boxes (100 × 100 × 160 mm) filled with 2 kg of white sand (150 μm diameter) moistened at 0.19 g/g by a nutrient solution (Saglio and Pradet, 1980). Water loss was compensated for by the addition of distilled water after weighing each box. The non-destructive measurements of hypocotyl lengths were carried out every 2 to 15 days, according to the temperature, in a dark room lit by yellow-green non-actinic lamplight.

Data fitting and statistical analysis

For each temperature and water potential, the relationship between cumulative germination and time from sowing was fitted to a Gompertz function:

$$G_t = G_{max} \times \exp[\exp(b - c \times t)]$$

where G_t is the cumulative germination at time t from sowing expressed in number of hours from sowing, G_{max} is the maximum cumulative germination expressed as the percentage of sown seeds, and b and c are shape parameters. From this fitted curve, the germination rate (i.e. $1/t(G)$) was calculated for several percentiles (i.e. 20%, 35%, 50%, 65% and 80%) as proposed by Gummerson (1986).

For each temperature, the relationship between hypocotyl length and time from germination was fitted to a Weibull function:

$$L_t = L_{max} \times [1 - \exp(-(b \times t)^c)]$$

where L_t is the hypocotyl length (in cm) at time t (hours from 80% germination), L_{max} is the maximum length expressed in cm, and b and c are shape parameters. The hypocotyl elongation rate (cm/h) was then calculated from the time needed to reach 5, 10, 15, 20 and 25 cm.

Germination or hypocotyl elongation rate of each percentile or length were thus plotted against temperature or water potential. In the range where this relationship was linear (i.e. 15°C to 25°C and −0.2 to −1.0 MPa), Tb and ψb could be calculated as the intercept of the linear regression with the *x*-axis. Finally, from the Tb or ψb of each percentile or length, the mean Tb or ψb value of each seed lot and its standard deviation were calculated. Mean values obtained for the different seed lots were statistically compared using Fisher's least significant difference (LSD) test at the 0.05 probability level.

Results and Discussion

The seed lots produced to undertake this study were as homogeneous as possible and were desiccated under known conditions in the field or under artificial conditions. This was in order to identify whether production conditions induce variations in seed germination or hypocotyl elongation responses. The seed lots differed in their response to both temperature and water potential, in germination, and to a lesser extent hypocotyl elongation.

The final germination percentages of the six seed lots were high (100%) regardless of temperature, except for 10°C (Fig. 35.1). At this temperature, final percentages differed between seed lots, ranging from 32% to 100%. Seeds that had not germinated after more than 240 h at 10°C were still alive, as shown by their ability to germinate quickly when placed back at 20°C (data not shown). This means that the temperature threshold above which germination becomes possible is not the same for all the seeds of a given lot.

Large differences in germination rate, between seed lots, were observed at extreme temperatures: at low temperatures in the suboptimal range (as shown at 10°C; Fig. 35.1) and temperatures above the optimum (i.e. 40°C). No differences were observed at 20°C. Regression curves of germination rate at 50% (i.e. $1/t$ (G = 50)) as a function of temperature (in the 10–30°C range) differed significantly between seed lots (T-test at α = 5%, data not shown). Germination responses to temperature therefore appeared to differ between seed lots. In the suboptimal range, germination rates differed between seed lots particularly at low temperatures (i.e. 10°C). In order to express these differences, a specific germination Tb was calculated for each seed lot. Calculated germination Tb values differed significantly between seed lots, and ranged from 3.5°C to 8.9°C (Table 35.2). Extreme Tb values were obtained for seed lots dried under natural conditions, in the field (F1 vs F4 and F5). This highlights the large variation in Tb values that can be generated by different natural production conditions.

Germination Tb is assumed to be a species specific characteristic (Ellis *et al.*, 1986, 1987), because it does not vary between cultivars. In this study, germination responses, especially to low and high temperatures, differed between seed lots of the same genotype that were produced under different conditions. Desiccation conditions, but not desiccation rate, appeared to affect the germination Tb value. In fact, few differences in germination Tb were observed between seed lots that were dried at different rates in chambers (i.e. seed lots C1 to C3). However, large differences were obtained between seed lots produced in the same conditions up until the desiccation phase, and then dried either in chambers or in the field (e.g. C1 vs F1). Moreover, other factors during seed development and before desiccation also affected the germination response to temperature, since the greatest differences were noticed between years (e.g. F2 vs F4).

Base water potential values ranged from −2.0 to −2.5 MPa between seed lots (Table 35.2). Germination Tb and Ψb values were correlated (Ψb = 0.11, Tb = 2.80; r^2 = 0.87, data not shown), so that seed lots, which were able to germinate at low temperature were also able to germinate at low water potentials (i.e. in harsher conditions). This relationship suggests that germination responses to cold temperatures and low water potentials would involve similar cellular processes. Such a

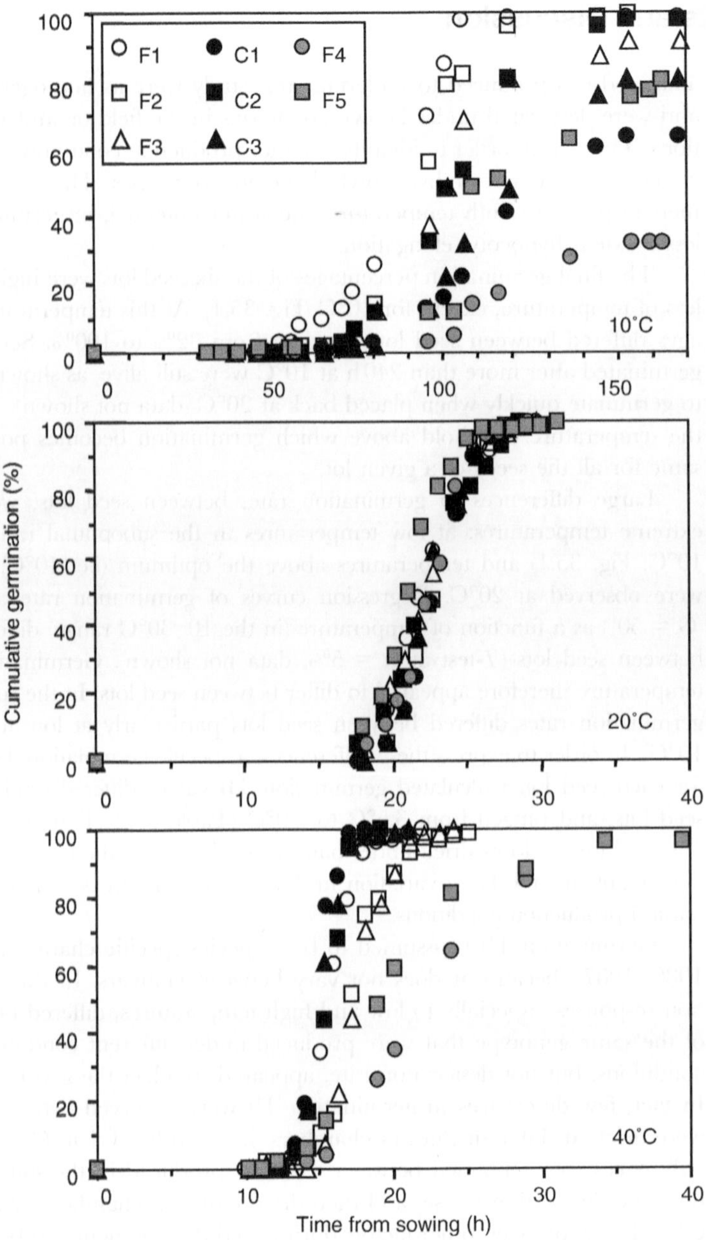

Fig. 35.1. Cumulative germination (percentage of sown seeds) of the eight seed lots described in Table 35.1 at three temperatures (i.e. 10°C, 20°C and 40°C). Note the different timescales.

correlation may reduce the number of measurements needed to characterize seed lots.

Hypocotyl elongation Tb varied less between seed lots than germination Tb (Table 35.2). This is in accordance with work by Lawlor *et al.* (1990), who showed a stronger effect of seed production conditions on germination Tb than on shoot elon-

Table 35.2. Seed lot characteristics determined in the laboratory.

	Germination		Hypocotyl elongation
Seed lot	Tb (°C)	ψb (MPa)	Tb (°C)
F1	3.5 ± 0.6 a	–2.5 ± 0.1 a	8.4 ± 0.3 bc
F2	5.6 ± 0.1 c	–2.3 ± 0.3 ab	8.6 ± 0.1 bc
F3	5.0 ± 0.5 b	–2.3 ± 0.1 ab	8.9 ± 0.2 cd
F4	8.9 ± 0.2 f	–1.9 ± 0.2 bc	8.4 ± 0.3 b
F5	8.9 ± 0.4 f	–1.9 ± 0.1 c	8.7 ± 0.2 bc
C1	6.2 ± 0.3 d	–2.0 ± 0.1 bc	8.4 ± 0.4 b
C2	6.4 ± 0.1 d	–2.0 ± 0.1 bc	7.8 ± 0.2 a
C3	7.0 ± 0.1 e	–2.1 ± 0.2 bc	9.1 ± 0.4 d

Germination base temperature (Tb), base water potential (ψb) and hypocotyl elongation Tb. Values are means ± standard deviation. Different letters in the same column indicate significant differences according to Fisher's least significant difference test (LSD) at the 0.05 probability level.

gation Tb in sorghum (*Sorghum bicolor* (L.) Moench). Variations in hypocotyl elongation Tb values were mainly due to the desiccation conditions, and particularly to the desiccation rate. In fact, the greatest difference between seed lots was observed for those dried in chambers at different rates (i.e. C1 to C3). However, the desiccation rate variations in the field were not sufficient to lead to differences in elongation Tb. No differences in hypocotyl elongation Tb were noticed for seed lots dried in field conditions from year to year in spite of great differences in micro-climate between years (i.e. F1 to F3 vs F4 and F5). Contrary to germination Tb, hypocotyl elongation Tb therefore appears not to be affected by production conditions before the desiccation phase.

Hypocotyl elongation was more sensitive to extreme temperatures than germination. Indeed, almost no hypocotyl elongation occurred at 10°C and 40°C for all the seed lots. Moreover, the calculated hypocotyl elongation Tb varied between seed lots from 7.8°C to 9.1°C (Table 35.2). This range is narrower than for germination and is up to 4.9°C higher than germination Tb. This higher Tb value for hypocotyl elongation is in agreement with several studies on other species (e.g. Lawlor *et al.*, 1990; Roman *et al.*, 1999).

In conclusion, as hypocotyl elongation is more sensitive to environmental factors and is a longer phase than germination, elongation differences may have a greater impact on emergence time differences. However, contrasting field conditions led to considerable differences in germination response to temperature, but to small differences in hypocotyl elongation response to temperature. The latter are principally affected by large variations in desiccation rates, and these variations are not observed under natural field conditions. Germination and hypocotyl elongation rates are lower at low temperatures, especially where seed lots differ the most. Hence, emergence time gaps between seed lots are more likely to occur at low temperatures. Further research is necessary to test whether these variables, introduced together with environmental sowing conditions into an emergence model, are able to reproduce differences, if any, in seed lot behaviours in the field.

Acknowledgements

We thank the staff of Ecole Supérieure d'Agriculture (ESA), Fédération Nationale des Agriculteurs Multiplicateurs de Semences (FNAMS) and Station Nationale d'Essais de Semences (SNES) for technical assistance, and the Conseil Régional des Pays de la Loire, the Conseil Général du Maine et Loire and Angers Agglomération (France) for funding this work. Many thanks to Dr Josephine Pithon for correcting the English.

References

Bouaziz, A. and Bruckler, L. (1989a) Modeling of wheat imbibition and germination as influenced by soil physical properties. *Soil Science Society of America Journal* 53, 219–227.

Bouaziz, A. and Bruckler, L. (1989b) Modeling wheat seedling growth and emergence: I. Seedling growth affected by soil water potential. *Soil Science Society of America Journal* 53, 1831–1838.

Bouaziz, A. and Bruckler, L. (1989c) Modeling wheat seedling growth and emergence: II. Comparison with field experiments. *Soil Science Society of America Journal* 53, 1838–1846.

Coste, F., Crozat, Y., Ladonne, F. and Wagner, M.H. (2002) Integrating seed age heterogeneity, desiccation rate and seed ageing rate for optimising both bean seed lot quality and seed yield. *Seed Science and Technology* 30, 586–596.

Coste, F., Raveneau, M.P. and Crozat Y. (2005) Spectrophotometrical pod colour measurement, a non-destructive method for monitoring seed drying? *Journal of Agricultural Science* 143, 1–10.

Dornbos Jr, D.L. (1995) Seed Vigour. In: Basra A.S. (ed.) *Seed Quality – Basic Mechanisms and Agricultural Implications.* The Haworth Press, New York, pp. 45–80.

Dürr, C., Aubertot, J.N., Richard, G., Dubrulle, P., Duval, Y. and Boiffin, J. (2001) Simple: a model for simulation of plant emergence predicting the effects of soil tillage and sowing operations. *Soil Science Society of America Journal* 65, 414–423.

Egli, D.B. and TeKrony, D.M. (1995) Soybean seed germination, vigor and field emergence. *Seed Science and Technology* 23, 595–607.

Ellis, R.H., Covell, S., Roberts, E.H. and Summerfield, R.J. (1986) The influence of temperature on seed germination rate in grain legumes. II. Intraspecific variation in chickpea (*Cicer arietinum* L.) at constant temperatures. *Journal of Experimental Botany* 37, 1503–1515.

Ellis, R.H., Hong, T.D. and Roberts, E.H. (1987) Comparison of cumulative germination and rate of germination of dormant and aged barley seed lots at different constant temperature. *Seed Science and Technology* 15, 717–727.

Finch-Savage, W.E., Steckel, J.R.A. and Phelps, K. (1998) Germination and post-germination growth to carrot seedling emergence: predictive threshold models and sources of variation between sowing occasions. *New Phytologist* 139, 505–516.

Gummerson, R.J. (1986) The effect of constant temperatures and osmotic potentials on the germination of sugar beet. *Journal of Experimental Botany* 37, 729–741.

Kulik, M.M. and Yaklich, R.W. (1982) Evaluation of vigor tests in soybean seeds: relationship of accelerated aging, cold, sand bench, and speed of germination tests to field performance. *Crop Science* 22, 766–770.

Lawlor, D.J., Kanemasu, E.T., Albrecht, W.C. and Johnson, D.E. (1990) Seed production environment influence on the base temperature for growth of sorghum genotypes. *Agronomy Journal* 82, 643–647.

Michel, B.E. and Kaufmann, M.R. (1973) The osmotic potential of polyethylene glycol 6000. *Plant Physiology* 51, 914–916.

Roman, E.S., Thomas, A.G., Murphy, S.D. and Swanton, C.J. (1999) Modeling germination and seedling elongation of common lambsquarters (*Chenopodium album*). *Weed Science* 47, 149–155.

Saglio, P.H. and Pradet, A. (1980) Soluble sugars, respiration and energy charge during aging of excised maize root tips. *Plant Physiology* 66, 516–519.

Sanhewe, A.J. and Ellis, R.H. (1996) Seed development and maturation in *Phaseolus vulgaris*. I. Ability to germinate and to tolerate desiccation. *Journal of Experimental Botany* 300, 949–958.

Zanakis, G.N., Ellis, R.H. and Summerfield, R.J. (1994) Seed quality in relation to seed development and maturation in 3 genotypes of soyabean (*Glycine max*). *Experimental Agriculture* 30, 139–156.

36 A Model of Seed Dormancy in Wild Oats (*Avena fatua*) for Investigating Genotype × Environment Interactions

S.K. Shepherd,[1] A.J. Murdoch,[1] J.M. Dunwell[2] and P.J. Lutman[3]

[1]*Seed Science Laboratory, Department of Agriculture, The University of Reading, Earley Gate, Reading RG6 6AR, UK;* [2]*School of Plant Sciences, The University of Reading, Whiteknights, Reading RG6 6AS, UK;* [3]*Rothamsted Research, Harpenden, Hertfordshire, AL5 2JQ, UK*

Abstract

Dormancy is an adaptive trait in seed populations that helps ensure that seed germination is distributed over time and occurs in environmental conditions suitable for seedling growth. Several genes, associated with seed dormancy in various plant species, have been integrated into a hypothetical dormancy model for *Avena fatua* L. (wild oats). Generally, the synthesis of, and sensitivity to, abscisic acid (ABA) during imbibition determines whether genes similar to those during maturation are expressed leading to a maintenance of dormancy during extended imbibition. Alternatively, there may be a shift towards expression of genes associated with gibberellins leading to germination. Environmental factors during maturation, after-ripening and imbibition are likely to interact with the genotype to affect gene expression and hence whether or not a seed germinates. In spite of the difficulties of working on a hexaploid species, *A. fatua* was selected for study because of its worldwide importance as a weed. Dormant and non-dormant genotypes of this species were also available. Gene expression studies are being carried out on three *A. fatua* genotypes produced under different environmental conditions to investigate the role of specific genes in dormancy and genotype × environment interactions in relation to dormancy.

Introduction

Dormancy is a complex trait thought to be controlled by a large number of genes. For example, in rice (*Oryza sativa* L.), six quantitative trait loci (QTLs) were found in the weedy rice strain SS18-2 with high order epistasis between several of

 Seeds: Biology, Development and Ecology (eds S. Adkins, S. Ashmore and S.C. Navie)

the QTLs, suggesting a complex genetic network in variation of seed dormancy (Gu *et al.*, 2004). Other investigations at the molecular level have compared gene expression or polypeptide levels between dormant and non-dormant genotypes, mutants and transgenic types, after-ripened seeds and seeds treated with gibberellic acid (GA_3) or ABA. Studies focused on ABA-responsive genes are dormancy or maturation-related, while GA-responsive genes are germination-related.

An early example of analysis of gene expression in wild oats (*Avena fatua* L.) is that of Dyer (1993) who used two-dimensional (2D) gel electrophoresis to monitor changes in soluble proteins and *in vitro* translation products of mRNA extracted from dry dormant and non-dormant line AN265 embryos. A year later, Li and Foley (1994) studied differential polypeptide patterns in imbibed dormant and non-dormant (i.e. after-ripened) line M73 *A. fatua* embryos. Li and Foley (1995) then focused on mRNA levels isolating complementary DNA (cDNA) clones (e.g. *Af30*), corresponding to genes that were differentially expressed in dormant and non-dormant (i.e. after-ripened) M73 embryos. Johnson *et al.* (1995) also characterized cDNA clones (e.g. *AfD1* and *AfD2*), which were differentially expressed in embryos of dormant and non-dormant (i.e. after-ripened) AN265 *A. fatua* caryopses during early imbibition (0–48 h). In 1997, Jones *et al.* used Northern blots to investigate gene expression of *Em1* (a late embryogenesis abundant (LEA) protein gene), α-*amylase 2-1* (*AMY2-1*) and *Viviparous 1* (*Vp1*) in embryos of mature seeds and after 48 h of imbibition. Further work using the dormant *A. fatua* line 'Bampton' was carried out to identify three proteins, Vp1 interacting protein 1 (VIP1), VIP2 and VIP3, which interact with Vp1 by a two-hybrid screen in yeast (Jones *et al.*, 2000).

In a diverse range of plant species, several other genes have been identified and associated with dormancy or germination, and expressed sequence tags (ESTs) from cDNA libraries have recently been sequenced. For example, EST libraries have been created for mature seed embryos from a dormant line of wheat (*Triticum aestivum* L. (Brevor)), produced in a cool, dormancy-inducing environment and also seeds imbibed in ABA (Zhang *et al.*, 2002). These treatments identify the genes that are expressed in dormant seeds. More recently, microarrays have been developed confirming that there is a complex network of genes involved in dormancy or regulation by ABA/GA (Suzuki *et al.*, 2003; Yazaki *et al.*, 2003).

Genes Associated with Maturation and Dormancy

Investigations into seed dormancy and germination have identified and explored, to some extent, the expression of many genes in several species. The main factors and genes from the literature have been summarized into a diagrammatic, hypothetical model of dormancy in *A. fatua* (Fig. 36.1). This model has been used to identify genes for further study of the interactions of genotype and environment on seed dormancy of *A. fatua*. The following account describes the factors and genes in the model.

Genes expressed during seed maturation and in dormant, imbibed seeds, are hypothesized to be important for the survival of dormant seeds in the soil to protect against desiccation and injury, rather than directly controlling seed dormancy. Seed maturation and dormancy is associated with a variety of genes, which are induced

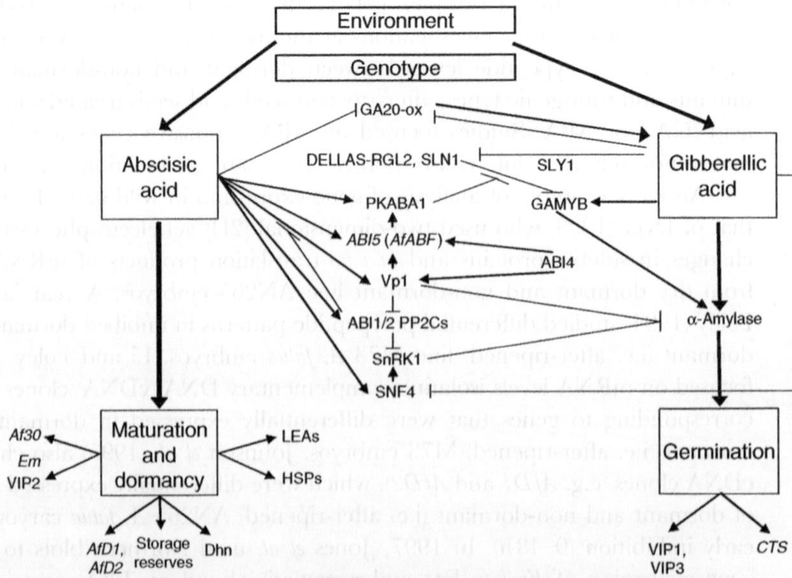

Fig. 36.1. Model of dormancy in wild oats (*Avena fatua*). Arrows indicate promotion or activation, bars indicate repression or inhibition.

by ABA and encode storage proteins and several LEA proteins that are thought to function as desiccation protectants (Dure *et al.*, 1989). In spite of their name, LEA proteins are not only found in seeds, but also occur in stress conditions during seedling development and are associated with tolerance to water stress in the nematode *Aphelenchus avenae* (Bastian) (Browne *et al.*, 2002). LEA proteins can suppress protein aggregation and inactivation under water stress (Goyal *et al.*, 2005). For example, dehydrins (Dhn) are a group of hydrophilic proteins, which can be induced by cold, water stress, freezing and ABA in plant mitochondria (Borovskii *et al.*, 2002). Also, heat shock proteins accumulate during late seed development and have been found in several plant species. They are thought to prevent aggregation of proteins during seed drying and assist in the refolding of proteins during imbibition (Lee *et al.*, 1995; Wehmeyer *et al.*, 1996).

Vp1 and *ABI3*

Vp1, or *ABI3* (*ABA insensitive 3*), from *Arabidopsis* (*Arabidopsis thaliana* (L.) Heynh.) are members of a large group of transcription factors that act as intermediates in regulating ABA-responsive genes during seed maturation and germination. *Vp1* also inhibits induction of germination-specific *AMY* genes in the aleurone (Hoecker *et al.*, 1995). Jones *et al.* (1997, 2000) studied *Vp1* in *A. fatua* in mature dry seed and during imbibition, and expression of *Vp1* was found to be higher in more dormant genotypes (Jones *et al.*, 1997). After-ripened seeds also had lower expression of *Vp1* during imbibition while seeds with secondary dormancy had higher expression. Similarly in wheat, *Vp1* expression in mature embryos of dormant and non-dormant wheat cultivars was positively correlated with the level of dormancy and sensitivity of the embryo to ABA (Nakamura and Toyama, 2001). However, sorghum (*Sorghum bicolor* (L.) Moench) embryos from a genotype susceptible to preharvest sprouting expressed a slightly higher level of *Vp1* mRNA than those from a line resistant to preharvest sprouting during embryogenesis (Carrari *et al.*, 2001). The timing of *Vp1* expression may, however, be important as it peaked earlier in seed development in the line susceptible to preharvest sprouting (Carrari *et al.*, 2001). During germination in the presence of fluridone, which inhibits ABA biosynthesis, *Vp1* expression was correlated with ABA sensitivity but not dormancy (Carrari *et al.*, 2001). However, inhibition of ABA biosynthesis may not be reliable in causing changes in *Vp1* expression to correlate with dormancy. In *Arabidopsis* expression of *Vp1*, together with *Leafy cotyledon1* (*Lec1*), *Fusca3* (*Fus3*) and *Em* revealed no correlation with dormancy in various ecotypes and mutants (Baumbusch *et al.*, 2004). They suggested that *ABI3* (*Vp1*) gene expression is regulated by cold and light, and not directly related to dormancy.

Preharvest sprouting can occur if there are cool and wet conditions during grain development of modern wheat varieties. In embryos excised from grains grown under such conditions, the majority of *Vp1* transcripts were not correctly spliced and produced transcripts of different sizes (McKibbin *et al.*, 2002). Consequently, the resultant non-functional proteins were unable to repress germination. So mis-splicing of this gene complicates expression studies as transcripts may not be functional.

ABI5 and basic leucine zipper genes

The *ABA insensitive 5* (*ABI5*) gene in *Arabidopsis* encodes a basic leucine zipper (bZIP) factor required for ABA response in seed and vegetative tissues (Finkelstein and Lynch, 2000). In rice, *ABI5* appears to interact with signalling effectors *ABI3*/*Vp1* and *ABI1* (Gampala *et al.*, 2002). Also, according to array work by Suzuki *et al.* (2003) in rice, *Vp1* activates *ABI5* and mediates ABA signalling. *Vp1* and *ABI5* bind ABA-responsive elements (ABREs) in promotion of ABA induced genes. The *ABI5* null mutant has a milder phenotype than *abi3* suggesting that other *ABI5*-related bZIPs have a role in the interaction of *Vp1* and ABA signalling (Finkelstein *et al.*, 2002). *ABI5* seems to be a positive regulator of ABA signalling in an ABA dependent manner (Suzuki *et al.*, 2003). Homologues from rice and wheat include *transcription factor responsible for ABA regulation 1* (TRAB1) (Hobo *et al.*, 1999) and *Triticum aestivum ABA response element-binding factor* (TaABF) (Johnson *et al.*, 2002), respectively.

PKABA1 and *TaABF*

ABA responsive protein kinase 1 (PKABA1) is an ABA induced unique serine/threonine protein kinase present in dormant seeds and in plants following 2 h of dehydration, and within 24 h of other stress treatments such as cold, osmotic stress and high salt concentration (Holappa and Walker-Simmons, 1995). In seeds, results suggest that PKABA1 is involved in the signal transduction pathway from ABA to suppress GA-inducible gene expression of *AMY* (Yamauchi *et al.*, 2002) and protease genes (Gómez-Cadenas *et al.*, 1999) in the aleurone layer of cereals.

TaABF and PKABA1 accumulated during wheat grain maturation and dormancy acquisition, and TaABF transcripts increased during 6–24 h of imbibition of dormant grains compared to after-ripened grains (Johnson *et al.*, 2002). TaABF is seed-specific and the TaABF protein is thought to serve as a physiological substrate for PKABA1 in the ABA signal transduction pathway during maturation and dormancy (Johnson *et al.*, 2002). It has been hypothesized that a high level of TaABF in the grain maintains dormancy (Johnson *et al.*, 2002).

ABI1 and *ABI2*

ABI1 and ABI2 are protein serine/threonine phosphatases (PP2Cs). Mutations leading to loss or decrease in function of ABI1/2, enhance responsiveness to ABA (Gosti *et al.*, 1999; Merlot *et al.*, 2001). The importance of ABI1 and ABI2 was also confirmed as expression was high in dormant *Arabidopsis* seeds (Lorenzo *et al.*, 2001). In beech (*Fagus sylvatica* L.) a PP2C similar to ABI1/2 was also detected in dormant seeds and increased after ABA treatment. However, it decreased and disappeared after dormancy-breaking treatment by stratification or GA, with no effect of drought stress, suggesting that this gene could be related to dormancy (Lorenzo *et al.*, 2001). Suzuki *et al.* (2003) showed that *ABI1/2* was induced with *Vp1* by ABA, while *Vp1* inhibited ABA induction of *ABI1/2*. This led to a model of 'feed forward regulation' of ABA signalling mediated by *Vp1* (Suzuki *et al.*, 2003).

ABI4

ABI4 is an AP2 domain transcription factor and is required during seed development but is not seed-specific (Finkelstein *et al.*, 1998). *ABI4* mutants have decreased sensitivity to ABA, salt and osmotic inhibition of germination and reduced expression of some LEA genes (Söderman *et al.*, 2000). Suzuki *et al.* (2003) hypothesize that *Vp1* and *ABI4* may balance ABA sensitivity for induction of some genes (e.g. *Em1*) through regulation of *ABI5* bZIPs.

Sugars and the SnRK1 complex

Sugars are hypothesized to be involved in responses mediated by ABA in seed maturation, dormancy and germination (Rock, 2000). Rook *et al.* (2001) proposed that interactions between ABA and sugar-signalling pathways shift a storage mode associated with seed development to a mobilization mode associated with germination. The sucrose non-fermenting 1 (SNF1)-related protein kinase (SnRK1) complex is hypothesized to be a mechanism in the 'cross-talk' between sugar, hormones and the environment during seed development and germination (Bradford *et al.*, 2003). In tomato (*Lycopersicon esculentum* Mill.), conditions that block germination such as dormancy, ABA, far-red light and low water potential maintain expression of *SNF4*. It is hypothesized that *SNF1* kinase is modulated by expression of *SNF4* (Bradford *et al.*, 2003). In cereals, *SnRK1* was expressed with *AMY1* and *2* in wheat embryos during grain development from 25 days post anthesis. SnRK1 was able to inhibit the AMY promoter activity (Laurie *et al.*, 2003). It is evident that sugar signalling interacts with *ABI3* (*Vp1*), *ABI4*, *ABI5* and *ABI1/2*. For example, *ABI4* and *ABI5* interact genetically in controlling several aspects of seed maturation and germination, as well as sugar metabolism and sensitivity to sugar (Brocard-Gifford *et al.*, 2003).

GA20-oxidase and DELLA proteins

Genes encoding GA20-oxidases have been found in many plant species, reviewed by Hedden and Kamiya (1997). Deficiency of this enzyme from the *GA20-ox2* gene leads to dwarfism in 'green revolution' rice (IR8) (Sasaki *et al.*, 2002). It is most highly expressed in mature leaves and so appears to be primarily involved in vegetative growth (Hedden, 2003). However, *GA20-ox1* is preferentially expressed in the reproductive organs (Sasaki *et al.*, 2002). In *Arabidopsis*, overproduction of GA20-oxidase led to a decrease in seed dormancy, increased stem elongation, earlier flowering and an increase in seed set (Huang *et al.*, 1998), possibly resulting from elevated levels in GA. In sorghum, expression was suppressed by ABA (Pérez-Flores *et al.*, 2003). *GA20-oxidase* was also identified as a candidate gene controlling the preharvest sprouting QTL on barley (*Hordeum vulgare* L.) chromosome 5H (Li *et al.*, 2004). When Calvo *et al.* (2004) isolated *GA20-ox1* from *F. sylvatica*, expression was undetectable in dry dormant seeds, low when imbibed under conditions that would break dormancy (i.e. stratified at 4°C or imbibed with GA), but significantly higher after treatments that inhibited germination – exposure to paclobutrazol (a GA_3 inhibitor) or amino oxyacetic acid (an ethylene biosynthesis inhibitor). Further treatments indicated that there was an interaction with gibberellins and ethylene (Calvo *et al.*, 2004).

On the other hand, the 'green revolution' gene from wheat, *reduced height* (*Rht*), is an allele associated with the GA signalling pathway via a different mechanism, which causes a reduced response to GA. It encodes a mutant form of a DELLA protein, a GA signalling repressor and so also causes GA malfunction (Peng *et al.*, 1999; Hedden, 2003; DELLA is an acronym of the single letter codes of five amino acids conserved in non-mutant 'DELLA' proteins). There are many *Rht* orthologues (e.g. *gibberellin insensitive* (*GAI*), *repressor-of-GA* (*RGA*), *RGA-like1-2* (*RGL1-2*), *slender1* (*SLN1*); Olszewski *et al.*, 2002). By studying various mutants, *RGL2* from *Arabidopsis* was found to be important in regulating seed germination and was enhanced by *GAI*, *RGL1* and *RGA* (Lee *et al.*, 2002). This is backed up by Tyler *et al.* (2004), who concluded that RGL2 is the most important protein controlling seed germination. It is suggested that DELLA proteins regulate the interaction of seed germination and the environment (Lee *et al.*, 2002).

AMY (*Amylase*)

AMY (*Amylase*) hydrolyses endosperm starch to provide energy for germination. ABA synthesized in the embryo interacts with *Vp1* to prevent expression of *AMY* genes in the aleurone layer during development (Hoecker *et al.*, 1999). High levels of germination-type *AMY* are produced during grain development when sensitive genotypes are exposed to cool temperatures (Mrva and Mares, 2001).

A Model of Key Dormancy and Germination Genes in *A. fatua*

From the earlier account of the association of variation in the expression of genes with germination and dormancy, a hypothetical model of these genes was developed. The model is based around the two hormones, ABA and GA. Levels of, and sensitivity to, these hormones may be influenced by the genotype and the environment. ABA is associated with maturation and dormancy, which in turn is associated with LEAs and other genes not directly controlling dormancy but have a role in, e.g. deposition of storage reserves or protecting the seed. Balancing ABA and GA are the proposed key genes involved in the switch between dormancy and germination of *A. fatua* seeds. In the centre is *Vp1*, which is a seed-specific gene linked with dormancy in *A. fatua* (Jones *et al.*, 1997, 2000). *Vp1* is induced by ABA and represses GA induced expression of germination genes such as *AMY* (Hoecker *et al.*, 1995). *Vp1* was also found to activate *ABI5* and together they bind ABA-responsive elements (ABREs) in promotion of ABA induced genes (Suzuki *et al.*, 2003). Vp1 interacts with the bZIP TRAB1 from rice (Hobo *et al.*, 1999); so Vp1 protein may also interact with the bZIP TaABF (Johnson *et al.*, 2002). *TaABF* is similar to *ABI5* of which there are many types from the same gene family, apparently with an overlapping function and perhaps individually important in different tissues. The homologue *Avena fatua ABF* (*AfABF*) appears to be the best candidate for a seed-specific bZIP gene involved in *A. fatua* dormancy. *TaABF* is ABA induced and is a physiological substrate for PKABA1 (i.e. PKABA1 phosphorylates TaABF) (Johnson *et al.*, 2002). *PKABA1* is ABA induced (Anderberg and Walker-Simmons, 1992) and overexpression represses GA-mediated expression of *AMY* (Gómez-Cadenas *et al.*, 1999). *ABI4* is hypothesized to balance ABA sensitivity via interaction with *ABI5*, *ABI4* and *ABI3* loci (Söderman *et al.*, 2000). *ABI1* and *ABI2* mRNA and enzyme activity increase in response to ABA (Leung *et al.*, 1997; Merlot *et al.*, 2001). These genes are negative regulators of ABA signalling and it is speculated that ABI1/ABI2 act as phosphatases in a negative feedback regulatory loop, allowing the cell to monitor ABA levels (Merlot *et al.*, 2001). Suzuki *et al.* (2003) presented a model of the feed forward regulation of ABA signalling mediated by *Vp1*. This shows that ABA induces *Vp1* and *ABI1/2*, while *Vp1* inhibits ABA induction of *ABI1/ABI2*. PP2Cs also phosphorylate and inactivate SNF1 protein kinases, SnRK1 (Sugden *et al.*, 1999), which may be a link between the SnRK1 complex and *ABI* genes.

High levels of *Lycopersicon esculentum SNF4* (*LeSNF4*) are associated with tomato seed dormancy and ABA, while a decrease in expression was observed during non-dormant seed germination or imbibition in GA (Bradford *et al.*, 2003). It is hypothesized that during maturation LeSNF4 (λ-subunit of the SnRK1 complex) binds to LeSNF1/SnRK1 promoting accumulation of storage reserves. *LeSNF1* transcripts were present in mature, dry and imbibed seeds, but ABA, GA, dormancy or germination status did not affect abundance. SnRK1 (with >80% amino acid identity to LeSNF1) is also present in wheat and was able to inhibit the AMY promoter in developing seeds (Laurie *et al.*, 2003).

Germination genes in the model include *GA20-oxidase*. ABA suppresses transcription of this gene in sorghum and beech (Pérez-Flores *et al.*, 2003; Calvo *et al.*, 2004). Expression was also low in GA_3 treated seeds (Calvo *et al.*, 2004). *Rht* orthologues such as the DELLA protein SLN1 from barley (Chandler *et al.*, 2002) and RGL2 from *Arabidopsis* (Dill and Sun, 2001) may also be important in seeds. RGL2 is enhanced by other DELLAs (Lee *et al.*, 2002) and protein degradation is induced by GA via Spindly-1 (SLY1) (Tyler *et al.*, 2004). ABA has no effect on GA enhanced SLN1 degradation (Gubler *et al.*, 2002). Also important in the GA response pathway is *GA-regulated myeloblastosis* (*GAMYB*), which is upregulated by GA (Gubler *et al.*, 1995) and binds to AMY and other promoters of genes encoding hydrolytic enzymes (Gubler *et al.*, 1999). SLN1 is thought to repress *GAMYB* via *PKABA1* (Zentella *et al.*, 2002). The *comatose* (*CTS*) locus also regulates germination potential by enhancing after-ripening, and increasing sensitivity to GA and pre-chilling (Russell *et al.*, 2000). *CTS*, an ATP binding cassette (ABC) transporter, does not increase in imbibed dormant seeds, but increases in non-dormant seeds and regulates transport of acyl-coenzyme As (acylCoAs) into the peroxisome (Footitt *et al.*, 2002).

Conclusions

Gene expression regulating the transition from dormancy to germination is interwoven, with importance of genotype determining the functional genes expressed, with the environment determining how these genes interregulate, and the balance of hormones. Interaction of genotype and environment determines the phenotype. From Fig. 36.1 it is apparent that several integrated genes have a key role in dormancy. Generally, it is likely that synthesis of, and sensitivity to, ABA during imbibition determines whether genes similar to those during maturation are expressed, particularly those genes that protect the seed during extended imbibition due to dormancy, or whether there is a shift towards GA associated genes and germination. The model will evolve as our knowledge improves. Further work is being carried out on *A. fatua* lines grown under different environmental conditions to produce seeds with a range of dormancy levels. Preliminary results suggest that *AfVp1* gene expression exhibits a genotype × environment interaction. Publicly available wheat and related species EST data has been used to identify homologues of the genes in the model to ascertain conserved regions for the design of primers to isolate these genes from *A. fatua*.

Acknowledgements

This project is funded by The University of Reading Research Endowment Trust Fund and The Lawes Agricultural Trust.

References

Anderberg, R.J. and Walker-Simmons, M.K. (1992) Isolation of a wheat cDNA clone for an abscisic acid-inducible transcript with homology to protein-kinases. *Proceedings of the National Academy of Sciences of the United States of America* 89, 10183–10187.

Baumbusch, L.O., Hughes, D.W., Galau, G.A. and Jakobsen, K.S. (2004) *LEC1*, *FUS3*, *ABI3* and *Em* expression reveals no correlation with dormancy in *Arabidopsis*. *Journal of Experimental Botany* 55, 77–87.

Borovskii, G.B., Stupnikova, I.V., Antipina, A.I., Vladimirova, S.V. and Voinikov, V.K. (2002) Accumulation of dehydrin-like proteins in the mitochondria of cereals in response to cold, freezing, drought and ABA treatment. *BMC Plant Biology* 11, 5.

Bradford, K.J., Downie, A.B., Gee, O.H., Alvarado, V., Yang, H. and Dahal, P. (2003) Abscisic acid and gibberellin differentially regulate expression of genes of the SNF1-related kinase complex in tomato seeds. *Plant Physiology* 132, 1560–1576.

Brocard-Gifford, I.M., Lynch, T.J. and Finkelstein, R.R. (2003) Regulatory networks in seeds integrating developmental, abscisic acid, sugar and light signaling. *Plant Physiology* 131, 78–92.

Browne, J., Tunnacliffe, A. and Burnell, A. (2002) Anhydrobiosis: plant desiccation gene found in a nematode. *Nature* 416, 38.

Calvo, A.P., Nicolas, C., Nicolas, G. and Rodriguez, D. (2004) Evidence of a cross-talk regulation of a GA 20-oxidase (*FsGA20ox1*) by gibberellins and ethylene during the breaking of dormancy in *Fagus sylvatica* seeds. *Physiologia Plantarum* 120, 623–630.

Carrari, F., Pérez-Flores, L., Lijavetzky, D., Enciso, S., Sanchez, R., Benech-Arnold, R. and Iusem, N. (2001) Cloning and expression of a sorghum gene with homology to maize *vp1*: its potential involvement in pre-harvest sprouting resistance. *Plant Molecular Biology* 45, 631–640.

Chandler, P.M., Marion-Poll, A., Ellis, P.M. and Gubler, F. (2002) Mutants at the *Slender1* locus of Barley cv. Himalaya: molecular and physiological characterization. *Plant Physiology* 129, 181–190.

Dill, A. and Sun, T. (2001) Synergistic derepression of gibberellin signaling by removing RGA and GAI function in *Arabidopsis thaliana*. *Genetics* 159, 777–785.

Dure, L. III, Crouch, M., Harada, J., Ho, T.H.D., Mundy, J., Quatrano, J., Thomas, T. and Sung, Z.R. (1989) Common amino acid domains among the LEA proteins of higher plants. *Plant Molecular Biology* 12, 475–486.

Dyer, W.E. (1993) Dormancy-associated embryonic messenger-RNAs and proteins in imbibing *Avena fatua* caryopses. *Physiologia Plantarum* 88, 201–211.

Finkelstein, R.R. and Lynch, T.J. (2000) The *Arabidopsis* abscisic acid response gene *ABI5* encodes a basic leucine zipper transcription factor. *Plant Cell* 12, 599–609.

Finkelstein, R.R., Wang, M.L., Lynch, T.J., Rao, S. and Goodman, H.M. (1998) The *Arabidopsis* abscisic-acid response locus *ABI4* encodes an APETALA 2 domain protein. *The Plant Cell* 10, 1043–1054.

Finkelstein, R.R., Gampala, S.S. and Rock, C.D. (2002) Abscisic acid signaling in seeds and seedlings. *The Plant Cell* 14, s15–s45.

Footitt, S., Slocombe, S.P., Larner, V., Kurup, S., Wu, Y., Larson, T., Graham, I., Baker, A. and Holdsworth, M. (2002) Control of germination and lipid mobilisation by *comatose*, the *Arabidopsis* homologue of human ALDP. *The EMBO Journal* 21, 2912–2922.

Gampala, S.S.L., Finkelstein, R.R., Sun, S.S.M. and Rock, C.D. (2002) ABI5 interacts with abscisic acid signaling effectors in rice protoplasts. *The Journal of Biological Chemistry* 277, 1689–1694.

Gómez-Cadenas, A., Verhey, S.D., Holappa, L.D., Shen, Q.X., Ho, T.H.D. and Walker-Simmons, M.K. (1999) An abscisic acid-induced protein kinase, PKABA1, mediates abscisic acid-suppressed gene expression in barley aleurone layers. *Proceedings of the National Academy of Sciences of the United States of America* 96, 1767–1772.

Gosti, F., Beaudoin, N., Serizet, C., Webb, A.A.R., Vartanian, N. and Giraudat, J. (1999) ABI1 protein phosphatase 2C is a negative regulator of abscisic acid signaling. *The Plant Cell* 11, 1897–1910.

Goyal, K., Walton, L.J. and Tunnacliffe, A. (2005) LEA proteins prevent protein aggregation due to water stress. *Biochemical Journal* 388, 151–157.

Gu, X.Y., Kianian, S.F. and Foley, M.E. (2004) Multiple loci and epistases control genetic variation for seed dormancy in weedy rice (*Oryza sativa*). *Genetics* 166, 1503–1516.

Gubler, F., Kalla, R., Roberts, J.K. and Jacobsen, J.V. (1995) Gibberellin-regulated expression of a *myb* gene in barley aleurone cells: evidence for Myb transactivation of a high-pI α-Amylase gene promoter. *The Plant Cell* 7, 1879–1891.

Gubler, F., Raventos, D., Keys, M., Watts, R., Mundy, J. and Jacobsen, J.V. (1999) Target genes and regulatory domains of the GAMYB transcriptional activator in cereal aleurone. *The Plant Journal* 17, 1–9.

Gubler, F., Chandler, P.M., White, R.G., Llewllyn, D.J. and Jacobsen, J.V. (2002) Gibberellin signalling in barley aleurone cells. Control of SLN1 and GAMYB expression. *Plant Physiology* 129, 191–200.

Hedden, P. (2003) The genes of the Green Revolution. *Trends in Genetics* 19, 5–9.

Hedden, P. and Kamiya, Y. (1997) Gibberellin biosynthesis: enzymes, genes and their regulation. *Annual Review of Plant Physiology and Plant Molecular Biology* 48, 431–460.

Hobo, T., Kowyama, Y. and Hattori, T. (1999) A bZIP factor, TRAB1, interacts with VP1 and mediates abscisic acid-induced transcription. *Proceedings of the National Academy of Sciences of the United States of America* 96, 15348–15353.

Hoecker, U., Vasil, I.K. and McCarty, D.R. (1995) Integrated control of seed maturation and germination programs by activator and repressor functions of *Viviparous-1* of maize. *Genes and Development* 9, 2459–2469.

Hoecker, U., Vasil, I.K. and McCarty, D.R. (1999) Signaling from the embryo conditions *Vp1*-mediated repression of alpha-amylase genes in the aleurone of developing maize seeds. *The Plant Journal* 19, 371–377.

Holappa, L.D. and Walker-Simmons, M.K. (1995) The wheat abscisic acid-responsive protein-kinase messenger-RNA, PKABA1, is upregulated by dehydration, cold temperature, and osmotic-stress. *Plant Physiology* 108, 1203–1210.

Huang, S., Raman, A.S., Ream, J.E., Fugiwara, H., Cerny, R.E. and Brown, S.M. (1998) Overexpression of 20-oxidase confers a gibberellin-overproduction phenotype in *Arabidopsis*. *Plant Physiology* 118, 773–781.

Johnson, R.R., Cranston, H.J., Chaverra, M.E. and Dyer, W.E. (1995) Characterization of cDNA clones for differentially expressed genes in embryos of dormant and nondormant *Avena fatua* L. caryopses. *Plant Molecular Biology* 28, 113–122.

Johnson, R.R., Wagner, R.L., Verhey, S.D. and Walker-Simmons, M.K. (2002) The abscisic acid-responsive kinase PKABA1 interacts with a seed-specific abscisic acid response element-binding factor, TaABF, and phosphorylates TaABF peptide sequences. *Plant Physiology* 130, 837–846.

Jones, H.D., Peters, N.C.B. and Holdsworth, M.J. (1997) Genotype and environment interact to control dormancy and differential expression of the *VIVIPAROUS 1* homologue in embryos of *Avena fatua*. *The Plant Journal* 12, 911–920.

Jones, H.D., Kurup, S., Peters N.C.B. and Holdsworth, M.J. (2000) Identification and analysis of proteins that interact with the *Avena fatua* homologue of the maize transcription factor *VIVIPAROUS 1*. *The Plant Journal* 21, 133–142.

Laurie, S., McKibbin, R.S. and Halford, N.G. (2003) Antisense SNF1-related (SnRK1) protein kinase gene represses transient activity of an alpha-amylase (α-*Amy2*) gene promoter in cultured wheat embryos. *Journal of Experimental Botany* 54, 739–747.

Lee, G.J., Pokala, N. and Vierling, E. (1995) Structure and *in vitro* molecular chaperone activity of cytosolic small heat shock proteins from pea. *Journal of Biological Chemistry* 270, 10432–10438.

Lee, S., Cheng, H., King, K.E., Wang, W., He, Y., Hussain, A., Lo, J., Harberd, N.P. and Peng, J. (2002) Gibberellin regulates *Arabidopsis* seed germination via *RGL2*, a *GAI/RGA*-like gene whose expression is upregulated following imbibition. *Genes and Development* 16, 646–658.

Leung, J., Merlot, S. and Giraudat, J. (1997) The *Arabidopsis ABSCISIC ACID-INSENSITIVE2* (*ABI2*) and *ABI1* genes encode homologous protein phosphatases 2C involved in abscisic acid signal transduction. *The Plant Cell* 9, 759–771.

Li, B.L. and Foley, M.E. (1994) Differential polypeptide patterns in imbibed dormant and after-ripened *Avena-fatua* embryos. *Journal of Experimental Botany* 45, 275–279.

Li, B.L. and Foley, M.E. (1995) Cloning and characterization of differentially expressed genes in imbibed dormant and after-ripened *Avena fatua* embryos. *Plant Molecular Biology* 29, 823–831.

Li, C., Ni, P., Francki, M., Hunter, A., Zhang, Y., Schibeci, D., Li, I., Tarr, A., Wang, J., Cakir, M., Yu, J., Bellgard, M., Lance, R. and Appels, R. (2004) Genes controlling seed dormancy and pre-harvest sprouting in a rice-wheat-barley comparison. *Functional and Integrative Genomics* 4, 84–93.

Lorenzo, O., Rodriguez, D., Nicolas, G., Rodriguez, P.L. and Nicolas, C. (2001) A new protein phosphatase 2C (*FsPP2C1*) induced by abscisic acid is specifically expressed in dormant beechnut seeds. *Plant Physiology* 125, 1949–1956.

McKibbin, R.S., Wilkinson, M.D., Bailey, P.C., Flintham, J.E., Andrew, L.M., Lazzeri, P.A., Gale, M.D., Lenton, J.R. and Holdsworth, M.J. (2002) Transcripts of *Vp-1* homeologues are misspliced in modern wheat and ancestral species. *Proceedings of the National Academy of Sciences of the United States of America* 99, 10203–10208.

Merlot, S., Gosti, F., Guerrier, D., Vavasseur, A. and Giraudat, J. (2001) The ABI1 and ABI2 protein phosphatases 2C act in a negative feedback regulatory loop of the abscisic acid signalling pathway. *The Plant Journal* 25, 295–303.

Mrva, K. and Mares, D.J. (2001) Induction of late maturity alpha-amylase in wheat by cool temperature. *Australian Journal of Agricultural Research* 52, 477–484.

Nakamura, S. and Toyama, T. (2001) Isolation of a *VP1* homologue from wheat and analysis of its expression in embryos of dormant and non-dormant cultivars. *Journal of Experimental Botany* 52, 875–876.

Olszewski, N., Sun, T.P. and Gubler, F. (2002) Gibberellin signaling: biosynthesis, catabolism and response pathways. *The Plant Cell* 14, s61–s80.

Peng, J.H., Richards, D.E., Moritz, T., Caño-Delgado, A. and Harberd, N.P. (1999) Extragenic suppressors of the *Arabidopsis gai* mutation alter the dose-response relationship of diverse gibberellin responses. *Plant Physiology* 119, 1119–1208.

Pérez-Flores, L., Carrari, F., Osuna-Fernandez, R., Rodriguez, M.V., Enciso, S., Stanelloni, R., Sanchez, R.A., Bottini, R., Iusem, N.D. and Benech-Arnold, R.L. (2003) Expression analysis of a *GA 20-oxidase* in embryos from two sorghum lines with contrasting dormancy: possible participation of this gene in the control of germination. *Journal of Experimental Botany* 54, 2071–2079.

Rock, C.D. (2000) Pathways to abscisic acid-regulated gene expression. *New Phytologist* 148, 357–396.

Rook, F., Corke, F., Card, R., Munz, G., Smith, C. and Bevan, M.W. (2001) Impaired sucrose-induction mutants reveal the modulation of sugar-induced starch biosynthetic gene expression by abscisic acid signalling. *The Plant Journal* 26, 421–433.

Russell, L., Larner, V., Kurup, S., Bougourd, S. and Holdsworth, M. (2000) The *Arabidopsis comatose* locus regulates germination potential. *Development* 127, 3759–3767.

Sasaki, A., Ashikari, M., Ueguchi-Tanaka, M., Itoh, H., Nishimura, A., Swapan, D., Ishiyama, K., Saito, T., Kobayashi, A., Khush, G.S., Kitano, H. and Matsouka, M. (2002) Green revolution: a mutant gibberellin-synthesis gene in rice. *Nature* 416, 701–702.

Söderman, E.M., Brocard I.M., Lynch T.J. and Finkelstein R.R. (2000) Regulation and function of the *Arabidopsis ABA-insensitive 4* gene in seed and abscisic acid response signaling networks. *Plant Physiology* 124, 1752–1765.

Sugden, C., Crawford, R.M., Halford, N.G. and Hardie, D.G. (1999) Regulation of spinach SNF1-related (SnRK1) kinases by protein kinases and phosphatases is associated with phosphorylation of the T loop and is regulated by 5-AMP. *The Plant Journal* 19, 433–439.

Suzuki, M., Ketterling, M.G., Li, Q.B. and McCarty, D.R. (2003) *Viviparous1* alters global gene expression patterns through regulation of abscisic acid signalling. *Plant Physiology* 132, 1664–1677.

Tyler, L., Thomas, S.G., Hu, J., Dill, A., Alonso, J.M., Ecker, J.R. and Sun, T.P. (2004) DELLA proteins and gibberellin-regulated seed germination and floral development in *Arabidopsis*. *Plant Physiology* 135, 1008–1019.

Wehmeyer, N., Hernandez, L.D., Finkelstein, R.R. and Vierling, E. (1996) Synthesis of small heat-shock proteins is part of the developmental program of late seed maturation. *Plant Physiology* 112, 747–757.

Yamauchi, D., Zentella, R. and Ho, T.H.D. (2002) Molecular analysis of the barley (*Hordeum vulgare* L.) gene encoding the protein kinase PKABA1 capable of suppressing gibberellin action in aleurone layers. *Planta* 215, 319–326.

Yazaki, J., Kishimoto, N., Nagata, Y., Ishikawa, M., Fujii, F., Hashimoto, A., Shimbo, K., Shimatani, Z., Kojima, K., Suzuki, K., Yamamoto, M., Honda1, S., Endo, A., Yoshida, Y., Sato, Y., Takeuchi, K., Toyoshima, K., Miyamoto, C., Wu, J., Sasaki, T., Sakata1, K., Yamamoto, K., Iba, K., Oda, T., Otomo, Y., Murakami, K., Matsubara, K., Kawai, J., Carninci, P., Hayashizaki, J. and Kikuchi, S. (2003) Genomics approach to abscisic acid- and gibberellin-responsive genes in rice. *DNA Research* 10, 249–261.

Zentella, R., Yamauchi, D. and Ho, T.H. (2002) Molecular dissection of the gibberellin/abscisic acid signaling pathways by transiently expressed RNA interference in barley aleurone cells. *The Plant Cell* 14, 2289–2301.

Zhang, D., Choi, D.W., Wanamaker, S., Fenton, R.D., Chin, A., Malatrasi, M., Turuspekov, Y., Walia, H., Akhunov, E.D., Kianian, P., Otto, C., Simons, K., Deal, K.R., Echenique, V., Stamova, B., Ross, K., Butler, G.E., Strader, L., Verhey, S.D., Johnson, R., Altenbach, S., Kothari, K., Tanaka, C., Shah, M.M., Laudencia-Chingcuanco, D., Han, P., Miller, R.E., Crossman, C.C., Chao, S., Lazo, G.R., Klueva, N., Gustafson, J.P., Kianian, S.F., Dubcovsky, J., Walker-Simmons, M.K., Gill, K.S., Dvorák, J., Anderson, O.D., Sorrells, M.E., McGuire, P.E., Qualset, C.O., Nguyen, H.T. and Close, T.J. (2002) Construction and evaluation of cDNA libraries for large-scale expressed sequence tag sequencing in wheat (*Triticum aestivum* L.). *Genetics* 168, 595–608.

37 Seed microRNA Research

R.C. Martin,[1] P.-P. Liu[2] and H. Nonogaki[2]

[1]*US Department of Agriculture–Agricultural Research Service (USDA–ARS), National Forage Seed Production Research Center (NFSPRC), 3450 SW Campus Way, Corvallis, OR 97331, USA;* [2]*Department of Horticulture, Oregon State University, Corvallis, OR 97331, USA*

Abstract

microRNAs (miRNAs) are key regulatory molecules that play critical roles in gene expression. The biological functions of miRNAs are important for developmental processes in plants and animals. Little is known about the functions of miRNAs in seeds. To gain a better understanding of the regulation of gene expression in seeds, it is necessary to characterize gene regulation by miRNAs, which are known to play significant roles in other plant organs and tissues. A simple method for isolating small RNAs from seeds and a non-radioactive detection system for seed-expressed miRNAs have been established earlier. In this study the established seed miRNA detection techniques, as well as a new method to isolate developing *Arabidopsis thaliana* (L.) Heynh. seeds from siliquae for miRNA analysis, are introduced. The information on miRNAs expressed in *Arabidopsis* and *Lycopersicon esculentum* Mill. and their target genes is also updated.

Introduction

microRNAs (miRNAs), small ~21 nucleotide regulatory molecules, were first discovered during developmental studies on *Caenorhabditis elegans* (Maupas) Dougherty and were found to be important in regulating the timing of larval development (Lee *et al.*, 1993). Recently there has been an explosion in research related to miRNAs. They have been found to be important in the regulation of developmental processes in a wide variety of animals and plants (reviewed by Ambros, 2004; Kidner and Martienssen, 2005). miRNAs are involved in diverse processes such as glucose-mediated insulin secretion (Poy *et al.*, 2004), brain morphogenesis (Giraldez *et al.*, 2005), cancer (Calin *et al.*, 2004; Gregory and Shiekhattar, 2005), as well as plant responses to dehydration, salinity, cold, gibberellic acid (GA) and abscisic acid (ABA) (Achard *et al.*, 2004; Sunkar and Zhu, 2004).

Interestingly, some of the plant miRNA regulatory pathways have been conserved between species as diverse as rice (*Oryza sativa* L.), *Arabidopsis thaliana* (L.)

Heynh. and lycopods (Floyd and Bowman, 2004; Axtell and Bartel, 2005), while some appear to be unique to monocots and dicots (Sunkar *et al.*, 2005). Plant miRNAs are processed from a stem loop precursor molecule by DICER-LIKE 1 (DCL1) (Kurihara and Watanabe, 2004), usually have near perfect complementarity to their target sequences and generally function to downregulate target gene expression by cleaving their target mRNA at this site (Llave *et al.*, 2002). There are also examples of plant miRNAs that act at the translational level (Aukerman and Sakai, 2003; Chen, 2004). miRNAs are involved in the regulation of many plant processes: floral development (Aukerman and Sakai, 2003; Chen, 2004), stem/leaf morphology (Palatnik *et al*, 2003; Laufs *et al.*, 2004), root development (Guo *et al.*, 2005; Wang *et al.*, 2005), hormone action (Achard *et al.*, 2004; Mallory *et al.*, 2005) and also in the production of miRNAs (Xie *et al.*, 2003; Vaucheret *et al.*, 2004). Plant miRNA research reports have focused mainly on floral, stem and leaf research. Seeds also need to be examined for miRNA expression and their potential function in development and germination. A simple miRNA isolation protocol and a system for screening multiple miRNA families expressed in seeds using non-radioactive probes and a miniblotter system have been developed (Martin *et al.*, 2005). All these established protocols, including detailed information on useful websites for miRNA research, will be summarized in this chapter. A method that was recently developed to extract developing seeds from *Arabidopsis* siliquae for miRNA detection is also introduced. Finally, updated information on the analysis of miRNAs and miRNA target genes in seeds, and potential problems in research will be discussed.

Semi-high Throughput miRNA Detection in Seeds

To develop a simple procedure for isolating miRNA from seeds, previous observations in preparing RNA from seeds were applied. While extracting RNA from seeds for gene expression analysis, such as RNA gel blotting and reverse transcriptase-polymerase chain reaction (RT-PCR), polysaccharide contamination is often a problem. During removal of polysaccharides from RNA samples using isopropanol, it was found that high molecular-weight (HMW) nucleic acids (i.e. genomic DNA and high molecular-weight RNA [mRNA and 28S and 18S rRNA]) could be separated from low molecular-weight (LMW) RNA (i.e. 5S rRNA, tRNA and small RNA) using isopropanol fractionation. This information was used to develop a relatively simple miRNA extraction procedure which has been described in detail in a recent publication (Martin *et al.*, 2005), and will only be described briefly here. The first step of extraction is essentially the same as the ordinary phenol-sodium dodecyl sulphate (SDS) RNA extraction method (Sambrook *et al.*, 1989). After lithium chloride precipitation, the miRNA remains in the supernatant and is further purified with a 40–50% (v/v) isopropanol precipitation. Much of the contaminating HMW RNA and DNA are removed by precipitation with 40% (v/v) isopropanol. The LMW RNA, which remains in the supernatant, is precipitated at a final isopropanol concentration of 50% (v/v). This precipitate is collected, washed with 80% (v/v) ethanol, dried and dissolved in formamide or water. The miRNA is then separated by electrophoresis on a 17% (w/v) polyacrylamide gel containing 7 M urea, transferred to a positively charged membrane (Hybond-N+, Amersham Biosciences, Piscataway, New Jersey, USA) using a semi-dry transfer unit (Bio-Rad Laboratories, Hercules, California, USA) and UV cross-linked to the membrane (for detailed instructions see Martin *et al.*, 2005).

Since commercial miRNA microarrays are not available for plant miRNA analysis, a method for semi-high throughput analysis of miRNAs expressed in seeds was developed. A miniblotter system (Immunetics, Boston, Massachusetts, USA), which is usually used for screening monoclonal antibodies, was applied for screening with multiple miRNA probes at the same time. This system has multiple wells that enable the application of 16 different probes to a preparative gel membrane at one time (Fig. 37.1a; Martin *et al.*, 2005). Essentially, LMW RNA is separated on a preparative gel and transferred to a positively charged membrane. The membrane is prehybridized in a container and then placed in the Immunetics system for hybridization. Approximately 170 μl of probe solution is added to the wells. Washing can be done in the same way as ordinary northern blotting using 2× SSC 0.2% (w/v) SDS at 65°C.

As use of radioactive probes is not feasible for this system, it is necessary to synthesize non-radioactive probes sensitive enough for this screening. This can be done using T7 RNA polymerase and digoxigenin (DIG) RNA labelling mix (Roche Applied Science, Indianapolis, Indiana, USA). For this purpose, DNA templates (oligomers) have to be designed based on the known miRNA sequences which can be found at the Sanger Institute miRBase web site (http://microrna.sanger.ac.uk/) (Table 37.1). A modified probe synthesis procedure using *mir*Vana™ miRNA probe construction kit (Ambion, Austin, Texas, USA) has been successfully used for the synthesis of non-radioactive miRNA probes (Martin *et al.*, 2005).

Immunetics screening allows identification of miRNAs expressed in seeds. The major miRNAs expressed in germinating tomato (*Lycopersicon esculentum* Mill.) seeds and seedlings have been identified using this method. Figure 37.1b shows an example of a preparative membrane, with miRNAs isolated from tomato seedlings, hybridized to different miRNA probes. Interestingly the same pattern of miRNA expression was observed in germinating *Arabidopsis* and tomato seeds, suggesting that these miRNAs may play an important role in germination (R.C. Martin, P.-P. Liu and H. Nonogaki, unpublished results).

The techniques described above can be applied for developing seeds. Immunetics screening has already been applied to identify miRNAs expressed in developing siliquae of *Arabidopsis*. However, using the whole siliqua for analysing miRNA expression in seeds could produce misleading results due to contamination with miRNAs expressed in siliquae. Moreover, manual dissection of developing seeds from siliquae is a time consuming process. Recently, a simple method to extract seeds from *Arabidopsis* siliquae has been developed (Fig. 37.2). The siliquae, which can be grouped based on stage/size, are carefully cut along the junction between the replum and the valves of the siliqua with a surgical blade. The opened siliquae are then placed in water containing fine sand (sand particle S-9887; Sigma, St. Louis, Missouri, USA) and shaken vigorously in a vortex several times, for a few seconds each time, to release the seeds into the water. As the particles settle, the seeds, which are lighter than sand, will accumulate on the surface of the sand and can be collected with a pipette. Both LMW and HMW RNAs were successfully extracted

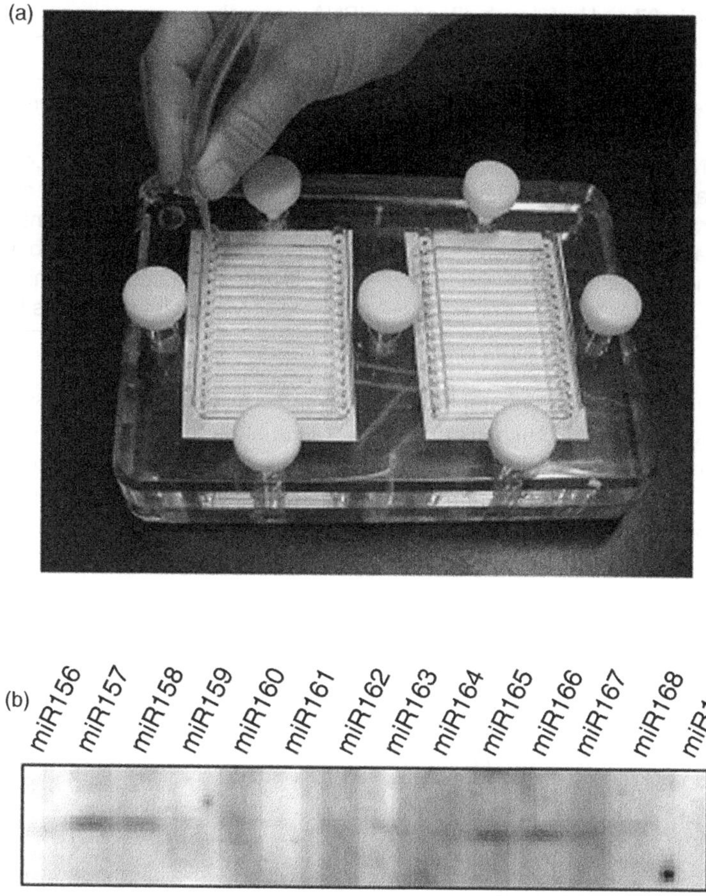

Fig. 37.1. (a) Photograph of Immunetics miniblotter system. Two wet prehybridized membranes are sealed between plastic plates. A pad and plastic overlay are placed between the membranes and the base of the plate. After removing excessive prehybridization solution from the wells, using a 200 μl pipette tip connected to a vacuum, 170 μl probe solution is applied to each well for hybridization. (b) Example of a membrane hybridized with the Immunetics screening system. Bands detected by chemiluminescence were visualized on X-ray film.

from developing seeds using this method (P.-P. Liu, R.C. Martin and H. Nonogaki, unpublished results).

miRNAs do Exist in Seeds, Now What?

miRNAs have already been shown to play critical roles in plant development. The identification of miRNAs expressed in seeds, by itself, does not provide much information towards a better understanding of the biology of seed development

Table 37.1. Useful web sites for miRNA research.

Web site name	Web site address
Memorial Sloan-Kettering Cancer Center, Computational Biology Center	http://www.microrna.org/
RNAi@elegansNet	http://c.elegans.tripod.com/RNAi.htm
Sanger Institute miRBase	http://microrna.sanger.ac.uk/
Arabidopsis Small RNA Project	http://asrp.cgrb.oregonstate.edu/db/
Bartel's Lab Homepage	http://web.wi.mit.edu/bartel/pub/
The McManus Lab	http://itsa.ucsf.edu/~micro/Faculty/McManus/miRarrays.html

and germination. Characterization of miRNA target gene expression patterns and their biological functions in seeds is essential as the next step. Fortunately, in the case of plants, the regions in miRNA target genes that are complementary to miRNA sequences are almost perfectly matched and are computationally predictable (Reinhart *et al.*, 2002; Rhoades *et al.*, 2002). For example, both miR156 and

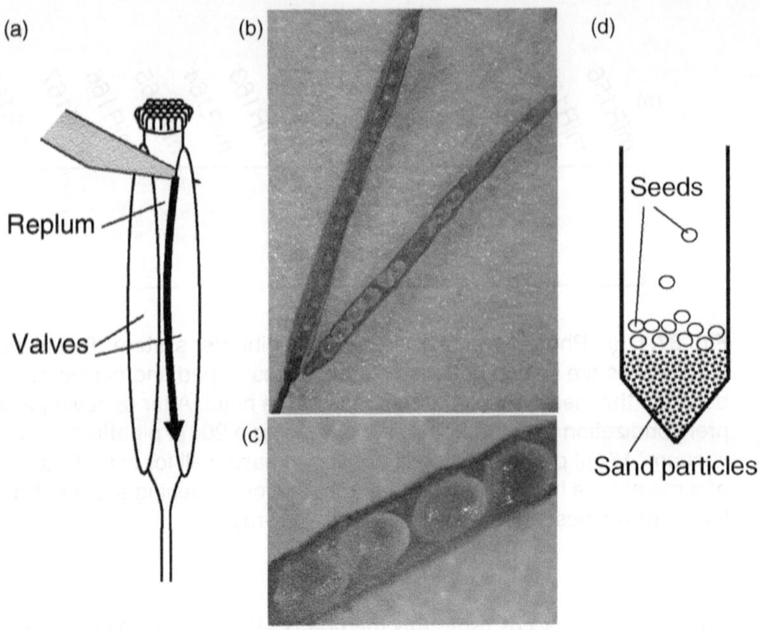

Fig. 37.2. Extraction of developing seeds from *Arabidopsis* siliquae. (a) Schematic representation of siliqua dissection as a preliminary step of seed extraction. Siliqua is gently opened at the valve-replum margin using a sharp surgical blade. (b) and (c) Opened siliquae containing intact developing seeds. (d) Schematic representation of seed extraction. Open siliquae are transferred to a plastic tube (15 or 50 ml) containing water and sand particles and briefly shaken using a vortex. Sand particles hit the opened siliquae and release the seeds into water. Siliquae can be removed by forceps. Since sand particles precipitate more rapidly than seeds, extracted seeds were recovered on the surface of the sand and can be collected with a pipette.

Table 37.2. Examples of miRNAs detected in germinating and germinated *Arabidopsis* seeds and their predicted targets. miRNAs are detected using Immunetics screening methods (data not shown).

miRNA	Predicted miRNA target genes	Function
miR156	*SQUAMOSA* promoter-binding protein-like (*SPL*)[a]	Flowering control
miR157	*SQUAMOSA* promoter-binding protein-like (*SPL*)[a]	Flowering control
miR162	*DICER-LIKE 1 (DCL1)*[b]	miRNA processing
miR165	Homeodomain-leucine zipper protein (*HD-ZIPIII*)[a]	Leaf polarity, vascular development
miR166	Homeodomain-leucine zipper protein (*HD-ZIPIII*)[a]	Leaf polarity, vascular development
miR173	*Trans*-acting siRNA generating transcript (*TAS1, TAS2*)[c]	ta-siRNA generation
miR319	TCP (basic-helix–loop-helix transcription factor)[d]	Leaf development
miR390	*Trans*-acting siRNA generating transcript (*TAS3*)[d]	ta-siRNA generation

[a]Rhoades *et al.* (2002); [b]Xie *et al.* (2003); [c]Allen *et al.* (2005); [d]Palatnik *et al.* (2003).

miR157, which are expressed in germinating *Arabidopsis* seeds (R.C. Martin, P.-P. Liu and H. Nonogaki, unpublished results), are predicted to target *SQUAMOSA* promoter-binding protein-like (*SPL*) genes (Table 37.2). These proteins belong to a group of transcription factors commonly recognized for their roles in floral development. *SQUAMOSA*-binding protein 1 and 2 were originally identified in *Antirrhinum majus* L. for their abilities to bind to *SQUAMOSA*, a floral meristem identity gene (Klein *et al.*, 1996). Sixteen *Arabidopsis* homologues of *SPL* have been identified, of which ten have miR156 complementary sites (Fig. 37.3). Therefore, it is necessary to determine which of the ten homologues is expressed during seed germination in order to understand the role of miR156 in germination (discussed later). miR162, which was also detected in germinating *Arabidopsis* and tomato seeds (data not shown), has been shown to target *DICER-LIKE 1* (*DCL1*) (Xie *et al.*, 2003). This is quite interesting since DCL1 is involved in the processing of pre-miRNA to form mature miRNA. This provides a negative feedback regulation for DCL1; as DCL1 protein increases, more miR162 is produced and *DCL1* mRNA is cleaved. With decreased levels of DCL1 protein, less miR162 is produced, and the stability of *DCL1* mRNA and its translation into proteins are promoted (Xie *et al.*, 2003). miR165/166 was also expressed in developing *Arabidopsis* seeds and is known to target the HD-ZIPIII transcription factor gene family, which is involved in various aspects of leaf, vascular tissue and flower development in *Arabidopsis* (McConnell *et al.*, 2001; Reinhart *et al.*, 2002; Rhoades *et al.*, 2002; Emery *et al.*, 2003). The TCP family of transcription factors, which contain a basic helix–loop–helix (bHLH) motif, are involved in meristem proliferation and are the target of miR319 (Cubas *et al.*, 1999; Palatnik *et al.*, 2003).

Although miRNAs are known for their negative regulatory roles, recent reports support a role for miRNAs in the production of *trans*-acting short interfering RNA (ta-siRNA). Both miR173 and miR390, which have also been detected in germinating *Arabidopsis* and tomato seeds and seedlings (R.C. Martin, P.-P. Liu and H. Nonogaki, unpublished results), have been shown to target ta-siRNA generating transcripts (*TAS1* or *TAS2* and *TAS3*, respectively) (Allen *et al.*, 2005). These

SPL2	GTG CTC TCT CTC TTC TGT CA
SPL3	TTG CTT ACT CTC TTC TGT CA
SPL4	CTG CTC TCT CTC TTC TGT CA
SPL5	CCG CTC TCT CTC TTC TGT CA
SPL6	GTG CTC TCT CTC TTC TGT CA
SPL9	GTG CTC TCT CTC TTC TGT CA
SPL10	GTG CTC TCT CTC TTC TGT CA
SPL11	GTG CTC TCT CTC TTC TGT CA
SPL13	GTG CTC TCT CTC TTC TGT CA
SPL15	GTG CTC TCT CTC TTC TGT CA

Fig. 37.3. Alignment of miRNA target sequences in the *SPL* gene family. Conserved bases are highlighted by black filling.

miRNAs target pre-ta-siRNA transcripts and lead to the production of ta-siRNA, which then negatively regulates other genes. Quite interestingly, *TAS3* targets *AUXIN RESPONSE FACTORS 3* and *4* (*ARF3* and *ARF4*) mRNA which encode transcription factors that are important in auxin signalling during plant growth and development. *TasiR-ARF*, which is the same as *TAS3*, was also identified through data base analysis and was shown to target *ARF2*, *ARF3* and *ARF4* (Williams *et al.*, 2005). These miRNAs and miRNA target genes are also summarized in Table 37.2.

After identifying the major families of miRNAs expressed in seeds, the pri- or pre-miRNAs can be amplified by RT-PCR and sequenced to identify the member of the miRNA gene family, which is transcribed in seeds (Schmittgen *et al.*, 2004). The scheme for primer design is shown in Fig. 37.4a and b. Alternatively, it may be possible to utilize 5' and 3' rapid amplification of cDNA end (RACE), with primers designed to bind the actual miRNA site, to isolate genes responsible for miRNA expression in seeds. This information will be needed, e.g. to characterize spatio-temporal expression of miRNA genes using promoter:reporter constructs. Since miRNAs often target a family of genes, it is also necessary to identify the specific target gene expressed particularly in seeds. For example, there are ten different *SPL* genes that are potential targets of miR156 (Fig. 37.3). In plants, most miRNAs negatively regulate target genes by mediating cleavage of target mRNA. A modified 5' RACE procedure can be used to amplify miRNA-mediated cleaved fragments, which can then be sequenced to identify the specific target gene (Llave *et al.*, 2002). It is, therefore, necessary to obtain the sequence of the expressed gene and the promoter region of the target gene for future experiments.

The ultimate goal of these experiments is to determine the biological roles of miRNAs and miRNA target genes in seed development and germination. Common approaches for analysing gene function include producing gene knockout and/or overexpression of mutants and examining the resulting phenotypes. Some miRNA target genes are redundant in their function (Millar and Gubler, 2005; Okushima *et al.*, 2005), therefore these approaches may not work. Target gene overexpression experiments also have potential pitfalls since miRNAs, which may be abundant in

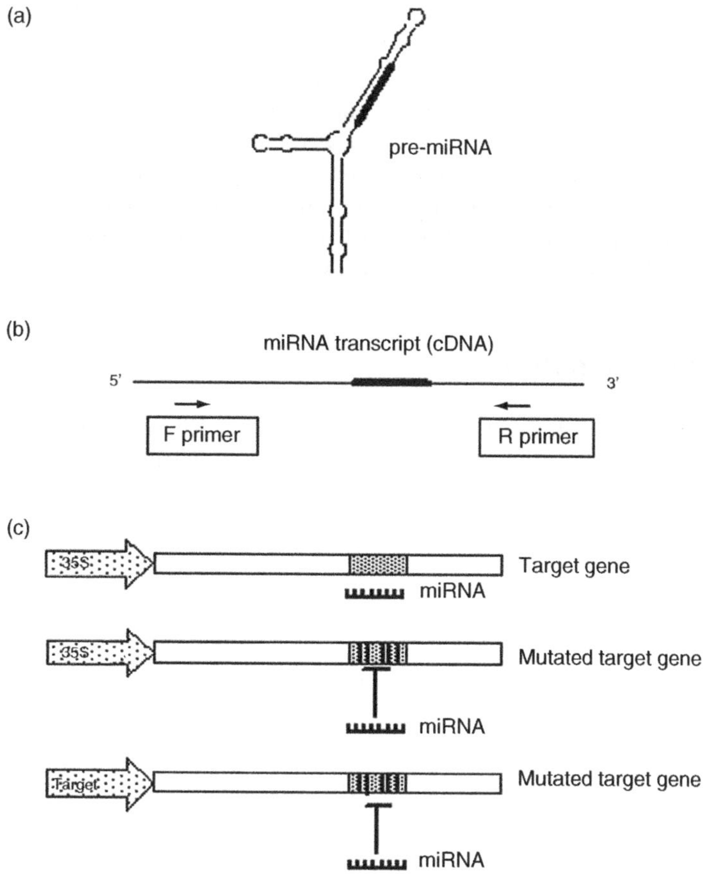

Fig. 37.4. (a) Schematic representation of secondary structure of a pre-miRNA molecule. (b) Schematic representation of primer design for RT-PCR to amplify miRNA transcripts or pri-miRNA. Forward and reverse primers can be designed in such a way that mature miRNA sequence (highlighted in thickened line) is present between two primers. (c) Schematic representation of construction of miRNA site-mutagenized target genes. Top: miRNA target gene without mutations driven by CaMV 35S constitutive promoter; middle: miRNA complementary site-mutagenized target gene driven by 35S promoter; bottom: miRNA complementary site-mutagenized target gene driven by its own promoter.

tissues, would still be able to repress the overexpressed target mRNAs. Also, overexpression, using a constitutive promoter such as CaMV35S, could cause pleiotropic effects if the gene is functioning in tissues where it is not normally expressed. A typical approach to characterize the function of the miRNA target gene is to mutagenize the miRNA complementary site in the target gene and express this mutant gene so that miRNA will no longer be able to cleave the target mRNAs (Fig. 37.4c). Constitutive expression of the mutated target gene has been informative in some instances (Achard *et al.*, 2004; Chen, 2004; Laufs *et al.*, 2004; Guo *et al.*, 2005), but can also produce pleiotropic effects on the plant phenotype in other cases (Wang *et al.*, 2005). However, if the authentic promoter and 3' non-coding region of the miRNA target gene is used, then the gene should be allowed to express precisely when it would normally be down-regulated by the miRNA (Palatnik *et al.*, 2003; Mallory *et al.*, 2004, 2005). This should better reflect the roles of miRNAs and target genes in plant growth and development. The approach is also promising for seed miRNA research.

The miRNA world is a rapidly advancing area of research. There are many groups studying various aspects of miRNA function and analysis. Several of these groups have web sites (Table 37.1) that include information, which could be very useful to people involved in miRNA research.

Acknowledgements

We are grateful to James Carrington, Zhixin Xie and Edward Allen, Center for Genome Research and Biocomputing at OSU, for their advice and helpful discussion. This work was supported by the National Science Foundation grant IBN-0237562 to H. Nonogaki.

References

Achard, P., Herr, A., Baulcombe, D.C. and Harberd, N.P. (2004) Modulation of floral development by a gibberellin-regulated microRNA. *Development* 131, 3357–3365.

Allen, E., Xie, Z., Gustafson, A.M. and Carrington, J.C. (2005) microRNA-directed phasing during trans-acting siRNA biogenesis in plants. *Cell* 121, 207–221.

Ambros, V. (2004) The functions of animal microRNAs. *Nature* 431, 350–355.

Aukerman, M.J. and Sakai, H. (2003) Regulation of flowering time and floral organ identity by a microRNA and its *APETALA2*-like target genes. *The Plant Cell* 15, 2730–2741.

Axtell, M.J. and Bartel, D.P. (2005) Antiquity of microRNAs and their targets in land plants. *The Plant Cell* 17, 1658–1673.

Calin, G.A., Sevignani, C., Dumitru, C.D., Hyslop, T., Noch, E., Yendamuri, S., Shimizu, M., Rattan, S., Bullrich, F., Negrini, M. and Croce, C.M. (2004) Human microRNA genes are frequently located at fragile sites and genomic regions involved in cancers. *Proceedings of the National Academy of Sciences of the United States of America* 101, 2999–3004.

Chen, X. (2004) A microRNA as a translational repressor of *APETALA2* in *Arabidopsis* flower development. *Science* 303, 2022–2025.

Cubas, P., Lauter, N., Doebley, J. and Coen, E. (1999) The TCP domain: a motif found in proteins regulating plant growth and development. *The Plant Journal* 18, 215–222.

Emery, J.F., Floyd, S.K., Alvarez, J., Eshed, Y., Hawker, N.P., Izhaki, A., Baum, S.F. and Bowman, J.L. (2003) Radial patterning of *Arabidopsis* shoots by class III *HDZIP* and *KANADI* genes. *Current Biology* 13, 1768–1774.

Floyd, S.K. and Bowman, J.L. (2004) Gene regulation: ancient microRNA target sequences in plants. *Nature* 428, 485–486.

Giraldez, A.J., Cinalli, R.M., Glasner, M.E., Enright, A.J., Thomson, J.M., Baskerville, S., Hammond, S.M., Bartel, D.P. and Schier, A.F. (2005) MicroRNAs regulate brain morphogenesis in zebrafish. *Science* 308, 833–838.

Gregory, R.I. and Shiekhattar, R. (2005) MicroRNA biogenesis and cancer. *Cancer Research* 65, 3509–3512.

Guo, H.S., Xie, Q., Fei, J.F. and Chua, N.H. (2005) MicroRNA directs mRNA cleavage of the transcription factor *NAC1* to downregulate auxin signals for *Arabidopsis* lateral root development. *The Plant Cell* 17, 1376–1386.

Kidner, C.A. and Martienssen, R.A. (2005) The developmental role of microRNA in plants. *Current Opinion in Plant Biology* 8, 38–44.

Klein, J., Saedler, H. and Huijser, P. (1996) A new family of DNA binding proteins includes putative transcriptional regulators of the *Antirrhinum majus* floral meristem identity gene *SQUAMOSA*. *Molecular and General Genetics* 259, 7–16.

Kurihara, Y. and Watanabe, Y. (2004) *Arabidopsis* micro-RNA biogenesis through Dicer-like 1 protein functions. *Proceedings of the National Academy of Sciences of the United States of America* 101, 12753–12758.

Laufs, P., Peaucelle, A., Morin, H. and Traas, J. (2004) MicroRNA regulation of the CUC genes is required for boundary size control in *Arabidopsis* meristems. *Development* 131, 4311–4322.

Lee, R.C., Feinbaum, R.L. and Ambros, V. (1993) The *C. elegans* heterochronic gene *lin-4* encodes small RNAs with anti-sense complementarity to *lin-14*. *Cell* 75, 843–854.

Llave, C., Xie, Z., Kasschau, K.D. and Carrington, J.C. (2002) Cleavage of *Scarecrowlike* mRNA targets directed by a class of *Arabidopsis* miRNA. *Science* 297, 2053–2056.

Mallory, A.C., Dugas, D.V., Bartel, D.P. and Bartel, B. (2004) MicroRNA regulation of NAC-domain targets is required for proper formation and separation of adjacent embryonic, vegetative, and floral organs. *Current Biology* 14, 1035–1046.

Mallory, A.C., Bartel, D.P. and Bartel, B. (2005) MicroRNA-directed regulation of Arabidopsis *AUXIN RESPONSE FACTOR17* is essential for proper development and modulates expression of early auxin response genes. *The Plant Cell* 17, 1360–1375.

Martin, R.C., Liu, P.-P. and Nonogaki, H. (2005) Seed microRNA techniques: simple purification of small RNAs from seeds and efficient detection of multiple microRNAs expressed in *Arabidopsis thaliana* and tomato (*Lycopersicon esculentum*) seeds. *Seed Science Research* 15, 319–328.

McConnell, J.R., Emery, J., Eshed, Y., Bao, N., Bowman, J. and Barton, M.K. (2001) Role of PHABULOSA and PHAVOLUTA in determining radial patterning in shoots. *Nature* 411, 709–713.

Millar, A.A. and Gubler, F. (2005) The Arabidopsis GAMYB-Like genes, MYB33 and MYB65, are microRNA-regulated genes that redundantly facilitate anther development. *The Plant Cell* 17, 705–721.

Okushima,Y., Overvoorde, P.J., Arima, K., Alonso, J.M., Chan, A., Chang, C., Ecker, J.R., Hughes, B., Lui, A., Nguyen, D., Onodera, C., Quach, H., Smith, A., Yu, G. and Theologis, A. (2005) Functional genomic anaylsis of the *AUXIN RESPONSE FACTOR* gene family members in *Arabidopsis thaliana*: unique and overlapping functions of *ARF7* and *ARF19*. *The Plant Cell* 17, 444–463.

Palatnik, J.F., Allen, E., Wu, X., Schommer, C., Schwab, R., Carrington, J.C. and Weigel, D. (2003) Control of leaf morphogenesis by microRNAs. *Nature* 425, 257–263.

Poy, M.N., Eliasson. L., Krutzfeldt, J., Kuwajima, S., Ma, X., Macdonald, P.E., Pfeffer, S., Tuschl, T., Rajewsky, N., Rorsman, P. and Stoffel, M. (2004) A pancreatic isletspecific microRNA regulates insulin secretion. *Nature* 432, 226–230.

Reinhart, B.J., Weinstein, E.G., Rhoades, M.W., Bartel, B. and Bartel, D.P. (2002) MicroRNAs in plants. *Genes and Development* 16, 1616–1626.

Rhoades, M.W., Reinhart, B.J., Lim, L.P., Burge, C.B., Bartel, B. and Bartel, D.P. (2002) Prediction of plant microRNA targets. *Cell* 110, 513–520.

Sambrook, J., Fritsch, E.F. and Maniatis, T. (1989) *Molecular Cloning: A Laboratory Manual*, 2nd edn. Cold Spring Harbor Laboratory Press, Cold Spring Harbor, New York.

Schmittgen, T.D., Jiang, J., Liu, Q. and Yang, L. (2004) A high-throughput method to monitor the expression of microRNA precursors. *Nucleic Acids Research* 32, e43.

Sunkar, R., Girke, T., Jain, P. and Zhu, J. (2005) Cloning and characterization of microRNAs from rice. *The Plant Cell* 17, 1397–1411.

Sunkar, R. and Zhu, J.K. (2004) Novel and stress-regulated microRNAs and other small RNAs from Arabidopsis. *The Plant Cell* 16, 2001–2019.

Vaucheret, H., Vazquez, F., Crete, P. and Bartel, D.P. (2004) The action of *ARGONAUTE1* in the miRNA pathway and its regulation by the miRNA pathway are crucial for plant development. *Genes and Development* 18, 1187–1197.

Wang, J., Wang, L., Mao, Y., Cai, W., Xue, H. and Chen, X. (2005) Control of root cap formation by microRNA-targeted auxin response factors in *Arabidopsis*. *The Plant Cell* 17, 2204–2216.

Williams, L., Carles, C.C., Osmont, K.S. and Fletcher, J.C. (2005) A database analysis method identifies an endogenous trans-acting short-interfering RNA that targets the *Arabidopsis ARF2*, *ARF3*, and *ARF4* genes. *Proceedings of the National Academy of Sciences of the United States of America* 102, 9703–9708.

Xie, Z., Kasschau, K.D. and Carrington, J.C. (2003) Negative feedback regulation of Dicer-Like1 in *Arabidopsis* by microRNA-guided mRNA degradation. *Current Biology* 13, 784–789.

38 Temporal Fulfilment of the Light Requirement for Seed Germination: an Example of its Use in Management of Rare Species

E.A. FITCH, J.L. WALCK AND S.N. HIDAYATI

Department of Biology, Middle Tennessee State University, Murfreesboro, TN 37132, USA

Abstract

Paysonia perforata (Rollins) O'Kane and Al-Shehbaz, and *P. stonensis* (Rollins) O'Kane and Al-Shehbaz are rare winter annuals found in frequently disturbed habitats, such as agricultural fields on floodplains. Both species are threatened by the loss of habitat, by its conversion to uses other than cultivation of annual crops. Considering the responses of the non-dormant portion of the seed set, *P. perforata* and *P. stonensis* seeds germinated to ≤18% in darkness with ≤10–12 weeks of light on alternating wet/dry substrate conditions (simulating current climatic conditions) and ≤26% with ≤4–6 weeks of light on continuously moist conditions (simulating predicted future conditions in a changed climate). This indicates that planting of crops along with any field preparation should be performed after seed dispersal (in early May) and before the seeds become photostimulated (i.e. by late July under current climatic conditions, but by late May under predicted future climatic conditions). Fields need to be harvested before seeds germinate and seedlings establish (in early September), and at very low disturbance levels since seeds are highly photostimulated in August under both climatic scenarios. Soybean (*Glycine max* (L.) Merr.) and maize (*Zea mays* L.) have been recommended as crops to provide an artificial means of maintaining the habitat of both *Paysonia* species. However, short-season forage crops should also be considered since the planting and harvesting times for them closely fit the life cycle and temporal dynamics of seed photoecology of the *Paysonia* species.

Introduction

A primary reason why non-dormant seeds of most species do not germinate while buried in soil is due to a requirement for light (Baskin and Baskin, 1989; Fenner and Thompson, 2005). However, seeds of some species receiving light during the dormancy-breaking period can germinate in darkness, whereas they would not have done so if they had not received light (e.g. Walck *et al.*, 1997a). If a soil disturbance buried these photostimulated seeds, they could germinate when environmental conditions became conducive. Subsequently, seedlings may fail to reach the soil's surface, and hence die and deplete the soil seed bank. Thus, the timing for the fulfilment of the light requirement during dormancy break in relation to disturbance plays a critical role in determining the germinability of non-dormant seeds in darkness and their capacity to form a soil seed bank.

Relatively few studies have examined the temporal dynamics for the fulfilment of the light requirement during dormancy break and considered their ecological consequences. Seeds of *Symphyotrichum pilosum* (Willd.) G.L. Nesom var. *pilosum* (syn. *Aster pilosus* Willd.), a dominant perennial plant during the early stages of old field succession in eastern USA, germinated to low percentages in darkness during simulated spring temperatures if they received ≤7 weeks of light during winter (cold stratification) and to moderate and high percentages if they received light for 8–12 weeks (Baskin and Baskin, 1985). In a comparative study, Walck *et al.* (1997b) found that seeds of the narrow-endemic *Solidago shortii* Torr. and Gray required no light to germinate to moderate and high percentages in darkness, whereas those of the geographically widespread *S. canadensis* L. var. *scabra* Torr. and Gray (syn. *S. altissima* L.) and *S. nemoralis* Ait. needed 10–12 weeks of light. The light requirement was fulfilled between 5 and 7 weeks for seeds of the weedy (winter) annual *Erophila verna* (L.) Chevall. (syn. *Draba verna* L.) (Baskin and Baskin, 1972). In contrast, seeds of other species apparently do not recall the light they received during dormancy break. Seeds of *Cyperus squarrosus* L. (syn. *Cyperus inflexus* Muhl.), a widely distributed summer annual sedge, germinated to low percentages in darkness at simulated spring temperatures regardless of the amount of light they received during cold stratification (Baskin and Baskin, 1971).

The implications for understanding the temporal aspects in fulfilling the light requirement for seed germination have so far not been addressed within a framework of management for a rare species. *Paysonia perforata* (Rollins) O'Kane and Al-Shehbaz (syn. *Lesquerella perforata* Rollins) occurs only in Wilson County, Tennessee (USA) and *P. stonensis* (Rollins) O'Kane and Al-Shehbaz (syn. *L. stonensis* Rollins) in Rutherford County, Tennessee (Rollins, 1993; Shea, 2001; O'Kane and Al-Shehbaz, 2002). *Paysonia perforata* is listed as an endangered species by the US Fish and Wildlife Service and by the state of Tennessee. Although *P. stonensis* was once a candidate for federal listing, it is still classified as endangered in Tennessee (US Fish and Wildlife Service, 1999, 2005; Bailey, 2004). These two species exhibit weedy tendencies by growing in disturbed habitats such as pastures, agricultural fields, along roadsides and on floodplains (Baskin and Baskin, 1990; Shea, 2001; US Fish and Wildlife Service, 2005). They are obligate winter annuals with their seeds being dormant at the time of dispersal and requiring warm (summer) temperatures for dormancy release (Baskin and Baskin, 1990; Fitch, 2004; US Fish and Wildlife Service, 2005). The longevity of the soil seed bank for *P. perforata* is unknown, but seeds of *P. stonensis* remain viable for at least 6 years (Baskin and Baskin, 1990).

Cultivation of annual crops, such as maize (*Zea mays* L.) or soybean (*Glycine max* (L.) Merr.), is recommended to provide an artificial way of maintaining the habitat of *P. perforata* and *P. stonensis* (Shea, 2001; US Fish and Wildlife Service, 2005).

Management advice was based on the above-ground life cycle of plants. Soil disturbance from ploughing, discing or harvesting should take place during late spring and summer: after seed maturation and dispersal in late April and early May, and before seed germination in September. Photoecology for germination has not been considered in the management protocols of these species. As such, we were concerned that disturbance, such as ploughing, that took place at an inopportune time (e.g. when seeds are photostimulated), could inadvertently deplete the soil seed bank.

The objective of the current research was to determine when the light requirement for germination was fulfilled under two types of substrate conditions: continuously moist vs alternating wet/dry. The information was then placed within a management framework, specifically to examine the planting and harvesting regimes of crops known to be grown in the same fields as both *Paysonia* species in light of global climatic change.

Materials and Methods

Plant material

Fresh seeds of *P. perforata* were collected in Wilson County, Tennessee on 27 April 2004, and those of *P. stonensis* in Rutherford County, Tennessee on 30 April 2004. They were stored dry in the laboratory between collection and the commencement of experiments.

General germination procedures

Germination tests were carried out in three light- and temperature-controlled incubators set at 12/12 h alternating temperatures of 25/15°C, 30/15°C and 35/20°C. These temperature regimes represent the mean daily maximum and minimum monthly air temperatures in central Tennessee: May and September, 25/15°C; June, 30/15°C; and July and August, 35/20°C (National Oceanic and Atmospheric Administration, 2002). A 14 h photoperiod extended 1 h before the beginning to 1 h after the end of the daily high temperature period. The white light source was 20 W cool white fluorescent tubes with a photon flux density (400–700 nm) at seed level ranging from 48 (top shelf) to 72 (bottom shelf) $\mu mol/m^2/s$.

Seeds were placed in 6 cm diameter plastic Petri dishes on limestone-derived topsoil, collected from an agricultural field in central Tennessee and moistened with distilled water. Three replicates of Petri dishes, each containing 50 seeds, were used for each test condition. Dishes were wrapped in plastic film to reduce water loss and those containing seeds incubated in darkness were wrapped additionally with two layers of aluminium foil. The criterion for germination was the emergence of the radicle. Viability of non-germinated seeds was determined by pinching them with forceps under a dissecting microscope to see if they contained firm, white (i.e. viable) embryos or soft, light brown (i.e. non-viable) ones. Tetrazolium tests confirmed that white embryos were viable and the brown ones were not.

Light requirements of seeds

Seeds of both species were incubated on topsoil starting on 10 May 2004. They were exposed to various light–dark treatments during a simulated sequence of natural seasonal temperature regimes. The temperature sequence was: May (4 week incubation period at 25/15°C) → June (4 weeks at 30/15°C) → July (4 weeks at 35/20°C) → August (4 weeks at 35/20°C). Seeds were given 0–16 weeks of light followed by 16–0 weeks of darkness. At 2-week intervals, germinated seeds were counted and removed from dishes and the remaining seeds were placed into darkness and left there for the rest of the 16-week period by wrapping the dishes with aluminium foil. At the end of the August regime, all seeds were incubated in darkness for 2 weeks at a September (25/15°C) temperature regime. Once seeds were moved to darkness, they were not checked for germination until the end of the treatment period. One set of controls was kept continuously in light during the temperature sequences (May → September), and a second set was kept in darkness.

The temperature and light–dark regimes were set up in duplicate: the topsoil substrate and seeds of one set were kept continuously moist throughout the 18-week experiment and those of a second set were alternately wet and dry. At the beginning of the experiment, the substrate for dishes in both treatments was moistened with distilled water. In the continuously moist treatment, the substrate was maintained wet by adding water as needed every time the dishes were checked. In the alternating wet/dry treatment, the substrate was allowed to dry for 13 days following the initial wetting until the end of the 2-week incubation period, and thereafter, they were given a 1-day wet/13-day dry cycle. Petri dish lids and plastic wraps were not used during the dry phase of the wet/dry treatment, but they were replaced during the wet phase. Seeds were fully imbibed within 2–4 h after wetting the substrate, and the moist substrate dried to its original weight within 24 h after the beginning of the dry phase of a cycle. The substrate in all dishes was moistened, and lids and plastic wraps were restored, before the seeds were placed into darkness. In addition, the set of controls that were in light was kept continuously moist during the September regime.

Statistical analyses

Germination percentages were determined based on the number of viable seeds. Means were compared by analysis of variance (ANOVA) followed by protected least significant difference tests (PLSDs, $P < 0.05$) (SPSS, 2003). Two ANOVAs were done: the first to detect species specific differences and the second to examine treatment effects (exposure time and moisture treatment) for each species. Data for percentages were arcsine square root transformed prior to statistical analysis. Germination percentages were also relativized by the following equation:

$$\text{relativized percentage} = \frac{\text{actual percentage} \times 100}{\text{highest percentage amongst a group of data}}$$

This standardization allowed comparisons among treatments to be equivalent when the amount of dormancy broken varied.

Results

Although germination responses were highly species specific regarding moisture treatment (species × treatment, $P = 0.011$), they were similar between the two species across the timed exposures (species factor, $P = 0.296$; species × time, $P = 0.052$; species × light × time, $P = 0.108$). Germination across the timed exposures was different between the two moisture treatments for *P. perforata* (treatment factor, $P = 0.006$; time factor, $P < 0.001$; treatment × time, $P < 0.001$) and *P. stonensis* (treatment factor, $P = 0.317$; time factor, $P < 0.001$; treatment × time, $P < 0.001$).

Seeds of *P. perforata* germinated to 4–15%, 40% and 54–61% in darkness when exposed to 0–10, 12 and 14–16 weeks of light, respectively, starting from May on an alternating wet/dry substrate, and they germinated to 3–8% and 19–29% in darkness when exposed to 0–4 and 6–16 weeks of light, respectively, on continuous moisture (Table 38.1). *P. stonensis* seeds germinated to 2–17% and 39–48% in darkness when exposed to 0–12 and 14–16 weeks of light, respectively, on an alternating wet/dry substrate, and they germinated to 3–13% and 23–37% in darkness when exposed to 0–6 and 8–16 weeks of light, respectively, on continuous moisture (Table 38.2). For both species, seeds in the light controls germinated to 67–86% and 42–47% on alternating wet/dry and constantly wet substrates, respectively.

Discussion

The timing and amount of light that seeds receive during the dormancy-breaking period can influence the germination of non-dormant seeds in darkness. Following dispersal in early May, seeds of *Paysonia* could be exposed to various periods of

Table 38.1. Germination percentages (mean ± SE) for seeds of *P. perforata*.

Month	Light	Dark	September	Alternating wet/dry			Continuously moist		
May	0	16	Dark	(0)	7 ± 2 ab	[8]	(0)	3 ± 1 a	[5]
	2	14	Dark	(0)	4 ± 2 a	[5]	(2)	6 ± 3 a	[9]
	4	12	Dark	(0)	5 ± 1 a	[6]	(0)	8 ± 2 a	[13]
June	6	10	Dark	(0)	8 ± 2 ab	[10]	(3)	23 ± 0 b	[36]
	8	8	Dark	(0)	9 ± 1 ab	[12]	(5)	21 ± 3 b	[33]
July	10	6	Dark	(0)	15 ± 1 b	[18]	(6)	29 ± 5 bc	[45]
	12	4	Dark	(0)	40 ± 5 c	[49]	(4)	22 ± 1 b	[35]
August	14	2	Dark	(0)	54 ± 3 d	[67]	(2)	19 ± 4 b	[30]
	16	0	Dark	(0)	61 ± 4 d	[76]	(5)	21 ± 1 b	[33]
	16	0	Light	(0)	67 ± 7 d	[83]	(3)	42 ± 11 c	[66]

Seeds were given 0–16 weeks of light followed by 16–0 weeks of darkness on an alternating wet/dry substrate or a continuously moist substrate. Exposure to the light–dark regimes occurred during a sequence of monthly temperatures: May (4 weeks, 25/15°C) → June (4 weeks, 30/15°C) → July (4 weeks, 35/20°C) → August (4 weeks, 35/20°C) → September (2 weeks, 25/15°C). Numbers in parentheses are percentages of germination that took place in light before placement into darkness, and those in brackets are relativized means. Values with dissimilar letters within columns are significantly different (PLSD, $P = 0.05$).

Table 38.2. Germination percentages (mean ± SE) for seeds of *P. stonensis*.

Month	Light	Dark	September	Alternating wet/dry			Continuously moist		
May	0	16	Dark	(0)	7 ± 3 abc	[7]	(0)	5 ± 1 a	[10]
	2	14	Dark	(0)	4 ± 3 ab	[4]	(0)	3 ± 2 ab	[6]
	4	12	Dark	(0)	2 ± 1 ab	[2]	(1)	9 ± 0 abc	[18]
June	6	10	Dark	(0)	3 ± 3 a	[3]	(5)	13 ± 5 acd	[26]
	8	8	Dark	(0)	9 ± 3 abc	[9]	(5)	23 ± 4 cde	[46]
July	10	6	Dark	(0)	10 ± 1 bc	[11]	(7)	30 ± 5 ef	[60]
	12	4	Dark	(0)	17 ± 5 c	[18]	(9)	24 ± 8 de	[48]
August	14	2	Dark	(0)	39 ± 9 d	[42]	(10)	37 ± 10 ef	[74]
	16	0	Dark	(0)	48 ± 2 d	[51]	(8)	33 ± 5 ef	[66]
	16	0	Light	(0)	86 ± 6 e	[91]	(5)	47 ± 1 f	[95]

Seeds were given 0–16 weeks of light followed by 16–0 weeks of darkness on an alternating wet/dry substrate or a continuously moist substrate. Exposure to the light–dark regimes occurred during a sequence of monthly temperatures: May (4 weeks, 25/15°C) → June (4 weeks, 30/15°C) → July (4 weeks, 35/20°C) → August (4 weeks, 35/20°C) → September (2 weeks, 25/15°C). Numbers in parentheses are percentages of germination that took place in light before placement into darkness, and those in brackets are relativized means. Values with dissimilar letters within columns are significantly different (PLSD, $P = 0.05$).

light on the soil surface until a burial event occurred sometime during summer. The photostimulated seeds would later germinate in soil when conditions became favourable, and seedlings from seeds covered by a few centimetres or more of soil would die before reaching the soil surface. Examining the temporal dynamics for the fulfilment of the light requirement is vital for understanding when disturbance, such as ploughing, could take place without inadvertently influencing the soil seed bank, particularly in the management of rare species in agroecosystems. However, management protocols also need to consider future environmental conditions.

For south-eastern USA, the Canadian Centre climate model shows a high degree of warming (5.5°C by 2100), which translates into lower soil moisture as higher temperatures increase evaporation (National Assessment Synthesis Team, 2000; Burkett *et al.*, 2001). In contrast, the Hadley Centre model simulates less warming (2.3°C) and a significant increase in precipitation by 2100, particularly during summer. The Canadian model shows precipitation decreasing in central Tennessee by ~20% in the next century, whereas the Hadley model predicts it increasing by about the same amount. Over the last 100 years in central Tennessee, annual rainfall trends show a very strong increase of ~25%. Our experimental design, which includes an alternating wet/dry treatment (current climatic scenario) and a continuously moist treatment (future scenario), allows an assessment on the impact of climatic change on seed ecology, particularly during the dormancy-break period, when precipitation is expected to increase the most.

Dormancy release was highly dependent on the moisture treatment (i.e. a higher percentage of seeds of both *Paysonia* species germinated on an alternating wet/dry substrate than on a continuously moist one) (Tables 38.1 and 38.2). Moreover, the timing for the fulfilment of the light requirement was shifted earlier under a continuously moist regime. Considering the non-dormant portion of the seed set (i.e. relativized means), seeds of *P. perforata* and *P. stonensis* germinated to low percentages in darkness (≤18%) with ≤10 and ≤12 weeks of light, respectively, on alternating wet/dry conditions. However, on continuously moist conditions *P. perforata* and *P. stonensis* seeds germinated to low percentages in darkness (≤26%), with ≤4 and ≤6 weeks of light, respectively. This would indicate that seeds of both species, dispersed in May, could lie on the soil surface for several weeks prior to being buried in July and still not germinate to high percentages in darkness during September under current climatic conditions, whereas the seeds that were buried in August would be expected to germinate to moderate to high percentages in darkness during September. On the other hand, seeds of both species dispersed in May would remain on the soil surface only until late May to early June under the predicted future climatic scenarios. Later, the seeds would be expected to germinate to relatively moderate to high percentages in darkness in September.

Planting along with any field preparation, such as ploughing, needs to be done following seed dispersal (in early May) but before seeds are photostimulated. Seeds fulfil their light requirement by late July under current climatic conditions and by late May under predicted future climatic scenarios. Crops would then need to be harvested before early September when seeds begin to germinate and seedlings form rosettes that overwinter until the following spring (Baskin and Baskin, 1990; Fitch, 2004). However, since seeds are highly photostimulated in August, under both present and future climatic scenarios, harvesting and other activities in the field have to be done at a very low disturbance level. Fields should remain fallow from September until May, when the above-ground life cycle of the plant is evident.

Maize and soybean have been grown primarily in the same fields as both *Paysonia* species, and are recommended for management purposes (Shea, 2001; US Fish and Wildlife Service, 2005). Maize for grain is planted usually from mid-April to early May in middle Tennessee and harvested from mid-September to mid-October, whereas maize for forage is planted from late April to mid-May and harvested from late August to mid-September (US Department of Agriculture, 1997; Tennessee Agricultural Statistics Service, 2002). Soybean for grain is planted typically from late May to late June and harvested from mid-October to mid-November. Planting and harvesting times of maize (grain and forage) overlap with the seed dispersal, and with the seedling and juvenile stages of the life cycle of both *Paysonia* species. Although the planting time for grain soybean would come after seed dispersal of *Paysonia*, harvesting could interfere with *Paysonia* seedling establishment.

An alternative agricultural practice would be to grow short-season crops for silage, haylage or hay. The planting and harvesting schedules for forage crops are: (i) not as strict as those for grain crops allowing adjustments to be made more easily depending on weather conditions; and (ii) fit closely the above-ground life cycle and seed photoecology of both *Paysonia* species. For example, a mixture of soybean, millet (*Setaria italica* (L.) Beauv.) and Johnson grass (*Sorghum halepense* (L.) Pers.) was grown for forage in a field with a population of *P. stonensis* during the 2004 growing season. The field was planted in early May following *Paysonia* seed dispersal and harvested in mid-July with relatively low disturbance at both times.

Acknowledgements

We thank the Tennessee Division of Natural Heritage for allowing access to study sites and collection of seeds. Support was provided by the Mary C. Dunn Graduate Scholarship from the Department of Biology and by the Faculty Research and Creative Activity Grant Program at Middle Tennessee State University.

References

Bailey, C. (2004) *Tennessee Natural Heritage Program Rare Plant List.* Tennessee Department of Environment and Conservation, Tennessee Division of Natural Heritage, Nashville, Tennessee.

Baskin J.M. and Baskin, C.C. (1971) The possible ecological significance of the light requirement for germination in *Cyperus inflexus. Bulletin of the Torrey Botanical Club* 98, 25–33.

Baskin, J.M. and Baskin, C.C. (1972) The light factor in the germination ecology of *Draba verna. American Journal of Botany* 59, 756–759.

Baskin, J.M. and Baskin, C.C. (1985) The light requirement for germination of *Aster pilosus* seeds: temporal aspects and ecological consequences. *Journal of Ecology* 73, 765–773.

Baskin, J.M. and Baskin, C.C. (1989) Physiology of dormancy and germination in relation to seed bank ecology. In: Leck, M.A., Parker, V.T. and Simpson, R.L. (eds) *Ecology of Soil Seed Banks.* Academic Press, San Diego, California, pp. 53–66.

Baskin, J.M. and Baskin, C.C. (1990) Seed germination biology of the narrowly endemic species *Lesquerella stonensis* (Brassicaceae). *Plant Species Biology* 5, 205–213.

Burkett, V., Ritschard, R., McNulty, S., O'Brien, J.J., Abt, R., Jones, J., Hatch, U., Murray, B., Jagtap, S. and Cruise, J. (2001) Potential consequences of climate variability and change for the southeastern United States. In: *National Assessment Synthesis Team. Climatic Change Impacts on the United States: The Potential Consequences of Climate Variability and Change. Foundation Report for the US Global Change Research Program.* Cambridge University Press, Cambridge, pp. 137–164.

Fenner, M. and Thompson, K. (2005) *The Ecology of Seeds.* Cambridge University Press, Cambridge.

Fitch, E.A. (2004) Photoecology of seed germination for two rare *Lesquerella* species: consequences for management. MSc thesis, Middle Tennessee State University, Murfreesboro, Tennessee.

National Assessment Synthesis Team (2000) *Climatic change impacts on the United States: the potential consequences of climate variability and change.* Overview report. US Global Change Research Program, Washington, DC.

National Oceanic and Atmospheric Administration (2002) *Comparative climatic data for the United States through 2002.* National Climatic Data Center, Asheville, North Carolina.

O'Kane, S.L. and Al-Shehbaz, I.A. (2002) *Paysonia,* a new genus segregated from *Lesquerella* (Brassicaceae). *Novon* 12, 379–381.

Rollins, R.C. (1993) *The Cruciferae of Continental North America: Systematics of the Mustard Family from the Arctic to Panama.* Stanford University Press, Stanford, California.

Shea, A.B. (2001) What's all the fuss about *Lesquerella*? *Tennessee Conservationist* 67(1), 16–19.

SPSS (2003) Version 12.0 for Windows. SPSS, Chicago, Illinois.

Tennessee Agricultural Statistics Service (2002) *Tennessee Agriculture.* Tennessee Department of Agriculture, Bulletin No. 37.

US Department of Agriculture (1997) *Usual Planting and Harvesting Dates for US Field Crops.* US Department of Agriculture, National Agricultural Statistics Service, Agricultural Handbook No. 628.

US Fish and Wildlife Service (1999) Endangered and threatened wildlife and plants; notice of reclassification of *Lesquerella stonensis* (Stones River bladderpod). *Federal Register* 64, 25216.

US Fish and Wildlife Service (2005) Technical/agency draft recovery plant for *Lesquerella perforata* (Spring Creek bladderpod). US Fish and Wildlife Service, Atlanta, Georgia.

Walck, J.L., Baskin, J.M. and Baskin, C.C. (1997a) A comparative study of the seed germination biology of a narrow endemic and two geographically widespread species of *Solidago* (Asteraceae). 1. Germination phenology and effect of cold stratification on germination. *Seed Science Research* 7, 47–58.

Walck, J.L., Baskin, J.M. and Baskin, C.C. (1997b) A comparative study of the seed germination biology of a narrow endemic and two geographically widespread species of *Solidago* (Asteraceae). 3. Photoecology of germination. *Seed Science Research* 7, 293–301.

39 Assisted Natural Recovery Using a Forest Soil Propagule Bank in the Athabasca Oil Sands

D.D. MACKENZIE AND M.A. NAETH

Department of Renewable Resources, University of Alberta, Edmonton, Alberta, T6G 2H1 Canada

Abstract

This research was undertaken to determine the natural regeneration potential of native boreal forest species following oil sands mining. The objective was to investigate the potential of two natural and locally available propagule sources (i.e. peat, and the litter and humus layer (LFH) of the forest floor) and two application depths (i.e. 10 and 20 cm) to enhance native plant establishment and diversity. A large-scale experiment was designed to assess treatment response at an operational level. In a controlled growth chamber, seeds and vegetative propagules were enumerated; in the field, plots were assessed for total canopy cover, canopy cover by species and species density. Preliminary growth chamber results indicate that application depth was the main influence on the number of viable propagules, while field results indicate that both propagule source and application depth are influencing vegetation establishments. LFH appears promising for oil sands revegetation, with a 20 cm application depth recommended for early community development.

Introduction

Surface mining in the Athabasca Oil Sands Region (AOSR) in the boreal forest of north-eastern Alberta, Canada, creates very large and intense disturbances, which must be reclaimed to self-sustaining ecosystems. The newly created landscapes lack plant propagules (i.e. seed, spores and vegetative parts) and require revegetation assistance from humans and through natural dispersal mechanisms (e.g. wind and birds). Oil sands operators use a peat–mineral mix as donor soil for reclamation. Peat is overstripped with underlying mineral soil to create more desirable physical and chemical properties than peat alone (Fung and Macyk, 2000). Revegetation efforts rely on transplanting trees and leaving native understorey species to establish through natural recovery. Low abundance of viable propagules (Fedkenheur and Heacock, 1979), dilution of total propagule density from overstripping (Putwain and Gillham, 1990), wind-dispersed weeds and poor representation of species adaptable to drier soil conditions (Box, 2003) make natural recovery problematic. The large disturbance size makes finding sources of native seed a major challenge. Native seed availability is limited to a few grasses; most boreal species are unavailable and unfeasible seed sources (Lanoue and Qualizza, 2000). Local seed collection is time-consuming and little is known about germination and establishment of understorey boreal species.

A source of native propagules is the LFH from upland forests. LFH is a thin organic horizon composed of fresh, intact, identifiable litter (L), fragmented and fermenting litter (F) and humus (H) with small amounts of moss (Paré *et al.*, 1993). Stripping a thin layer from the surface for large-scale operations is considered uneconomical. Research has shown that this horizon contains extensive numbers of propagules for revegetation after natural disturbances, such as fires and tree throw (Whittle *et al.*, 1997). Using LFH as a donor soil could add ecological benefits that far outweigh the cost of salvaging through additions of upland species tolerant of drier conditions and a more concentrated density of propagules at the surface (Leck *et al.*, 1989; Qi and Scarratt, 1998).

Effective use of donor soils as a propagule source for revegetation relies on application depth (Rokich *et al.*, 2000). The few studies assessing application depth of topsoil on mine sites have concluded that respreading at thin vs thick depths results in similar seedling density and species richness outcomes (Rokich *et al.*, 2000; Zhang *et al.*, 2001). Research on seedling establishment from a Jarrah forest in Australia found that the majority of seeds emerged from the upper 2 cm of surface soil and indicated that respreading topsoil at depths of <2 cm could maximize its use (Grant *et al.*, 1996).

Using donor soils as a revegetation propagule source on mine sites is well documented in subtropical, temperate and arid regions, however its applicability in the boreal forest has not been well researched (Iverson and Wali, 1981; Koch *et al.*, 1996; Holmes, 2001; Zhang *et al.*, 2001). The current research focuses on increasing our knowledge of vascular plants in the soil propagule bank from peat and LFH, two common organic horizons found in the AOSR. Differences in soil propagule banks from a postharvested lowland and upland forest, dominated by black spruce (*Picea mariana* (Mill.) BSP) and trembling aspen (*Populus tremuloides* Michx.), respectively, were assessed prior to salvaging at the donor site and after placement on the receiver site. Donor soils were applied at thin (~10 cm) and thick (~20 cm) application depths. The research was conducted at an operational scale, with large treatments to incorporate responses to equipment used in normal operations.

Materials and Methods

The research area is located at the Syncrude mine site (57°05.95' N, 11°38.90' W) north of Fort McMurray, Alberta, Canada. Mean annual temperature and precipitation for the region are 1.5°C and 435 mm, respectively (Hackbarth and Nastasa, 1979). Soils are organic in lowlands and fine textured minerals in uplands. Predisturbance vegetation is representative of the mixed wood boreal forest. Organic soils are dominated by black spruce and tamarack (*Larix laricina* (Du Roi) K. Koch), and mineral soils are dominated by a coniferous–deciduous forest mix. Uplands typically consist of trembling aspen, balsam poplar (*Populus balsamifera* L.) and white spruce (*Picea glauca* Moench) (Fung and Macyk, 2000).

Donor sites

In late September 2003 donor sites were divided into adjacent peat and LFH areas for salvaging. Organic horizons at the peat donor site were >40 cm and the LFH donor soil had a thin LFH layer (7.5 cm), both sites had an underlying fine textured mineral soil. Two 300 m transects and one 240 m transect spread 75 m apart were used to sample the soil propagule bank and presence or absence of above-ground vegetation at 10 m intervals. A 0.1 m^2 quadrat was used to assess species composition adjacent to the soil samples. Adjacent to the quadrat a 10 × 10 cm area of surface organic horizon was sampled at 0–10 cm. A serrated knife was used to cut dimensions of the sample. Core samples did not allow separation of upper and lower strata in the field. The full depth of the organic horizons <10 cm was sampled. Samples were split in half and depths recorded as upper and lower strata. Salvaging donor soils occurred in late October 2004 using large bulldozers. Peat and LFH donor soils were overstripped with 5–20 cm of underlying mineral soil. Both donor soils were stockpiled until treatment set up at the receiver site in late February 2004.

Receiver site

A complete randomized design consisting of two factors (i.e. donor soil type) and two levels (i.e. application depth) was applied. Treatments were replicated three times and experimental units sized to accommodate equipment and material availability. Each experimental unit was 25 m wide by 150 m long. Treatments were applied with standard size mining haul trucks, hoes and bulldozers. The frozen peat lumps were respread in early June to fill in bare areas and the desired application depth was obtained using a mid-size bulldozer pulling a 15 m wide pipe with one pass-down slope and one pass-up slope.

All treatments were sampled immediately after the peat treatment redistribution. Two randomly located transects were placed within each treatment and samples were taken along each transect every 10 m using a 7.5 cm diameter core at a 7.5 cm depth. Core size was based on a sampling area sufficient to include seed and plant vegetative parts. Vegetation was sampled in late July using three randomly located transects per replicate. A 0.01 m^2 quadrant was used to assess emergent density.

Growth chamber methods

The 10 × 10 cm samples from the donor site were resampled with a 5 cm diameter core to reduce space required in the controlled growth chamber. The samples were thoroughly mixed by hand and spread to a depth of 1–2 cm in 8 × 12 cm plastic containers lined with 1 cm of vermiculite. Samples from the receiver site were spread on 12.5 × 12.5 cm plastic containers to a depth of 2 cm using the same mixing procedures described above. Samples were placed randomly throughout the growth chamber. Soils within the containers were kept moist. Growth chamber temperatures mimicked growing conditions at Fort McMurray (i.e. 21°C during the day for 16 h and 15°C at night for 8 h). Emergence was monitored for 7 months to confirm no new seedlings emerged for each enumeration period. Samples were remixed after each enumeration period to promote emergence by bringing up buried seeds and reducing thickness of the moss layer to promote light penetration (Thompson *et al.*, 1997). Some species were not identifiable due to death between enumeration periods or prior to emergence of identifiable structures.

Data analyses

Two-way analysis of variance was used to determine significant effects between treatments for the response variables. Response variables were total density, species richness (R), species diversity (H'), evenness (E) and Sorenson's qualitative similarity index (S) to above-ground vegetation at the donor sites. Species richness was calculated as the total number of species in each donor site and each replicate at the receiver site. Species richness excluded unidentified individuals (i.e. *Salix* and *Carex*). Diversity was calculated using the Shannon–Wiener index and evenness was calculated using the formula $E = H'/\log_{10}R$ (Magurran, 1988). Sorenson's similarity index was calculated using the formula $S = 2 \times N/(2 \times N + R1 + R2)$, where N = number of species found in both sites, and R1 and R2 are species richness in donor site and receiver site, respectively. Response variables that did not meet homogeneity of variances were log-transformed or square-root-transformed. When transformations did not improve heterogeneity of variance a non-parametric alternative, the Scheirer-Ray Hare test, was conducted (Scheirer *et al.*, 1976). A P value of 0.05 was used to detect differences between donor soil treatments and a P value of 0.10 was used to compare application depths.

Results and Discussion

Soil propagule bank at donor sites

A total of 1215 germinants emerged from the 172 samples, 1002 from upland and 213 from lowland donor soils. More than twice as many viable propagules/m^2 emerged from the LFH than the peat donor site (Table 39.1). This supports results from Moore and Wein (1977) and Hill and Stevens (1981) where higher propagule density was found on upland soils of deciduous forests vs organic soils of bogs. The lower abundance of propagules in bogs is not fully understood. Hill and Stevens (1981) postulated lower abundance in peat due to its impermeability, leaving seeds on the surface and hence, higher susceptibility to seed predation and germination. However, Jauhiainen (1998) and Mcgraw (1987) found viable seeds at depths >40 cm in bogs. Several researchers found high viable seed numbers in peat soils, often associated with increased seed longevity caused by cold, wet anaerobic conditions (Champness and Morris, 1948; Ødum, 1965; Granström, 1988). The high seed numbers are often represented by few species that may be tolerant of soil conditions leading to increased seed longevity (Warr *et al.*, 1993). Differences in species reproductive strategies between donor sites may contribute to soil propagule abundance differences. Species found in harsh soil conditions, such as bogs, rely more on vegetative reproduction, resulting in less seed inputs, whereas species in more suitable growing conditions may rely more upon seed regeneration, therefore leaving more seed dispersers on upland sites, thus contributing to a higher seed bank pool (Jauhiainen, 1998; Grime, 2001).

Propagule density, species richness, diversity and similarity decreased with increasing depth in both donor sites (Table 39.1). Evenness was higher in the lower stratum of the peat donor site. Propagule abundance and species richness generally decreased with increasing depth (Leck *et al.*, 1989; Jackson *et al.*, 1996). Implications of salvaging a donor soil too deep will reduce propagule availability at the receiver site surface due to dilution effects (Tacey and Glossop, 1980; Rokich *et al.* 2000).

Soil propagule bank at receiver site

From 336 samples a total of 786 germinants was enumerated. No significant differences were found between donor soils for total propagule density (Table 39.2). The large reduction in soil propagules compared to what was originally at the donor site may be explained through effects from materials handling and dry spring conditions. Koch *et al.* (1996) found major losses in seed density throughout stripping (26%), stockpiling (69%) and respreading (87%) in Jarrah forest rehabilitation. Overstripping can dilute propagules at the surface with the lower layer (Tacey and Glossop, 1980; Rokich *et al.*, 2000). Stockpiling can cause propagule death *in situ* and respreading can physically bury or injure propagules. Less than 1% of the propagules emerged from plant vegetative parts in both donor soils; the large loss from physical injury or death during the dry spring.

Unexpectedly, peat resulted in slightly higher soil propagule density than LFH and there were considerably fewer species, which contradicts findings at the donor sites. The peat treatments were expected to have a larger propagule loss at the surface due to increased dilution because the source material was thicker. Since peat treatments were respread with a pipe, adequate mixing may have brought propagules to the surface, with effects similar to ripping donor soils (Rokich *et al.*, 2000).

The larger reduction in total density in LFH vs peat treatments may be related to increased clay content from overstripping mineral soil at similar depths. During the second growth chamber study, the clay seemed to create a smooth, light impermeable layer when watered, which may have reduced emergence. High clay content at the surface can cause reductions in light penetration at depths <2 mm, thus restricting light availability for seed germination (Galinato and van der Valk, 1986; Fenner, 2000).

Application depth significantly altered total density at the surface with the thick treatments having more propagules in the upper 7.5 cm of donor soil, contradicting results from other research (Rokich *et al.*, 2000; Zhang *et al.*, 2001). Application depth, however, did not significantly affect species richness, diversity,

Table 39.1. Total density and diversity indices from the two donor sites.

	Peat			LFH		
Attribute	Upper	Lower	Entire	Upper	Lower	Entire
Total density	2511.1 ± 518.8	1102.8 ± 333.8	3613.9 ± 614.7	5480.8 ± 953.4	3626.6 ± 609.8	9107.5 ± 1203.9
Richness	19	10	19	35	29	37
Diversity	2.29	2.14	2.42	2.57	2.54	2.62
Evenness	0.78	0.93	0.72	0.72	0.75	0.82
Similarity	0.30	0.26	0.37	0.4	0.37	0.65

Total density is mean emergents/m^2 ± standard error (SE). Peat n = 30; LFH n = 56.

evenness or similarity to the donor site vegetation, which is consistent with a similar study using forest soil for rehabilitation of bauxite mines in Australia (Rokich *et al.*, 2000). The reduced densities in thin application treatments may be attributed to increased surface clay content and uneven distribution of donor soil throughout the treatment. The increased mineral content at the surface buried the organic layer, containing most of the seeds, in various locations throughout the treatment and mixed more mineral material with donor soil creating a dilution effect. When the bulldozer lowered its blade to apply a thin layer of donor soil, chunks of mineral soil on the surface of the overburdened pile were brought up and placed over the applied donor soil and/or mixed with the donor soil. The thin application depth was patchier than the thick depth creating areas with no donor soil.

The LFH treatment was significantly more similar to above-ground vegetation prior to salvaging than the lowland treatment. Other diversity measures were not significantly different, but higher in LFH treatments. A large reduction of species dependent on regeneration from plant vegetative parts (34% of total propagule density) in peat treatments reduced overall number of species emerging on the receiver site.

Vegetation

The LFH treatment was significantly greater than peat treatments for all response variables except evenness (Table 39.2). Thick application depths had significantly higher plant densities. Application depths were not significantly different for species richness, diversity, evenness and similarity. Thin peat treatments had higher diversity indices compared to thick peat treatments, and no significant interaction effects were found. This trend was similar to that in the soil propagule bank at the receiver site.

These results contrast with growth chamber results from the soil propagule bank at the receiver site. The higher densities in the LFH treatments can be attributed to higher abundance and number of species at the LFH donor site with more species adaptable to drier soil conditions. Most species in the peat were hydrophilic and unable to adapt to drier conditions, leaving few elastic species capable of tolerating the drier soil conditions. Additional benefits from increased species richness,

Table 39.2. Treatment effects on total density and diversity indices from the soil propagule bank and vegetation at the receiver site.

	Peat		LFH	
Attribute	Thin	Thick	Thin	Thick
Soil propagules				
Total density	309.4 ± 74.9 b	828.7 ± 163.5 a	341.7 ± 136.4 b	527.3 ± 234.1 a
Richness	5.6 ± 0.7	7.7 ± 1.2	8.3 ± 2.9	13.0 ± 4.0
Diversity	1.51 ± 0.19	1.61 ± 0.15	1.51 ± 0.36	2.09 ± 0.15
Evenness	0.88 ± 0.07	0.82 ± 0.12	0.76 ± 0.02	0.85 ± 0.04
Similarity	0.14 ± 0.04 b	0.14 ± 0.01 b	0.27 ± 0.06 a	0.27 ± 0.08 a
Vegetation				
Total density	1.51 ± 0.51 bd	2.54 ± 0.86 bc	8.57 ± 1.83 ad	16.75 ± 3.38 ac
Richness	5.33 ± 1.33 b	4.67 ± 0.67 b	17.67 ± 2.19 a	19.33 ± 3.18 a
Diversity	1.44 ± 0.13 b	1.17 ± 0.14 b	2.26 ± 0.13 a	2.34 ± 0.19 a
Evenness	0.90 ± 0.04	0.78 ± 0.10	0.79 ± 0.02	0.80 ± 0.05
Similarity	0.13 ± 0.03 b	0.16 ± 0.02 b	0.34 ± 0.01 a	0.32 ± 0.03 a

Total density is in plants/m^2 and the values for all attributes are means ± standard error (SE). Different letters in the same row indicate significant difference at $P \leq 0.05$ for donor soil type and $P \leq 0.10$ for application depth. $n = 3$.

species diversity, similarity to the donor site and evenness all contribute to a more self-sustaining ecosystem when using LFH. The majority of species added to the receiver site with LFH are not available from commercial seed suppliers. The higher diversity and species richness in LFH treatments will play an important role in nutrient cycling and energy flow on reclaimed landscapes (Roberts, 2004). Salvaging LFH in the Oil Sands Region has already started to be implemented in mining companies' reclamation plans.

Conclusions

The thin LFH layer from upland forests has great potential as surface soil for reclamation in the oil sands region. It contains many propagules and species that will increase revegetation success. In early plant community development stages, thin applications created plant communities similar in diversity with similar species numbers, however, not as dense as thick applications. Due to the fragile nature of donor soils in initial stages of development higher densities would result in more resistance to wind and water erosion, but costs utilizing more donor soil may outweigh the benefits of adding less donor soil with equal numbers of species. To get shallower application depths of LFH without problems associated with mixing or burial of underlying clay, different equipment may be needed. While peat will continue to be overstripped with mineral soil, overstripping the LFH layer may detrimentally affect the efficiency of the use of native propagules. Further research on optimal stripping depth will be required if this valuable resource is to be used efficiently in the future.

Acknowledgements

The project was funded by Syncrude Canada Ltd. Thanks to Clara Qualizza for assistance and support during this project.

References

Box, J. (2003) Critical factors and evaluation criteria for habitat translocation. *Journal of Environmental Planning and Management* 46, 839–856.

Champness, S.S. and Morris, K. (1948) The population of buried seeds in relation to contrasting pasture and soil types. *Journal of Ecology* 364, 149–173.

Fedkenheur, A.W. and Heacock, H.M. (1979) Potential of soil amendments as sources of native plants for revegetation of Athabasca oil sands tailings. *Canadian Land Reclamation Association Fourth Annual General Meeting.* Edmonton, Alberta, Canada, pp. 223–236.

Fenner, M. (2000) *Seeds: The Ecology of Regeneration in Plant Communities,* 2nd edn. CAB International, Wallingford, UK.

Fung, M.Y.P. and Macyk, T.M. (2000) Reclamation of oil sands mining areas. *Reclamation of Drastically Disturbed Lands. Agronomy Monograph* 41, 755–774.

Galinato, M.I. and van der Valk, A.G. (1986) Seeds germination traits of annuals and emergents recruited during drawdowns in the Delta Marsh, Manitoba, Canada. *Aquatic Botany* 26, 89–102.

Granström, A. (1988) Seed banks at six open and afforested heathland sites in southern Sweden. *Journal of Applied Ecology* 25, 297–306.

Grant, C.D., Bell, D.T., Koch, J.M. and Loneragan, W.A. (1996) Implications of seedling emergence to site restoration following bauxite mining in Western Australia. *Restoration Ecology* 4, 146–154.

Grime, J.P. (2001) *Plant Strategies, Vegetation Processes, and Ecosytem Properties,* 2nd edn. Wiley, Chichester, UK.

Hackbarth, D.A. and Nastasa, N. (1979) The hydrogeology of the Athabasca oil sands area, Alberta. *Alberta Energy and Natural Resources and Alberta Research Council Bulletin* 28, 39.

Hill, M.O. and Stevens, P.A. (1981) The density of viable seed in soils of forest plantations in upland Britain. *Journal of Ecology* 69, 693–709.

Holmes, P.M. (2001) Shrubland restoration following woody alien invasion and mining: effects of topsoil depth, seed sources and fertilizer addition. *Restoration Ecology* 9, 71–84.

Iverson, L.R. and Wali, M.K. (1981) Buried, viable seeds and their relation to revegetation after surface mining. *Journal of Range Management* 35, 648–652.

Jackson, R.B., Canadell, J., Ehleringer, J.R., Mooney, H.A., Sala, O.E. and Schulze, E.D. (1996) A global analysis of root distributions for terrestrial biomes. *Oecologia* 108, 389–411.

Jauhiainen, S. (1998) Seed and spore banks of two boreal mires. *Annales Botanici Fennici* 35, 197–201.

Koch, J.M., Ward, S.C., Grant, C.D. and Ainsworth, G.L. (1996) Effects of bauxite mine restoration operations on topsoil seed reserves in the jarrah forest of Western Australia. *Restoration Ecology* 4, 368–376.

Lanoue, A. and Qualizza, C. (2000) LFH and shallow mineral horizons as inoculants on reclaimed areas to improve native species catch. 1999 Status Report. Syncrude Canada Ltd.

Leck, M.A., Parker, T.V. and Simpson, R.L. (1989) *Ecology of Soil Seed Banks.* Academic Press, San Diego, California.

Magurran, A. (1988) *Ecological Diversity and its Measurement.* Princeton University Press, New Jersey.

Mcgraw, J.B. (1987) Seed-bank properties of an Appalachian sphagnum bog and a model of the depth distribution of viable seeds. *Canadian Journal of Botany* 65, 2028–2035.

Moore, J.M. and Wein, R.W. (1977) Viable seed populations by soil depth and potential site recolonization after disturbance. *Canadian Journal of Botany* 55, 2408–2412.

Ødum, S. (1965) Germination of ancient seeds: floristical observation and archaeologically dated soil samples. *Danish Botanical Archive* 24, 1–70.

Paré, D., Bergeron, Y. and Camire, C. (1993) Changes in the forest floor of Canadian southern boreal forest after disturbance. *Journal of Vegetation Science* 4, 811–818.

Putwain, P.D. and Gillham, D.A. (1990) The significance of viable seed bank in the restoration of heathlands. *Biological Conservation* 52, 1–16.

Qi, M. and Scarratt, J.B. (1998) Effect of harvesting method on seed bank dynamics in a boreal mixedwood forest in northwestern Ontario. *Canadian Journal of Botany* 76, 872–883.

Roberts, M.R. (2004) Response of the herbaceous layer to natural disturbance in North American forests. *Canadian Journal of Botany* 82, 1273–1283.

Rokich, D.P., Dixon, K.W., Sivasithamparam, K. and Meney, K.A. (2000) Topsoil handling and storage effects on woodland restoration in Western Australia. *Restoration Ecology* 8, 196–208.

Scheirer C.J., Ray, W.S. and Hare, N. (1976) The analysis of ranked data derived from completely randomized factorial designs. *Biometrics* 32, 429–434.

Tacey, W.H. and Glossop, B.L. (1980) Assessment of topsoil handling techniques for rehabilitation of sites mined for bauxite within the Jarrah forest of Western Australia. *Journal of Applied Ecology* 17, 195–201.

Thompson, K., Bakker, J. and Bekker, R. (1997) *The Soil Seed Banks of North West Europe: Methodology, Density and Longevity.* Cambridge University Press, Cambridge.

Warr, S.J., Thompson, K. and Kent, M. (1993) Seed banks as a neglected area of biogeographic research: a review of literature and sampling techniques. *Progress in Physical Geography* 17, 329–347.

Whittle, C.A., Duchesne, L.C. and Needham, T. (1997) The importance of buried seed and vegetative propagation in the development of post fire plant communities. *Environmental Review* 5, 79–87.

Zhang, Z.Q., Shu, W.S., Lan, C.Y. and Wong, M.H. (2001) Soil seed bank as an input of seed source in revegetation of lead/zinc mine tailings. *Restoration Ecology* 9, 378–385.

40 The Effect of Light Intensity on Seed Production and Quality in a Number of Australian Wild Oat (*Avena fatua* L.) Lines

S.W. Adkins[1] and L.J. Armstrong[2]

[1]*Integrated Seed Research Unit, School of Land and Food Sciences, The University of Queensland, St Lucia, Queensland 4072, Australia;*
[2]*School of Computer and Information Science, Edith Cowan University, Mt Lawley, Western Australia, 6050 Australia*

Abstract

Several aspects of the maternal environment (e.g. air temperature, soil moisture, soil nutritional status and photoperiod) are known to influence seed production and quality in a range of plant species. However, little is known about the effect of other environmental factors, such as the light intensity perceived by the developing plant, on these seed characters. This is particularly important when elucidating the persistence mechanisms of annual weeds that may produce seeds under a crop canopy. Therefore, the aim of the present study was to determine the effect of light intensity on the reproductive characteristics of a number of wild oat (*Avena fatua* L.) lines originating from a range of locations in Australia. Under a reduced light intensity and in the absence of competition, the overall plant growth, seed production, seed weight and depth of seed dormancy (but not seed viability) were all reduced compared with that seen in plants growing in full sunlight. In addition, the reduced light intensity conditions significantly increased plant height and reduced tiller production, but did not affect leaf area production. In general, these responses were the same for all six lines studied and indicate that the reduced light intensity perceived by late germinating weeds present within a crop canopy causes the production of seeds with traits that are likely to result in rapid loss from the soil seed bank. In addition, all lines grown under the reduced light intensity conditions had an extended development time (up to 70 days), that it would be unlikely that they would reach maturity before the crop was harvested, and therefore would not be able to return any seed to the soil seed bank.

 Seeds: Biology, Development and Ecology (eds S. Adkins, S. Ashmore and S.C. Navie)

Introduction

Earlier studies have helped to elucidate how the photoperiod experienced by wild oat (*Avena fatua* L.) influence some of the important persistence characteristics of weeds. An increase in the day length under which the plants were grown led to a decrease in seed production and a decrease in the depth of seed dormancy possessed by these seeds (Armstrong, 1994). However, little is known about how other components, such as light intensity and quality, modulate the effect of the light environment on the persistence of this serious annual weed of cereal crops worldwide.

In the environment of a growing crop, *A. fauta* plants are exposed to a range of light intensities and qualities. For example, when the crop is in its early stages of growth, newly emerged *A. fatua* seedlings are exposed to a low amount of shading. However, as the crop canopy develops, shading will increase, reducing the light level by as much as 90–95% of that experienced in direct sunlight (Puckridge and Ratkowsky, 1971; Fisher and Wilson, 1976; Steven *et al.*, 1986; Cudney *et al.*, 1991). Such shading could be expected to influence plant morphology, phenology and reproductive ability. Shading is known to reduce plant growth because it reduces photosynthesis (Monteith and Elson, 1969), reduces leaf thickness and length, reduces root to shoot ratios, increases apical dominance, reduces tiller production (Holmes and Smith, 1975; Ludlow, 1978; Ballare *et al.*, 1988), increases stem elongation (Carpenter and Hopen, 1985), reduces dry matter production and reduces flower and seed production (Dall'armellina and Zimdahl, 1988; Stoller and Myers, 1989; Stirling *et al.*, 1990). Such developmental changes could be expected to occur in *A. fatua* plants, especially in those emerging late from the seed bank, into a closing crop canopy.

Although some studies have investigated the effect of shading on *A. fatua* development (Barnes *et al.*, 1990; Beyschlag *et al.*, 1990) and others have concentrated at the emergence of *A. fatua* seedlings in a crop (McBeath *et al.*, 1970; Peters and Wilson, 1983; Martin and Field, 1988), the effects of shade alone have not been determined. In order to separate the effects of shade from that of soil water and nutrient status, it might be more appropriate to assess the influence of shade under partially controlled conditions where the plants are grown individually, away from other competition effects. This approach can be problematical, as the quality of light experienced within a crop canopy is different to that produced under controlled conditions using shade cloth. However, assessment of light intensity effects upon plant growth and development, using shade cloth (Yates, 1989), has been undertaken in the past and found to provide significant insight into the effect of light intensity on plant growth and development under plant canopies.

The aim of the present study was to determine whether the light intensity experienced by growing *A. fatua* plants could have an effect on their reproductive capacity (i.e. seed production, seed weight, seed viability and depth of dormancy) and on features that may determine their long-term persistence in the soil seed bank. The light intensities used represent those that *A. fatua* would experience if it emerged before or into a closing crop canopy. In order to analyse any changes in reproductive capacity that may occur, a number of phenological and morphological characters were recorded. The *A. fatua* lines used in this study came from a number of locations within the northern grain growing region of Australia and were used to determine if the genetic variation in natural populations of *A. fatua* at different locations affected this response.

Materials and Methods

Plant material

A. fatua seeds were hand-collected from six locations in the northern wheat belt of Australia (i.e. from near Springsure, Toowoomba and Warwick in Queensland, and near Narrabri, Coonabarabran and Moree in New South Wales). The collected seeds were used to generate six near isogenic lines (one per collection; Armstrong, 1994) using the single seed desent method. In this procedure, each line underwent at least three generations of self-pollination to produce partially homozygous (at least 88% homozygous; Allard, 1960) lines.

Plant growth conditions

Seeds from the lines were germinated on two Whatman filter papers in plastic Petri dishes (9 cm diameter) and when seedlings were 3 days old they were planted, one per pot, into 30 cm diameter plastic pots containing a heavy black soil/potting compost mix (1:1, v/v). The soil was watered and maintained at field capacity and the plants grown outside the field site at the University of Queensland during the months of June–August (the normal time for *A. fatua* growth and development in the northern wheat belt). When the seedlings were 20 days old, they were placed under three light intensity conditions (i.e. 0%, 40% and 80% reductions in natural sunlight) created by using a non-spectral shift, black Saron shade cloth (Yates, 1989) obtained from KMart Pty., Ltd, Brisbane, Australia. The shade cloth was attached to metal frames (3 × 4 × 4 m) that were set up 5 m apart. The plants were grown under these shade conditions until they reached maturity.

Measurement of plant characters

For each of the light intensity treatments applied to the six lines, a number of phenological, morphological (i.e. tiller production, plant height, shoot biomass and leaf area) and reproductive (i.e. seed production and seed weight) development features were measured. The phenological development was assessed by measuring the duration of the three major plant developmental phases: (i) the vegetative (from initial germination to floral initiation); (ii) reproductive (from floral initiation to anthesis); and (iii) seed maturation (from anthesis to seed maturity) phases, as described previously (Armstrong and Adkins, 1998). The morphological and reproductive features were measured at maturity (i.e. at the conclusion of the experiment). Seed viability and depth of dormancy were measured by using eight different concentrations (0–10 mM) of gibberellic acid (GA_3) applied to batches of caryopses (lots of ten per replicate) taken from plant harvests (Armstrong and Adkins, 1998).

Statistical design and data analysis

A completely randomized block design was used, with each light treatment consisting of three replicated blocks of six near isogenic lines (each having three replicate plants). For the study of seed weight, three replicate samples (each with ten caryopses) were used. For the study of seed dormancy, four replicate samples (each with ten caryopses) were assessed using GA_3. Analysis of variance (ANOVA) and linear correlation (Pearson's rank correlation) were carried out on the data-sets using the SAS statistical program (SAS Institute Inc., Cary, North Carolina, USA) (SAS Institute, 1987).

Results and Discussion

Phenological development

When considering all the lines together, the average time taken by the plants to mature increased from 127 to 165 days, as the light intensity was reduced from full sunlight to 20% of that value (Fig. 40.1). This trend was obvious in four of the lines (i.e. Springsure, Toowoomba, Warwick and Coonabarabran), less apparent in the Narrabri line and not apparent in the Moree line. The delay in plant development time, observed under the reduced light intensities, may be the result of reduced photosynthesis, which in turn led to a reduction in the plant growth rate (Monteith and Elson, 1969). The delay in reaching maturity under the 80% light intensity reduction conditions could have a bearing on the survival of *A. fatua* in the field. Such delays would extend the life of the plant to a period well beyond the time of crop harvest, thus not allowing seed from these plants to be returned to the seed bank. The response observed between the lines was variable, with the Narrabri line showing the shortest delay in plant maturity time (10 days at 80% light intensity reduction), compared with the Toowoomba, Warwick and Coonabarabran lines (~70 days). The overall time taken to mature in full sunlight (~127 days) was comparable with that seen in earlier field-grown plants of the same lines (Armstrong, 1994).

Reproductive development

When considering all the lines together, the average number of primary seeds produced per plant was reduced from 187 to 109, as the light intensity was reduced from full sunlight to 20% of that value (Fig. 40.2). The reduction in the numbers of seeds produced under shade may have been due to the observed reduction in growth rate (Fig. 40.1), which in turn reduced flower production and/or promoted flower abortion. This result would suggest that plants growing under reduced light intensities in the field may produce less seeds than plants growing outside the crop canopy, and such effects have been previously reported (McBeath *et al.*, 1970; Peters and Wilson, 1983; Martin and Field, 1988). This response was only observed in the Springsure line, where seed production in full sunlight was considerably higher than in all other lines.

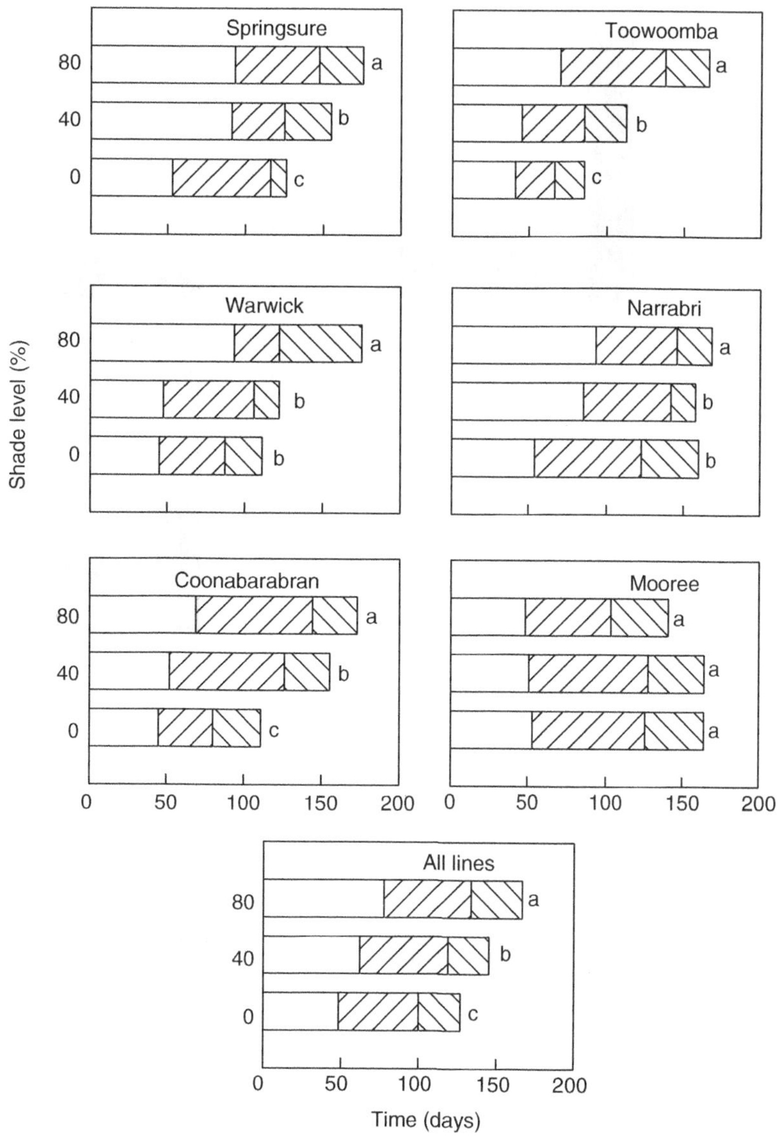

Fig. 40.1. The influence of shade on the duration of the vegetative (❑), reproductive (▨) and seed maturation (⧅) growth phases of six near isogenic lines of *Avena fatua* grown under simulated field conditions at the University of Queensland from April to December 1991. Total development time values within each line followed by the same letter are not significantly different ($P < 0.05$).

The mean seed weight of all the lines declined (from 253 to 204 mg) as the light intensity was reduced (Fig. 40.3). This would suggest that later-emerging seedlings, exposed to a closing crop canopy, would produce smaller seeds. However, this trend was not statistically significant in every one of the individual lines (e.g. Narrabri and Moree). The overall reduction in seed weight, under reduced light

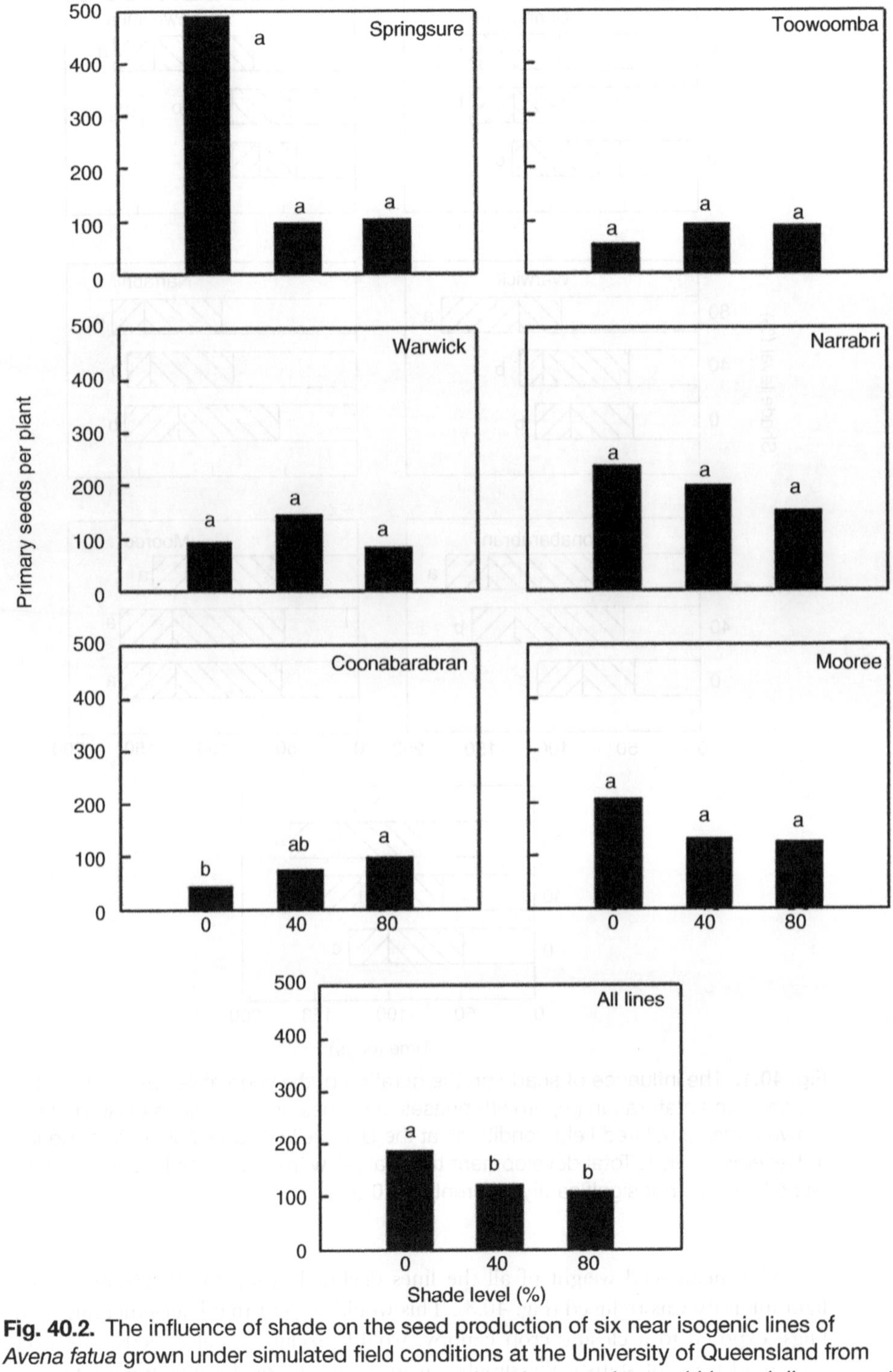

Fig. 40.2. The influence of shade on the seed production of six near isogenic lines of *Avena fatua* grown under simulated field conditions at the University of Queensland from April to December 1991. Seed numbers followed by the same letter within each line are not significantly different ($P < 0.05$).

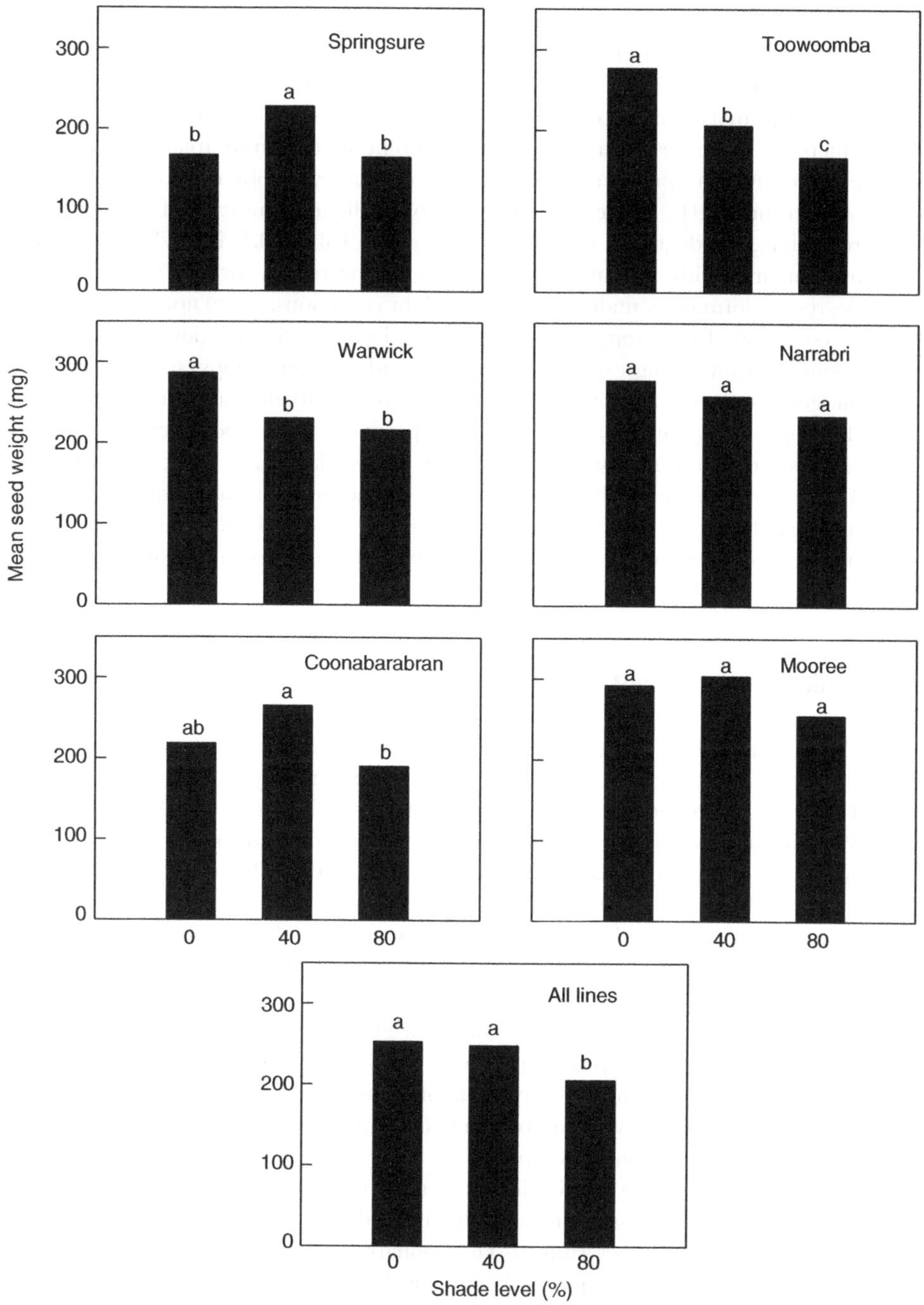

Fig. 40.3. The influence of shade on the mean seed weight of six near isogenic lines of *Avena fatua* grown under simulated field conditions at the University of Queensland from April to December 1991. Seed weight values followed by the same letter within each line are not significantly different ($P < 0.05$).

intensity, may be because of a reduced growth rate of these plants (Fig. 40.1). The mean seed weight produced under full sunlight (253 mg) is comparable with that produced in field studies (Armstrong, 1994) on the same lines.

All primary seeds produced in these studies were >95% viable; therefore, no differences were observed between the different light intensity treatments or different lines. All lines produced seeds that had dormancy, as observed in earlier studies (Armstrong, 1994). However, under the lowest light intensity used (i.e. 80% light reduction), the degree of dormancy was lowest (Table 40.1; i.e. 37% germination as compared with 11% under full sunlight). The reason for this decrease in the degree of dormancy, under the reduced light conditions, is unknown and is under investigation. In a cropping situation, the decrease in seed dormancy, brought about by plants being exposed to shading inside a crop canopy, may not occur in all seeds produced on a plant since panicles are able to elongate and emerge above the crop canopy. Seeds produced on these panicles may not be affected in the same way as those produced within the canopy (Beyschlag *et al.*, 1990; Cousens *et al.*, 1991). However, seeds that are produced on the non-elongating secondary tillers are likely to have a decreased level of dormancy. Thus, *A. fatua* plants establishing into a developing crop canopy will have an increased proportion of non-dormant seeds entering the soil seed bank.

Morphological development

When considering all the lines together, the average number of fertile tillers produced per plant declined from 15 to 7 as the light intensity was reduced (Fig. 40.4). However, this reduction in tiller production was only significant for two lines (i.e. Warwick and Narrabri). This finding may explain the reduction in seed numbers produced under the reduced light intensity (Fig. 40.2). There was also a decrease in the total number of tillers (reproductive and vegetative) produced under reduced light intensity in most lines, with the average number of total tillers declining from 23 to 10 per plant. The number of both fertile and total tillers produced in full sunlight was similar to that shown by plants in earlier growth studies (Armstrong, 1994). However, some variability was noticed between lines, with the Springsure and Toowoomba lines producing more tillers per plant than all the other lines under the reduced light intensity situation. This suggests that the Springsure and Toowoomba lines may be more competitive than all of the other lines studied, when emerging later in the crop.

When considering all the lines together, the average plant height increased from 74 to 98 cm as light intensity was reduced (Fig. 40.5), and this was also true for three of the six lines studied (i.e. Coonabarabran, Narrabri and Springsure). Increased plant height, under shading, is believed to be an important escape response (Grime, 1979), allowing plants to quickly avoid the effects of shade by emerging above the crop canopy. This effect has been observed many times in the field, where *A. fatua* plants are known to emerge up above the crop canopy (Beyschlag *et al.*, 1990; Cousens *et al.*, 1991).

Reduced light intensity did not significantly influence the total leaf area produced, which ranged from 616 cm^2 per plant (in 80% light reduction) to 705 cm^2

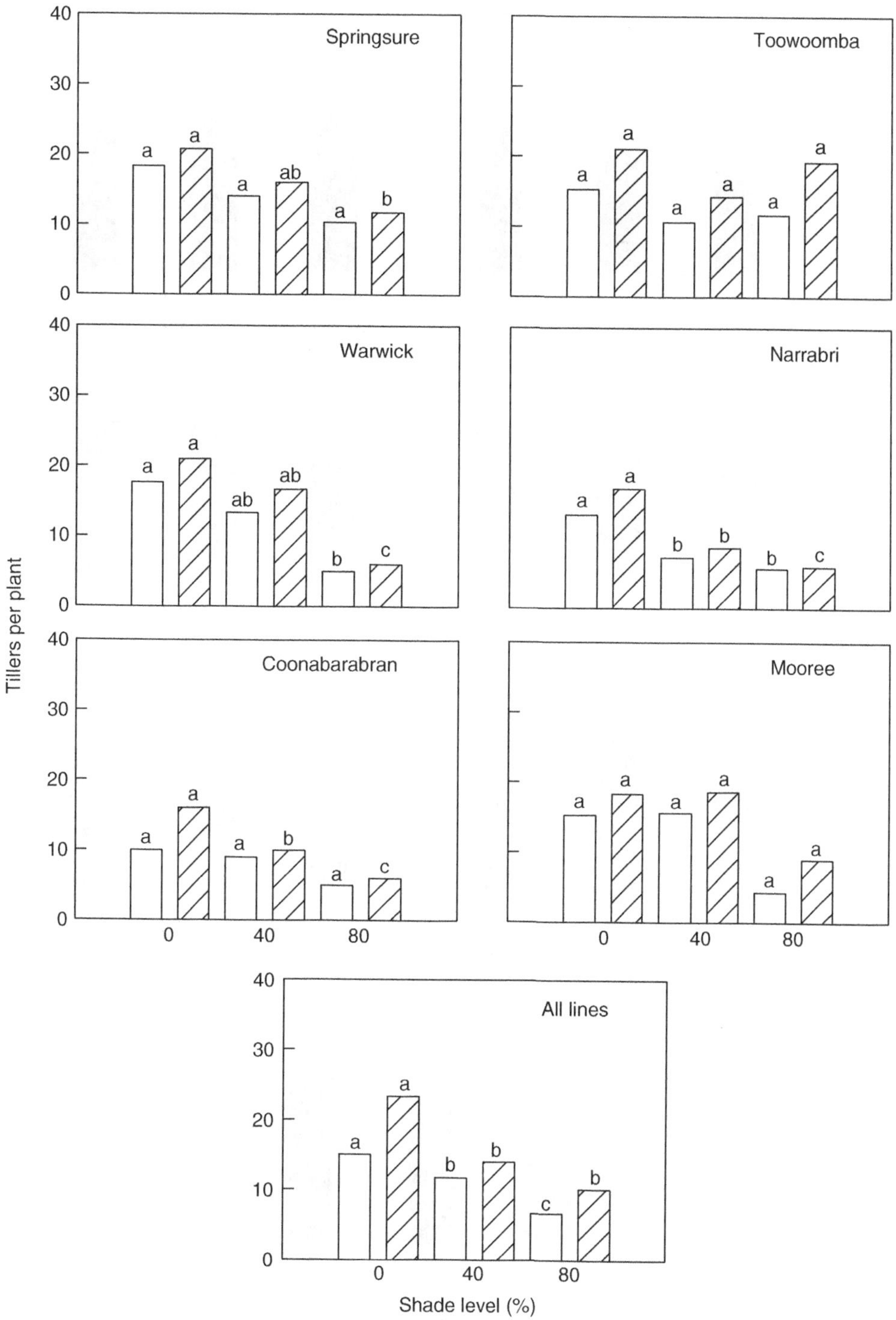

Fig. 40.4. The influence of photoperiod on the fertile (❑) and total (▨) tiller production of six near isogenic lines of *Avena fatua* grown under simulated field conditions at the University of Queensland from April to December 1991. Tiller values followed by the same letter within each line are not significantly different ($P < 0.05$).

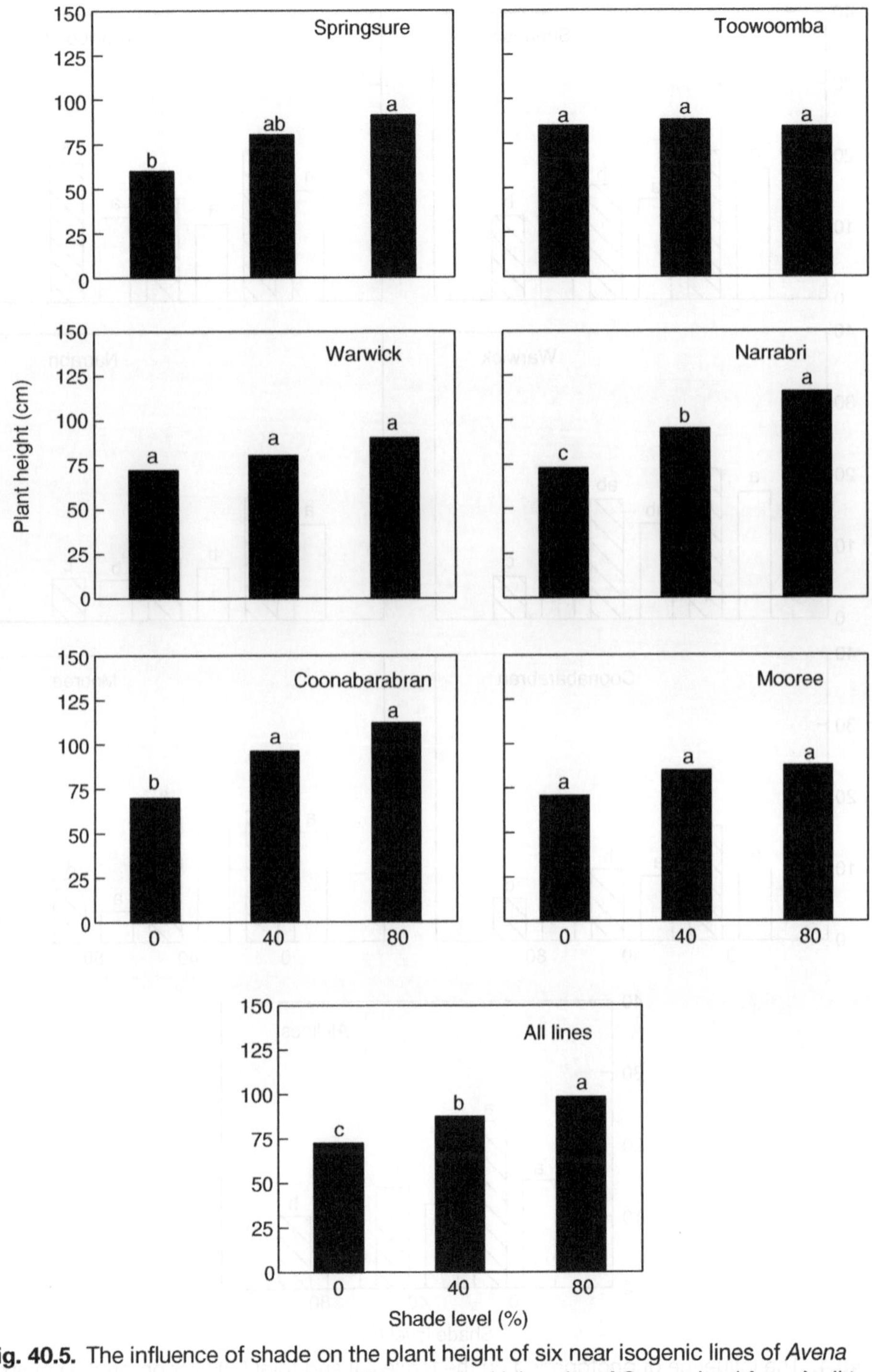

Fig. 40.5. The influence of shade on the plant height of six near isogenic lines of *Avena fatua* grown under simulated field conditions at the University of Queensland from April to December 1991. Plant height values followed by the same letter within each line are not significantly different ($P < 0.05$).

Table 40.1. The influence of gibberellic acid (GA_3) (50 µM) on the germination of six lines of *Avena fatua*.

Line	Germination (%)		
	0% light reduction	40% light reduction	80% light reduction
Springsure	8 c	68 a	30 b
Toowoomba	18 b	15 b	90 a
Warwick	28 a	38 a	53 a
Narrabri	0 b	3 a	0 b
Coonabarabran	5 b	8 b	35 a
Moree	5 b	0 b	13 a
Mean of all lines	11 b	22 b	37 a

Seeds were harvested from plants that were grown outdoors under three levels of light intensity at the University of Queensland. There were four replicates of ten caryopses in each treatment, for each line. Germination percentages followed by the same letter within each row are not significantly different ($P < 0.05$).

per plant (in 40% light reduction) (Fig. 40.6). This observation would suggest that, under a crop canopy, *A. fatua* is capable of maintaining a comparable amount of leaf area to that produced under full sunlight, allowing for an increased efficiency of light capture in these shaded conditions. However, this observation is in contrast to an earlier study, where significant reductions in *A. fatua* leaf growth were observed in the field when plants were shaded by a wheat (*Triticum aestivum* L.) crop (Friend *et al.*, 1962). This would suggest that other competition factors (e.g. for water and nutrients) may be determining leaf area production in a cropping situation and this needs further clarification. Some variability was observed for leaf area between the lines, with the Moree line producing the greatest leaf area under the moderate reduction in light intensity. This greater capacity of the Moree line to produce leaf area, under this reduced light level, indicates that it is less affected by shade than other lines, as was seen in its phenological development (Fig. 40.1).

Plants grown under a reduced light intensity produced less shoot biomass than those grown under full sunlight, with average shoot biomass declining from 28 to 16 g per plant; however, this trend was not significant (Fig. 40.7). This reduction in shoot biomass may be due to the lower photosynthetic capacity and reduced growth of these plants under shade, as has already been described. The shoot biomass values of plants produced in full sunlight were comparable with those observed in earlier studies (Armstrong, 1994).

Conclusions

A. fatua does not display the growth characteristics of an extremely 'shade-tolerant' species (Grime, 1979), with reduced light intensity conditions (applied in the absence of any other kind of competition) reducing shoot biomass (Fig. 40.7), seed production (Fig. 40.2), seed weight (Fig. 40.3) and the degree of seed dormancy (Table 40.1), but greatly extending the time of plant development (Fig. 40.1). However, *A. fatua* did display some evidence of compensating for the reduced light levels,

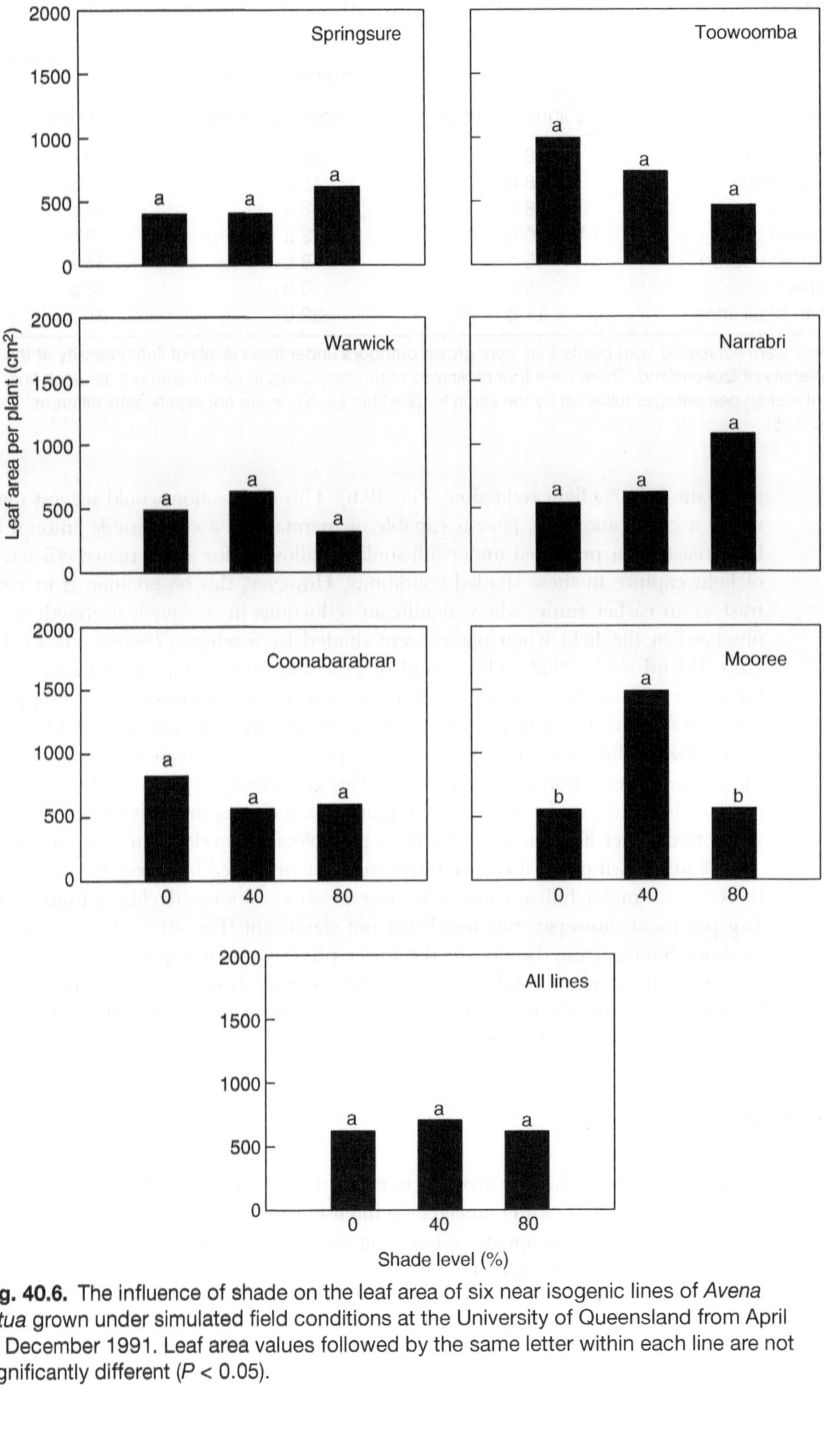

Fig. 40.6. The influence of shade on the leaf area of six near isogenic lines of *Avena fatua* grown under simulated field conditions at the University of Queensland from April to December 1991. Leaf area values followed by the same letter within each line are not significantly different ($P < 0.05$).

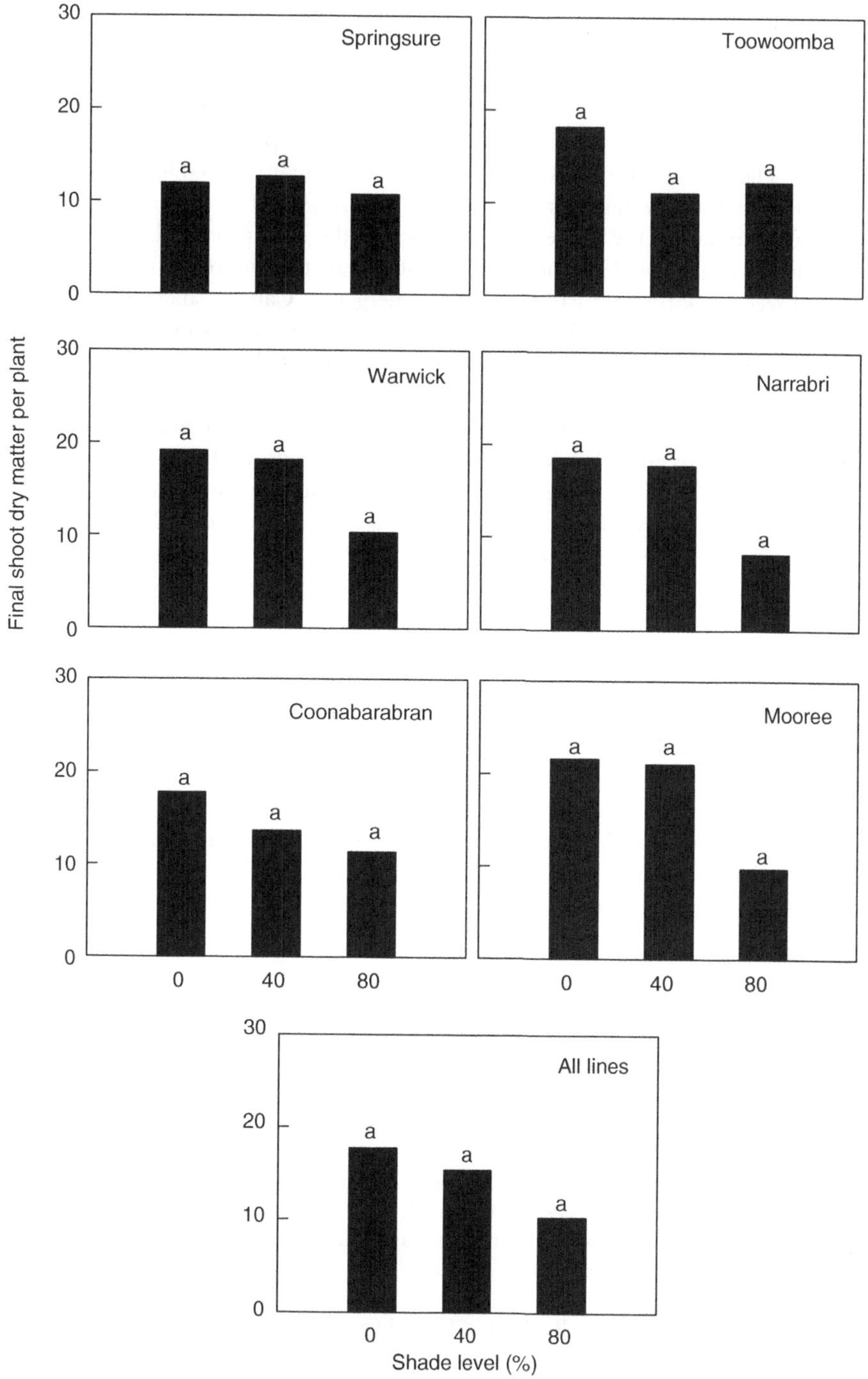

Fig. 40.7. The influence of shade on the shoot dry weight of six near isogenic lines of *Avena fatua* grown under simulated field conditions at the University of Queensland from April to December 1991. Shoot weight values followed by the same letter within each line are not significantly different ($P < 0.05$).

by increasing shoot height (Fig. 40.5), which may allow its upper leaves to avoid shading in the crop canopy in a cropping situation. Reduced light intensity caused similar detrimental effects to all of the lines studied and although some variability in response was observed in these lines, this was small and indicates that few differences would be detectable among lines in a field population.

The findings from this study have confirmed that reduced light intensity can be detrimental to the persistence mechanisms of *A. fatua*, by reducing seed production and by reducing the degree of dormancy in the seed produced. Fewer seeds with a lower degree of dormancy will probably lead to a more rapid depletion of the soil seed bank through earlier seedling emergence. Canopy shading would effect all *A. fatua* plants emerging into the crop, but would have its greatest influence on those emerging later in the season when the crop canopy is closing. An interesting effect of reduced light intensity, observed in this study, is the delay it causes in plant development (up to ~70 days). This could result in some plants not reaching maturity before the crop is harvested. While in others, only a few seeds will be produced by this time. Consequently, this would prevent most of the late emerging plants from returning large numbers of seeds to the soil seed bank. In addition, reduced light intensity may dramatically reduce the competitive ability of *A. fatua* in the crop as it drastically reduces shoot and tiller production. Thus, one aspect of a highly competitive crop might be its ability to shade the weeds within it, reducing their growth, extending their development, reducing their seed set and lowering their seed dormancy.

Acknowledgements

The authors wish to acknowledge some financial assistance from the Grains Research and Development Corporation, Australia, in the form of a scholarship for L.J. Armstrong and additional operating funding. We are also grateful for the technical support provided by W. Bean.

References

Allard, R.W. (1960) *Principles of Plant Breeding*. Wiley, New York.

Armstrong, L.J. (1994) Studies on the biology of wild oats (*Avena fatua* L.): the effect of selected environmental factors on growth and reproduction. PhD thesis, The University of Queensland, St. Lucia, Australia.

Armstrong, L.J. and Adkins, S.W. (1998) Variation in photoperiod response of different isogenic lines of wild oats (*Avena fatua* L.). *Weed Science* 46, 39–47.

Ballare, C.L., Sanchez, R.A., Scopel, A.L. and Ghersa, C.M. (1988) Morphological response of *Datura ferox* L. seedlings to the presence of neighbours: their relationships with canopy microclimate. *Oecologia* 76, 288–293.

Barnes, P.W., Beyschlag, W., Ryel, R., Flint, S.D. and Caldwell, M.M. (1990) Plant competition for light analysed with a multispecies canopy model. III. Influence of canopy structure in mixed and monocultures of wheat and wild oat. *Oecologia* 82, 560–566.

Beyschlag, W., Barnes, P.W., Ryel, R., Caldwell, M.M. and Flint, S.D. (1990) Plant competition for light analysed with a multispecies canopy model. II. Influence of photosynthesis characteristics on mixtures of wheat and wild oat. *Oecologia* 82, 374–380.

Carpenter, J.L. and Hopen, H.J. (1985) A comparison of the biology of wild and cultivated proso millet (*Panicum miliaceum*). *Weed Science* 33, 795–799.

Cousens, R.D., Weaver, S.E., Martin, T.D., Blair, A.M. and Wilson, J. (1991) Dynamics of competition between wild oats (*Avena fatua* L.) and winter cereals. *Weed Research* 31, 203–210.

Cudney, D.W., Jordon, L.S. and Hall, A.E. (1991) Effect of wild oat (*Avena fatua*) infestations of light interception and growth rate of wheat (*Triticum aestivum*). *Weed Science* 39, 175–179.

Dall'armellina, A.A. and Zimdahl, R.L. (1988) Effect of light on growth and development of field bindweed (*Convolvulus arvensis*) and Russian knapweed (*Centaurea repens*). *Weed Science* 36, 779–783.

Fisher, R.A. and Wilson, G.L. (1976) Studies on grain production in *Sorghum bicolor* (L.) Moench. VI. Profiles of photosynthesis, illuminance and foliage arrangement. *Australian Journal of Agricultural Research* 27, 35–44.

Friend, D.J.C., Helson, V.A. and Fisher, J.E. (1962) Leaf growth in Marquis wheat, as regulated by temperature, light intensity, and day length. *Canadian Journal of Botany* 40, 1299–1311.

Grime, J.P. (1979) *Plant Strategies and Vegetation Processes.* Wiley, New York.

Holmes, M.G. and Smith, H. (1975) The function phytochrome in plants growing in the natural environment. *Nature* 254, 512–514.

Ludlow, M.M. (1978) Light relations of pastures. In: Wilson, J.R. (ed.) *Plant Relations in Pastures.* CSIRO, Melbourne, Australia, pp. 35–49.

Martin, M.P.L.D. and Field, R.J. (1988) Influence of time of emergence of wild oat on competition with wheat. *Weed Research* 28, 111–116.

McBeath, D.K., Dew, D.A. and Friesen, H.A. (1970) Competition between barley and wild oats as affected by nitrogen, barban and time of seeding. *Canadian Journal of Plant Science* 50, 541–550.

Monteith, J.L. and Elson, J. (1969) Light interception and radiative exchange in crop stands. In: Eastin, J.D. (ed.) *Physiological Aspects of Crop Yield.* American Society of Agronomy, Madison, Wisconsin, pp. 89–115.

Peters, N.C.B. and Wilson, B.J. (1983) Some studies on the competition between *Avena fatua* L. and spring barley. II. Variations of *A. fatua* emergence and development and its influence of crop yield. *Weed Research* 23, 305–311.

Puckridge, D.W. and Ratkowsky, D.A. (1971) Photosynthesis of wheat under field conditions. IV. The influence of density and leaf area index on the response to radiation. *Australian Journal of Agricultural Research* 22, 11–20.

SAS Institute (1987) *Sas/Stat Guide for Personal Computers.* Version 6. SAS Institute Inc., Cary, North Carolina.

Steven, M.D., Biscoe, P.V., Jaggerd, K.W. and Paruntu, J. (1986) Foliage cover and radiation interception. *Field Crops Research* 13, 75–87.

Stirling, C.M., Williams, J.H., Black, C.R. and Ong, C.K. (1990) The effect of timing of shade on development, dry matter production and light-use efficiency in groundnut (*Arachis hypogaea* L.) under field conditions. *Australian Journal of Agricultural Research* 41, 633–644.

Stoller, E.W. and Myers, R.A. (1989) Effects of shading and soybean, *Glycine max* (L.) interference on *Solanum ptycanthum* (Dun.) (eastern black nightshade) growth and development. *Weed Research* 29, 307–316.

Yates, D.J. (1989) Shade factors of a range of shade cloth materials. *Acta Horticulturae* 257, 201–217.

41 Seed Ecology of *Apiaceae* Weeds in Pyrethrum

R.P. RAWNSLEY,[1] A.J. GRACIE,[2] P.A. LANE,[2] P.H. BROWN[3] AND T. GROOM[4]

[1]*Tasmanian Institute of Agricultural Research, Burnie, Tasmania, Australia;* [2]*School of Agricultural Science, University of Tasmania, Hobart, Tasmania, Australia;* [3]*Tasmanian Institute of Agricultural Research, Private Bag 98 Hobart, Tasmania 7001, Australia;* [4]*Botanical Resources Australia Pty. Ltd, Industrial Drive, Ulverstone, Tasmania, Australia*

Abstract

Anthriscus caucalis M. Bieb. (burr chervil) and *Torilis nodosa* (L.) Gaertn. (knotted hedge-parsley) are two widely occurring weed species in pyrethrum (*Tanacetum cinerariifolium* (Trevir.) Sch. Bip.) crops in northern Tasmania. Little is known about the seed ecology of these two *Apiaceae* weed species. This study examined the level of seed dormancy of both species and their germination response to fluctuating temperatures, dark and light, stratification and planting depth. Both *A. caucalis* and *T. nodosa* displayed morphological dormancy (i.e. an underdeveloped embryo) that was overcome by warm stratification. In addition, seeds of *A. caucalis* displayed physical exogenous dormancy that was broken by scarification and a physiological endogenous dormancy that was overcome by dry storage at 20°C. More than 90% of freshly collected, mature seeds of *T. nodosa* germinated within 14 days of imbibition at 20°C. Optimum seed germination temperature was between 6°C and 15°C for *A. caucalis* and between 18°C and 23°C for *T. nodosa*. Optimum seed burial depth for seedling emergence was between 0 and 30 mm for both species, with emergence reduced at 50 mm and prevented if seed was planted at 70 mm. Removal of a light stimulus was found to restrict the germination of *A. caucalis* and *T. nodosa*. The results of this study provide valuable information for the development of an effective integrated weed management strategy for control of these weeds in pyrethrum.

Introduction

Anthriscus caucalis M. Bieb. (burr chervil) and *Torilis nodosa* (L.) Gaertn. (knotted hedge-parsley) are two problematic weeds of pyrethrum (*Tanacetum cinerariifolium* (Trevir.) Sch. Bip.). Pyrethrum is a perennial daisy, belonging to the family *Asteraceae*, grown for the production of the natural insecticide pyrethrin. Pyrethrum

 Seeds: Biology, Development and Ecology (eds S. Adkins, S. Ashmore and S.C. Navie)

production is a relatively new industry to Tasmania, but in a short time Tasmania has become the world's second highest producer of pyrethrum behind Kenya. Both *A. caucalis* and *T. nodosa* are considered a threat to the pyrethrum industry due to an absence of appropriate chemical control options, and a paucity of information on the aspects of their biology (including their seed dormancy and germination characteristics), required for the development of effective control strategies.

Both *A. caucalis* and *T. nodosa* are winter annuals and like other weed species depend on seed ecological adaptations to enable plants to be dispersed in space and time. Seed dormancy is considered one of the most important adaptations contributing to the success of weeds and, almost without exception, seeds of weed species possess one or more factors contributing to seed dormancy (Anderson, 1996). There are many different types of seed dormancy and such dormancy classification is quite complex. Over the last century, there have been a number of seed dormancy classification reviews. The terms and definitions used in this chapter are consistent with those documented by Geneve (1999).

A wide range of seed dormancy mechanisms have been reported for members of the family *Apiaceae* (Chaturvedi and Muralia, 1975). Morphological dormancy (i.e. underdeveloped embryos) appears to be common in this family and has been reported in *Apium graveolens* L. (Jacobsen and Pressman, 1979), *Conium maculatum* L. (Baskin and Baskin, 1990) and *Pastinaca sativa* L. (Baskin and Baskin, 1979). *Thaspium pinnatifidum* (Buckl.) Gray has been shown to have deep complex morphophysiological dormancy (MPD) (Baskin *et al.*, 1992). From the genus *Anthriscus*, Baskin *et al.* (2000) reported that *Anthriscus sylvestris* (L.) Hoffm. only required a long cold stratification treatment for loss of physiological dormancy and for growth of the embryo (deep complex MPD). Within the genus *Torilis*, seed of *Torilis japonica* (Houtt.) DC. has been found to be dormant at time of dispersal (Roberts, 1979; Grime *et al.*, 1981) and chilling was required to break dormancy (Grime *et al.*, 1981). In contrast, Baskin and Baskin (1975) found that seeds of *T. japonica* were non-dormant at the time of dispersal and were induced into dormancy by low temperatures.

There is little information on the environmental factors involved in controlling seed germination of both *A. caucalis* and *T. nodosa*. The objective of this series of studies was to assess the level of seed dormancy of both *A. caucalis* and *T. nodosa* and examine some of the major environmental factors affecting their seed germination and seedling emergence.

Materials and Methods

Unless stated otherwise, individual treatments for all laboratory experiments consisted of 50 seeds placed on two sheets of Postlip filter paper moistened with distilled water in 85 mm diameter Petri dishes. Petri dishes were incubated in a Contherm™ incubator at a constant temperature (20°C) with an illumination period of 12 h per day. A light intensity at seed level of ~140 μmol/m^2/s was produced by two 15 W fluorescent tubes. A seed was considered germinated when the radicle protruded a minimum of 1 mm from the seed coat (ISTA, 1996). Embryos were excised from seeds with a scalpel and their lengths measured under a dissecting microscope equipped with a micrometer.

Dormancy determination of *A. caucalis*

Seeds of *A. caucalis* were collected on 15 January 2003 from a commercial pyrethrum field at Don, Tasmania. The seed moisture content was 7.5% and seed viability, as determined by a tetrazolium (TZ) test (ISTA, 1996), was 92%. Seeds were packaged into sealed zip-lock plastic bags and dry-stored at constant temperatures of 4°C or 20°C.

Seed was scarified with a scalpel where the seed coat of *A. caucalis* seed was nicked to expose the endosperm. Four replicates of 25 seeds were germinated for each treatment. Seeds were incubated at 12 h alternating day/night temperatures of 15/10°C. The number of germinated seeds was recorded 14 days after incubation (DAI). Seeds were also germinated in complete darkness by applying a double wrapping of aluminium foil around the Petri dishes containing the imbibed seeds. An untreated control was included in which the seeds were not scarified.

Dormancy determination of *T. nodosa*

Mature seeds of *T. nodosa* were collected from a commercial pyrethrum field in February 2002. Seed viability, as determined by a TZ test (ISTA, 1996), was 96%. Moistened seeds of *T. nodosa* were stored at 4°C for 0, 1, 2, 3, 4, 6, 8, 12 and 16 weeks with a 12 h light/dark photoperiod. Following stratification treatment, seeds were transferred to a constant temperature of 20°C with a 12 h light/dark photoperiod. Seeds were also kept in complete darkness for the prechilling treatment of 0, 2, 4, 8 and 16 weeks and following transfer to 20°C, by applying a double wrapping of reflective aluminium foil around each Petri dish. Seeds incubated in light were counted every 7 days following placement at 20°C for 28 days and germinated seeds were recorded and removed. Seeds incubated in complete darkness were counted only once after incubation at 20°C for 21 days.

Germination characteristics of *A. caucalis* and *T. nodosa*

Temperature

A Terratec™ thermogradient table with ten thermocouples was used to determine the effect of temperature on germination of *T. nodosa* and *A. caucalis*. The thermogradient plate had thermocouples spaced 15 cm apart, resulting in regions with constant temperatures ranging from 0°C to 40°C. One hundred seeds of *T. nodosa* and 50 seeds of *A. caucalis* were exposed to each temperature treatment under constant light. Seeds were observed daily for 28 days and germination counts were recorded every 7 days. Temperatures above 25°C caused localized drying, and consequently seeds were remoistened twice daily throughout the experiment.

Planting depth

Fifty seeds of both *T. nodosa* and *A. caucalis* were sown at one of seven soil depths: 0, 5, 10, 20, 30, 50 and 70 mm in a red ferrosol soil (Isbell, 1996), a soil type on which the majority of pyrethrum crops are cultivated in Tasmania, in 12.5 cm diameter pots. The soil was heat sterilized at 100°C for 24 h to ensure that any weed seeds present were destroyed, ground with a mortar and pestle, and then passed through a 5 mm sieve and rewetted prior to use. Pots were placed in a glasshouse (20°C ± 5°C) for a period of 49 days and watered twice daily for 2 min to replace transpiration losses. Emergence counts were recorded weekly.

Osmotic stress

Solutions of polyethylene glycol 8000 (PEG 8000) with osmotic potentials of 0.00, −0.25, −0.50, −0.75 and −1.00 MPa were prepared (Michel, 1983). The osmotic potential of each PEG solution was confirmed using an osmometer. Fifty seeds of both *T. nodosa* and *A. caucalis* were imbibed on two sheets of Postlip filter paper moistened with 5 ml of each solution in separate 85 mm diameter Petri dishes. Germination counts were recorded every 7 days for a period of 28 days.

Statistical analysis

All experiments were arranged as randomized complete block designs with four replications, except for the temperature and soil depth study, where three replications were used. Percentage germination data were subjected to analysis of variance (ANOVA) with Fisher's protected least significant difference (LSD) test ($P < 0.05$) for mean separation using the statistical analysis package SAS (SAS Institute, 1998).

Results and Discussion

Dormancy determination of *A. caucalis*

The initial germination percentage of the *A. caucalis* seed lot was 15.0% ± 4.7%. This increased significantly ($P < 0.05$) with dry storage at 20°C, with germination reaching 91% after 52 weeks in storage (Fig. 41.1a). There was also an increase in the percentage of seed germinating in complete darkness following storage at 20°C. The germination percentage in darkness was significantly ($P < 0.05$) lower than the untreated seed with an illumination period of 12 h per day at all storage durations, except at 12 and 16 weeks after storage. Scarification ($P < 0.05$) increased germination, which was significantly ($P < 0.05$) higher than the untreated seed at all storage durations prior to 39 weeks of storage.

Storage of *A. caucalis* seeds at 4°C resulted in no discernible pattern of change in germination percentage (Fig. 41.1b). Under complete darkness, seeds failed to germinate or germination was restricted to levels below 5%. The initial germination of scarified seeds was 41% and this did not significantly ($P > 0.05$) change with storage duration at 4°C. Similarly, for the untreated seeds there was no significant ($P > 0.05$) change over the first 38 weeks of storage, although there was a significant ($P < 0.05$) increase after 52 weeks of storage. Mean (± one standard deviation (SD)) embryo length prior to imbibition was 0.56 ± 0.07 mm and increased to 0.82 ± 0.13 mm 6 days after imbibition of warm stratification. Like many *Apiaceae* species, *A. caucalis* was found to have a morphological dormancy.

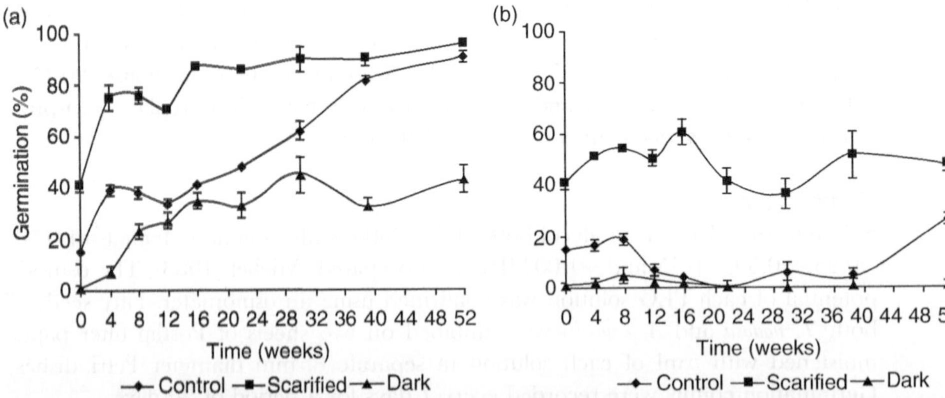

Fig. 41.1. Germination response of *Anthriscus caucalis* seeds dry-stored at 20°C (a) and 4°C (b) for differing periods up to 52 weeks and germinated at 15/10°C. Seed treatments were scarification, incubation in complete darkness or untreated control. Mean values ± standard error (SE) of mean are shown.

The differences in germination response to scarification from the seed lots dry-stored at 20°C and 4°C indicated that there is combinational seed dormancy associated with *A. caucalis*. In addition to a morphological dormancy, seeds of *A. caucalis* have an exogenous dormancy, which is broken by scarification and a physiological endogenous dormancy, which is overcome by dry storage. This finding is consistent with the winter annual behaviour of *A. caucalis*. Maturation and seed dispersal occurs in early summer and emergence of seedlings is not observed until mid-autumn/early winter. It is likely that during this dry warm period from seed dispersal to emergence, the requirements for after-ripening and some loss of seed coat dormancy will occur.

Dormancy determination of *T. nodosa*

Although a morphological dormancy was observed for *T. nodosa*, freshly collected seeds of *T. nodosa* exhibited no other innate dormancy behaviour with an initial germination percentage of 93% at 20°C. The germination of seeds of *T. nodosa* was 79% in complete darkness, which was significantly ($P < 0.05$) lower than with a 12h daily illumination period (Fig. 41.2). There was a significant ($P < 0.05$) reduction in the germination of *T. nodosa* seeds when stratified in complete darkness. Increases in the duration of stratification resulted in no significant ($P > 0.05$) difference in the germination percentage for those seeds exposed to a 12h daily illumination period.

Seeds of *T. nodosa* remain attached to the senescing adult plant until dispersal occurs in early autumn and therefore an exogenous or a physiological endogenous dormancy is unnecessary as conditions for germination are favourable at this time. Seed germination of *T. nodosa* was unaffected by stratification at 4°C and inducement of dormancy over the winter period was therefore considered unlikely. This is in contrast to the related species *T. japonica*, where field populations have been found to be induced into dormancy by low temperatures (Baskin and Baskin, 1975).

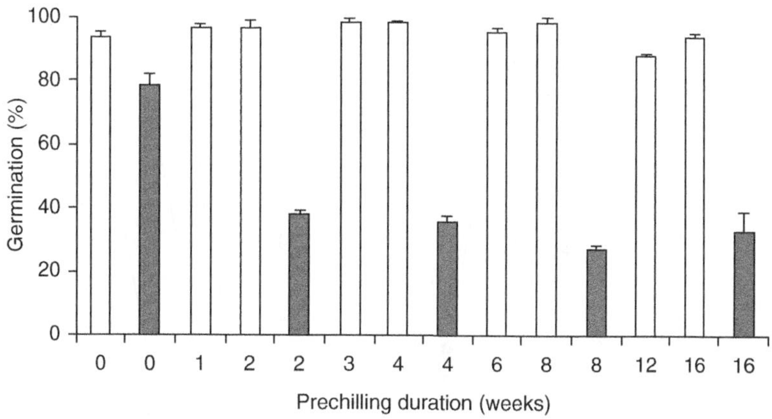

Fig. 41.2. Effect of stratification (4°C) duration on the germination of *Torilis nodosa* incubated in complete darkness (shaded bars) or with a 12 h daily illumination period (open bars). Mean values ± standard error (SE) of mean are shown.

Dispersed seeds of *T. nodosa* were not induced into dormancy by cold stratification and are unlikely to be exposed to dark conditions in a non-tillage production system like pyrethrum. Therefore, *T. nodosa* has the ability to behave predominantly as a facultative winter annual with germination spread over autumn and spring.

Temperature

T. nodosa seeds required a temperature above 4.2°C, but below 35°C, to germinate (Fig. 41.3a). The highest level of germination of this species occurred at 18.2°C and 23.4°C. At 28 DAI, the mean germination for seed imbibed at 18.2°C and 23.4°C was 91% and 94%, respectively, which was significantly ($P < 0.05$) higher than all other temperatures. *A. caucalis* seeds required a temperature above 2°C, but below 29.5°C, for germination. The maximum germination percentage occurred at 6°C, 10.5°C and 14.5°C (Fig. 41.3b). At 28 DAI, the mean germination percentage at 6°C, 10.5°C and 14.5°C was 66.7%, 71.3% and 74.7%, respectively. Initially (i.e. at 7 DAI) the highest level of germination was achieved at temperatures of 14.5°C, 18°C and 22°C, but at these temperatures the final germination at 28 DAI was only 56% and 42% at 18°C and 22°C, respectively.

The results suggest that both *T. nodosa* and *A. caucalis* are capable of behaving as facultative winter annuals with some germination occurring in spring, but with the majority taking place in autumn following seed dispersal. The optimum temperature for the germination of *A. caucalis* was lower than that for *T. nodosa. A. caucalis* may therefore behave more like a strict winter annual than *T. nodosa.* One could expect that a mild spring would result in minimal germination of *A. caucalis.* Only the cool temperatures experienced during the winter periods in northern Tasmania would restrict germination of *T. nodosa.*

Planting depth

The optimum planting depth of *T. nodosa* and *A. caucalis* seeds for seedling emergence occurred between 0 and 30 mm (Fig. 41.4a). At a planting depth of 50 mm,

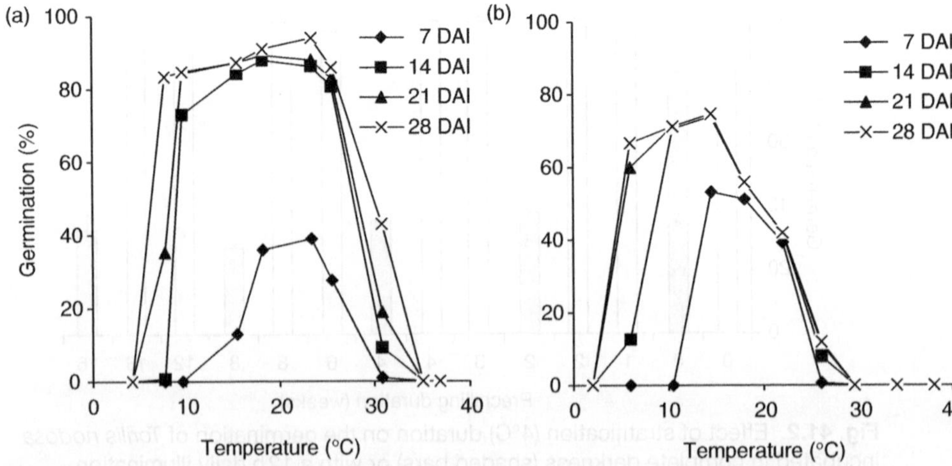

Fig. 41.3. Effect of temperature on the germination of *Torilis nodosa* (a) and *Anthriscus caucalis* (b).

emergence of both species was significantly ($P < 0.05$) reduced, with emergence totally inhibited at 70 mm. As *T. nodosa* and *A. caucalis* have seeds of similar size it was not surprising that their emergence response to planting depths was similar.

The results suggest that a non-tillage production system, such as that for a short-lived perennial crop like pyrethrum, would favour the germination of *T. nodosa* and *A. caucalis* through accumulation of seeds at or near the surface (Dorado *et al.*, 1999). Slight burial of *T. nodosa* and *A. caucalis* seeds just below the surface would most likely occur as a result of the movement of machinery, grazing animals and abiotic factors during the growing season, favouring the germination of these seeds. In an attempt to reduce seedling emergence of these weeds in the following crops, conventional tillage may bury a large proportion of the seed to depths that inhibit emergence. The restriction in emergence of *T. nodosa* and *A. caucalis*, when planted at depths of 50 and 70 mm, was a result of the small seed size that restricted epigeal emergence from greater depths.

Osmotic stress

There was no significant ($P < 0.05$) difference in germination of *T. nodosa* and *A. caucalis* at osmotic potentials between 0 and −0.5 MPa (Fig. 41.4b). However, there was a highly significant ($P < 0.001$) reduction in germination at an osmotic potential >−0.75 MPa for both species, and germination was totally inhibited at an osmotic potential of −1.0 MPa.

The results indicate that both *T. nodosa* and *A. caucalis* are sensitive to osmotic stress. During the summer period in north-western Tasmania, the upper 5 cm layer of the red ferrosol soils becomes dry, especially under a non-tillage system, and soil water potentials commonly fall to levels near permanent wilting point (i.e. −1.5 MPa) (McLaren and Cameron, 1996). Germination of any newly dispersed *T. nodosa* and *A. caucalis* seeds at such time would be restricted and would be unlikely to occur until moisture availability increased in autumn. In addition to the

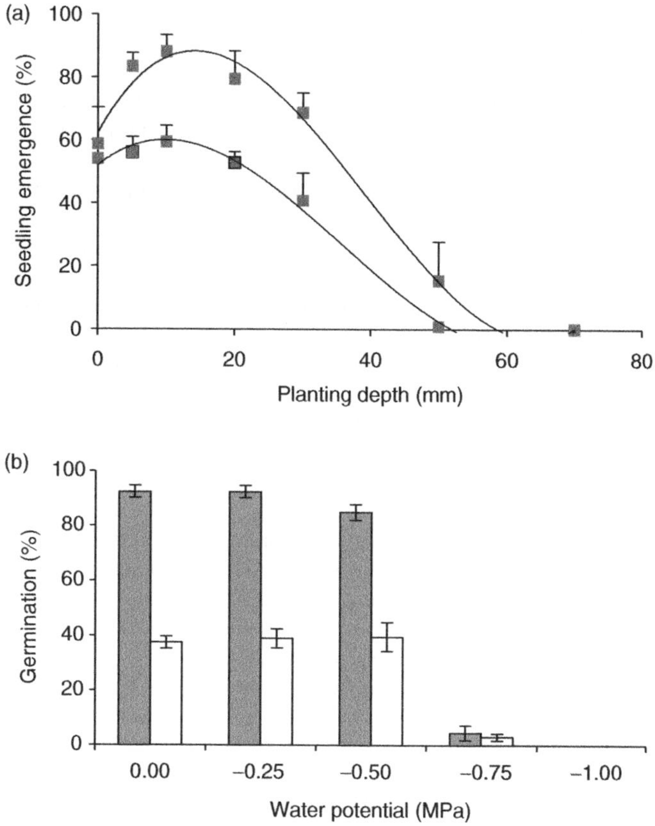

Fig. 41.4. Effect of (a) planting depth and (b) osmotic potential on the emergence and germination of *Torilis nodosa* (■) and *Anthriscus caucalis* (□). Mean values ± standard error (SE) of mean are shown.

low initial germination of dispersed *A. caucalis* seeds, due to an imposed seed coat dormancy and a requirement for after-ripening, low levels of moisture availability during the summer period would also restrict emergence. Seed coat dormancy and a requirement for after-ripening are viewed as survival strategies of *A. caucalis*, in that they limit the number of false breaks that will occur during the summer period. Although *T. nodosa* has been shown to have a high initial germination percentage, susceptibility to false breaks in the summer is overcome by a delayed dispersal mechanism of the mature seeds, which are held tightly in compact umbels and have been observed to remain on the parent until senescence has been completed, generally in late summer/early autumn.

References

Anderson, W.P. (1996) *Weed Science, Principles and Applications.* West Publishing, St. Paul, Minnesota.

Baskin, C.C., Chester, E.W. and Baskin, J.M. (1992) Deep complex morphophysiological dormancy in seeds of *Thaspium pinnatifidum* (Apiaceae). *International Journal of Plant Sciences* 153, 565–571.

Baskin, C.C., Milberg, P., Andersson, L. and Baskin, J.M. (2000) Deep complex morphophysiological dormancy in seeds of *Anthriscus sylvestris* (Apiaceae). *Flora* 195, 245–251.

Baskin, J.M. and Baskin, C.C. (1975) Ecophysiology of seed dormancy and germination in *Torilis japonica* in relation to its life cycle. *Bulletin of the Torrey Botanical Club* 102, 67–72.

Baskin, J.M. and Baskin, C.C. (1979) Studies on the autecology and population biology of the weedy monocarpic perennial *Pastinaca sativa*. *Journal of Ecology* 67, 601–610.

Baskin, J.M. and Baskin, C.C. (1990) Seed germination ecology of poison hemlock, *Conium maculatum*. *Canadian Journal of Botany* 68, 2018–2024.

Chaturvedi, S.N. and Muralia, R.N. (1975) Germination inhibitors in some Umbelliferae seeds. *Annals of Botany* 39, 1125–1129.

Dorado, J., Del Monte, J.P. and Lopez-Fando, C. (1999) Weed seedbank response to crop rotation and tillage in semiarid agroecosystems. *Weed Science* 47, 67–73.

Geneve, R.L. (1999) Seed dormancy in vegetable and flower seeds. *Combined Proceedings International Plant Propagator's Society* 49, 248–254.

Grime, J.P., Mason, G., Curits, A.V., Rodman, J., Band, S.R., Mowforth, M.A.G., Neal, A.M. and Shaw, S. (1981) A comparative study of germination characteristics in a local flora. *Journal of Ecology* 69, 1017–1059.

Isbell, R.F. (1996) *The Australian Soil Classification.* CSIRO Australia, Collingwood, Australia.

ISTA (1996) *International Rules for Seed Testing.* International Seed Testing Association, Switzerland.

Jacobsen, J.V. and Pressman, E. (1979) A structural study of germination in celery (*Apium graveolens* L.) seed with emphasis on endosperm breakdown. *Planta* 144, 241–248.

McLaren, R.G. and Cameron, K.C. (1996). *Soil Science Sustainable Production and Environmental Protection*, 2nd edn. Oxford University Press, Auckland, New Zealand.

Michel, B.E. (1983) Evaluation of the water potentials of solutions of polyethylene glycol 8000 both in the absence and presence of other solutes. *Plant Physiology* 72, 66–70.

Roberts, H.A. (1979) Periodicity of seedling emergence and seed survival in some Umbelliferae. *Journal of Applied Ecology* 16, 195–201.

SAS Institute (1998) *User Guide*, release 6.03 edn. SAS Institute, Cary, North Carolina.

42 Plant Dispersal Strategies, Seed Bank Distribution and Germination of Negev Desert Species

Y. GUTTERMAN,[1] T. GENDLER[1] AND Z.-Y. HUANG[1,2]

[1]Ben-Gurion University of the Negev, Desert Research Institutes, Sede Boker Campus, 84990 Israel; [2]Institute of Botany, The Chinese Academy of Sciences, Beijing 100093, Peoples Republic of China

Abstract

In a natural flat area of deep loess soil in the Negev Desert of Israel, near Sede Boker, the distribution of seeds of different species, in the seed bank, is a result of their strategies of seed dispersal, seed size and structure and seed germination. Samples were collected in summer from 39 plots measuring 350 cm^2. In *Carrichtera annua* (L.) DC., which has a protection strategy, seeds are enclosed in lignified structures as aerial seed banks on the dry and lignified plant until they are dispersed by rain. Of 3244 *C. annua* seeds, 3110 were found in the aerial seed bank. In *Spergularia diandra* (Guss.) Heldr. & Sart., which has an escape strategy, the tiny dust-like mature seeds are dispersed into cracks in the soil by wind. Of 1043 seeds, 896 were found in the upper 0–1 cm layer of soil and only 99 at a depth of 1–2 cm. No seeds remained on the dry plants. *Neotorularia torulosa* (Desf.) Hedge & J. Léonard has developed a dual dispersal strategy. Of 954 *N. torulosa* seeds, 630 were found protected in the aerial seed bank. The rest were dispersed by wind, 177 escaped into the upper 0–1 cm layer of soil and 147 were at a depth of 1–2 cm.

Introduction

The desert biome: unpredictable quantity and distribution of rainfall

The Negev Desert highlands, a part of the northern belt of the Saharo-Arabian phytogeographical region (which is ~6000 km long in total), receive an annual average rainfall of 100 mm. This rainfall occurs in winter along with mild temperatures. The short growing season usually occurs between December and April. The

 Seeds: Biology, Development and Ecology (eds S. Adkins, S. Ashmore and S.C. Navie)

quantity of rainfall is unpredictable, ranging from 25 to 180 mm per annum. On a single day, the amount of rainfall may reach 50 mm, which is significantly higher than the total amount of rain that may fall during another year. The distribution of the rainfall during the season is also unpredictable, and so is the number of days with rain per season. The summers are long, very hot, without rain and with very low relative humidity during the day (Evenari and Gutterman, 1976; Evenari *et al.*, 1982; Gutterman, 2002). Similar environmental conditions also exist in other deserts within the northern belt of the Saharo-Arabian phytogeographical region, and they also receive an annual average rainfall of about 100 mm (Batanouny, 1981, 2001).

Seed collection

In large areas of the Negev Desert highlands, there is an extremely high level of seed collection. In a natural desert habitat, in the Negev Desert near Sede Boker, seeds that were dispersed on to the soil surface at the end of the summer were collected by ants within 7 min, and within 2 h, all the 2500 seeds from several common annual plant species were collected, mainly by three species of 'harvesting ants' (Gutterman, 2002).

Massive seed collection has also been observed in other parts of the Negev and in other deserts (Bernstein, 1974; Reichman, 1979; Andersen, 1982, 1991a,b; Abramsky, 1983; Medel and Vasquez, 1994; Valone *et al.*, 1994; Gutterman, 2002).

Seed dispersal strategies

Under such extreme desert conditions, and intense seed collection pressure, different groups within the approximately 250 annual plant species occurring in the Negev Desert highlands have developed different seed dispersal strategies (Fig. 42.1) (Gutterman, 2002).

Escape strategy

Some of the very common annual plant species, belonging to different families and genera, produce many tiny dust-like seeds, e.g. *Schismus arabicus* Nees (*Poaceae*) and *Spergularia diandra* (Guss.) Heldr. & Sart. (*Caryophyllaceae*). These seeds are dispersed by wind after their maturation at the beginning of the summer. They fall and escape into small cracks in the loess soil crust a few millimetres below the soil surface, and become a part of the long-lasting soil seed banks of these plant species (Gutterman, 2002).

Protection strategy

AERIAL-PROTECTED SEED BANKS. In other very common annual plant species belonging to different families and genera, seeds are relatively larger and produced in much lesser numbers. They are covered by dry lignified tissues of the dead mother plant. These seeds become a long-lived protected aerial seed bank. Each of these plant species has developed its own mechanism of seed dispersal by rain. The

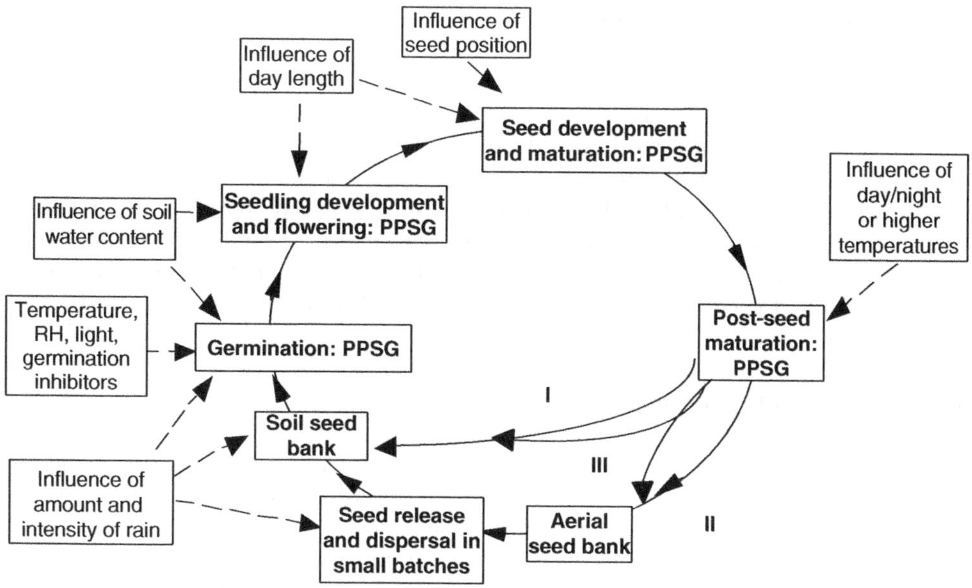

Fig. 42.1. The influences of environmental and maternal factors on phenotypic plasticity of seed germination (PPSG) during the different stages of the annual life cycle of ephemerals in the Negev Desert of Israel. Seed dispersal strategies: escape (I), protection (II) and dual strategy of protection and escape on the same plant (III).

seeds are dispersed in small portions, year after year, by any rainfall event that is long and strong enough. In some of these plant species, the seeds develop a mucilaginous layer when they touch the wet surface of the soil and germinate within a few hours, as in *Blepharis* species (*Acanthaceae*) and *Anastatica hierochuntica* L. (*Brassicaceae*). The seeds adhere to the soil crust by their mucilaginous layer, which may also delay their collection by ants. Ants do not collect seeds while it is raining or after the seeds have germinated (Witztum *et al.*, 1969; Gutterman *et al.*, 1973; Gutterman and Witztum, 1977; Zaady *et al.*, 1997; Gutterman, 2002).

SUBTERRANEAN-PROTECTED SEED BANKS. In some plant species, a portion of the seeds are located a few millimetres below the soil surface and are covered by the dry lignified tissues of the dead mother plant. These seeds germinate *in situ*, such as the subterranean 'seeds' of the amphicarpic plant species *Emex spinosa* (L.) Campd. (*Polygonaceae*) (Evenari *et al.*, 1977) and *Gymnarrhena micrantha* Desf. (*Asteraceae*) (Koller and Roth, 1964).

Dual strategy: escape and protection strategies on the same plant

In some plant species, there is a dual strategy of seed dispersal. Some of the seeds are dispersed by wind after their maturation at the beginning of the summer, while the remainder are protected by the tissues of the dead, dry and lignified mother plant and are dispersed by rain in the following winter. This dual strategy is found in *Neotorularia torulosa* (Desf.) Hedge & J. Léonard (*Brassicaceae*) (Gutterman, 2002).

Seed banks in desert loess soils

Desert loess soils are covered by a biological soil crust. The depth of this crust depends on the vegetation cover as well as the average rainfall. In uncovered areas between the shrubs, the higher the average rainfall, the deeper the biological soil crust, i.e. a depth of 1 mm in areas with 50 mm of rainfall and 16 mm in areas with 250 mm of rainfall (Shem-Tov *et al.*, 2002). Near Sede Boker, seeds are mostly located in the upper 2 cm layer of the soil. Only in depressions and diggings, such as those of porcupines (*Hystrix indica* Kerr) and ibex (*Capra ibex nubiana* Cuvier), are seeds found in deeper layers, after being covered by soil erosion (Gutterman, 2001b).

Seed germination strategies

The germination strategies used by a particular plant species are a part of the complementary set of adaptations made to suit a particular habitat (Went, 1948, 1949, 1953; Gutterman, 2002). During the period of seed development and maturation, seed germination is affected by environmental factors as well as maternal factors. These influences may increase the phenotypic plasticity of seed germination (PPSG). Therefore, only a small portion of the seeds of the plant species in the seed bank may be ready for germination after a particular rainfall event. The phenomenon of PPSG was observed in the seeds of certain plant species with seeds dispersed by wind, such as *S. arabicus* and *S. diandra*, as well as with seeds dispersed by rain, such as in *Carrichtera annua* (L.) DC. (Gutterman, 2001a, 2002).

The location of the seeds within the soil profile may affect their germination pattern. There are two very common annual plant species occurring in the Negev Desert that have tiny seeds that are dispersed by wind after maturation (i.e. *S. arabicus* and *S. diandra*). The caryopses of *S. arabicus* may begin to germinate after 4 h and will reach much higher percentages of germination in darkness than in light. In *S. diandra*, a higher percentage of seeds germinate in light than in darkness. A higher percentage of the caryopses of *S. arabicus* located close to the soil surface germinate when the rain that engenders their germination starts in the evening rather than during the daytime. The opposite may be true for the seed germination of *S. diandra.* Some of the plant species occurring in the Negev Desert have seeds that develop a mucilaginous layer on their seed coats when wetted. *C. annua* seeds are dispersed by rainfall events that last for ~1.5 h. The seeds fall on to the wet soil and adhere to the soil surface by this mucilaginous layer. This process may delay the collection of the seeds by ants and keep them in a moist condition for a longer period, and possibly increase their germination percentage. Ants do not collect seeds once radicles have protruded from their seed coats. Therefore, the faster the seed germinates, the higher the chance that ants will not collect it (Gutterman, 2002).

The aims of this study are: (i) to analyse the distribution of seeds in the seed bank of some of the most common plant species that occurred in the research area; and (ii) to find out whether there are significant differences in the distribution of seeds in the different local seed banks (i.e. between plant species with different strategies of seed dispersal as well as with different strategies of seed germination).

Materials and Methods

Research site, soil sampling and seed banks

The research site is located near Sede Boker on the Zin plateau (34°46'E, 30°51'N, 460 m asl) in the Negev Desert highlands of Israel. Sede Boker is located in the middle of the Negev Desert and the average annual rainfall is ~100 mm. The research site is a natural habitat on a flat area of scrubland of deep loess soil in which *Hammada scoparia* (Pomel) Iljin is the dominant shrub. The shrubs are distributed a few metres apart and geophytes and annual plant species appear between these shrubs.

On 3 August 2004, soil samples of 350 cm^2 from 39 plots located in the same natural habitat were collected from 0 to 1 cm and 1 to 2 cm depths before the first rains of the following growing season. The two layers of the soil samples were progressively sieved through several mesh sizes (i.e. 1000, 750, 500 and 250 μm). More than 98% of this soil was thin loess soil, which easily passed through the 250 μm mesh. None of the seeds of plants existing in this area could pass through this fine mesh. The seeds and the larger particles of the soil were separated through larger mesh, and manually under a stereomicroscope. This dry isolation method was easy and did not require wetting the samples, which might have changed the viability and germinability of the seeds.

For the determination of the number of mucilaginous seeds present, the larger particles of the soil were poured on the 250 μm mesh, washed in distilled water, the remaining particles air-dried and the mucilaginous seeds counted under a stereomicroscope. The number of seeds of each species was then determined and recorded.

Seed samples of the most common species that were separated from the soil were dry-stored at room temperature before being tested for germination on 19 May 2005. Four replicates of 50 seeds from each species were placed in 52 mm diameter Petri dishes, on a No. 1 filter paper, and wetted with 1.5 ml of distilled water. Seeds were germinated at 25°C in darkness. The experiment was terminated after the germination percentage reached a plateau, and no additional germination was encountered.

Results

After seed maturation, but before the first rains of the following growing season, the soil is dry and the soil crust (about 1 cm deep) has a pattern of cracks and splits of different widths and depths. At this time, seeds are collected from the different plots. There is a correlation between the location of seed collection of the different species in the seed bank and the different strategies of seed dispersal of the four most common species (Table 42.1). The tiny seeds that are dispersed by wind in summer escape into the cracks formed in the soil crust, mostly in the upper 1 cm layer of soil. In other plant species, seeds remain in lignified structures on the mother plant during summer and are dispersed during rain in winter. The seeds of many of these species are mucilaginous, and hence adhere to the wet soil surface immediately after their dispersal. These seeds may then germinate from this position.

Table 42.1. The seed bank distribution of four common plant species found in the Negev Desert.

	Number of seeds			
Seed bank location	*Carrichtera annua*	*Spergularia diandra*	*Schismus arabicus*	*Neotorularia torulosa*
Aerial	3110	0	10	630
Crust (0–1 cm)	123	896	515	177
Soil (1–2 cm)	10	99	152	147
Total	3243 (2375)	995 (730)	677 (496)	954 (699)

The numbers in brackets after the totals are the estimated number of seeds per square metre.

S. diandra has an 'escape' strategy of seed dispersal. The great majority of its tiny seeds were found in the upper layer of the soil crust (i.e. in the upper 0–1 cm) and a relatively small amount reached the 1 cm layer below this. None of the seeds remained in the dry capsules of the dead plant. The opposite was found in *C. annua,* which has a protection strategy of seed dispersal. All the seeds of the last season remained enclosed in the dry lignified siliquae (i.e. fruit) on the dead plants. Relatively small numbers of seeds were found in the upper soil layer, and even less were found in the lower soil layer. Most of the seeds in the lower layer of the soil were the seeds that remained from previous years. The seeds of this plant are dispersed by rain and adhere to the soil surface by the aid of their mucilaginous seed coats or the seeds may be washed into cracks in the soil. In *N. torulosa* there is a dual strategy of seed dispersal: (i) seeds from the unlignified upper fruit disperse their seeds by wind during the summer after maturation and escape into cracks in the soil; and (ii) seeds that mature in the lower lignified fruit remain enclosed and protected during summer and are dispersed by rain in the following winter.The majority of the seeds are located in the aerial seed bank, enclosed in the lignified fruit, while a higher proportion of seeds with escape strategy are located in the upper 1 cm layer of the soil with the rest in the lower level (Table 42.1). In these soils, it was found that the number of seeds below the 2 cm level is negligible. When samples of seeds from the soil and aerial seed bank collected on 3 August 2004 were germinated, it was found that 60% of *C. annua* seeds from the aerial seed bank germinated, but none of the seeds that were separated from the upper (0–1 cm) or the lower (1–2 cm) layer of the soil germinated (Fig. 42.2). In contrast, in *S. diandra,* ~70% of the seeds from both soil layers germinated. In *S. arabicus,* only a few seeds germinated. In *N. torulosa,* a small percentage of seeds germinated from the seed samples were separated from the aerial seed bank as well as from the upper layer of the soil. No germination was observed in the seed samples separated from the lower layer of the soil.

Discussion

In the present study it was found that the location of the seeds of different plant species in the aerial or soil seed bank is a result of the different strategies used by

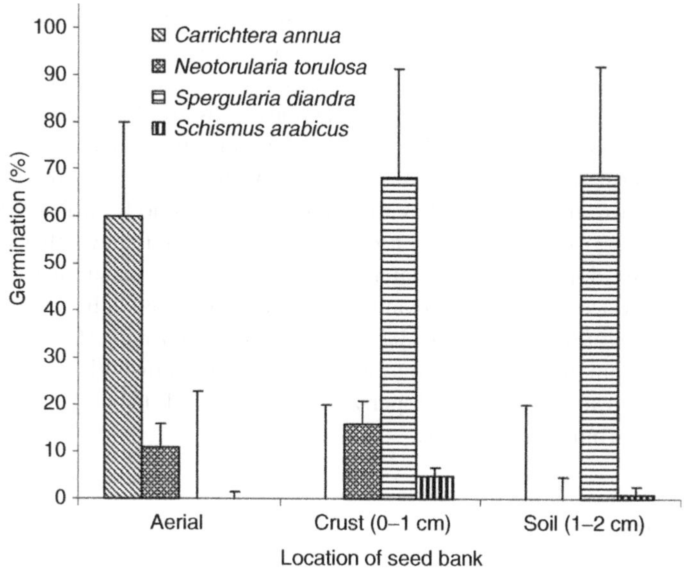

Fig. 42.2. Germination of seeds from the seed banks of four common Negev Desert plant species. Seeds were germinated on 19 May 2005 in Petri dishes on filter paper in darkness at 25°C. Germination was terminated after 48 h. These seed samples were separated from the aerial seed bank, the soil crust (0–1 cm) or the soil layer underneath (1–2 cm). They were collected on 3 August 2004 from a natural desert habitat (see Table 42.1). Bars indicate ± standard error (SE).

these plant species for seed dispersal and germination. In *S. diandra* and *S. arabicus*, as well as in some of the seeds of *N. torulosa* that are dispersed from their mother plants during summer, the seeds are located mainly in the upper 1 cm layer of the soil with relatively few seeds present in the 1 cm layer below this. Seeds that are located in the aerial seed bank still enclosed in their siliquae, as seen with *C. annua*, or in lignified lower fruit, as seen with *N. torulosa*, are dispersed by rains during the following year(s) (Fig. 42.1; Table 42.1). The cracks and splits in the loess soil surface, which are 1 cm, or even 2 cm, deep are crucial for the penetration of the seeds into the soil profile. Nearly, all of the seeds in these soils are located in the cracks or splits in the soil crust and in the upper 2 cm layer of the profile (Gutterman, 2001b, 2002). Plants like *Blepharis* species keep their seeds protected in lignified structures on the plants for many years and during rain a small portion of these seeds are released, adhere to the soil surface by their mucilaginous hairs and rapidly germinate (Gutterman *et al.*, 1967; Gutterman and Witztum, 1977). In other plants, such as *Plantago coronopus* L., seeds are protected for many years in the lignified inflorescences on plants. During some rainfall events, small portions of these seeds are released and fall on to the wet soil surface. The mucilaginous layer develops following wetting and the seeds adhere to the soil surface. In *Asteriscus pygmaeus* (DC.) Coss. & Dur. the capitula are opened by rain and release some achenes (i.e. seeds); in addition the pappus of the disconnected achenes (i.e. seeds) can act to disperse the seed in wind (Gutterman, 2002).

During summer, many plants using the escape strategy of seed dispersal disperse their tiny seeds into the open cracks appearing in the soil surface crust. The smaller the seed, the higher the chances are that it will enter the smaller cracks in the soil surface. In contrast, mucilaginous seeds are dispersed by rain during winter when rain events may adhere them to the soil surface and allow them to germinate (Gutterman *et al.*, 1967; Gutterman and Witztum, 1977; Zaady *et al.*, 1997), or if the seeds are small enough they may adhere to the muddy surface even without mucilage (e.g. *Mesembryanthemum nodiflorum* L.). The result shows that many of the seeds that are dispersed by rain later germinate from the soil surface, or very close to the soil surface, which is an advantage for seeds that need light for germination, such as *Artemisia sieberi* Bess. (*Asteraceae*) (Evenari and Gutterman, 1976; Gutterman, 2002).

From the studies undertaken it has been shown that there is a correlation between a species' seed dispersal strategy, seed distribution in the seed bank and seed germination strategy.

References

Abramsky, Z. (1983) Experiments on seed predation by rodents and ants in the Israeli desert. *Oecologia* 57, 328–332.

Andersen, A.N. (1982) Seed removal by ants in the mallee of northwestern Victoria. In: Buckley, R.C. (ed.) *Ant–Plant Interactions in Australia.* Dr W. Junk Publishers, The Hague, The Netherlands, pp. 31–43.

Andersen, A.N. (1991a) Seed harvesting by ants in Australia. In: Huxley, C.R. and Cutler, D.F. (eds) *Ant–Plant Interactions.* Oxford University Press, Oxford, UK pp. 493–503.

Andersen, A.N. (1991b) Parallels between ants and plants: implications for community ecology. In: Huxley, C.R. and Cutler, D.F. (eds) *Ant–Plant Interactions.* Oxford University Press, Oxford, UK pp. 539–558.

Batanouny, K.H. (1981) *Ecology and Flora of Qatar.* Centre for Applied and Scientific Research, University of Qatar, Doha, Qatar.

Batanouny, K.H. (2001) *Plants in the Deserts of the Middle East: Adaptations of Desert Organisms.* Springer, Berlin, Germany.

Bernstein, R.A. (1974) Seasonal food abundance and foraging activity in some desert ants. *The American Naturalist* 108, 490–498.

Evenari, M. and Gutterman, Y. (1976) Observations on the secondary succession of three plant communities in the Negev Desert, Israel. I. *Artemisietum herbae albae.* In: Jacques, R. (ed.) *Hommage au Prof. P. Chouard Etudes de Biologie Vegetale, CNRS.* Gif sur Yvette, Paris, pp. 57–86.

Evenari, M., Kadouri, A. and Gutterman, Y. (1977) Eco-physiological investigations on the amphicarpy of *Emex spinosa* (L.) Campd. *Flora* 166, 223–238.

Evenari, M., Shanan, L. and Tadmor, N. (1982). *The Negev: The Challenge of a Desert,* 2nd edn. Harvard University Press, Cambridge, Massachusetts.

Gutterman, Y. (2001a) Phenotypic germination plasticity related to caryopsis size in *Schismus arabicus. Seed Science Research* 11, 173–178.

Gutterman, Y. (2001b) *Regeneration of Plants in Arid Ecosystems Resulting from Patch Disturbance. Geobotany 27.* Kluwer Academic Publishers, Dordrecht, The Netherlands.

Gutterman, Y. (2002) *Survival Strategies of Annual Desert Plants. Adaptations of Desert Organisms.* Springer, Berlin, Germany, pp. 348.

Gutterman, Y. and Witztum, A. (1977) The movement of integumentary hairs in *Blepharis ciliaris* L. Burtt. *Botanical Gazette* 138, 29–34.

Gutterman, Y., Witztum, A. and Evenari, M. (1967) Seed dispersal and germination in *Blepharis persica* (Burm.) Kuntze. *Israel Journal of Botany* 16, 213–234.

Gutterman, Y., Witztum, A. and Heydecker, W. (1973) Studies of the surfaces of desert plant seeds. II. Ecological adaptations of the seeds of *Blepharis persica. Annals of Botany* 37, 1051–1055.

Koller, D. and Roth, N. (1964) Studies on the ecological and physiological significance of amphicarpy in *Gymnarrhena micrantha* (Compositae). *American Journal of Botany* 51, 26–35.

Medel, R.G. and Vasquez, R.A. (1994) Comparative analysis of harvester ant assemblages of Argentinian and Chilean arid zones. *Journal of Arid Environments* 26, 363–371.

Reichman, O.J. (1979) Desert granivore foraging and its impact on seed densities and distributions. *Ecology* 60, 1085–1092.

Shem-Tov, S., Zaady, E. and Gutterman, Y. (2002) Germination of *Carrichtera annua* (Brassicaceae) seeds on soil samples collected along a rainfall gradient in the Negev Desert of Israel. *Israel Journal of Plant Sciences* 50, 115–120.

Valone, T.J., Brown, J.H. and Heske, E. J. (1994) Interactions between rodents and ants in the Chihuahuan desert: an update. *Ecology* 75, 252–254.

Went, F.W. (1948) Ecology of desert plants. I. Observations on germination in the Joshua Tree National Monument, California. *Ecology* 29, 242–253.

Went, F.W. (1949) Ecology of desert plants. II. The effect of rain and temperature on germination and growth. *Ecology* 30, 1–13.

Went, F.W. (1953) The effects of rain and temperature in plant distribution in the desert. *Proceedings International Symposium Desert Research,* May 1952. *Research Council of Israel* 2, 230–240.

Witztum, A., Gutterman, Y. and Evenari, M. (1969) Integumentary mucilage as an oxygen barrier during germination of *Blepharis persica* (Burm.) Kuntze. *Botanical Gazette* 130, 238–241.

Zaady, E., Gutterman, Y. and Boeken, B. (1997) The germination of mucilaginous seeds of *Plantago coronopus, Reboudia pinnata* and *Carrichtera annua* on cyanobacterial soil crust from the Negev Desert. *Plant and Soil* 190, 247–252.

43 Seed Biology of Tropical Australian Plants

S.M. BELLAIRS[1] AND N. ASHWATH[2]

[1]*School of Science and Primary Industries, Charles Darwin University, Darwin, Northern Territory 0909, Australia;* [2]*School of Biological and Environmental Sciences, Central Queensland University, Rockhampton, Queensland 4700, Australia*

Abstract

Australia has a diverse tropical flora. In this chapter we review some aspects of the seed biology of Australian tropical forest and woodland communities. Unfortunately, there is a lack of knowledge of seed biology of most species and some habitats in tropical Australia. Although much less research has been carried out on the tropical flora, in comparison with the temperate flora of Australia, some general trends begin to emerge from published studies. The *Eucalyptus*-dominated forest and woodland flora have a bimodal peak in seed production, with the woody vegetation flowering towards the end of the dry winter season, and the herb and grass understorey flowering around the end of the wet season. The rainforest communities show extended fruit production with a peak around early summer (i.e. December). These communities do not maintain a large seed bank. Many species produce dormant seeds, but for most species seed dormancy is broken at the beginning of the following summer wet season.

Introduction

The environment of tropical Australia is very different to that of temperate Australia. Seasonal conditions that favour successful seedling establishment are quite different, and this has resulted in differences in germination and dormancy responses. Most Australian native seed biology research has been focused on temperate species, and both taxonomic understanding and knowledge of the seed biology of tropical species have lagged well behind. For example, of the 130 papers published between 1994 and 2005 in the Australian native seed biology workshop series (http://www.acmer.com.au), only 15 papers have dealt with tropical species specifically, and a further 19 papers have discussed examples of both temperate and tropical floras.

This review will focus on some aspects of seed biology of the forest and woodland/savannah communities of northern Australia (i.e. north of the Tropic of Capricorn), including rainforest and *Eucalyptus*-dominated communities. Seed production, soil seed storage, dormancy and germination responses to controlled storage conditions are also discussed.

Vegetation Communities and Environment

Savannah woodlands and open forests are the most widespread vegetation types within the wet–dry tropics of northern Australia (Dunlop and Webb, 1991) and are dominated by *Eucalyptus* species with a tall grass understorey. Dense *Eucalyptus* forest occurs along the northern and eastern coastal regions of Australia and tree density reduces to woodland and then to scattered trees as average annual rainfall decreases along southern and western gradients.

Rainforest occurs in some areas along the east coast of Australia, and in numerous small patches through the northernmost portion of the Northern Territory (NT) and Western Australia (WA). In the northern region of the NT and the Kimberley region of WA, there are 16,500 rainforest patches, which average <5 ha in area (Russell-Smith *et al.*, 1992). These include moist rainforests of evergreen species and seasonally dry rainforests dominated by deciduous or semi-deciduous species. Many tree species in these rainforests have bat- or bird-dispersed seeds and are likely to have expanded into the NT from South-east Asia or north Queensland (Russell-Smith and Lee, 1992).

Rainfall is summer-dominated in northern Australia (Fig. 43.1). Darwin has an annual rainfall of 1700 mm, with >90% falling between November and April. On the north-east coast, Cairns also has a strongly summer-dominated rainfall pattern, but some rain still tends to occur during the winter months. The rainfall decreases to the south of Darwin and to the west of Cairns. Tennant Creek has a shorter summer wet season and a much lower annual rainfall, averaging 371 mm. Temperatures show only a slight decrease during the winter months at Darwin, but there is substantial variation between higher summer and lower winter temperatures further inland towards Katherine.

Fire is frequent in the wet–dry tropical areas of northern Australia and many areas are burnt annually. Most fire events are human-lit, either by pastoralists, fire managers or aboriginal traditional landowners. Fire intensity is low during the early dry season, but the high-intensity fires during the middle to late dry season can affect the seed production of trees (Setterfield, 1997). Fire in the late wet season and early dry season can destroy seeds of grasses and herbaceous plants before they are mature and incorporated in the soil seed bank.

Eucalyptus Forest and Woodland/Savannah Vegetation

Seed production and dispersal

Trees and shrubs tend to produce mature seeds during the late dry season, from August to October (Brennan, 1996; Williams *et al.*, 1999; Brock, 2001). Tree seed production can be affected by fire. Mild fires at the beginning of the dry season have

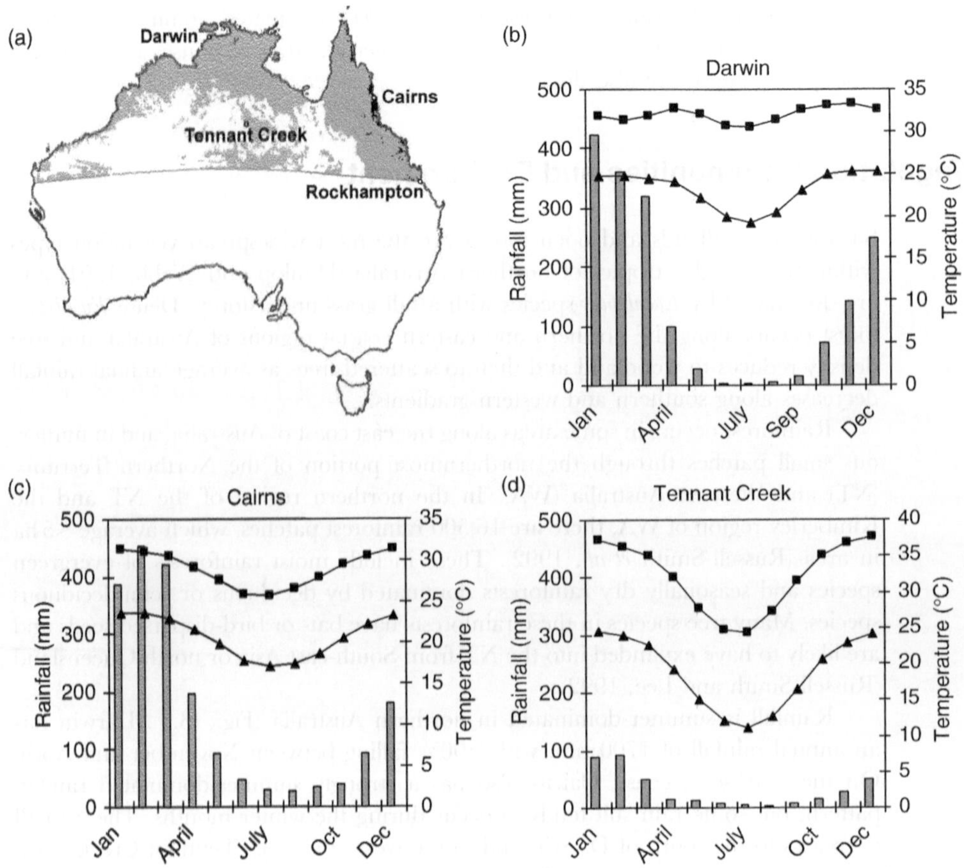

Fig. 43.1. (a) Map of *Eucalyptus* forest/woodland (grey shading) and rainforest (black) vegetation in tropical Australia. Ombrothermic diagrams of (b) Darwin, (c) Cairns and (d) Tennant Creek. Mean monthly rainfall, mean daily minimum and mean daily maximum temperature data are taken from Bureau of Meteorology, Australia. (From Bureau of Meteorology, Australia. With permission.)

little effect on seed production of dominant *Eucalyptus* species, but the more intense fires later in the dry season greatly reduce seed production (Setterfield, 1997).

Most of the tropical woody species have vertebrate-dispersed seeds, which are larger when compared with those of woody plants of temperate floras (Lord *et al.*, 1997). Ant dispersal of seeds is also common (Andersen *et al.*, 2000) and only ~25% of species have unassisted seed dispersal, which is less than the proportion of subtropical or temperate floras (Lord *et al.*, 1997). The small tree *Petalostigma pubescens* Domin has a particularly interesting seed dispersal mechanism (Clifford and Monteith, 1989). The first step involves the consumption of the drupaceous fruit by emus (*Dromaius novaehollandiae* Latham), during which the flesh is removed from the woody endocarp. On drying in the faeces the endocarp cracks and explodes, further dispersing the segments and seeds. Finally the seeds contain an ant-attracting elaiosome, which attract the ants to carry the seeds further away. Clifford and Monteith (1989) suggest that the later dispersal mechanisms are of considerable importance as up to 1200 seeds may be deposited in a single emu scat. In contrast, *Jedda multicaulis* JR Clarkson (*Thymelaeaceae*) displays minimal dispersal of seeds. The indehiscent fruit is about 6 cm diameter and encloses a single large seed in a coriaceous pericarp about 3 mm thick. At maturity the heavy fruit weighs down the branch tips until the fruit rests on the soil surface and germination often occurs while the fruit is still attached to the plant (Clarkson and Clifford, 1987).

Grasses and herbs produce seeds from February to May, which is the end of the wet season and the early dry season (Andrew and Mott, 1983; Brennan, 1996; Crowley and Garnett, 1999). They are using available soil moisture for seed production before the soil profile dries out, whereas the woody vegetation with deeper roots can access groundwater sources to produce seeds during the dry season.

Most tropical Australian grasses either have seeds adapted for adhering to the coats of animals or do not have any structures to assist dispersal (Lord *et al.*, 1997). Many of the dominant grass species have sharp-pointed calluses with hygroscopically active spirally twisted awns that drill the floret containing the seed into the soil (Mott, 1978). Seed dispersal has been investigated for two grasses with this type of floret, *Sorghum intrans* Benth. and *Sorghum stipoideum* (Ewart & Jean White) CA Gardner & CE Hubb. Most of the seeds of these species were dispersed within 2 m of the parent plant and then drilled themselves into the soil to depths up to 2 cm (Andrew and Mott, 1983). Ant consumption of grass seeds is considerable, but most of the seeds that are dispersed by ants are also consumed (Andersen *et al.*, 2000). Most forbs do not have adaptations for seed dispersal, but ~20% are adapted for ant dispersal and some species are adapted for wind or vertebrate dispersal (Lord *et al.*, 1997). Forb species of the Australian tropics also tend to produce larger seeds than subtropical or temperate floras, while the average grass seed sizes are more similar (Lord *et al.*, 1997).

Soil seed banks

In the *Eucalyptus* woodland communities of northern Australia, the soil seed bank tends to be lower in size than that found in southern Australia. The size ranges from 49 to 792 germinable seeds/m^2 and is composed mainly of grass and forb seeds with few shrub seeds (Manning, 2004; Williams *et al.*, 2005). The germinable seed bank increased from the late wet season (February) and early dry season (May) to a maximum towards the end of the dry season (October). Neither of these studies detected germinable tree seeds in the soil seed bank. Early season fires are likely to kill seeds only if the seeds are on the soil surface, whereas later season fires can result in lethal temperatures at several millimetres below the surface (Williams *et al.*, 2004). Late season fires are also likely to stimulate legume germination from greater soil depths. Fire reduces the seed bank of perennial and annual native grasses, but Mott and Andrew (1985) found >250 seeds/m^2 of the annual *S. intrans* germinating from sites that had been burnt. Seed production by the perennial *Themeda triandra* Forssk. was much lower, and this grass relied on resprouting of the parent plants with very poor survival of seedlings to maturity.

Introduced grass weeds are a major problem in the Darwin region and can produce larger soil seed banks than native species. In June, the seed density of gamba grass (*Andropogon gayanus* Kunth.) averaged 2100 ± 1100 seeds/m^2 and that of mission grass (*Pennisetum polystachion* (L.) J.A. Schultes) averaged 3365 ± 1607 seeds/m^2 across several sites, which reduced to 21 ± 21 and 350 ± 249 seeds/m^2 by December prior to wet season rains. During this period, the germinable native grass seed density averaged 300 ± 10 seeds/m^2 across several sites in June and increased to 595 ± 445 seeds/m^2 by December as seed dormancy decreased (Setterfield *et al.*, 2004).

Seed dormancy and germination

Tropical Australian species tend to have an optimum germination temperature ranging from 20°C to 30°C, whereas southern Australian species tend to have maximum germination at ~15°C. This trend is demonstrated by *Callitris intratropica* RT Baker & HG Smith compared with southern *Callitris* species (Scott, 1970) and by *Allocasuarina littoralis* (Salisb.) LAS Johnson and *Allocasuarina torulosa* (Aiton) LAS Johnson (Crowley and Jackes, 1990) compared with *Allocasuarina humilis* (Otto & F. Dietr.) LAS Johnson, *Allocasuarina fraseriana* (Miq.) LAS Johnson and *Allocasuarina campestris* (Diels) LAS Johnson (Bellairs and Bell, 1990; Bell *et al.*, 1995). However, populations of *T. triandra* from Lae (in Papua New Guinea) and northern Australia had similar germination responses to incubation temperature as populations from southern Australia (Groves *et al.*, 1982).

Seed dormancy has been investigated in few woody species. However, *Acacia* species generally respond to a heat shock treatment, with the exception of a few species that germinate readily due to lack of a hard seed coat (e.g. *Acacia harpophylla* F. Muell.). Some woody species have increased germination following scarification of the endocarp or testa including *Alphitonia*, *Brachychiton*, *Cochlospermum*, *Erythrophleum* and *Grevillea* (Ashwath *et al.*, 1994, 2003; McIntyre *et al.*, 1994; King, 2005). Leaching or removal of fruit has been shown to increase germination of *Melia*, *Terminalia* and the palm *Livistona* (Moncur and Gunn, 1989; Ashwath *et al.*, 1994). Removal or leaching of the fleshy fruit combined with scarification can increase germination of *Jacksonia* and *Terminalia* (McIntyre *et al.*, 1994; King, 2005).

Seed dormancy has been studied in a range of native grass species. Most of the native grasses that have been investigated, including the annual species *Brachiaria subquadripara* (Trin.) RD Webster, *Dactyloctenium radulans* (R. Br.) P. Beauv., *S. intrans* and *S. stipoideum* and the perennial species *Aristida latifolia* Domin, *Bothriochloa decipiens* (Hack.) CE Hubb., *Chrysopogon fallax* ST Blake, *C. latifolius* ST Blake, *Sorghum plumosum* (R. Br.) P. Beauv., *Sporobolus indicus* (L.) R. Br. and *T. triandra* have a very high proportion of seeds with innate dormancy at seed fall (Mott, 1978; Andrew and Mott, 1983; Lane, 2000; McIvor and Howden, 2000). Dormancy is broken during the dry season and the seeds are non-dormant at the beginning of the following wet season. Warm storage for several months is required for breaking dormancy in *Sorghum* species, *Chrysopogon* species and *T. triandra* (Mott, 1978). Removal of the husk from the caryopsis and the application of gibberellic acid increased germination of dormant seeds. Other genera including *Brachyachne*, *Eulalia* and *Sorghum* also show substantial dormancy release when the husk is removed (Fesuk and Ashwath, 2002). Smoke has been found to increase germination of some tropical grasses including *Digitaria breviglumis* (Domin) Henrard, *Heteropogon triticeus* (R. Br.) Stapf and *Triodia bitextura* Lazarides (Fesuk and Ashwath, 2002; Williams *et al.*, 2005). McIvor and Howden (2000) show that the germination rate increases as the length of storage increases, with the seeds germinating faster if stored on soil compared with storage in similar temperature conditions in an oven.

Some legume forbs have dormancy overcome by a heat shock treatment (e.g. *Chamaecrista*, *Crotalaria*, *Indigofera*, *Pycnospora* and *Tephrosia*), but two native perennial legume forbs (*Galactia* and *Glycine*) did not respond (Williams *et al.*, 2003, 2005). *Triumfetta* (*Tiliaceae*) also has dormancy broken by the application of a heat treatment. Williams *et al.* (2005) also noted an overall increase in soil seed bank emergence of forbs when smoke was applied. A combined treatment of a heat shock plus the application of nitrate was the optimum fire-related treatment for breaking dormancy of *Chamaecrista* (Williams *et al.*, 2003).

Seed longevity

Many tropical trees and shrubs have short-lived seeds in soil, as indicated by the soil seed bank data. Commercial seed companies and researchers use various types of storage conditions to maintain seed viability, with the majority using cold rooms (CRs). Seed storage properties vary with species, provenance, seed processing techniques and many other factors. Ashwath *et al.* (2003) carried out an extensive study to test germination, storage and responses to smoke treatment of >800 Central Queensland species/provenances. Seeds were collected from the field, tested for germination soon after collection and representative samples of these seed lots (102 species representing 83 genera of the Central Queensland flora) were stored in cloth/paper bags contained in metal tins in a: (i) garden shed (GS) (8–60°C, 48 species); (ii) air-conditioned (AC) laboratory (8–29°C, 51 species); (iii) CR (8–10°C, 800 species/provenances); and (iv) freezer (F) (−18°C, 53 species).

Germination tests were conducted following storage for 12–24 months, with the majority of species tested around 18 months. The CR germination data were compared with initial germination data to determine which species showed an increase in germination, associated with the breaking of dormancy. The germination data of GS, AC and F were compared with those of CR to reveal the effects of different storage conditions on subsequent germination. Selected seed lots (84) were also treated with smoke water/aerosol (SW) following storage, and the effects of SW treatment were determined by comparing the SW data with those of CR. The resulting ratios were grouped into four classes: (i) treatments showing a substantial improvement in germination compared with initial germination (>125%); (ii) those having little effect; (iii) those showing a substantial reduction (<75%); and (iv) those completely inhibiting germination while their counterparts germinated successfully in other treatments.

Among the 92 species tested for dormancy, 31 species showed improved germination, 19 species showed little change and 35 species showed reduced germination (Fig. 43.2a). Seven species showed complete loss of viability during storage. Overall, >50% of the tested species either improved or remained unchanged in their germination when stored in a CR.

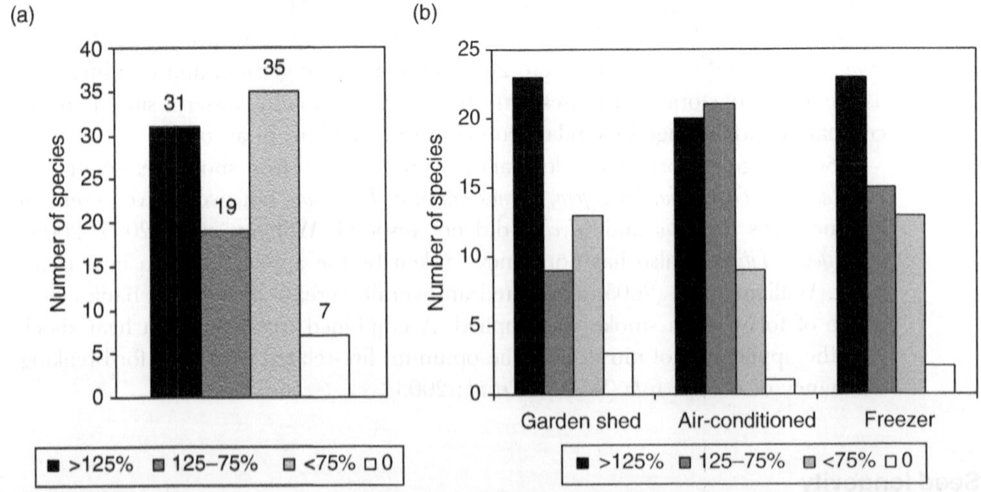

Fig. 43.2. The number of central Queensland species showing improved, unaffected, reduced or inhibited germination: (a) when stored in a cold room (CR) for 12–24 months, as compared with initial germination at collection (92 species); and (b) when stored in a garden shed (GS), air-conditioned room or freezer, as compared with germination following storage in a CR (i.e. >125% indicates a proportionate increase in germination percentage of >25% compared with that of the comparative germination result).

Of the 53 species stored in different conditions (Fig. 43.2b), 20 species showed improved germination in all three storage conditions compared with CR storage and the overall responses were similar among the three storage conditions. The seeds were air-dried and contained <10% moisture and hence they were not adversely affected by storage in a freezer. Storage in warm conditions, such as in a GS, may be essential for after-ripening some species, such as grasses (Fesuk and Ashwath, 2005), whereas storage in a freezer may be needed to prevent ageing of certain other species. For short-term storage (up to 2 years), all storage conditions appear to yield similar results when we consider a wide range of tropical native species. Testing of stored seeds at 5 and 10 years will be required to assess the best long-term storage conditions.

The responses of 84 species to smoke treatment (after being stored in a CR for 12–24 months) were also assessed. Smoke treatment improved germination in 37 species, but reduced in 29 species, and there was little change in 16 species. Two species showed no germination. Smoke improved germination in >40% of the tested species and it either reduced or inhibited germination in about one-third of the species. The response to smoke treatment often mimicked the effects of storage (e.g. freezer or GS; Fig. 43.2) suggesting that the storage conditions may also induce similar germination responses as smoke treatments.

The responses of tropical native species to storage were variable and no single storage condition is optimal for all or the majority of tropical native species, at least for 2 years of storage. Therefore, it is necessary to undertake species-specific storage and/or seed dormancy studies to maximize either storability or germinability of native plant seeds.

Rainforests

Seed production and dispersal

In rainforests, seed production occurs throughout the year but peaks in the wet season. In north-eastern Australia, this peak in fruit production occurs between August and April (Hopkins and Graham, 1989). In the rainforest patches in the NT, the peak is more distinct and occurs in December, while fruit production was low in May and June (Price, 2004). Many species have episodes of very high seed production (i.e. masts), with large seed crops produced every 2–6 years (Hopkins *et al.*, 1990; Connell and Green, 2000).

Many rainforest species have vertebrate-dispersed seeds. In the north-eastern Australian rainforest 83–93% of the tree species have bird-dispersed seeds (Hopkins *et al.*, 1990). Cassowary (*Casuarius casuarius*) dung in north Queensland rainforest was found to contain diaspores of 78 plant species and germination was observed for 70 species. The success of germination was variable, with 397 of the 400 recovered *Beilschmiedia oligandra* LS Sm. seeds germinating, but only 6 of 1851 *Elaeocarpus foveolatus* F. Muell. seeds germinating (Stocker and Irvine, 1983). Dispersal of seeds by birds and flying foxes (i.e. large bats) is important for maintaining genetic diversity in the small patches. White *et al.* (2004) investigated seed dispersal into revegetated patches of rainforest and found much higher numbers of native species (in higher densities) dispersed by flying fauna, ground-dwelling mammals and wind to the adjacent site than to sites that were 600 m or 2 km away.

At least two tropical species are long-lived and semelparous. The palm *Corypha elata* Roxb. can grow 20 m tall with a 1 m diameter trunk before it flowers and then dies (Brock, 2001). *Bambusa arnhemica* F. Muell. is an arborescent, clumping bamboo that is endemic to riparian vegetation in the northern part of NT. It has an estimated lifespan of 40–50 years after which it flowers gregariously in clumps that range in size from a few hectares to several kilometres across, and then dies (Franklin, 2004).

Soil seed banks

Rainforest soil seed banks range from 400 to 5000 seeds/m^2 (Table 43.1), with size and composition of the soil seed bank highly dependent on the forest type. Primary forests had low quantities of seeds, and this increased for forests with pioneer species present. The largest soil seed bank occurred in a forest that had been disturbed and had a substantial weed seed bank (Hopkins *et al.*, 1990). Russell-Smith and Lucas (1994) also showed that the dormant seeds in monsoon forest seed banks were dominated by woody pioneer species and exotic weeds. Savannah taxa were represented in the seed bank of drier monsoon forest types.

Table 43.1. Seed density and life form composition of rainforest soil seed banks.

Size (seeds/m2)	Herbs (%)	Pioneer trees (%)	Primary trees (%)	Reference
434	5	75	20	Hopkins *et al.* (1990)
537	20	67	13	Hopkins *et al.* (1990)
4758	98	1	<1	Hopkins *et al.* (1990)
25–144	35	65	Absent	Russell-Smith and Lucas (1994)

Russell-Smith and Lucas (1994) measured the dormant seed bank only.

Seed dormancy

Seed dormancy mechanisms occur in many pioneer species. Scarification increases germination of *Abrus, Alphitonia, Cathormion* and *Dodonaea* (McIntyre *et al.*, 1994).

Sun *et al.* (1995) report that application of boiling water and soaking for 48h increase germination of *Alphitonia petriei* Braid (*Rhamnaceae*). Heat shock breaks physical dormancy in the hard-seeded *Acacia mangium* Willd. (Hopkins and Graham, 1984). Removal of the fleshy fruit promotes germination of a rare undescribed *Ryparosa* species (Webber and Woodrow, 2004) and also of *Melia azedarach* L. var. *australasica* (Juss.) C. DC, but incubation at 30°C or higher is necessary (Moncur and Gunn, 1989). Removal of the fleshy fruit and scarification is required to germinate *Grewia* seeds (McIntyre *et al.*, 1994). Hopkins and Graham (1987) report that seeds of many pioneer species are dormant in the forest soil, but germinate readily once placed in soil in a nursery. This may indicate inhibition by low red light levels as observed in some South-east Asian rainforest species (Metcalfe, 1996).

Primary rainforest species tend to have quick germinating, short-lived seeds that do not have any dormancy (Hopkins and Graham, 1983, 1987).

Seed longevity

Hopkins and Graham (1987) buried seeds of 50 species in forest soil in nylon bags and found that pioneer and early secondary species survived up to 2 years of burial. Primary forest species had very short longevity in soil, with a mean viability of 10% after 6 months of soil storage. Primary forest species with soft coats were obligate immediate germinators.

Other Communities

In comparison with the southern Australian flora, there is a lack of knowledge of seed biology for most species and some habitats in northern Australia. Sandstone escarpment areas contain a diverse range of vegetation communities, rich in endemic species, which are threatened by changing fire regimes. Since most northern Australian communities are subject to frequent fires, these communities tend to be protected from fire and include many obligate seeder species. Likewise, the unique serpentine communities of Central Queensland also contain endemic serpentine floras. We have little information on seed production, viability, longevity or dormancy of any of these species.

Acknowledgements

N. Ashwath acknowledges the financial support of the National Heritage Trust in the seed storage research and S.M. Bellairs acknowledges support from the Environmental Research Institute of the Supervising Scientist for his seed biology research, but the views expressed are not necessarily those of the sponsoring organizations.

References

Andersen, A.N., Azcarate, F.M. and Cowie, I.D. (2000) Seed selection by an exceptionally rich community of harvester ants in the Australian seasonal tropics. *Journal of Animal Ecology* 69, 975–984.

Andrew, M.H. and Mott, J.J. (1983) Annuals with transient seed banks: the population biology of indigenous *Sorghum* species of tropical north-west Australia. *Australian Journal of Ecology* 8, 265–276.

Ashwath, N., McLaughlin, M., McIntyre, W., Plummer, J. and Slee, M. (1994) Seed germination in a selection of tree, shrub, forb and grass species native to Kakadu National Park and its environs. In: Bellairs, S.M. and Bell, L.C. (eds) *Proceedings National Workshop on Native Seed Biology for Revegetation.* Australian Centre for Minesite Rehabilitation Research, Kenmore and Chamber of Mines and Energy, Perth, Australia, pp. 81–86.

Ashwath, N., Fesuk, S., Heard, B., Dodd, B. and Hood, B. (2003) Native Plant Seedbank for Central Queensland. Central Queensland University, Rockhampton, Australia. Available at: http://www.cqu.edu.au/seedbank/

Bell, D.T., Rokich, D.P., McChesney, C.J. and Plummer, J.A. (1995) Effects of temperature, light and gibberellic acid on the germination of seeds of 43 species native to the southwest of Western Australia. *Journal of Vegetation Science* 6, 797–806.

Bellairs, S.M. and Bell, D.T. (1990) Temperature effects on the seed germination of ten Kwongan species from Eneabba, Western Australia. *Australian Journal of Botany* 38, 451–458.

Brennan, K. (1996) Flowering and fruiting phenology of native plants in the Alligator Rivers Region with particular reference to the Ranger uranium mine lease area. Supervising Scientist Report 107. Supervising Scientist, Canberra, ACT, Australia.

Brock, J. (2001) *Native Plants of Northern Australia.* Reed New Holland, New South Wales, Australia.

Clarkson, J.R. and Clifford, H.T. (1987) Germination of *Jedda multicaulis* J.R. Clarkson (Thymelaeaceae): an example of cryptogeal germination in the Australian flora. *Australian Journal of Botany* 35, 715–720.

Clifford, H.T. and Monteith, G.B. (1989) A three stage seed dispersal mechanism in Australian Quinine Bush (*Petalostigma pubescens* Domin.). *Biotropica* 21, 284–286.

Connell, J.H. and Green, P.T. (2000) Seedling dynamics over thirty-two years in a tropical rain forest tree. *Ecology* 81, 568–584.

Crowley, G. and Garnett, S. (1999) Seeds of the annual grasses *Schizachyrium* spp. as a food resource for tropical granivorous birds. *Australian Journal of Ecology* 24, 208–220.

Crowley, G.M. and Jackes, B.R. (1990) Germination characteristics in tropical provenances of *Allocasuarina littoralis* and *A. torulosa*. *Forest Ecology and Management* 35, 227–238.

Dunlop, C.R. and Webb, L.J. (1991) Flora and vegetation. In: Haynes, C.D., Ridpath, M.G. and Williams, M.A.J. (eds) *Monsoonal Australia Landscape, Ecology and Man in the Northern Lowlands.* Balkema, Rotterdam, The Netherlands, pp. 41–60.

Fesuk, S. and Ashwath, N. (2002) Tropical native grasses: strategies for seed germination and storage. In: Adkins, S.W., Bellairs, S.M. and Bell, L.C. (eds) *Proceedings of the Fourth Australian Workshop on Native Seed Biology for Revegetation.* Australian Centre for Mining Environmental Research, Kenmore, Australia, pp. 135–142.

Fesuk, S. and Ashwath, N. (2005) Optimising seed treatment and storage conditions for tropical grasses of Australia. In: Adkins, S.W., Ainsley, P.J., Bellairs, S.M., Coates, D.J. and Bell, L.C. (eds) *Proceedings of the Fifth Australian Workshop on Native Seed Biology.* Australian Centre for Minerals Extension and Research, Kenmore, Australia, pp. 81–88.

Franklin, D.C. (2004) Synchrony and asynchrony: observations and hypotheses for the flowering wave in a long-lived semelparous bamboo. *Journal of Biogeography* 31, 773–786.

Groves, R.H., Hagon, M.W. and Ramakrishnan, P.S. (1982) Dormancy and germination of seed of eight populations of *Themeda australis. Australian Journal of Botany* 30, 373–386.

Hopkins, M.S. and Graham, A.W. (1983) The species composition of soil seed banks beneath lowland tropical rainforest in North Queensland, Australia. *Biotropica* 15, 90–99.

Hopkins, M.S. and Graham, A.W. (1984) Viable soil seed banks in disturbed lowland tropical rainforest sites in North Queensland. *Australian Journal of Ecology* 9, 71–79.

Hopkins, M.S. and Graham, A.W. (1987) The viability of seeds of rainforest species after experimental soil burials under tropical wet lowland rainforest in north eastern Australia. *Australian Journal of Ecology* 12, 97–108.

Hopkins, M.S. and Graham, A.W. (1989) Community phenological patterns of a lowland tropical rainforest in north-eastern Australia. *Australian Journal of Ecology* 14, 399–414.

Hopkins, M.S., Tracey, J.G. and Graham, A.W. (1990) The size and composition of soil seed-banks in remnant patches of three structural rainforest types in north Queensland. *Australian Journal of Ecology* 15, 43–50.

King, M. (2005) Optimising seed germination in the field of six Top End species used for minesite rehabilitation. Graduate Diploma. Tropical Environmental Management thesis, Charles Darwin University, Darwin, Australia.

Lane, A. (2000) Germination behaviour of selected tropical native grasses and Rhodes grass: implications for revegetation. In: Asher, C.J. and Bell, L.C. (eds) *Proceedings of the Third Australian Workshop on Native Seed Biology for Revegetation.* Australian Centre for Mining Environmental Research, Kenmore, Australia, pp. 135–144.

Lord, J., Egan, J., Clifford, T., Jurado, E., Leishman, M., Williams, D. and Westoby, M. (1997) Larger seeds in tropical floras: consistent patterns independent of growth form and dispersal mode. *Journal of Biogeography* 24, 205–211.

Manning, J. (2004) Sustainability of vegetation development on Nabarlek mine rehabilitation areas: an assessment of nutrient cycling and seed bank ecology. BSc (Hons) thesis, Charles Darwin University, Darwin, Australia.

McIntyre, W., Plummer, J. and Ashwath, N. (1994) Seed germination in a selection of tree, shrub, forb and grass species native to Kakadu National Park and its environs. In: Bellairs, S.M. and Bell, L.C. (eds) *Proceedings National Workshop on Native Seed Biology for Revegetation.* Australian Centre for Minesite Rehabilitation Research, Kenmore and Chamber of Mines and Energy, Perth, Australia, p. 129.

McIvor, J.G. and Howden, S.M. (2000) Dormancy and germination characteristics of herbaceous species in the seasonally dry tropics of northern Australia. *Austral Ecology* 25, 213–222.

Metcalfe, D.J. (1996) Germination of small-seeded tropical rain forest plants exposed to different spectral compositions. *Canadian Journal of Botany* 74, 516–520.

Moncur, M.W. and Gunn, B.V. (1989) Seed development and germination responses of *Melia azedarach* var. *australasica*. In: Turnbull, J.W. (ed.) *Tropical Tree Seed Research*. ACIAR Proceedings No. 28, pp. 24–28.

Mott, J.J. (1978) Dormancy and germination in five native grass species from savannah woodland communities of the Northern Territory. *Australian Journal of Botany* 26, 621–631.

Mott, J.J. and Andrew, M.H. (1985) The effect of fire on the population dynamics of native grasses in tropical savannas of north-west Australia. *Proceedings of the Ecological Society of Australia* 13, 231–239.

Price, O.F. (2004) Indirect evidence that frugivorous birds track fluctuating fruit resources among rainforest patches in the Northern Territory, Australia. *Austral Ecology* 29, 137–144.

Russell-Smith, J. and Lee, A. (1992) Plant populations and monsoon rainforest in the Northern Territory, Australia. *Biotropica* 24, 471–487.

Russell-Smith, J. and Lucas, D.E. (1994) Regeneration of monsoon rain-forest in northern Australia – the dormant seed bank. *Journal of Vegetation Science* 5, 161–168.

Russell-Smith, J., McKenzie, N.L. and Woinarski, J.C.Z. (1992) Conserving vulnerable habitat in northern and northwestern Australia: the rainforest archipelago. In: Moffatt, I. and Webb, A. (eds) *Conservation and Development Issues in Northern Australia*. North Australia Research Unit, Darwin, Australia, pp. 63–68.

Scott, L. (1970) Temperatures for testing germination of *Callitris* seeds. *Australian Forestry Research* 4, 37–40.

Setterfield, S.A. (1997) The impact of experimental fire regimes on seed production of two tropical eucalypt species in northern Australia. *Australian Journal of Ecology* 22, 279–287.

Setterfield, S., Bellairs, S.M., Douglas, M. and Calnan, T. (2004) Seed bank dynamics of two exotic grass species in Australia's northern savannas. In: Sindel, B.M. and Johnson, S.B. (eds) *Proceedings of the Fourth Australian Weed Society Conference*. Weed Society of New South Wales, Wagga Wagga, Australia, pp. 555–557.

Stocker, G.C. and Irvine, A.K. (1983) Seed dispersal by Cassowaries (*Casuarius casuarius*) in north Queensland's rainforests. *Biotropica* 15, 170–176.

Sun, D., Dickinson, G.R. and Bragg, A.L. (1995) Direct seeding of *Alphitonia petriei* (Rhamnaceae) for gully revegetation in tropical northern Australia. *Forest Ecology and Management* 73, 249–257.

Webber, B.L. and Woodrow, I.E. (2004) Cassowary frugivory, seed defleshing and fruit fly infestation influence the transition from seed to seedling in the rare Australian rainforest tree, *Ryparosa* sp. Nov. 1 (Achariaceae). *Functional Plant Biology* 31, 505–516.

White, E., Tucker, N., Meyers, N. and Wilson, J. (2004) Seed dispersal to revegetated isolated rainforest patches in North Queensland. *Forest Ecology and Management* 192, 409–426.

Williams, P.R., Congdon, R.A., Grice, A.C. and Clarke, P.J. (2003) Fire-related cues break seed dormancy of six legumes of tropical eucalypt savannas in north-eastern Australia. *Austral Ecology* 28, 507–514.

Williams, P.R., Congdon, R.A., Grice, A.C. and Clarke, P.J. (2004) Soil temperature and depth of legume germination during early and late season fires in a tropical savanna of north-eastern Australia. *Austral Ecology* 29, 258–263.

Williams, P.R., Congdon, R.A., Grice, A.C. and Clarke, P.J. (2005) Germinable soil seed banks in a tropical savanna: seasonal dynamics and effects of fire. *Austral Ecology* 30, 79–90.

Williams, R.J., Myers, B.A., Eamus, D. and Duff, G.A. (1999) Reproductive phenology of woody species in a north Australian tropical savanna. *Biotropica* 31, 626–636.

Index

Page numbers in **bold** refer to illustrations and tables.

Printed and bound by CPI Group (UK) Ltd, Croydon, CR0 4YY

16/09/2026

14963056-0007